MANUEL A. ARCINIEGA
6205 ARAPAHO
772-7325

INTRODUCTION TO LINEAR ALGEBRA

LEE W. JOHNSON
R. DEAN RIESS

Virginia Polytechnic Institute
and State University

ADDISON-WESLEY PUBLISHING COMPANY
Reading, Massachusetts • Menlo Park, California
London • Amsterdam • Don Mills, Ontario • Sydney

This book is in the
Addison-Wesley Series in Mathematics

Lynn H. Loomis
Consulting Editor

Library of Congress Cataloging in Publication Data

Johnson, Lee W
Introduction to linear algebra.

Includes index.
1. Algebras, Linear. I. Riess, Ronald Dean, 1940- joint author. II. Title.
QA184.J63 512′.5 80-19984
ISBN 0-201-03392-5

ISBN 0-201-03392-5
BCDEFGHIJK-HA-898765432

PREFACE

Linear algebra is an important component of undergraduate mathematics, particularly for students majoring in the scientific, engineering, and social science disciplines. At the practical level, matrix theory and the related vector space concepts provide a language and a powerful computational framework for posing and solving important problems. Beyond this, elementary linear algebra is a valuable introduction to mathematical abstraction and logical reasoning because the theoretical development is self-contained, consistent, and accessible to most students.

Therefore this book stresses both practical computation and theoretical principles and centers around

matrix theory and systems of linear equations,

elementary vector-space concepts, and

the eigenvalue problem.

The text is designed for a one-term course at the late-freshman or sophomore level and is organized so that these three topics can be covered in even a short (10-week) course. However there is enough material for a more leisurely paced course or for a sophomore/junior course at the level of the usual "advanced engineering mathematics" sequence.

To provide a measure of flexibility, we have written Chapters 3, 4, 5, and 6 so that they are essentially independent and can be taken in any order once Chapters 1 and 2 are covered. In particular Chapter 5 (Determinants) can be covered before Chapter 3(Eigenvalues) if desired, all or portions of Chapter 4 (Abstract Vector Spaces) can be covered at any time, etc. The chapter dependencies are given schematically by the following diagram.

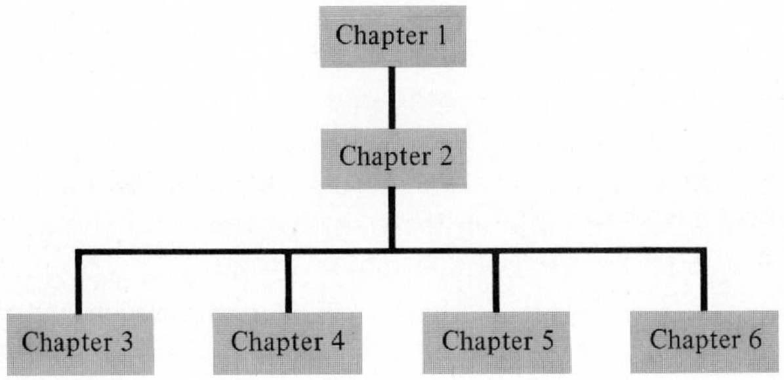

For example, a short course at the beginning level can be built around Sections 1.1–1.8, 2.1–2.4, 3.1–3.6 and can be supplemented by other topics such as least-squares approximations, abstract vector spaces, determinant theory, etc. (Also a brief review of vector geometry is given in the appendix.)

We have introduced the idea of linear independence very early (in Section 1.6) and use it extensively thereafter. Linear independence is one unifying thread running throughout the text. In addition, the early introduction of linear independence will ease the transition from the concrete material in Chapter 1 to the more abstract ideas in Chapter 2, such as subspace, basis, and dimension in R^n. The treatment of the eigenvalue problem in Chapter 3 is more detailed than that in most books at this level, and we emphasize computation and application of eigenvalues as well as the theoretical foundations. In Sections 3.2 and 3.3, eigenvalues are introduced in the traditional way, including a brief discussion of determinants. In Sections 3.7–3.9 we use similarity transformations to explain reduction to Hessenberg form, and then we show the parallels between Hessenberg form for the eigenvalue problem and echelon form for the problem of solving $A\mathbf{x} = \mathbf{b}$. The material in Sections 3.7–3.9 is theoretically complete, computationally relevant, and developmentally important for ideas such as the Cayley–Hamilton theorem and generalized eigenvectors. Finally if a more comprehensive treatment of determinants is desired before introducing eigenvalues, Chapter 5 can be covered before Chapter 3.

Our applications have been drawn from differential equations and difference equations and from problems of interpolating data and finding best least-squares fits to data. In particular in their major curriculum most students have been exposed to problems of drawing curves that fit experimental or empirical data, and they can appreciate the techniques from linear algebra that can be applied to these problems. In Chapter 6 we also discuss numerical methods for linear algebra and include a number of simple computer programs that can be used to implement the numerical methods. These programs are in the form of well-documented subroutines and include, for example, programs that solve $(n \times n)$ systems $A\mathbf{x} = \mathbf{b}$ where A and $\mathbf{b}$ may contain complex entries, programs that use Householder transformations to reduce A to Hessenberg form, and programs that use the Givens algorithm to find the eigenvalues and eigenvectors for a symmetric matrix.

As far as the nature of the material permits, we have tried to organize each section so that it can be covered in one period; where such organization is not possible, we have tried to follow with a short section. The individual sections contain a number of examples, many worked in extreme detail so that the student can read them unassisted. Some sections also contain optional material, marked with a dagger (†); this material can be omitted without any loss of continuity. In particular, entire sections, such as †2.5 and †2.6, may be skipped. In keeping with the introductory nature of the text, many of the exercises are routinely computational while many of the theoretical exercises contain fairly strong hints and should be relatively easy. Some of the exercises are designed also to motivate and give concrete illustrations of topics in later sections. We have made a determined effort to construct the exercises so that the numerical calculations proceed smoothly and do not obscure the point of the problem, and we have tried also to be certain that the answer key is correct.

We have attempted to make the text theoretically complete and self-contained although the inherent nature of the material means that some of the topics demand more mathematical maturity. In particular, Section †3.10 and some of the later sections in Chapter 4 are probably not appropriate for most beginning students.

Finally we would like to express our appreciation and gratitude to the following reviewers for their assistance during the development of the manuscript: Betty J. Barr from the University of Houston, Monte Boisen from Virginia Polytechnic Institute and State University, William A. Brown from the University of Southern Maine, Alexander Hahn from the University of Notre Dame, Steven K. Ingram from Norwich University, Robert M. McConnel from The University of Tennessee, and Gerald J. Roskes from Queens College of The City University of New York.

Blacksburg, Virginia L.W.J.
January 1981 R.D.R.

CONTENTS

1

MATRICES AND SYSTEMS OF LINEAR EQUATIONS

1.1 INTRODUCTION AND GAUSS ELIMINATION

In science, engineering, and the social sciences, one of the most important and frequently occurring mathematical problems is finding a simultaneous solution to a set of linear equations involving several unknowns. A simple example is the problem of finding values for x_1 and x_2 that simultaneously satisfy the equations

$$\begin{aligned} x_1 + x_2 &= 3 \\ x_1 - x_2 &= 1. \end{aligned}$$

This system of equations is easily solved by observing that the first equation requires $x_1 = 3 - x_2$. Inserting that relationship in the second equation gives $(3 - x_2) - x_2 = 1$, which yields $x_2 = 1$; and hence $x_1 = 2$. An alternative approach is to add the two equations together finding $(x_1 + x_2) + (x_1 - x_2) = 4$, or $2x_1 = 4$. Thus again $x_1 = 2$ and $x_2 = 1$ is the solution. This simple system of two equations in two unknowns is quite easy to solve, but obviously some sort of systematic procedure is needed to handle the more complicated problems that arise in practice. To answer this need, the first topic of this chapter is the general problem of solving a system of m linear equations in n unknowns; and the goal is to develop a *systematic* and computationally efficient method for solving such systems. Whenever it is appropriate, we will also comment on the practical aspects of solving systems of linear equations on a digital computer.

An $(m \times n)$ ***system of linear equations*** is the system of m linear equations in n unknowns:

$$\begin{array}{lcl} a_{11}x_1 + a_{12}x_2 + \dots + a_{1n}x_n &=& b_1 \\ a_{21}x_1 + a_{22}x_2 + \dots + a_{2n}x_n &=& b_2 \\ \quad\vdots & & \vdots \\ a_{m1}x_1 + a_{m2}x_2 + \dots + a_{mn}x_n &=& b_m. \end{array} \tag{1.1}$$

In (1.1) the coefficients a_{ij} for $1 \leqslant i \leqslant m$, $1 \leqslant j \leqslant n$ are known constants as are $b_1, b_2, \dots, b_m$. The unknowns in the system are designated $x_1, x_2, \dots, x_n$; and a

solution to (1.1) is a set of n numbers $x_1, x_2, \ldots, x_n$ that satisfies each of the m equations in (1.1)*. The subscript notation in (1.1) is necessary to identify the variables and the constants. For example in the general (3×3) system

$$\begin{aligned} a_{11}x_1 + a_{12}x_2 + a_{13}x_3 &= b_1 \\ a_{21}x_1 + a_{22}x_2 + a_{23}x_3 &= b_2 \\ a_{31}x_1 + a_{32}x_2 + a_{33}x_3 &= b_3, \end{aligned}$$

the subscript notation signifies that a_{32} is the coefficient of the second variable, x_2, in the third equation. The system (1.1) is called linear since each equation is of the first degree in each of the variables $x_1, x_2, \ldots, x_n$. That is, none of the equations contains terms such as x_1^2, x_1x_2, $\sin x_3$, etc. An example of a system of *nonlinear* equations is

$$\begin{aligned} x_1^2 + x_2^2 &= 9 \\ x_1 + x_2 &= 1. \end{aligned}$$

(The solutions of this system are the points on the intersection of the circle $x_1^2 + x_2^2 = 9$ and the line $x_1 + x_2 = 1$.)

The linear system (1.1) is called ***consistent*** if it has at least one solution and is called ***inconsistent*** if there is no solution. In (1.1) we have placed no restriction on the relative sizes of m and n; so we may have more equations than unknowns $(m > n)$, more unknowns than equations $(m < n)$, or an equal number of equations and unknowns. Before discussing methods for solving (1.1), we show by example (even when $m = n$) that a linear system may have no solution, a unique solution, or infinitely many solutions. As one simple example, the (2×2) system

$$\begin{aligned} x_1 + x_2 &= 2 \\ 2x_1 + 2x_2 &= 4 \end{aligned}$$

is consistent but has infinitely many solutions. Clearly any values of x_1 and x_2 that are related by $x_2 = 2 - x_1$ (for example, $x_1 = 1$, $x_2 = 1$ or $x_1 = -3$, $x_2 = 5$) will satisfy both equations. On the other hand the (2×2) system

$$\begin{aligned} x_1 + x_2 &= 2 \\ x_1 + x_2 &= 1 \end{aligned}$$

is clearly inconsistent. Finally the first example given,

$$\begin{aligned} x_1 + x_2 &= 3 \\ x_1 - x_2 &= 1, \end{aligned}$$

is a consistent system with the unique solution $x_1 = 2$, $x_2 = 1$.

The examples above are typical because a consistent system of linear equations

---

* For clarity of presentation, we will assume throughout this chapter that the constants a_{ij} and b_i are real numbers although all statements are equally valid for complex constants. When we consider eigenvalue problems, we will occasionally encounter linear systems having complex coefficients; but the solution technique is no different.

always has either exactly one solution or infinitely many solutions. Each of these three systems has a simple geometric interpretation, as shown in Fig. 1.1. Geometrically a solution to a (2×2) system of linear equations represents a point of intersection of two lines; and there are three possibilities: the two lines may be coincident (the same line), they may be parallel (having no points of intersection), or the lines may intersect in exactly one point.

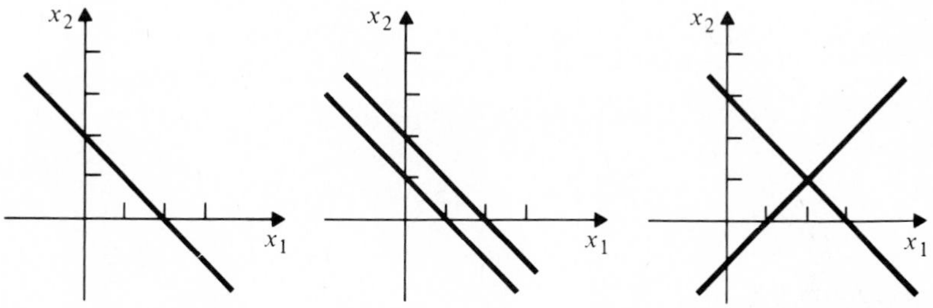

Fig. 1.1 Geometry of (2×2) linear systems.

The simplest and best-known method for solving a general $(m \times n)$ system as in (1.1) is the familiar variable-elimination technique known as Gauss elimination. This technique is computationally practical and is the procedure most widely used in computer-software packages that are designed to solve systems of linear equations. We shall illustrate Gauss elimination with an example, and then follow with a more careful description of this solution process. Consider the (3×3) linear system

$$\begin{aligned} x_1 + 2x_2 + x_3 &= 4 \\ x_1 + x_2 + 3x_3 &= 0 \\ 2x_1 - x_2 - x_3 &= 1. \end{aligned}$$

If we multiply the first equation by -1 and add the result to the second equation, and then multiply the first equation by -2 and add the result to the third equation, we obtain the following system where x_1 has been eliminated from the second and third equations:

$$\begin{aligned} x_1 + 2x_2 + x_3 &= 4 \\ -x_2 + 2x_3 &= -4 \\ -5x_2 - 3x_3 &= -7. \end{aligned}$$

For the moment we ignore the first equation and eliminate x_2 by multiplying the second equation by -5 and adding the result to the third equation. In this way we eliminate x_2 from the third equation and obtain a system having a "triangular form":

$$\begin{aligned} x_1 + 2x_2 + x_3 &= 4 \\ -x_2 + 2x_3 &= -4 \\ -13x_3 &= 13. \end{aligned}$$

From the third equation we find $x_3 = -1$. Using $x_3 = -1$ in the second equation yields $-x_2 - 2 = -4$, or $x_2 = 2$. Finally in the first equation, $x_1 + 4 - 1 = 4$, or $x_1 = 1$. We can easily verify that $x_1 = 1$, $x_2 = 2$, $x_3 = -1$ is a solution of the original system. (Geometrically the solution $x_1 = 1$, $x_2 = 2$, $x_3 = -1$ is the point of intersection of the three planes: $x_1 + 2x_2 + x_3 = 4$, $x_1 + x_2 + 3x_3 = 0$, $2x_1 - x_2 - x_3 = 1$.)

In general the steps of Gauss elimination follow almost exactly as in the example above. The objective of Gauss elimination is to transform the system (1.1)

$$\begin{array}{ccccccc} a_{11}x_1 & + & a_{12}x_2 & + \dots + & a_{1n}x_n & = & b_1 \\ a_{21}x_1 & + & a_{22}x_2 & + \dots + & a_{2n}x_n & = & b_2 \\ \vdots & & & & \vdots & & \vdots \\ a_{m1}x_1 & + & a_{m2}x_2 & + \dots + & a_{mn}x_n & = & b_m \end{array}$$

systematically into an equivalent linear system that is easier to solve. By a system that is ***equivalent*** to (1.1), we mean a system with unknowns $x_1, x_2, \dots, x_n$ that has exactly the same solution set as (1.1). To be more precise, consider the $(r \times n)$ system:

$$\begin{array}{ccccccc} c_{11}x_1 & + & c_{12}x_2 & + \dots + & c_{1n}x_n & = & d_1 \\ c_{21}x_1 & + & c_{22}x_2 & + \dots + & c_{2n}x_n & = & d_2 \\ \vdots & & & & \vdots & & \vdots \\ c_{r1}x_1 & + & c_{r2}x_2 & + \dots + & c_{rn}x_n & = & d_r. \end{array} \tag{1.2}$$

Then (1.1) and (1.2) are equivalent systems if any solution of (1.1) is a solution of (1.2) and any solution of (1.2) is a solution of (1.1).

Gauss elimination [that is, the transformation of (1.1) into an easily solved but equivalent system] is carried out using the three ***elementary operations*** listed below:

1. interchange of two equations,
2. multiplication of an equation by a nonzero scalar, and
3. addition of a constant multiple of one equation to another.

In Exercise 11, Section 1.1, the reader is asked to prove that the application of any of these elementary operations to a system will produce an equivalent system. A precise description of Gauss elimination is most easily given in the language of matrix theory; so we defer this precise description until Section 1.2. However in order at least to describe the essence of Gauss elimination and to allow the reader some practice in solving linear systems, we will indicate briefly how the basic steps are carried out.

The first stage of Gauss elimination as applied in solving (1.1) is to use the first equation and the three elementary operations to eliminate x_1 from equations 2, 3, ..., m. This elimination is easily done (if $a_{11} \neq 0$) by multiplying the first equation by $-a_{i1}/a_{11}$ and adding the result to the ith equation, for $i = 2, 3, \dots, m$.

For each i, $i \geqslant 2$, the effect is that of replacing the original ith equation by a new equation in which x_1 does not appear, and of producing a system equivalent to (1.1):

$$\begin{aligned} a_{11}x_1 + a_{12}x_2 + \ldots + a_{1n}x_n &= b_1 \\ a'_{22}x_2 + \ldots + a'_{2n}x_n &= b'_2 \\ \vdots \qquad\qquad \vdots \quad &\quad \vdots \\ a'_{m2}x_2 + \ldots + a'_{mn}x_n &= b'_m. \end{aligned} \tag{1.3}$$

In (1.3) the primed terms indicate the equations that result when multiples of the first equation are added to equations 2, 3, . . . , m of (1.1). In particular the primed terms in (1.3) are

$$a'_{ij} = a_{ij} - \frac{a_{i1}}{a_{11}} a_{1j} \qquad \text{and} \qquad b'_i = b_i - \frac{a_{i1}}{a_{11}} b_1.$$

If $a_{11} = 0$ in (1.1), then we would have to interchange equations in (1.1) in order to eliminate x_1. We will consider complications of this sort more thoroughly in the next section. Also note that if we can solve the last $m-1$ equations of (1.3) for $x_2, x_3, \ldots, x_n$, then we can determine x_1 from the first equation of (1.3). Thus in (1.3) we have essentially reduced the problem of solving (1.1) to the smaller problem of solving a system of $m-1$ equations in $n-1$ unknowns.

The next step in Gauss elimination is to use the second equation of (1.3) to eliminate x_2 from equations 3, 4, . . . , m. If $a'_{22} \neq 0$, this elimination is accomplished by multiplying the second equation of (1.3) by $-a'_{i2}/a'_{22}$ and adding the result to the ith equation of (1.3), for $i = 3, 4, \ldots, m$. Carrying this out, we obtain

$$\begin{aligned} a_{11}x_1 + a_{12}x_2 + a_{13}x_3 + \ldots + a_{1n}x_n &= b_1 \\ a'_{22}x_2 + a'_{23}x_3 + \ldots + a'_{2n}x_n &= b'_2 \\ a''_{33}x_3 + \ldots + a''_{3n}x_n &= b''_3 \\ \vdots \qquad\qquad \vdots \quad &\quad \vdots \\ a''_{m3}x_3 + \ldots + a''_{mn}x_n &= b''_m. \end{aligned} \tag{1.4}$$

Moreover (1.4) is equivalent to (1.3), and (1.3) is equivalent to (1.1) since they were derived by elementary operations. Thus any solution of (1.4) is a solution of (1.1), and any solution of (1.1) is a solution of (1.4).

Although Gauss elimination may proceed not quite so neatly as we are describing it, our objective is to produce a linear system that is equivalent to (1.1) and that has the following form (in the case $m \leqslant n$):

$$\begin{aligned} c_{11}x_1 + c_{12}x_2 + \ldots + c_{1m}x_m + \ldots + c_{1n}x_n &= d_1 \\ c_{22}x_2 + \ldots + c_{2m}x_m + \ldots + c_{2n}x_n &= d_2 \\ \vdots \qquad\qquad \vdots \quad &\quad \vdots \\ c_{mm}x_m + \ldots + c_{mn}x_n &= d_m. \end{aligned} \tag{1.5}$$

If the coefficients $c_{11}, c_{22}, \ldots, c_{mm}$ are all nonzero, then we can solve (1.5) relatively easily; in the special case that $m = n$, the solution will also be unique. The solution process for (1.5) is called ***backsolving*** and proceeds in the obvious fashion. That is, if we allow $x_{m+1}, \ldots, x_n$ to be variables (or free parameters), then x_m is determined from the last equation of (1.5) by

$$x_m = (d_m - c_{m,\,m+1}x_{m+1} - \ldots - c_{mn}x_n)/c_{mm}.$$

This formula for x_m is then inserted in the $(m-1)$st equation of (1.5) to determine x_{m-1}. In general, x_i is determined from the ith equation of (1.5) using $x_{i+1}, \ldots, x_n$. We will treat the case that some $c_{ii} = 0$ and the case that $m > n$ in the next section, using the idea of an echelon matrix. (Echelon matrices will serve also as convenient notational devices allowing us to organize the steps of Gauss elimination in a compact fashion.) We conclude this section with a number of simple examples that illustrate Gauss elimination.

EXAMPLE 1.1 To demonstrate Gauss elimination, we use the procedure to solve the system

$$\begin{aligned} x_1 - 2x_2 + x_3 &= 2 \\ 2x_1 + x_2 - x_3 &= 1 \\ -3x_1 + x_2 - 2x_3 &= -5. \end{aligned} \tag{1.6}$$

Following the description given above, we multiply the first equation by -2 and add the result to the second, and then multiply the first equation by 3 and add the result to the third (in the previous notation, $-a_{21}/a_{11} = -2$ and $-a_{31}/a_{11} = 3$). These steps yield an equivalent system of the form (1.3):

$$\begin{aligned} x_1 - 2x_2 + x_3 &= 2 \\ 5x_2 - 3x_3 &= -3 \\ -5x_2 + x_3 &= 1. \end{aligned} \tag{1.6a}$$

The final step for this simple example is to eliminate x_2 from the third equation by adding the second equation to the third (in the notation used above, $-a'_{32}/a'_{22} = 1$). We thus arrive at a "triangular" system [of the form (1.5)], which is equivalent to (1.6):

$$\begin{aligned} x_1 - 2x_2 + x_3 &= 2 \\ 5x_2 - 3x_3 &= -3 \\ -2x_3 &= -2. \end{aligned} \tag{1.6b}$$

If we backsolve, the last equation of this triangular system yields $x_3 = 1$; and by having $x_3 = 1$, the second equation requires $5x_2 = 0$, or $x_2 = 0$. Given that $x_3 = 1$ and $x_2 = 0$, then the first equation yields $x_1 = 1$. Clearly the only solution of the triangular system (1.6b) is $x_1 = 1$, $x_2 = 0$, $x_3 = 1$; and since the triangular system and the original system (1.6) are equivalent, we know that the only solution of (1.6) is also given by $x_1 = 1$, $x_2 = 0$, $x_3 = 1$.

* for proper signs, multiply the first with reference to the second and third

EXAMPLE 1.2 In this example we use Gauss elimination to show that the system (1.7) has no solution:

$$\begin{aligned} x_1 - 2x_2 - x_3 &= -2 \\ 2x_1 + x_2 + 3x_3 &= 1 \\ -3x_1 + x_2 - 2x_3 &= 1.1. \end{aligned} \tag{1.7}$$

Proceeding as in Example 1.1, we multiply the first equation by -2 and add the result to the second equation, and then multiply the first equation by 3 and add to the third equation. These steps give

$$\begin{aligned} x_1 - 2x_2 - x_3 &= -2 \\ 5x_2 + 5x_3 &= 5 \\ -5x_2 - 5x_3 &= -4.9. \end{aligned}$$

Next adding the second equation to the third, we obtain a system *equivalent* to (1.7):

$$\begin{aligned} x_1 - 2x_2 - x_3 &= -2 \\ 5x_2 + 5x_3 &= 5 \\ 0x_3 &= .1. \end{aligned} \tag{1.7a}$$

Because there is no number x_3 such that $0x_3 = .1$, the system (1.7a) is an inconsistent system. Since (1.7a) and (1.7) are equivalent systems, then (1.7) has no solution. In a later section we will develop a procedure for finding a "best" approximate solution to (1.7), that is, a procedure for finding a set of values x_1, x_2, x_3 that comes closest to solving (or "best fits") the three equations in (1.7).

EXAMPLE 1.3 This example shows how Gauss elimination can be used to exhibit all the solutions of a consistent system that has infinitely many solutions. Applying Gauss elimination to

$$\begin{aligned} x_1 - 2x_2 - x_3 &= -2 \\ 2x_1 + x_2 + 3x_3 &= 1 \\ -3x_1 + x_2 - 2x_3 &= 1, \end{aligned} \tag{1.8}$$

we find after eliminating x_1 that (1.8) is equivalent to

$$\begin{aligned} x_1 - 2x_2 - x_3 &= -2 \\ 5x_2 + 5x_3 &= 5 \\ -5x_2 - 5x_3 &= -5. \end{aligned}$$

Multiplying the second equation by 1 and adding the result to the third, we next obtain

$$\begin{aligned} x_1 - 2x_2 - x_3 &= -2 \\ 5x_2 + 5x_3 &= 5 \\ 0x_3 &= 0. \end{aligned} \tag{1.8a}$$

This case is unlike (1.7a) because the last equation $0x_3 = 0$ can be satisfied by any value of x_3. In fact we may as well dispense completely with the last equation of (1.8a); then we see that (1.8) is equivalent to the (2×3) system

$$\begin{aligned} x_1 - 2x_2 - x_3 &= -2 \\ 5x_2 + 5x_3 &= 5. \end{aligned} \tag{1.8b}$$

Backsolving (1.8b), we find $x_2 = 1 - x_3$ from the second equation of (1.8b); and then $x_1 = -2 + x_3 + 2x_2 = -2 + x_3 + 2(1 - x_3) = -x_3$ from the first equation. Thus the solution of (1.8) is

$$\begin{aligned} x_1 &= \quad -x_3 \\ x_2 &= 1 - x_3 \end{aligned}$$

where x_3 is a free parameter. For example $x_3 = 0$, $x_2 = 1$, $x_1 = 0$ is a solution as is $x_3 = 4$, $x_2 = -3$, $x_1 = -4$.

EXAMPLE 1.4 In this example we solve a simple "rectangular" system, the (3×4) system

$$\begin{aligned} x_2 + x_3 - x_4 &= 0 \\ x_1 - x_2 + 3x_3 - x_4 &= -2 \\ x_1 + x_2 + x_3 + x_4 &= 2. \end{aligned}$$

We immediately observe that the first equation cannot be used to eliminate x_1 from the second and third equations. This problem is easily overcome by interchanging the first and second equations, and we obtain an equivalent system

$$\begin{aligned} x_1 - x_2 + 3x_3 - x_4 &= -2 \\ x_2 + x_3 - x_4 &= 0 \\ x_1 + x_2 + x_3 + x_4 &= 2. \end{aligned}$$

Eliminating x_1, we have

$$\begin{aligned} x_1 - x_2 + 3x_3 - x_4 &= -2 \\ x_2 + x_3 - x_4 &= 0 \\ 2x_2 - 2x_3 + 2x_4 &= 4. \end{aligned}$$

Finally we eliminate x_2 from the third equation by multiplying the second equation by -2 and adding the result to the third:

$$\begin{aligned} x_1 - x_2 + 3x_3 - x_4 &= -2 \\ x_2 + x_3 - x_4 &= 0 \\ -4x_3 + 4x_4 &= 4. \end{aligned}$$

We can solve the last equation for x_3 in terms of x_4 (and find $x_3 = x_4 - 1$) or for x_4 in terms of x_3 (and find $x_4 = 1 + x_3$). Selecting $x_3 = x_4 - 1$, we have $x_2 = 1$ from the second equation and $x_1 = 2 - 2x_4$. If we had chosen to write the solution in terms of x_3 instead of x_4, we would have used $x_4 = 1 + x_3$, then found $x_2 = 1$ and finally $x_1 = -2x_3$. Thus for this example we can give two forms for the solution set:

$$\begin{aligned} x_1 &= 2 - 2x_4 \\ x_2 &= 1 \\ x_3 &= x_4 - 1 \end{aligned} \qquad \text{or} \qquad \begin{aligned} x_1 &= \quad -2x_3 \\ x_2 &= 1 \\ x_4 &= 1 + x_3. \end{aligned}$$

We could have rearranged variables in the original system so that Gauss elimination resulted in solving for x_3 and x_4 in terms of x_1. For this particular example we will always obtain $x_2 = 1$, and hence it would not have been possible to solve for x_1, x_3, and x_4 in terms of x_2. What is invariant about these solution sets is that $x_2 = 1$ and that any two of the remaining variables may be expressed in terms of the third. Hence we have one independent (unconstrained) variable and three dependent (constrained) variables in any case, and we can choose whichever form of the solution set best suits our purposes.

EXAMPLE 1.5 This example illustrates how some of the "diagonal" terms, c_{ii} in (1.5), might be zero. The example demonstrates also that we can still solve the system in this case, but that some additional mathematical terminology is necessary to describe the situation precisely. (In fact the equivalent system that we find from Gauss elimination will be termed "echelon" in the next section.)

Given the (3×4) system

$$\begin{aligned} x_1 + x_2 + 3x_3 + x_4 &= 12 \\ 2x_1 + 2x_2 + 7x_3 - 9x_4 &= 16 \\ x_1 + x_2 + x_3 - 4x_4 &= 1, \end{aligned}$$

the first phase of Gauss elimination happens to elimiate x_2 as well as x_1:

$$\begin{aligned} x_1 + x_2 + 3x_3 + x_4 &= 12 \\ x_3 - 11x_4 &= -8 \\ -2x_3 - 5x_4 &= -11. \end{aligned}$$

We now accept the fact that x_2 is not constrained by the last two equations, and we proceed to eliminate x_3 from the third equation:

$$\begin{aligned} x_1 + x_2 + 3x_3 + x_4 &= 12 \\ x_3 - 11x_4 &= -8 \\ -27x_4 &= -27. \end{aligned}$$

Backsolving, we find $x_4 = 1$, $x_3 = 3$, $x_1 = 2 - x_2$, which shows that the original system is consistent and has an infinite set of solutions (although the variables x_4 and x_3 must be fixed at 1 and 3 respectively in any solution).

EXAMPLE 1.6 This example treats a (3×2) system, which is an example of an ***overdetermined system***, that is, a system with more constraints (or equations) than variables. We normally do not expect an overdetermined system to have a solution, but such a system may have a unique solution, or even infinitely many solutions (see Exercise 6, Section 1.1). The example below is inconsistent:

$$\begin{aligned} x_1 - x_2 &= 3 \\ 2x_1 + x_2 &= 6 \\ x_1 + x_2 &= 1. \end{aligned}$$

Eliminating x_1, we obtain the equivalent system

$$\begin{aligned} x_1 - x_2 &= 3 \\ 3x_2 &= 0 \\ 2x_2 &= -2; \end{aligned}$$

and this system is clearly inconsistent, for the third equation requires $x_2 = -1$ while the second equation requires $x_2 = 0$.

1.1 EXERCISES

1. Solve each system of linear equations or state that the system is inconsistent:

a) $\begin{aligned} x_1 + 2x_2 &= -5 \\ 2x_1 - x_2 &= 5 \end{aligned}$

b) $\begin{aligned} 2x_1 - 3x_2 &= 5 \\ -4x_1 + 6x_2 &= -10 \end{aligned}$

c) $\begin{aligned} x_1 + 2x_2 &= 4 \\ 2x_1 + 6x_2 &= 12 \end{aligned}$

d) $\begin{aligned} 2x_1 - 3x_2 &= 5 \\ -4x_1 + 6x_2 &= 10 \end{aligned}$

e) $\begin{aligned} x_1 + x_2 &= 2 \\ 3x_1 + 3x_2 &= 6 \end{aligned}$

f) $\begin{aligned} x_1 - x_2 + x_3 &= 3 \\ 2x_1 + x_2 - 4x_3 &= -3 \end{aligned}$

g) $\begin{aligned} x_1 - x_2 + x_3 &= 4 \\ 2x_1 - 2x_2 + 3x_3 &= 2 \end{aligned}$

h) $\begin{aligned} x_1 + x_2 - x_3 &= 2 \\ -x_1 - x_2 + x_3 &= -2 \end{aligned}$

i) $\begin{aligned} 2x_1 + 2x_2 + 4x_3 &= 6 \\ x_1 + x_2 + 2x_3 &= 4 \end{aligned}$

j) $\begin{aligned} x_1 - x_2 + x_3 &= 1 \\ x_1 + x_2 - x_3 &= 1 \\ 2x_1 + 3x_3 &= 8 \end{aligned}$

k) $\begin{aligned} x_1 + 2x_2 + 4x_3 &= 1 \\ x_1 + 2x_3 &= 0 \\ 2x_1 + 3x_2 + 7x_3 &= 0 \end{aligned}$

l) $\begin{aligned} 3x_1 + 2x_2 + x_3 &= 2 \\ x_1 - x_2 - 2x_3 &= -3 \\ -4x_1 - 2x_2 + 2x_3 &= 6 \end{aligned}$

m) $\begin{aligned} x_1 + x_2 - x_3 &= 1 \\ 2x_1 - x_2 + 7x_3 &= 8 \\ -x_1 + x_2 - 5x_3 &= -5 \end{aligned}$

n) $\begin{aligned} 2x_1 + 3x_2 - 4x_3 &= 3 \\ x_1 - 2x_2 - 2x_3 &= -2 \\ -x_1 + 16x_2 + 2x_3 &= 16 \end{aligned}$

2. Solve each system of linear equations or state that the system is inconsistent.

a) $\begin{aligned} x_1 - x_2 &= 3 \\ 2x_1 + x_2 &= 9 \end{aligned}$

b) $\begin{aligned} x_1 + 2x_2 &= 3 \\ 2x_1 + 4x_2 &= 6 \end{aligned}$

c) $\begin{aligned} x_1 + 2x_2 &= 3 \\ 2x_1 + 4x_2 &= 5 \end{aligned}$

d) $\begin{aligned} x_1 + x_2 + x_3 &= 4 \\ 2x_1 + x_2 - x_3 &= 1 \\ x_1 - x_2 - 2x_3 &= 4 \end{aligned}$

e) $\begin{aligned} x_1 - x_2 - x_3 &= 1 \\ x_1 + x_3 &= 2 \\ x_2 + 2x_3 &= 3 \end{aligned}$

f) $\begin{aligned} x_1 - x_2 + 2x_3 &= 2 \\ 2x_1 - 2x_2 + 6x_3 &= 6 \\ 3x_1 - 3x_2 + 8x_3 &= 8 \end{aligned}$

3. Solve the (3×3) system of linear equations where $a_{11} = 1$, $a_{21} = 2$, $a_{31} = 3$, $a_{12} = -1$, $a_{22} = -2$, $a_{32} = -2$, $a_{13} = 1$, $a_{23} = 1$, $a_{33} = 3$, $b_1 = -1$, $b_2 = 0$, and $b_3 = -2$.

4. In Exercise 1, parts (a), (b), (c), and (d), interpret each system as representing two lines in the plane. Sketch the graphs of these lines and give a geometric interpretation of the solution set.

5. Find all values of b for which the systems are consistent. Are the solutions unique?

a) $\begin{aligned} 2x_1 + 3x_2 - 4x_3 &= 3 \\ x_1 - 2x_2 - 2x_3 &= -2 \\ -x_1 + 16x_2 + 2x_3 &= b \end{aligned}$

b) $\begin{aligned} x_1 - 2x_2 + x_3 &= 3 \\ 2x_1 - x_2 + x_3 &= 0 \\ x_1 + 4x_2 - x_3 &= b \end{aligned}$

6. For each of the following conditions, give one example of a system of three linear equations in two unknowns for which

a) the system has a unique solution,
b) the system has infinitely many solutions,
c) the system is inconsistent.

7. Consider the system

$$\begin{aligned} x_1 - x_2 + x_3 + x_4 &= 4 \\ 2x_1 - x_2 - x_3 - 2x_4 &= -8. \end{aligned}$$

Find a form of the solution set where x_3 and x_4 are arbitrary, and find another form where x_1 and x_2 are arbitrary. Verify that $x_1 = 1, x_2 = 2, x_3 = 2, x_4 = 3$ is a member of both solution sets.

8. Consider the (2×2) system

$$\begin{aligned} a_{11}x_1 + a_{12}x_2 &= b_1 \\ a_{21}x_1 + a_{22}x_2 &= b_2. \end{aligned}$$

Show that if $a_{11}a_{22} - a_{12}a_{21} \neq 0$, then this system is equivalent to a system of the form

$$\begin{aligned} c_{11}x_1 + c_{12}x_2 &= d_1 \\ c_{22}x_2 &= d_2 \end{aligned}$$

where $c_{11} \neq 0, c_{22} \neq 0$. Note that the second system is always consistent and has a unique solution. (Hint: First suppose $a_{11} \neq 0$, and then consider the special case that $a_{11} = 0$.)

9. Find all values of a for which the system given is inconsistent.

a) $\begin{aligned} x_1 + x_2 &= 5 \\ 2x_1 + ax_2 &= 4 \end{aligned}$ b) $\begin{aligned} x_1 + 2x_2 &= -3 \\ ax_1 - 2x_2 &= 5 \end{aligned}$ c) $\begin{aligned} x_1 + 3x_2 &= 4 \\ 2x_1 + 6x_2 &= a \end{aligned}$

10. In Exercise 8 suppose $a_{11}a_{22} - a_{12}a_{21} = 0$ and $a_{11} \neq 0$. Give conditions on b_1 and b_2 that ensure the system is consistent. Can solutions ever be unique?

11. Prove that either of the elementary operations (2) or (3) applied to (1.1) produces an equivalent system. [Hint: To simplify this proof, represent the ith equation in (1.1) as $f_i(x_1, x_2, \ldots, x_n) = b_i$; so $f_i(x_1, x_2, \ldots, x_n) = a_{i1}x_1 + a_{i2}x_2 + \ldots + a_{in}x_n$ for $i = 1, 2, \ldots, m$. With this notation, (1.1) has the form of (A) below. Next for example if a multiple of c times the jth equation is added to the kth equation, a new system of the form (B) is produced:

$$\text{(A)}\quad \begin{aligned} f_1(x_1, x_2, \ldots, x_n) &= b_1 \\ &\vdots \\ f_j(x_1, x_2, \ldots, x_n) &= b_j \\ &\vdots \\ f_k(x_1, x_2, \ldots, x_n) &= b_k \\ &\vdots \\ f_m(x_1, x_2, \ldots, x_n) &= b_m \end{aligned} \qquad \text{(B)}\quad \begin{aligned} f_1(x_1, x_2, \ldots, x_n) &= b_1 \\ &\vdots \\ f_j(x_1, x_2, \ldots, x_n) &= b_j \\ &\vdots \\ g(x_1, x_2, \ldots, x_n) &= r \\ &\vdots \\ f_m(x_1, x_2, \ldots, x_n) &= b_m \end{aligned}$$

where $g(x_1, x_2, \ldots, x_n) = f_k(x_1, x_2, \ldots, x_n) + cf_j(x_1, x_2, \ldots, x_n)$ and $r = b_k + cb_j$. To show that operation (3) gives an equivalent system, show that any solution of (A) is a solution of (B) and vice versa.]

12. Solve the system of two *nonlinear* equations in two unknowns

$$\begin{aligned} x_1^2 - 2x_1 + x_2^2 &= b_1 \\ 9x_1^2 \qquad - 4x_2^2 &= b_2 \end{aligned}$$

for (a) $b_1 = 0, b_2 = 36$; and (b) $b_1 = -3/4, b_2 = 36$. (Hint: Complete the square in the first equation.)

1.2 MATRICES AND ECHELON FORM

In this section we begin our introduction to matrix theory by relating matrices to the problem of solving systems of linear equations. Initially we will show that matrix theory provides a convenient and natural symbolic language to describe linear systems. Later we will show that matrix theory is also an appropriate and powerful framework within which to analyze and solve more general linear problems such as least-squares approximations, the representation of linear operations, eigenvalue problems etc.

To begin, consider the (3×3) system of linear equations

$$\begin{aligned} x_1 + 2x_2 + x_3 &= 4 \\ 2x_1 - x_2 - x_3 &= 1 \\ x_1 + x_2 + 3x_3 &= 0. \end{aligned}$$

If we write the coefficients of this system in a square array

$$\begin{bmatrix} 1 & 2 & 1 \\ 2 & -1 & -1 \\ 1 & 1 & 3 \end{bmatrix},$$

then we have expressed compactly and naturally all of the information contained in the left-hand side of the three equations. An array of this sort is an example of a "matrix." In general let us consider an $(m \times n)$ system of linear equations as in (1.1):

$$\begin{aligned} a_{11}x_1 + a_{12}x_2 + \dots + a_{1n}x_n &= b_1 \\ a_{21}x_1 + a_{22}x_2 + \dots + a_{2n}x_n &= b_2 \\ \vdots \qquad\qquad\qquad\qquad & \quad \vdots \\ a_{m1}x_1 + a_{m2}x_2 + \dots + a_{mn}x_n &= b_m. \end{aligned}$$

The coefficients of this linear system can be written in a rectangular array having m rows and n columns, and we designate this array as A:

$$A = \begin{bmatrix} a_{11} & a_{12} & \dots & a_{1n} \\ a_{21} & a_{22} & \dots & a_{2n} \\ \vdots & & & \vdots \\ a_{m1} & a_{m2} & \dots & a_{mn} \end{bmatrix}. \tag{1.9}$$

Such a rectangular array of numbers, having m rows and n columns, is called an $(m \times n)$ ***matrix***. We will use subscripts i and j to designate the position of an entry a_{ij} in a matrix A. In particular a_{ij} is the number in the ith row and jth column of A. For example the matrix A given by

$$A = \begin{bmatrix} 6 & 3 & 7 \\ 2 & 1 & 4 \end{bmatrix}$$

is a (2×3) matrix (A has 2 rows and 3 columns) where $a_{11} = 6, a_{12} = 3, a_{23} = 4$, etc. We will frequently use the notation $A = (a_{ij})$ to denote a matrix A with entries a_{ij}. There is a natural arithmetic associated with matrices that makes them useful in a variety of practical problems, but in this section we will focus on the use of matrices as a convenient and precise notation to express the steps of Gauss elimination.

Returning to the $(m \times n)$ system (1.1), we call the $(m \times n)$ matrix A given in (1.9) the ***coefficient matrix*** for the system (1.1). If we now form a matrix that includes the constants from the right-hand side of (1.1), $b_1, b_2, \ldots, b_m$, as well as the coefficients a_{ij}, we will have a matrix that expresses compactly all the relevant information contained in (1.1). Such a matrix is called the ***augmented matrix*** for (1.1), and it is usually denoted as $[A \vdots b]$. Precisely if $B = [A \vdots b]$ is the augmented matrix for the system (1.1), then B is the $(m \times (n+1))$ matrix given by:

$$B = \begin{bmatrix} a_{11} & a_{12} & \cdots & a_{1n} & b_1 \\ a_{21} & a_{22} & \cdots & a_{2n} & b_2 \\ \vdots & & & & \vdots \\ a_{m1} & a_{m2} & \cdots & a_{mn} & b_m \end{bmatrix}. \tag{1.10}$$

EXAMPLE 1.7 For the system in Example 1.1

$$\begin{aligned} x_1 - 2x_2 + x_3 &= 2 \\ 2x_1 + x_2 - x_3 &= 1 \\ -3x_1 + x_2 - 2x_3 &= -5, \end{aligned}$$

the coefficient matrix A and the augmented matrix B are given by

$$A = \begin{bmatrix} 1 & -2 & 1 \\ 2 & 1 & -1 \\ -3 & 1 & -2 \end{bmatrix} \qquad B = \begin{bmatrix} 1 & -2 & 1 & 2 \\ 2 & 1 & -1 & 1 \\ -3 & 1 & -2 & -5 \end{bmatrix}$$

An initial explanation of why the notion of an augmented matrix is useful is found in the next example, and we will formalize the procedures in Example 1.8 shortly.

EXAMPLE 1.8 In this example we will use the augmented matrix B of Example 1.7 to solve the system (1.6). The basic idea is to use matrices as a shorthand notation, relieving us of the necessity of keeping track of the unknowns x_1, x_2, x_3. We will perform essentially the same operations on the augmented matrix B as we did when solving (1.6) in Example 1.1. To be specific, we start with

$$B = \begin{bmatrix} 1 & -2 & 1 & 2 \\ 2 & 1 & -1 & 1 \\ -3 & 1 & -2 & -5 \end{bmatrix},$$

which "represents" the system (1.6). We next multiply the first row of B by -2 and add the result to the second row of B, then multiply the first row of B by 3 and add the result to the third row of B. This operation produces the (3×4) matrix B_1 where

$$B_1 = \begin{bmatrix} 1 & -2 & 1 & 2 \\ 0 & 5 & -3 & -3 \\ 0 & -5 & 1 & 1 \end{bmatrix}.$$

We note that B_1 is the augmented matrix for the system (1.6a) where (1.6a) is equivalent to (1.6) (again, see Example 1.1). Thus B_1 represents (in matrix form) the first step of Gauss elimination applied to (1.6). We next add the second row of B_1 to the third row of B_1, and obtain a new matrix B_2 where

$$B_2 = \begin{bmatrix} 1 & -2 & 1 & 2 \\ 0 & 5 & -3 & -3 \\ 0 & 0 & -2 & -2 \end{bmatrix}.$$

Again we note that B_2 is the augmented matrix for the system (1.6b) of Example 1.1. Thus B_2 represents the system

$$\begin{aligned} x_1 - 2x_2 + x_3 &= 2 \\ 5x_2 - 3x_3 &= -3 \\ -2x_3 &= -2, \end{aligned}$$

which we found before by the direct use of Gauss elimination on (1.6).

This example demonstrates how matrices can be used to display conveniently the steps of Gauss elimination and shows also that solving systems of linear equations amounts essentially to recording, shifting, and modifying arrays of data. To formalize the ideas inherent in Example 1.8, we use a concept similar to the elementary operations defined in the last section. For matrices the ***elementary row operations*** are

1. interchange of two rows,
2. multiplication of a row by a nonzero constant, and
3. addition of a constant multiple of one row to another.

We next say that two $(m \times n)$ matrices, B and C, are ***row equivalent*** if we can obtain one from the other by a sequence of elementary row operations. Now if B is the augmented matrix for the system (1.1), and if C is a matrix that is row equivalent to B, then clearly C is the augmented matrix for a system of equations that is equivalent to (1.1). This observation follows because the elementary row operations for matrices exactly duplicate the elementary operations for systems of equations. As a particular example, the matrix B_2 in Example 1.8 is row equivalent to B; moreover B_2 is the augmented matrix for a linear system that is equivalent to the system represented by the augmented matrix B.

Thus we can solve a linear system by forming the augmented matrix, B, for the system and then using elementary row operations to transform the augmented matrix to a row-equivalent matrix, C, which represents a "simpler" system. We are now ready to specify this simpler form, which is called echelon form. Informally a matrix C is in echelon form if all the nonzero entries of C lie in a staircase-shaped region on or above the diagonal. Two such matrices in echelon form are illustrated in Fig. 1.2, in which the entries denoted by $*$ are nonzero while the entries denoted by $\times$ may or may not be zero. The purpose of echelon form should be evident, for if either C_1 or C_2 is the augmented matrix for a linear system, then the system can be solved easily. The formal definition of echelon form (see Definition 1.1) may seem unduly complicated, but the definition expresses mathematically the key properties of echelon form.

1. All rows with nonzero entries are grouped together at the top of the matrix.

2. The first nonzero entry of row 1 is to the left of the first nonzero entry of row 2, which is to the left of the first nonzero entry of row 3, etc.

$$C_1 = \begin{bmatrix} * & \times & \times & \times & \times & \times & \times \\ 0 & 0 & * & \times & \times & \times & \times \\ 0 & 0 & 0 & * & \times & \times & \times \\ 0 & 0 & 0 & 0 & 0 & * & \times \\ 0 & 0 & 0 & 0 & 0 & 0 & 0 \end{bmatrix} \qquad C_2 = \begin{bmatrix} 0 & * & \times & \times & \times \\ 0 & 0 & * & \times & \times \\ 0 & 0 & 0 & 0 & * \end{bmatrix}$$

Fig. 1.2 Examples of echelon form.

DEFINITION 1.1.

An $(m \times n)$ matrix C is in ***echelon form*** if

1. There is an integer k such that all the entries of any row of C below the kth row are zero.
2. If the first (counting from left to right) nonzero entry of the ith row is in column p_i and the first nonzero entry of the $(i+1)$st row is in column p_{i+1}, then $p_i < p_{i+1}$ (for $1 \leqslant i \leqslant k-1$).

For example in Fig. 1.2, the (5×7) matrix C_1 has, in the notation of Definition 1.1,

$$k = 4,\ p_1 = 1,\ p_2 = 3,\ p_3 = 4,\ p_4 = 6.$$

For C_2, the values are $k = 3$, $p_1 = 2$, $p_2 = 3$, and $p_3 = 5$.* Clearly if the augmented matrix for a system is in echelon form, then we can recognize immediately whether or not solutions to the system exist; and if so, they can be found immediately by backsolving. Thus the aim of Gauss elimination is to produce a row-equivalent matrix that is in echelon form. This can always be done as we shall show in the next section.

EXAMPLE 1.9 The matrices A, B, and D are in echelon form while C and E are not.

$$A = \begin{bmatrix} 2 & 1 & 6 & 4 & -2 \\ 0 & 1 & 0 & 2 & 4 \\ 0 & 0 & 3 & 1 & 5 \\ 0 & 0 & 0 & 1 & 4 \end{bmatrix} \quad B = \begin{bmatrix} 1 & 5 & 0 & 2 & 3 \\ 0 & 0 & 6 & 1 & 1 \\ 0 & 0 & 0 & 2 & 4 \\ 0 & 0 & 0 & 0 & 0 \end{bmatrix} \quad C = \begin{bmatrix} 3 & 1 & 4 & 6 \\ 0 & 5 & 2 & 4 \\ 0 & 1 & 3 & 6 \\ 0 & 2 & 1 & 1 \end{bmatrix}$$

---

* Many definitions of echelon form require that the first nonzero entry in each row be a 1, but we do not insist on this. A related concept is that of "reduced echelon form," which we discuss later.

$$D = \begin{bmatrix} 2 & 1 & 6 & 0 & 3 & 4 \\ 0 & 0 & 2 & 0 & 4 & 3 \\ 0 & 0 & 0 & 0 & 1 & 1 \\ 0 & 0 & 0 & 0 & 0 & -1 \end{bmatrix} \qquad E = \begin{bmatrix} 0 & 5 & 3 & 1 \\ 2 & 6 & 1 & 5 \\ 0 & 0 & 2 & 1 \\ 0 & 0 & 0 & 0 \end{bmatrix}$$

If these matrices are augmented matrices for linear systems, then A represents the (4×4) system

$$\begin{aligned} 2x_1 + x_2 + 6x_3 + 4x_4 &= -2 \\ x_2 \qquad\quad + 2x_4 &= 4 \\ 3x_3 + x_4 &= 5 \\ x_4 &= 4 \end{aligned}$$

while C represents the (4×3) overdetermined system

$$\begin{aligned} 3x_1 + x_2 + 4x_3 &= 6 \\ 5x_2 + 2x_3 &= 4 \\ x_2 + 3x_3 &= 6 \\ 2x_2 + x_3 &= 1. \end{aligned}$$

Note that the system represented by A has a nice "triangular" form, and we can see at a glance that this system is consistent and has a unique solution. For A, in the notation of Definition 1.1, $k = 4, p_1 = 1, p_2 = 2, p_3 = 3, p_4 = 4$. The matrix C is not in echelon form, for the first nonzero entry in the second row is in column two, but there are also nonzero entries in column two in the third and fourth rows. In fact without further eliminations we cannot tell whether the system represented by C is consistent or inconsistent. (With further eliminations it is easy to verify that the system is inconsistent.) As we continue, B clearly represents a consistent system of four equations in four unknowns with infinitely many solutions, and D represents an inconsistent system of four equations in five unknowns. While E is not in echelon form, the interchange of rows one and two will transform E to a row-equivalent matrix in echelon form. In the terminology of Definition 1.1, we see for B that $k = 3$, $p_1 = 1, p_2 = 3, p_3 = 4$ while for D, $k = 4, p_1 = 1, p_2 = 3, p_3 = 5, p_4 = 6$.

A slightly more unusual matrix in echelon form is

$$F = \begin{bmatrix} 0 & 1 & 2 & -1 & 3 \\ 0 & 0 & 2 & 4 & 1 \\ 0 & 0 & 0 & 1 & 3 \\ 0 & 0 & 0 & 0 & 0 \end{bmatrix},$$

and we see that the definition of echelon form allows the possibility that the first several columns may consist entirely of zeros.

EXAMPLE 1.10 Solve the (3×4) system of linear equations

$$\begin{aligned} x_1 + 2x_2 + x_3 - x_4 &= 1 \\ x_1 + 2x_2 - x_3 + x_4 &= 5 \\ -x_1 + x_2 + 2x_3 \qquad &= 0. \end{aligned}$$

The augmented matrix for this system is

$$B = \begin{bmatrix} 1 & 2 & 1 & -1 & 1 \\ 1 & 2 & -1 & 1 & 5 \\ -1 & 1 & 2 & 0 & 0 \end{bmatrix}.$$

To introduce zeros in the (2, 1) and (3, 1) locations, we multiply the first row of B by -1 and add the result to the second row, and then add the first row of B to the third row of B. The result is a new (3×5) matrix, B_1, which is row equivalent to B:

$$B_1 = \begin{bmatrix} 1 & 2 & 1 & -1 & 1 \\ 0 & 0 & -2 & 2 & 4 \\ 0 & 3 & 3 & -1 & 1 \end{bmatrix}.$$

Interchanging rows two and three of B_1, we get a matrix B_2, which is in echelon form:

$$B_2 = \begin{bmatrix} 1 & 2 & 1 & -1 & 1 \\ 0 & 3 & 3 & -1 & 1 \\ 0 & 0 & -2 & 2 & 4 \end{bmatrix}.$$

Since B_2 was obtained from B by elementary row operations, B_2 and B are row equivalent. Hence as augmented matrices, they represent equivalent linear systems. Since B_2 represents the system

$$\begin{aligned} x_1 + 2x_2 + x_3 - x_4 &= 1 \\ 3x_2 + 3x_3 - x_4 &= 1 \\ -2x_3 + 2x_4 &= 4, \end{aligned}$$

then the original system is seen to have the solution

$$\begin{aligned} x_3 &= x_4 - 2 \\ x_2 &= (7 - 2x_4)/3 \\ x_1 &= (4x_4 - 5)/3. \end{aligned}$$

EXAMPLE 1.11 Below is an example of a homogeneous system. As we note in the next section, a system of the form (1.1) is called homogeneous when all the terms $b_1, b_2, \ldots, b_m$ on the right-hand side are zero.

$$\begin{aligned} x_1 + 2x_2 + x_3 + 3x_4 &= 0 \\ 2x_1 + 4x_2 + 3x_3 + x_4 &= 0 \\ 3x_1 + 6x_2 + 6x_3 + 2x_4 &= 0 \end{aligned}$$

The associated augmented matrix for this homogeneous system is

$$B = \begin{bmatrix} 1 & 2 & 1 & 3 & 0 \\ 2 & 4 & 3 & 1 & 0 \\ 3 & 6 & 6 & 2 & 0 \end{bmatrix}.$$

The augmented matrix is row equivalent to

$$C = \begin{bmatrix} 1 & 2 & 1 & 3 & 0 \\ 0 & 0 & 1 & -5 & 0 \\ 0 & 0 & 0 & 8 & 0 \end{bmatrix}.$$

Since C is in echelon form, we can read off the solution to the original system: $x_4 = 0$, $x_3 = 0$, $x_1 = -2x_2$. It should be clear from this example that if B is an augmented matrix for *any* homogeneous linear system, and if C is row equivalent to B, then C is also the augmented matrix for a homogeneous system. That is, the last column of zeros will be maintained under elementary row operations.

We close this section with a brief discussion of reduced echelon form, which is a simple extension of the idea of echelon form. We introduce reduced echelon form with an example, beginning with the (4×5) matrix A, which is in echelon form:

$$A = \begin{bmatrix} 2 & 1 & 6 & 4 & -2 \\ 0 & 1 & 0 & 2 & 4 \\ 0 & 0 & 3 & 1 & 5 \\ 0 & 0 & 0 & 1 & 4 \end{bmatrix}.$$

If A is the augmented matrix for a linear system, then we can backsolve the system to find the solution. An alternative to backsolving is to use elementary row operations to transform A into a row-equivalent matrix of the form

$$B = \begin{bmatrix} * & 0 & 0 & 0 & \times \\ 0 & * & 0 & 0 & \times \\ 0 & 0 & * & 0 & \times \\ 0 & 0 & 0 & * & \times \end{bmatrix}.$$

Clearly it is a trivial matter to solve any linear system for which B is the augmented matrix, and B is an example of a matrix in reduced echelon form. (In the notation of Definition 1.1, a matrix C that is in echelon form is in ***reduced echelon form*** if all the entries in column p_i are zero in rows $1, 2, \ldots, i-1$, for $1 \leqslant i \leqslant k$.)

It is easy to see that any matrix in echelon form can be transformed by elementary row operations to a row-equivalent matrix in reduced echelon form as we illustrate with the (4×5) matrix A given above. To transform A to reduced echelon form, we essentially perform Gauss elimination, this time moving from right to left. The sequence of elementary row operations, beginning with A, follows. First multiply the fourth row of A by -1, add the result to the third row, and obtain

$$A_1 = \begin{bmatrix} 2 & 1 & 6 & 4 & -2 \\ 0 & 1 & 0 & 2 & 4 \\ 0 & 0 & 3 & 0 & 1 \\ 0 & 0 & 0 & 1 & 4 \end{bmatrix}.$$

Multiply row four of A_1 by -2 and add the result to the second row:

$$A_2 = \begin{bmatrix} 2 & 1 & 6 & 4 & -2 \\ 0 & 1 & 0 & 0 & -4 \\ 0 & 0 & 3 & 0 & 1 \\ 0 & 0 & 0 & 1 & 4 \end{bmatrix}.$$

Multiply row four by -4 and add the result to the first row:

$$A_3 = \begin{bmatrix} 2 & 1 & 6 & 0 & -18 \\ 0 & 1 & 0 & 0 & -4 \\ 0 & 0 & 3 & 0 & 1 \\ 0 & 0 & 0 & 1 & 4 \end{bmatrix}.$$

(Note that the echelon form of A ensures that these elementary row operations do not disturb any of the entries in columns 1 to 3.) Continuing, we now introduce zeros into the third column of A_3, above the third row, by multiplying row three by -2, adding the result to row one, and obtaining

$$A_4 = \begin{bmatrix} 2 & 1 & 0 & 0 & -20 \\ 0 & 1 & 0 & 0 & -4 \\ 0 & 0 & 3 & 0 & 1 \\ 0 & 0 & 0 & 1 & 4 \end{bmatrix}.$$

Next multiply row two by -1 and add the result to row one:

$$A_5 = \begin{bmatrix} 2 & 0 & 0 & 0 & -16 \\ 0 & 1 & 0 & 0 & -4 \\ 0 & 0 & 3 & 0 & 1 \\ 0 & 0 & 0 & 1 & 4 \end{bmatrix}.$$

The matrix A_5 is in reduced echelon form and as an augmented matrix, represents the system

$$\begin{aligned} 2x_1 \qquad\qquad\qquad &= -16 \\ x_2 \qquad\qquad &= -4 \\ 3x_3 \qquad &= 1 \\ x_4 &= 4. \end{aligned}$$

1.2 EXERCISES

1. Find the coefficient matrix and the augmented matrix for each of the systems (a), (c), (g), and (k) in Exercise 1, Section 1.1.
2. Find the coefficient matrix and the augmented matrix for the systems (b), (d), and (e) in Exercise 1, Section 1.1.

3. Suppose $A = (a_{ij})$ is a (2×3) matrix where $a_{11} = 2$, $a_{12} = 1$, $a_{13} = 6$, $a_{21} = 4$, $a_{22} = 3$, $a_{23} = 8$.

a) Reduce A to echelon form.

b) If A is the augmented matrix for a linear system, what is the system?

c) Using the echelon form found in (a), solve the system in (b).

4. Reduce the matrix C in Example 1.9 to echelon form.

5. Let

$$A = \begin{bmatrix} 1 & 2 & -1 & -2 \\ 0 & 2 & -2 & -3 \\ 0 & 0 & 0 & 1 \end{bmatrix} \qquad B = \begin{bmatrix} 2 & -1 & 3 \\ 0 & 1 & 1 \\ 0 & 2 & -1 \\ 0 & 0 & -3 \end{bmatrix}$$

$$C = \begin{bmatrix} -1 & 4 & -3 & 4 & 6 \\ 0 & 2 & 1 & -3 & -3 \\ 0 & 0 & 0 & 1 & 2 \end{bmatrix} \qquad D = \begin{bmatrix} 1 & -3 & 4 & 1 & -3 \\ 0 & 1 & 2 & -2 & 0 \\ 0 & 2 & 0 & -1 & 5 \\ 0 & 0 & 4 & -2 & 10 \end{bmatrix}.$$

a) Determine if the matrices above are in echelon form; and if not, reduce them to echelon form.

b) Consider each matrix above as an augmented matrix for a linear system. Identify the system and solve it using the echelon form found in (a).

6. Solve each of the systems (b), (d), (f), (h), and (j) of Exercise 1, Section 1.1, by reducing the augmented matrix for the system to echelon form.

7. Form the augmented matrix for each system, reduce the augmented matrix to echelon form, and give the solution of the system.

a) $\begin{aligned} x_1 - x_2 &= -1 \\ x_1 + x_2 &= 3 \end{aligned}$

b) $\begin{aligned} x_1 + x_2 - x_3 &= 2 \\ 2x_1 \qquad - x_3 &= 1 \end{aligned}$

c) $\begin{aligned} x_1 + 3x_2 - x_3 &= 1 \\ 2x_1 + 5x_2 + x_3 &= 5 \\ x_1 + x_2 + x_3 &= 3 \end{aligned}$

d) $\begin{aligned} x_1 + x_2 + 2x_3 &= 6 \\ 3x_1 + 4x_2 - x_3 &= 5 \\ -x_1 + x_2 + x_3 &= 2 \end{aligned}$

e) $\begin{aligned} x_1 + x_2 - 3x_3 &= -1 \\ x_1 + 2x_2 - 5x_3 &= -2 \\ -x_1 - 3x_2 + 7x_3 &= 3 \end{aligned}$

f) $\begin{aligned} x_1 + x_2 + x_3 &= 1 \\ 2x_1 + 3x_2 + x_3 &= 2 \\ x_1 - x_2 + 3x_3 &= 2 \end{aligned}$

8. Put each of the matrices in Example 1.9 into reduced echelon form.

9. Put the matrices A, B, and D of Exercise 5 into reduced echelon form.

10. Show that A and B are row-equivalent matrices. (That is, apply a sequence of elementary row operations to A to produce B; see Exercise 12.)

$$A = \begin{bmatrix} 1 & 2 \\ 2 & 3 \end{bmatrix} \qquad B = \begin{bmatrix} 1 & 2 \\ 3 & 4 \end{bmatrix}.$$

11. Regard each matrix below as the augmented matrix for a linear system. Determine by inspection (without solving the system) which of these represent consistent systems and which of the consistent systems have unique solutions.

$$A = \begin{bmatrix} 1 & 2 & 1 & 4 \\ 0 & 4 & 3 & 1 \\ 0 & 0 & 0 & 2 \end{bmatrix} \qquad B = \begin{bmatrix} 1 & 2 & 1 & 4 \\ 0 & 4 & 3 & 1 \\ 0 & 0 & 0 & 0 \end{bmatrix}$$

$$C = \begin{bmatrix} 1 & 2 & 1 \\ 0 & 1 & 3 \\ 0 & 0 & 4 \end{bmatrix} \qquad D = \begin{bmatrix} 1 & 4 & 3 & 7 & 6 \\ 0 & 0 & 8 & 2 & 1 \\ 0 & 0 & 0 & 1 & 3 \end{bmatrix}$$

$$E = \begin{bmatrix} 1 & 2 & 6 & 3 \\ 0 & 2 & 1 & 4 \\ 0 & 0 & 3 & 0 \end{bmatrix} \qquad F = \begin{bmatrix} 4 & 7 & 1 \\ 0 & 2 & 0 \\ 0 & 0 & 0 \end{bmatrix}$$

12. Let A and I be as below. Prove that if $b - cd \neq 0$, then A is row equivalent to I.

$$A = \begin{bmatrix} 1 & d \\ c & b \end{bmatrix} \qquad I = \begin{bmatrix} 1 & 0 \\ 0 & 1 \end{bmatrix}.$$

13. Each matrix is the augmented matrix for a system of linear equations. Find the solution, or state that the system is inconsistent.

a) $\begin{bmatrix} 1 & 1 & 0 \\ 0 & 1 & 0 \end{bmatrix}$ b) $\begin{bmatrix} 1 & 2 & 1 & 0 \\ 0 & 1 & 3 & 0 \end{bmatrix}$ c) $\begin{bmatrix} 2 & 4 & 2 & 0 \\ 0 & 3 & 1 & 0 \\ 0 & 0 & 4 & 0 \end{bmatrix}$

d) $\begin{bmatrix} 2 & 4 & 2 & 2 & 0 \\ 0 & 1 & 2 & 2 & 0 \\ 0 & 0 & 1 & 0 & 0 \end{bmatrix}$ e) $\begin{bmatrix} 1 & 1 & 1 & 0 \\ 0 & 0 & 1 & 0 \\ 0 & 0 & 1 & 0 \end{bmatrix}$ f) $\begin{bmatrix} 1 & 1 & 2 & 0 & 2 & 0 \\ 0 & 0 & 1 & 1 & 0 & 0 \\ 0 & 0 & 0 & 1 & 1 & 0 \end{bmatrix}$

14. Repeat Exercise 13 for

a) $\begin{bmatrix} 1 & 1 & 0 \\ 0 & 0 & 0 \end{bmatrix}$ b) $\begin{bmatrix} 1 & 2 & 2 & 0 \\ 0 & 1 & 0 & 0 \end{bmatrix}$ c) $\begin{bmatrix} 1 & -1 & 1 & 0 \\ 0 & 1 & 1 & 0 \\ 0 & 0 & 2 & 0 \end{bmatrix}$

d) $\begin{bmatrix} 1 & 2 & 1 & 2 & 0 \\ 0 & 2 & 2 & 4 & 0 \\ 0 & 0 & 1 & 1 & 0 \end{bmatrix}$ e) $\begin{bmatrix} 1 & 0 & 1 & 0 & 0 \\ 0 & 0 & 1 & 1 & 0 \\ 0 & 0 & 0 & 1 & 0 \end{bmatrix}$ f) $\begin{bmatrix} 1 & 0 & 0 & 0 & 0 \\ 0 & 1 & 1 & 1 & 0 \\ 0 & 0 & 1 & 1 & 0 \end{bmatrix}$

1.3 REDUCTION TO ECHELON FORM AND SOLUTIONS OF HOMOGENEOUS SYSTEMS

This brief section contains two important theoretical results relating to the solution of a system of linear equations. The first result, found in Theorem 1.1 and its corollary, gives a thorough description of the relation between Gauss elimination and reduction to echelon form. The second result (Theorem 1.2) is important because in linear algebra many other concepts (such as dimension, see Chapter 2) depend on its validity. Furthermore both of these results play important roles in connecting the theory of linear algebra to practical computation.

Theorem 1.1

Let B be an $(m \times n)$ matrix. Then there is an $(m \times n)$ matrix C such that

1. C is in echelon form, and
2. B is row equivalent to C.

Corollary

Any $(m \times n)$ system of linear equations is equivalent to some $(m \times n)$ system of linear equations whose augmented matrix is in echelon form.

The proof of Theorem 1.1 is rather tedious and amounts to merely applying the elementary row operations, discussed in Section 1.2, to a general matrix. Since the proof contains no really new ideas, we defer it to the end of this section. The corollary to Theorem 1.1 is immediate since as we observed in the last section, the elementary row operations used to transform the augmented matrix of an arbitrary linear system to echelon form exactly duplicate the elementary operations for systems of equations. The computational significance of the corollary is that any system of linear equations can be reduced quickly to an equivalent system whose augmented matrix is in echelon form. The echelon form of the augmented matrix allows immediate recognition of whether the system is consistent or not; and if it is consistent, the reduced system can be backsolved to find the solution or solutions.

As we have mentioned, Gauss elimination is more than just a theoretical tool. Indeed most of the general-purpose computer-software packages currently used for solving systems of linear equations employ either Gauss elimination or a closely related variant called LU-decomposition. In Chapter 6 we discuss a few of the practical considerations associated with the use of Gauss elimination in computer programs. The practical aspects mostly concern the role of "pivoting" strategies, which are used in an attempt to control roundoff errors. As we shall see in Chapter 6, roundoff errors result from the fact that a computer (or a hand calculator) can represent a given number with only a fixed and finite number of digits. Thus a digital computer cannot exactly represent irrational numbers (such as π or $\sqrt{3}$) or rational numbers (such as 1/3) that have an infinite decimal expansion. Moreover a computer cannot usually add numbers of different orders of magnitude without an error or a loss of significant figures. For these reasons we are well advised to use pivoting strategies in Gauss elimination to combat the effects of roundoff error.

The other result of this section, which we prove below, is really the theoretical underpinning for a variety of important concepts in linear algebra and matrix theory, concepts such as rank, dimension, and eigenvalues of matrices. In order to state this result, we need the definition of a homogeneous system: an $(m \times n)$ system of linear equations of the form

$$\begin{array}{c} a_{11}x_1 + a_{12}x_2 + \ldots + a_{1n}x_n = 0 \\ a_{21}x_1 + a_{22}x_2 + \ldots + a_{2n}x_n = 0 \\ \vdots \qquad\qquad\qquad\qquad \vdots \quad \vdots \\ a_{m1}x_1 + a_{m2}x_2 + \ldots + a_{mn}x_n = 0 \end{array} \qquad (1.11)$$

is a ***homogeneous*** system of linear equations. Thus (1.11) is just a special case of (1.1) where $b_1 = b_2 = \ldots = b_m = 0$. We note that (1.11) is *always* a consistent system since the choice $x_1 = x_2 = \ldots = x_n = 0$ is a solution to (1.11)—this solution is called the ***trivial solution*** or ***zero solution***. We will establish the important result that when (1.11) has more unknowns than equations (that is, when $n > m$), then (1.11) has other solutions besides the trivial solution. This result is suggested by our intuition since $n > m$ means that (1.11) has fewer constraints (equations) than freedoms (unknowns, or variables).

EXAMPLE 1.12 An example of a homogeneous system with fewer equations than unknowns is

$$\begin{aligned} x_1 - x_2 + 2x_3 &= 0 \\ x_1 - 2x_2 + x_3 &= 0. \end{aligned}$$

Clearly the choice $x_1 = 0, x_2 = 0, x_3 = 0$ is a solution of this homogeneous system. Theorem 1.2, which we prove below, asserts that this (2×3) system has nontrivial solutions also. This assertion is easy to verify in this example since we see that the system is equivalent to

$$\begin{aligned} x_1 - x_2 + 2x_3 &= 0 \\ -x_2 - x_3 &= 0. \end{aligned}$$

So for instance if we set $x_3 = 1$, then we find $x_2 = -1$ and $x_1 = -3$, which is a nontrivial solution.

While a homogeneous system is always consistent, a nonhomogeneous system of m equations in n unknowns may have no solution regardless of whether m is less than n, m is greater than n, or $m = n$. For instance the (2×3) nonhomogeneous system.

$$\begin{aligned} x_1 + x_2 + x_3 &= 1 \\ x_1 + x_2 + x_3 &= 7 \end{aligned}$$

is obviously inconsistent.

Theorem 1.2 always nontrivial solutions when rows < columns

A homogeneous $(m \times n)$ system of linear equations always has nontrivial solutions when $m < n$.

Proof. Before giving the complete proof, we wish to consider the special case of (1.11) where $m = 2$ and $n = 3$. The principle employed in proving Theorem 1.2 for this special case is exactly the same as that used for the general case; so this example should serve to illustrate and clarify the general proof. Given a (2×3) homogeneous system

$$\begin{aligned} b_{11}x_1 + b_{12}x_2 + b_{13}x_3 &= 0 \\ b_{21}x_1 + b_{22}x_2 + b_{23}x_3 &= 0, \end{aligned} \qquad (1.12)$$

trivial solution – zero solution

the augmented matrix for this system has the form

$$B = \begin{bmatrix} b_{11} & b_{12} & b_{13} & 0 \\ b_{21} & b_{22} & b_{23} & 0 \end{bmatrix}.$$

According to Theorem 1.1, we can use elementary row operations on B to derive a matrix C that is row equivalent to B and that is in echelon form. Since C is in echelon form, we can assume C has the form below:

$$C = \begin{bmatrix} c_{11} & c_{12} & c_{13} & 0 \\ 0 & c_{22} & c_{23} & 0 \end{bmatrix}.$$

For C we do not specify whether $c_{11}, c_{12}, c_{13}, c_{22}, c_{23}$ are zero or not—the important thing to note is that the fourth column consists of zeros (since elementary row operations applied to B cannot disturb the zeros in the fourth column of B). Thus C serves as a general model for a (2×4) matrix in echelon form where the fourth column consists of zeros. That is, (1.12) is equivalent to a reduced system of the form

$$\begin{aligned} c_{11}x_1 + c_{12}x_2 + c_{13}x_3 &= 0 \\ c_{22}x_2 + c_{23}x_3 &= 0. \end{aligned} \tag{1.13}$$

To argue that (1.13) has a nontrivial solution (the point of Theorem 1.2) is now an easy matter. If $c_{11} = 0$, then the choice $x_1 = 1$, $x_2 = 0$, $x_3 = 0$ gives a nontrivial solution to (1.13). If $c_{11} \neq 0$ but $c_{22} = 0$, then we can choose $x_2 = 1, x_3 = 0$ [which solves the second equation of (1.13)] and choose x_1 from

$$x_1 = -c_{12}/c_{11}$$

to solve the first equation. The only remaining situation is $c_{11} \neq 0$ and $c_{22} \neq 0$. In this case we set $x_3 = 1$ and determine x_2 and x_1 by backsolving (1.13).

The proof for the general $(m \times n)$ case of Theorem 1.2 follows these same lines. Let B be the $(m \times (n+1))$ augmented matrix for the system (1.11) where $m < n$. By Theorem 1.1 and its corollary, the system represented by B is equivalent to a system of the form

$$\begin{aligned} c_{11}x_1 + c_{12}x_2 + \ldots + c_{1m}x_m + \ldots + c_{1n}x_n &= 0 \\ c_{22}x_2 + \ldots + c_{2m}x_m + \ldots + c_{2n}x_n &= 0 \\ \ddots \quad c_{mm}x_m + \ldots + c_{mn}x_n &= 0. \end{aligned} \tag{1.14}$$

The augmented matrix representing (1.14) is in echelon form; so the coefficients c_{ij}, $i > j$, are zero and are not shown in (1.14). However we are *not* asserting that the diagonal coefficients, c_{ii}, are nonzero. As in the special (2×3) case above if $c_{11} = 0$, then $x_1 = 1, x_2 = x_3 = \ldots = x_n = 0$ is a nontrivial solution to the original system. On the other hand let us suppose that $c_{11}, c_{22}, \ldots, c_{rr}$ are nonzero, but $c_{r+1,r+1} = 0$ where $r+1 \leqslant m$. In this case we choose $x_{r+1} = 1$, $x_{r+2} = \ldots = x_n = 0$ and backsolve in (1.14) for $x_1, x_2, \ldots, x_r$. The last case to consider is that in which c_{11}, $c_{22}, \ldots, c_{mm}$ are all nonzero. In this case we set $x_{m+1} = \ldots = x_n = 1$, backsolve (1.14) for $x_1, \ldots, x_m$, and again obtain a nontrivial solution. Thus since any system of the form (1.14) has a nontrivial solution, Theorem 1.2 is established. ■

We conclude this section by sketching the proof of Theorem 1.1. Let B be an $(m \times n)$ matrix. If B has only zero entries, then B is already in echelon form. If B contains nonzero entries, we find the first column of B that contains a nonzero entry. Suppose $b_{rs} \neq 0$ but $b_{ij} = 0$ for $1 \leqslant i \leqslant m$ and $j = 1, 2, \ldots, s-1$. Interchange row r and row 1 to produce a row-equivalent matrix B_1 where the $(1, r)$ entry of B_1 is nonzero. Add multiples of the first row of B_1 to rows 2, 3, . . ., m to produce a row-equivalent matrix B_2 with zeros in the $(2, r), (3, r), \ldots, (m, r)$ positions. Now fix the first row of B_2 and repeat the process on the remaining $m-1$ rows of B_2. Ultimately we arrive at an $(m \times n)$ matrix C that is row equivalent to B and is in echelon form.

1.3 EXERCISES

1. Which of the following are homogeneous systems? Which systems have a nontrivial solution?

a) $\begin{aligned} x_1 + x_2 &= 0 \\ x_1 - x_2 &= 0 \end{aligned}$ b) $\begin{aligned} x_1 + x_2 - x_3 &= 0 \\ x_1 + 2x_2 + x_3 &= 4 \end{aligned}$ c) $\begin{aligned} x_1 + x_2 &= 0 \\ x_2 &= 1 \end{aligned}$ d) $\begin{aligned} x_1 - x_2 + 2x_3 &= 0 \\ x_1 + x_2 - 3x_3 &= 0 \end{aligned}$ nontrivial solution because more unknowns than equations Th. 1.2

2. Give an example of a (2×2) homogeneous system that has a nontrivial solution. Does this contradict Theorem 1.2?

3. Find a nontrivial solution of the following systems such that $x_3 = 1$. Is it possible to find a solution where $x_2 = 2$?

a) $\begin{aligned} x_1 + x_2 - x_3 &= 0 \\ x_1 - x_2 + x_3 &= 0 \end{aligned}$ b) $\begin{aligned} x_1 + x_2 - x_3 &= 0 \\ x_1 - x_2 - x_3 &= 0 \end{aligned}$

c) $\begin{aligned} x_1 + x_2 + x_3 &= 0 \\ 2x_1 + 2x_2 + 2x_3 &= 0 \end{aligned}$ d) $\begin{aligned} x_1 + 2x_2 + x_3 + 3x_4 &= 0 \\ 2x_1 + 5x_2 + 2x_3 + x_4 &= 0 \\ x_1 + 3x_2 + x_3 - x_4 &= 0 \end{aligned}$

4. a) Are each of the systems in Exercise 1 consistent?
   b) Is the following system consistent?

$$\begin{aligned} x_1 + x_2 - x_3 &= 0 \\ x_1 + x_2 - x_3 &= 1 \end{aligned}$$

   c) Does the answer to (b) contradict Theorem 1.2?

5. Let B be the the (3×4) matrix given below. Find a (3×4) matrix C where C is in echelon form and where B is row equivalent to C.

$$B = \begin{bmatrix} 0 & 1 & 2 & 1 \\ 1 & -1 & 3 & 4 \\ 2 & -1 & 8 & 1 \end{bmatrix}$$

If B is the augmented matrix for a (3×3) system, is the system consistent?

6. Prove Theorem 1.1 for the special case:

$$B = \begin{bmatrix} b_{11} & b_{12} & b_{13} \\ b_{21} & b_{22} & b_{23} \end{bmatrix}.$$

(Hint: If $b_{11} = 0$ but $b_{21} \neq 0$, interchange rows 1 and 2. If $b_{11} = b_{21} = 0$ and also $b_{12} = 0$, again interchange rows 1 and 2, etc.)

7. From Exercise 8, Section 1.1, if $a_{11}a_{22} - a_{21}a_{12} \neq 0$, then the matrix

$$\begin{bmatrix} a_{11} & a_{12} & 0 \\ a_{21} & a_{22} & 0 \end{bmatrix}$$

is row equivalent to a matrix of the form

$$\begin{bmatrix} c_{11} & c_{12} & 0 \\ 0 & c_{22} & 0 \end{bmatrix}$$

where $c_{11} \neq 0$ and $c_{22} \neq 0$.

a) Can the system

$$\begin{aligned} a_{11}x_1 + a_{12}x_2 &= 0 \\ a_{21}x_1 + a_{22}x_2 &= 0 \end{aligned}$$

have a nontrivial solution when $a_{11}a_{22} - a_{21}a_{12} \neq 0$?

b) Find all values λ for which

$$\begin{aligned} (3-\lambda)x_1 + 2x_2 &= 0 \\ 2x_1 + (3-\lambda)x_2 &= 0 \end{aligned}$$

has nontrivial solutions. [Hint: Use part (a).]

1.4 MATRIX OPERATIONS

Augmented matrices can be used to represent compactly systems of linear equations, and matrices can be used as a shorthand notation to describe the steps of Gauss elimination. Moreover the language and notation of matrix theory were convenient to describe the objective of Gauss elimination—a system whose augmented matrix is in echelon form. In this section we show that a natural arithmetic associated with matrices makes them a useful computational and theoretical tool, as well as a notational device. In fact this arithmetic gives matrices a structure that makes them a mathematical system having wide applicability. Before defining the arithmetic operations for matrices, we must first define what we mean by equality of matrices.

DEFINITION 1.2.

Let $A = (a_{ij})$ be an $(m \times n)$ matrix, and let $B = (b_{ij})$ be an $(r \times s)$ matrix. We say A and B are ***equal*** (and write $A = B$) if $m = r$, $n = s$ and if $a_{ij} = b_{ij}$ for all i and j, $1 \leqslant i \leqslant m$, $1 \leqslant j \leqslant n$.

Thus two matrices are equal if they both have the same number of rows and columns and, moreover, all of their corresponding entries are equal.

The first matrix operation we consider is the product of two matrices. The form of the product operation is suggested by the form of an $(m \times n)$ system of linear equations. We begin by recalling that a point in three-dimensional space can be represented as a three-dimensional vector (an ordered triple of real numbers):

$$\mathbf{x} = \begin{bmatrix} x_1 \\ x_2 \\ x_3 \end{bmatrix};$$

and in general a point in n-dimensional space is an ordered n-tuple of real numbers (an n-dimensional vector):

$$\mathbf{x} = \begin{bmatrix} x_1 \\ x_2 \\ x_3 \\ \vdots \\ x_n \end{bmatrix}.$$

We will use R^n to denote Euclidean n-space (the set of all n-dimensional vectors with real components), and we will denote vectors by boldface type. Thus R^n is the set defined by

$$R^n = \left\{\mathbf{x} : \mathbf{x} = \begin{bmatrix} x_1 \\ x_2 \\ \vdots \\ x_n \end{bmatrix}, x_1, x_2, \ldots, x_n \text{ real numbers}\right\}.$$

Frequently we will find it convenient to regard a $(1 \times m)$ matrix as a row vector, an $(n \times 1)$ matrix as a column vector, and a (1×1) matrix as a scalar.

Given these preliminaries, the multiplication operation for matrices is suggested by the system (1.15)

$$\begin{aligned} a_{11}x_1 + a_{12}x_2 + \ldots + a_{1n}x_n &= b_1 \\ a_{21}x_1 + a_{22}x_2 + \ldots + a_{2n}x_n &= b_2 \\ \vdots \qquad\qquad\qquad\qquad & \quad \vdots \\ a_{m1}x_1 + a_{m2}x_2 + \ldots + a_{mn}x_n &= b_m. \end{aligned} \tag{1.15}$$

In (1.15) the ith equation is $a_{i1}x_1 + a_{i2}x_2 + \ldots + a_{in}x_n = b_i$, which can be written in a briefer form as

$$\sum_{j=1}^{n} a_{ij}x_j = b_i. \tag{1.16}$$

In vector calculus the scalar product (or dot product) of two vectors $\mathbf{y}$ and $\mathbf{z}$

$$\mathbf{y} = \begin{bmatrix} y_1 \\ y_2 \\ \vdots \\ y_n \end{bmatrix} \qquad \mathbf{z} = \begin{bmatrix} z_1 \\ z_2 \\ \vdots \\ z_n \end{bmatrix}$$

is defined as the number $y_1z_1 + y_2z_2 + \ldots + y_nz_n = \sum_{j=1}^{n} y_jz_j$. So if A is the coefficient matrix for (1.15),

$$A = \begin{bmatrix} a_{11} & a_{12} & \dots & a_{1n} \\ a_{21} & a_{22} & & a_{2n} \\ \vdots & & & \vdots \\ a_{m1} & a_{m2} & \dots & a_{mn} \end{bmatrix},$$

and if $\mathbf{x}$ is a vector in R^n,

$$\mathbf{x} = \begin{bmatrix} x_1 \\ x_2 \\ \vdots \\ x_n \end{bmatrix},$$

then the left-hand side of (1.16) is precisely the scalar product of the ith row of A with the vector $\mathbf{x}$. Thus if we define the product of A and $\mathbf{x}$ to be the $(m \times 1)$ vector $A\mathbf{x}$ where the ith component of $A\mathbf{x}$ is the scalar product of the ith row of A with $\mathbf{x}$, then $A\mathbf{x}$ is given by

$$A\mathbf{x} = \begin{bmatrix} \sum_{j=1}^{n} a_{1j}x_j \\ \sum_{j=1}^{n} a_{2j}x_j \\ \vdots \\ \sum_{j=1}^{n} a_{mj}x_j \end{bmatrix}$$

Next let $\mathbf{b}$ be the vector defined by the right-hand side of (1.15):

$$\mathbf{b} = \begin{bmatrix} b_1 \\ b_2 \\ \vdots \\ b_m \end{bmatrix}.$$

Using the definition of equality (Definition 1.2), we see that the simple equation

$$A\mathbf{x} = \mathbf{b}$$

is equivalent to the system (1.15).

In a natural fashion we can extend the idea of the product of a matrix and a vector to the product, AB, of an $(m \times n)$ matrix A and an $(n \times s)$ matrix B by defining the ijth entry of AB to be the scalar product of the ith row of A with the jth column of B. Formally we have the following definition.

DEFINITION 1.3.

Let $A = (a_{ij})$ be an $(m \times n)$ matrix, and let $B = (b_{ij})$ be an $(r \times s)$ matrix. If $n = r$, then the ***product*** AB is defined to be the $(m \times s)$ matrix $C = (c_{ij})$ where

$$c_{ij} = \sum_{k=1}^{n} a_{ik} b_{kj}.$$

If $n \neq r$, then the product AB is not defined.

If $C = AB$, then the definition above provides a formula for the ijth entry of C, $1 \leqslant i \leqslant m$, $1 \leqslant j \leqslant s$. This definition is easy to visualize by thinking of the ijth entry of AB as being given by the scalar product of the ith row of A and the jth column of B. Furthermore for the scalar product to make sense, a row from A must have the same number of entries as a column from B, and this equality is enforced in Definition 1.3 by the requirement that $n = r$. Also the product AB should have dimension $(m \times s)$ since there are m rows in A and s columns in B. Finally we note that this definition is used for computation as well as for theoretical purposes. For example any computer program that requires the formation of a matrix product must use Definition 1.3 in forming the product. The following example gives some examples of matrix products. Note particularly that the matrix product is not commutative; that is, normally AB and BA are *different* matrices.

EXAMPLE 1.13 Let the matrices A, B, C, D, and I be given by

$$A = \begin{bmatrix} 1 & 2 & 1 \\ 3 & -1 & 6 \end{bmatrix} \qquad B = \begin{bmatrix} 3 & 4 \\ 0 & -2 \\ 1 & 1 \end{bmatrix}$$

$$C = \begin{bmatrix} 2 & 1 \\ 1 & 3 \end{bmatrix} \qquad D = \begin{bmatrix} 1 & 1 \\ 2 & 0 \end{bmatrix} \qquad I = \begin{bmatrix} 1 & 0 \\ 0 & 1 \end{bmatrix}.$$

Since A is a (2×3) matrix and B is a (3×2) matrix, we see according to Definition 1.3 that AB is a (2×2) matrix and BA is a (3×3) matrix. The entries of the products AB and BA are easily found from Definition 1.3. For example the $(1, 2)$ entry of AB is the scalar product of the first row of A and the second column of B:

$$\sum_{k=1}^{3} a_{1k} b_{k2} = a_{11}b_{12} + a_{12}b_{22} + a_{13}b_{32} = 1 \cdot 4 + 2 \cdot (-2) + 1 \cdot 1 = 1.$$

In detail AB and BA are given by

$$AB = \begin{bmatrix} 4 & 1 \\ 15 & 20 \end{bmatrix} \qquad BA = \begin{bmatrix} 15 & 2 & 27 \\ -6 & 2 & -12 \\ 4 & 1 & 7 \end{bmatrix}.$$

Note that $AB \neq BA$; in fact AB and BA do not even have the same dimension. We see that CD and DC are both (2×2) matrices where

$$CD = \begin{bmatrix} 4 & 2 \\ 7 & 1 \end{bmatrix} \qquad DC = \begin{bmatrix} 3 & 4 \\ 4 & 2 \end{bmatrix}.$$

Again we see that $CD \neq DC$. Occasionally the matrix product is commutative; for example $IC = CI$ as the reader may verify. As an example where matrix multiplication is not defined, we observe that the product CB cannot be formed since C is (2×2) and B is (3×2). However BC is defined, and we find

$$BC = \begin{bmatrix} 10 & 15 \\ -2 & -6 \\ 3 & 4 \end{bmatrix}.$$

Since a vector can be regarded as a special type of matrix, Definition 1.3, of course, also applies to define the product of a matrix and a vector. For example let

$$\mathbf{w} = \begin{bmatrix} 1 \\ 1 \end{bmatrix};$$

then $B\mathbf{w}$ is the product of a (3×2) matrix and a (2×1) matrix; so $B\mathbf{w}$ is a (3×1) matrix (a column vector) and $B\mathbf{w}$ is given by

$$B\mathbf{w} = \begin{bmatrix} 7 \\ -2 \\ 2 \end{bmatrix}.$$

The next example emphasizes the natural equivalence between a system of linear equations and a matrix equation $A\mathbf{x} = \mathbf{b}$ where A is the coefficient matrix of the system. This equivalence is a direct result, of course, of the way matrix multiplication was defined.

EXAMPLE 1.14 We again use the system (1.6) as an example:

$$\begin{aligned} x_1 - 2x_2 + x_3 &= 2 \\ 2x_1 + x_2 - x_3 &= 1 \\ -3x_1 + x_2 - 2x_3 &= -5. \end{aligned}$$

Using the system above to define A, $\mathbf{x}$, and $\mathbf{b}$, we have

$$A = \begin{bmatrix} 1 & -2 & 1 \\ 2 & 1 & -1 \\ -3 & 1 & -2 \end{bmatrix}, \qquad \mathbf{x} = \begin{bmatrix} x_1 \\ x_2 \\ x_3 \end{bmatrix}, \qquad \mathbf{b} = \begin{bmatrix} 2 \\ 1 \\ -5 \end{bmatrix};$$

so that $A\mathbf{x}$ is the vector

$$A\mathbf{x} = \begin{bmatrix} x_1 - 2x_2 + x_3 \\ 2x_1 + x_2 - x_3 \\ -3x_1 + x_2 - 2x_3 \end{bmatrix}.$$

Thus from the definition of matrix equality we see that $A\mathbf{x} = \mathbf{b}$ is solved if and only if the components of the solution vector, $\mathbf{x}$, satisfy the system (1.6). That is, the system (1.6) and the matrix equation $A\mathbf{x} = \mathbf{b}$ are equivalent problems.

Two further arithmetic operations that are necessary for theory and computation are matrix addition and the multiplication of a matrix by a scalar.

DEFINITION 1.4.

Let $A = (a_{ij})$ and $B = (b_{ij})$ both be $(m \times n)$ matrices. The ***sum***, $A + B$, is defined to be the $(m \times n)$ matrix $C = (c_{ij})$ where

$$c_{ij} = a_{ij} + b_{ij}.$$

Note that this definition requires A and B to have the *same* dimension before their sum is defined. The product of a scalar and a matrix is given in Definition 1.5.

DEFINITION 1.5.

Let $A = (a_{ij})$ be an $(m \times n)$ matrix, and let r be a scalar. The product, rA, is defined to be the $(m \times n)$ matrix $D = (d_{ij})$ where

$$d_{ij} = ra_{ij}.$$

EXAMPLE 1.15 Let the matrices A, B, and C be given by

$$A = \begin{bmatrix} 1 & 3 \\ -2 & 7 \end{bmatrix}, \quad B = \begin{bmatrix} 6 & 1 \\ 2 & 4 \end{bmatrix}, \text{ and } C = \begin{bmatrix} 1 & 2 & -1 \\ 3 & 0 & 5 \end{bmatrix}.$$

Then

$$A + B = \begin{bmatrix} 7 & 4 \\ 0 & 11 \end{bmatrix}, \quad 3C = \begin{bmatrix} 3 & 6 & -3 \\ 9 & 0 & 15 \end{bmatrix}, \quad A + 2B = \begin{bmatrix} 13 & 5 \\ 2 & 15 \end{bmatrix},$$

and $$(A + 2B)(3C) = \begin{bmatrix} 84 & 78 & 36 \\ 141 & 12 & 219 \end{bmatrix}$$

whereas $A + C$ and $B + C$ are not defined.

In matrix manipulations we frequently find it convenient and useful to express an $(m \times n)$ matrix $A = (a_{ij})$ in the form

$$A = [\mathbf{A}_1, \mathbf{A}_2, \ldots, \mathbf{A}_n] \tag{1.17a}$$

where for each j, $1 \leqslant j \leqslant n$, $\mathbf{A}_j$ denotes the jth column of A. Hence $\mathbf{A}_j$ is the $(m \times 1)$

column vector

$$\mathbf{A}_j = \begin{bmatrix} a_{1j} \\ a_{2j} \\ \vdots \\ a_{mj} \end{bmatrix}. \tag{1.17b}$$

For example if A is the (2×3) matrix

$$A = \begin{bmatrix} 1 & 3 & 6 \\ 2 & 4 & 0 \end{bmatrix}, \tag{1.18}$$

then $A = [\mathbf{A}_1, \mathbf{A}_2, \mathbf{A}_3]$ where

$$\mathbf{A}_1 = \begin{bmatrix} 1 \\ 2 \end{bmatrix}, \quad \mathbf{A}_2 = \begin{bmatrix} 3 \\ 4 \end{bmatrix}, \quad \mathbf{A}_3 = \begin{bmatrix} 6 \\ 0 \end{bmatrix}.$$

Our reason for writing A in the form of (1.17a) is found in Theorems 1.3 and 1.4 below. These two theorems provide alternate ways of expressing the vector $A\mathbf{x}$ and the matrix AB; these alternate ways are extremely useful in developing essential concepts in matrix theory. We will use these results repeatedly in all that follows.

Theorem 1.3

Let $A = [\mathbf{A}_1, \mathbf{A}_2, \ldots, \mathbf{A}_n]$ be an $(m \times n)$ matrix whose jth column is $\mathbf{A}_j$, and let $\mathbf{x}$ be the $(n \times 1)$ column vector

$$\mathbf{x} = \begin{bmatrix} x_1 \\ x_2 \\ \vdots \\ x_n \end{bmatrix}.$$

Then the product $A\mathbf{x}$ can be expressed as

$$A\mathbf{x} = x_1\mathbf{A}_1 + x_2\mathbf{A}_2 + \ldots + x_n\mathbf{A}_n. \tag{1.19}$$

The proof of this theorem is not difficult and uses only Definitions 1.2, 1.3, 1.4, and 1.5; we have left the proof as an exercise for the reader. As an example of (1.19), let A be the matrix in (1.18)

$$A = \begin{bmatrix} 1 & 3 & 6 \\ 2 & 4 & 0 \end{bmatrix},$$

and let $\mathbf{x}$ be the vector in R^3

$$\mathbf{x} = \begin{bmatrix} x_1 \\ x_2 \\ x_3 \end{bmatrix}.$$

It is then easy to verify that

$$Ax = x_1 \begin{bmatrix} 1 \\ 2 \end{bmatrix} + x_2 \begin{bmatrix} 3 \\ 4 \end{bmatrix} + x_3 \begin{bmatrix} 6 \\ 0 \end{bmatrix};$$

so that $A\mathbf{x} = x_1\mathbf{A}_1 + x_2\mathbf{A}_2 + x_3\mathbf{A}_3$. Similarly for the vector $\mathbf{v}$

$$\mathbf{v} = \begin{bmatrix} 2 \\ 1 \\ 3 \end{bmatrix},$$

it is easy to check that $A\mathbf{v} = 2\mathbf{A}_1 + \mathbf{A}_2 + 3\mathbf{A}_3$. Finally we see from (1.19) that the equation $A\mathbf{x} = \mathbf{b}$ corresponding to the $(m \times n)$ system (1.1) can be expressed as

$$x_1\mathbf{A}_1 + x_2\mathbf{A}_2 + \ldots + x_n\mathbf{A}_n = \mathbf{b}. \tag{1.20}$$

Thus (1.20) says that solving $A\mathbf{x} = \mathbf{b}$ amounts to showing that $\mathbf{b}$ can be written as a weighted sum (or linear combination) of the columns of A. For example the equation $A\mathbf{x} = \mathbf{b}$ given by

$$\begin{bmatrix} 1 & 2 & 1 \\ 2 & -1 & -1 \\ 1 & 1 & 3 \end{bmatrix} \begin{bmatrix} x_1 \\ x_2 \\ x_3 \end{bmatrix} = \begin{bmatrix} 4 \\ 1 \\ 0 \end{bmatrix},$$

which corresponds to the system

$$\begin{aligned} x_1 + 2x_2 + x_3 &= 4 \\ 2x_1 - x_2 - x_3 &= 1 \\ x_1 + x_2 + 3x_3 &= 0, \end{aligned}$$

can be written as the vector equation $x_1\mathbf{A}_1 + x_2\mathbf{A}_2 + x_3\mathbf{A}_3 = \mathbf{b}$:

$$x_1 \begin{bmatrix} 1 \\ 2 \\ 1 \end{bmatrix} + x_2 \begin{bmatrix} 2 \\ -1 \\ 1 \end{bmatrix} + x_3 \begin{bmatrix} 1 \\ -1 \\ 3 \end{bmatrix} = \begin{bmatrix} 4 \\ 1 \\ 0 \end{bmatrix}.$$

So solving $A\mathbf{x} = \mathbf{b}$ amounts to expressing $\mathbf{b}$ as a linear combination of the columns of A. We see that $x_1 = 1$, $x_2 = 2$, $x_3 = -1$ is the solution since

$$\begin{bmatrix} 1 \\ 2 \\ 1 \end{bmatrix} + 2 \begin{bmatrix} 2 \\ -1 \\ 1 \end{bmatrix} - \begin{bmatrix} 1 \\ -1 \\ 3 \end{bmatrix} = \begin{bmatrix} 4 \\ 1 \\ 0 \end{bmatrix}.$$

Although (1.20) is not particularly efficient as a computational tool, it is useful for understanding how the internal structure of the coefficient matrix affects the possible solutions of the linear system $A\mathbf{x} = \mathbf{b}$.

Another important observation, which we will have occasion to use later, is an alternate way of expressing the product of two matrices as given in Theorem 1.4.

Theorem 1.4

Let A be an $(m \times n)$ matrix, and let $B = [\mathbf{B}_1, \mathbf{B}_2, \ldots, \mathbf{B}_s]$ be an $(n \times s)$ matrix whose kth column is $\mathbf{B}_k$. Then the jth column of AB is $A\mathbf{B}_j$ so that

$$AB = [A\mathbf{B}_1, A\mathbf{B}_2, \ldots, A\mathbf{B}_s].$$

Proof. This theorem is really just an observation as is Theorem 1.3. To see that the jth column of AB is $A\mathbf{B}_j$, we merely note that the jth column of AB contains the entries

$$\begin{bmatrix} \sum_{k=1}^{n} a_{1k}b_{kj} \\ \sum_{k=1}^{n} a_{2k}b_{kj} \\ \vdots \\ \sum_{k=1}^{n} a_{mk}b_{kj} \end{bmatrix};$$

and these are precisely the components of the column vector $A\mathbf{B}_j$ where

$$\mathbf{B}_j = \begin{bmatrix} b_{1j} \\ b_{2j} \\ \vdots \\ b_{nj} \end{bmatrix}.$$

It follows that we can write AB in the form $AB = [A\mathbf{B}_1, A\mathbf{B}_2, \ldots, A\mathbf{B}_s]$. ∎

To illustrate Theorem 1.4, let A and B be given by

$$A = \begin{bmatrix} 2 & 6 \\ 0 & 4 \\ 1 & 2 \end{bmatrix}, \quad B = \begin{bmatrix} 1 & 3 & 0 & 1 \\ 4 & 5 & 2 & 3 \end{bmatrix}.$$

Thus

$$\mathbf{B}_1 = \begin{bmatrix} 1 \\ 4 \end{bmatrix}, \quad \mathbf{B}_2 = \begin{bmatrix} 3 \\ 5 \end{bmatrix}, \quad \mathbf{B}_3 = \begin{bmatrix} 0 \\ 2 \end{bmatrix}, \quad \mathbf{B}_4 = \begin{bmatrix} 1 \\ 3 \end{bmatrix}$$

and

$$A\mathbf{B}_1 = \begin{bmatrix} 26 \\ 16 \\ 9 \end{bmatrix}, \quad A\mathbf{B}_2 = \begin{bmatrix} 36 \\ 20 \\ 13 \end{bmatrix}, \quad A\mathbf{B}_3 = \begin{bmatrix} 12 \\ 8 \\ 4 \end{bmatrix} \quad A\mathbf{B}_4 = \begin{bmatrix} 20 \\ 12 \\ 7 \end{bmatrix}.$$

Calculating AB, we see immediately that AB is a (3×4) matrix with columns $A\mathbf{B}_1$, $A\mathbf{B}_2$, $A\mathbf{B}_3$ and $A\mathbf{B}_4$; or equivalently $AB = [A\mathbf{B}_1, A\mathbf{B}_2, A\mathbf{B}_3, A\mathbf{B}_4]$.

1.4 EXERCISES

1. Determine whether the following matrix products are defined. When the product is defined, give the size of the product.

   a) AB and BA when A is (2×3) and B is (3×4)
   b) AB and BA when A is (2×3) and B is (2×4)
   c) AB and BA when A is (3×7) and B is (6×3)
   d) AB and BA when A is (2×3) and B is (3×2)
   e) AB and BA when A is (3×3) and B is (3×1)
   f) $A(BC)$ and $(AB)C$ when A is (2×3), B is (3×5), and C is (5×4)
   g) AB and BA when A is (4×1) and B is (1×4)

2. What is the size of the product $(AB)(CD)$ where A is (2×3), B is (3×4), C is (4×4), and D is (4×2)? Also calculate the size of $A(B(CD))$ and $((AB)C)D$.

3.
$$A = \begin{bmatrix} 2 & 3 \\ 1 & 4 \end{bmatrix} \quad B = \begin{bmatrix} 1 & 2 \\ 1 & 4 \end{bmatrix} \quad \mathbf{u} = \begin{bmatrix} 1 \\ 3 \end{bmatrix} \quad \boldsymbol{v} = [2 \quad 4]\cdot$$

$$C = \begin{bmatrix} 2 & 1 \\ 4 & 0 \\ 8 & -1 \\ 3 & 2 \end{bmatrix} \quad D = \begin{bmatrix} 2 & 1 & 3 & 6 \\ 2 & 0 & 0 & 4 \\ 1 & -1 & 1 & -1 \\ 1 & 3 & 1 & 2 \end{bmatrix} \quad \mathbf{w} = \begin{bmatrix} 2 \\ 3 \\ 1 \\ 1 \end{bmatrix}$$

   Find the following.

   a) AB b) BA c) DC d) $A\mathbf{u}$
   e) $\boldsymbol{v}B$ f) $\boldsymbol{v}\mathbf{u}$ g) $\boldsymbol{v}B\mathbf{u}$ h) $D\mathbf{w}$
   i) $A+B$ j) $(A+2B)\mathbf{u}$ k) CA l) $D\mathbf{w}+4CB\mathbf{u}$

4. $A = \begin{bmatrix} 1 & 1 \\ 1 & 0 \end{bmatrix} \quad B = \begin{bmatrix} 2 & 1 \\ 3 & 2 \end{bmatrix} \quad C = \begin{bmatrix} 2 & 0 \\ 3 & 1 \end{bmatrix}$

   Find the following.

   a) $A+B$ b) $(A+2B)C$ c) $BC+2A$ d) $A(BC)$

5. Refer to the matrices and vectors in Exercise 3.

   a) Identify the column vectors in $A = [\mathbf{A}_1, \mathbf{A}_2]$ and $D = [\mathbf{D}_1, \mathbf{D}_2, \mathbf{D}_3, \mathbf{D}_4]$.
   b) In part (a), is $\mathbf{A}_1$ in R^2, R^3, or R^4? Is $\mathbf{D}_1$ in R^2, R^3, or R^4?
   c) Form the (2×2) matrix with columns $[A\mathbf{B}_1, A\mathbf{B}_2]$, and verify that this matrix is the product AB.
   d) Verify that the vector $D\mathbf{w}$ is the same as $2\mathbf{D}_1 + 3\mathbf{D}_2 + \mathbf{D}_3 + \mathbf{D}_4$.

6. Let $\mathcal{O}$ and A be the matrices given below. [$\mathcal{O}$ is called the (2×2) zero matrix.]

$$\mathcal{O} = \begin{bmatrix} 0 & 0 \\ 0 & 0 \end{bmatrix} \quad A = \begin{bmatrix} 1 & 2 \\ 2 & 4 \end{bmatrix}.$$

   Find a (2×2) matrix B

$$B = \begin{bmatrix} b_{11} & b_{12} \\ b_{21} & b_{22} \end{bmatrix}$$

   such that $B \neq \mathcal{O}$, $AB = \mathcal{O}$, and $BA \neq \mathcal{O}$.

7. a) Express the linear systems (i) and (ii) in the form $A\mathbf{x} = \mathbf{b}$.

i) $\begin{aligned} 2x_1 - x_2 &= 3 \\ x_1 + x_2 &= 3 \end{aligned}$ ii) $\begin{aligned} x_1 - 3x_2 + x_3 &= 1 \\ x_1 - 2x_2 + x_3 &= 2 \\ x_2 - x_3 &= -1 \end{aligned}$

b) Express systems (i) and (ii) in the form of (1.20).

c) Solve the systems (i) and (ii) by Gauss elimination. For each system $A\mathbf{x} = \mathbf{b}$, determine how to represent $\mathbf{b}$ as a linear combination of the columns of the coefficient matrix.

8. Solve $A\mathbf{x} = \mathbf{b}$ where A and $\mathbf{b}$ are given by

$$A = \begin{bmatrix} 1 & 1 \\ 1 & 2 \end{bmatrix}, \quad \mathbf{b} = \begin{bmatrix} 2 \\ 3 \end{bmatrix}.$$

9. Let A and I be the matrices

$$A = \begin{bmatrix} 1 & 1 \\ 1 & 2 \end{bmatrix}, \quad I = \begin{bmatrix} 1 & 0 \\ 0 & 1 \end{bmatrix}.$$

a) Find a (2×2) matrix B such that $AB = I$. (Hint: Use Theorem 1.4 to determine the column vectors of B.)

b) Show that $AB = BA$ for the matrix B found in part (a).

10. Prove Theorem 1.3 by showing that the ith component of $A\mathbf{x}$ is equal to the ith component of $x_1\mathbf{A}_1 + x_2\mathbf{A}_2 + \ldots + x_n\mathbf{A}_n$ where $1 \leqslant i \leqslant m$.

11. a) Let A and C be the matrices

$$A = \begin{bmatrix} 1 & 3 \\ 1 & 4 \end{bmatrix}, \quad C = \begin{bmatrix} 2 & 6 \\ 3 & 6 \end{bmatrix}.$$

If possible, find (2×2) matrices Q and B such that $A + Q = C$ and $AB = C$. (Hint: Use Theorem 1.4 to find B.)

b) Let A and C be the matrices

$$A = \begin{bmatrix} 1 & 1 & 1 \\ 0 & 2 & 1 \\ 2 & 4 & 3 \end{bmatrix} \quad C = \begin{bmatrix} 1 & 0 & 0 \\ 1 & 2 & 0 \\ 1 & 3 & 5 \end{bmatrix}.$$

If possible, find (3×3) matrices Q and B such that $A + Q = C$ and $AB = C$.

12. A (3×3) matrix $T = (t_{ij})$ is called an ***upper-triangular*** matrix if T has the form

$$T = \begin{bmatrix} t_{11} & t_{12} & t_{13} \\ 0 & t_{22} & t_{23} \\ 0 & 0 & t_{33} \end{bmatrix}.$$

Formally T is upper triangular if $t_{ij} = 0$ whenever $i > j$. If A and B are upper-triangular (3×3) matrices, verify that the product AB is also upper triangular.

13. An $(n \times n)$ matrix $T = (t_{ij})$ is called upper triangular if $t_{ij} = 0$ whenever $i > j$. Suppose A and B are $(n \times n)$ upper-triangular matrices. Use Definition 1.3 to prove that the product AB is upper triangular. That is, show the ijth entry of AB is zero when $i > j$.

1.5 ALGEBRAIC PROPERTIES OF MATRIX OPERATIONS

In the last section we defined the matrix operations of addition, multiplication, and multiplication of a matrix by a scalar. For these operations to be useful, the basic rules they obey must be determined. For example we have already shown that given two matrices, A and B, the products AB and BA are usually *not* equal even if both products are defined. By contrast, the real number system has $ab = ba$ for any two real numbers a and b. Observing that AB and BA are normally not equal emphasizes the fact that the algebraic properties of matrix operations are not the same as the familiar algebraic properties of the real number system. Another example of where the properties of matrix arithmetic differ from the arithmetic properties of the real numbers is the following. If a, b, and c are real numbers such that $ab = ac$, and if $a \neq 0$, then $b = c$. (This property of the real number system is usually called the cancellation law.) Now consider the (2×2) matrices A, B, and C where

$$A = \begin{bmatrix} 1 & 1 \\ 1 & 1 \end{bmatrix}, \quad B = \begin{bmatrix} 1 & 4 \\ 2 & 1 \end{bmatrix}, \quad C = \begin{bmatrix} 2 & 2 \\ 1 & 3 \end{bmatrix}.$$

Clearly $AB = AC$, but $B \neq C$. Therefore the cancellation law does not hold in general for matrices.

Given that the familiar properties of the real number system do not carry over directly to matrix operations, the objective of this section is to state precisely the rules that matrix operations do obey. Most of these rules will seem obvious; their proofs will probably seem trivial; and indeed most of the proofs are quite simple. However certain subtleties should be noted. For example the two matrix products $(AB)C$ and $A(BC)$ are formed differently, and it is only through Definitions 1.2 and 1.3 that we can establish that the products are equal (see Theorem 1.6). As another example Theorem 1.7 asserts that $(r+s)A = rA + sA$ where r and s are scalars and A is an $(m \times n)$ matrix. A moment's reflection reveals that there is actually something to prove here, for two different addition operations are involved. To be specific, we form $(r+s)A$ by first adding two scalars together and then performing a multiplication of a scalar and a matrix. On the other hand, forming $rA + sA$ involves the sum of two matrices rather than two scalars; so the same symbol $+$ is being used to represent two distinct operations. (We do not bother to distinguish between the symbols for scalar addition and matrix addition since it is normally clear from the context which addition is intended.) Our first theorem lists some of the properties that matrix addition satisfies.

Theorem 1.5

If A, B, and C are $(m \times n)$ matrices, then the following are true.

1. $A + B = B + A$.
2. $(A + B) + C = A + (B + C)$.
3. There exists a unique $(m \times n)$ matrix $\mathcal{O}$ (called the ***zero matrix***) such that $A + \mathcal{O} = A$ for any $(m \times n)$ matrix A.

4. Given any $(m \times n)$ matrix A, there exists a unique $(m \times n)$ matrix P such that $A + P = \mathcal{O}$.

These properties are easily established, and we leave the proofs of (2)–(4) to the reader and prove only (1). We anticipate the results of the exercises, however, and observe for (3) that the zero matrix, $\mathcal{O}$, is the $(m \times n)$ matrix all of whose entries are zero. In (4) the matrix P is usually called the additive inverse for A, and the reader can show that $P = (-1)A$. In terms of notation we usually write $A + (-1)B$ as $A - B$; so part (4) reduces to the statement $A - A = \mathcal{O}$.

Proof of part (1). If $A = (a_{ij})$ and $B = (b_{ij})$ are $(m \times n)$ matrices, then by Definition 1.4, the ijth entry of $A + B$ is $a_{ij} + b_{ij}$, and the ijth entry of $B + A$ is $b_{ij} + a_{ij}$. However the real numbers $a_{ij} + b_{ij}$ and $b_{ij} + a_{ij}$ are equal; so $A + B = B + A$. This is an instance in which an algebraic property of the real numbers is valid also for matrices; and in fact the commutativity of matrix addition is derived from the commutativity of real number addition. ■

Three associative properties involving scalar and matrix multiplication are given by Theorem 1.6.

Theorem 1.6

1. If A, B, and C are $(m \times n)$, $(n \times p)$, and $(p \times q)$ matrices, respectively, then $(AB)C = A(BC)$.
2. If r and s are scalars then $r(sA) = (rs)A$.
3. $r(AB) = (rA)B = A(rB)$.

The proof is again left to the reader, but we shall give one example to illustrate the theorem.

EXAMPLE 1.16 Let A, B, and C be given by

$$A = \begin{bmatrix} 1 & 2 \\ -1 & 3 \end{bmatrix}, \quad B = \begin{bmatrix} 2 & -1 & 3 \\ 1 & -1 & 1 \end{bmatrix}, \quad C = \begin{bmatrix} 3 & 1 & 2 \\ -2 & 1 & -1 \\ 4 & -2 & -1 \end{bmatrix}.$$

Then

$$AB = \begin{bmatrix} 4 & -3 & 5 \\ 1 & -2 & 0 \end{bmatrix} \quad \text{and} \quad BC = \begin{bmatrix} 20 & -5 & 2 \\ 9 & -2 & 2 \end{bmatrix}.$$

Therefore $(AB)C$ is the product of a (2×3) matrix with a (3×3) matrix while $A(BC)$ is the product of a (2×2) matrix with a (2×3) matrix. Forming these products, we find

$$(AB)C = \begin{bmatrix} 38 & -9 & 6 \\ 7 & -1 & 4 \end{bmatrix} \quad \text{and} \quad A(BC) = \begin{bmatrix} 38 & -9 & 6 \\ 7 & -1 & 4 \end{bmatrix};$$

so the case above is a special case of (1). As an example of (2), let $r = 2$ and $s = 3$: then

$$2(3A) = 2\begin{bmatrix} 3 & 6 \\ -3 & 9 \end{bmatrix} = \begin{bmatrix} 6 & 12 \\ -6 & 18 \end{bmatrix} = (2\cdot 3)A = 6A.$$

Finally the distributive properties connecting addition and multiplication are given by Theorem 1.7.

Theorem 1.7

1. If A and B are $(m \times n)$ matrices, and C is an $(n \times p)$ matrix, then $(A+B)C = AC + BC$.
2. If A is an $(m \times n)$ matrix, and B and C are $(n \times p)$ matrices, then $A(B+C) = AB + AC$.
3. If r and s are scalars, and A is an $(m \times n)$ matrix, then $(r+s)A = rA + sA$.
4. If r is a real number, and A and B are $(m \times n)$ matrices, then $r(A+B) = rA + rB$.

Proof. We shall prove (1) and leave the others to the reader. First we observe that $(A+B)C$ and $AC+BC$ are both $(m \times p)$ matrices. To show that these two matrices are equal componentwise, let $Q = A+B$ where $Q = (q_{ij})$. Then $(A+B)C = QC$, and the rsth entry of QC is given by

$$\sum_{k=1}^{n} q_{rk}c_{ks} = \sum_{k=1}^{n} (a_{rk}+b_{rk})c_{ks} = \sum_{k=1}^{n} a_{rk}c_{ks} + \sum_{k=1}^{n} b_{rk}c_{ks}.$$

Because

$$\sum_{k=1}^{n} a_{rk}c_{ks} + \sum_{k=1}^{n} b_{rk}c_{ks}$$

is precisely the rsth entry of $AC+BC$, we have shown that $(A+B)C = AC+BC$. ■

One further, quite important matrix operation is the transpose of a matrix. Stated informally, the transpose of an $(m \times n)$ matrix A is a new $(n \times m)$ matrix formed by interchanging the rows and columns of A. The precise definition is as follows

DEFINITION 1.6.

If $A = (a_{ij})$ is an $(m \times n)$ matrix, then the ***transpose*** of A (denoted A^{T}) is an $(n \times m)$ matrix, $A^{T} = (b_{ij})$, where $b_{ij} = a_{ji}$ for all i and j where $1 \leqslant i \leqslant n$ and $1 \leqslant j \leqslant m$.

EXAMPLE 1.17

Let
$$A = \begin{bmatrix} 1 & 3 & 7 \\ 2 & 1 & 4 \end{bmatrix}.$$

Then

$$A^{\mathrm{T}} = \begin{bmatrix} 1 & 2 \\ 3 & 1 \\ 7 & 4 \end{bmatrix}.$$

As a second example involving vectors, let $\mathbf{x}$ be the vector in R^3 given by

$$\mathbf{x} = \begin{bmatrix} 1 \\ -3 \\ 2 \end{bmatrix}.$$

Then $\mathbf{x}^{\mathrm{T}}$ is the (1×3) vector

$$\mathbf{x}^{\mathrm{T}} = [1, -3, 2].$$

It is of interest to observe that both the products $\mathbf{x}^{\mathrm{T}}\mathbf{x}$ and $\mathbf{x}\mathbf{x}^{\mathrm{T}}$ are defined and that $\mathbf{x}^{\mathrm{T}}\mathbf{x}$ is a scalar [or a (1×1) matrix]:

$$\mathbf{x}^{\mathrm{T}}\mathbf{x} = [1, -3, 2] \begin{bmatrix} 1 \\ -3 \\ 2 \end{bmatrix} = 1 + 9 + 4 = 14.$$

On the other hand, $\mathbf{x}\mathbf{x}^{\mathrm{T}}$ is a (3×3) matrix:

$$\mathbf{x}\mathbf{x}^{\mathrm{T}} = \begin{bmatrix} 1 \\ -3 \\ 2 \end{bmatrix} [1, -3, 2] = \begin{bmatrix} 1 & -3 & 2 \\ -3 & 9 & -6 \\ 2 & -6 & 4 \end{bmatrix}.$$

If $\mathbf{y}$ is the vector

$$\mathbf{y} = \begin{bmatrix} 1 \\ 2 \\ 1 \end{bmatrix},$$

then $\mathbf{y}^{\mathrm{T}}\mathbf{x}$ is another way of writing the familiar scalar product of $\mathbf{y}$ and $\mathbf{x}$:

$$\mathbf{y}^{\mathrm{T}}\mathbf{x} = [1, 2, 1] \begin{bmatrix} 1 \\ -3 \\ 2 \end{bmatrix} = 1 - 6 + 2 = -3.$$

Furthermore we should note that $\mathbf{y}^{\mathrm{T}}\mathbf{x} = \mathbf{x}^{\mathrm{T}}\mathbf{y}$, but we should also be careful to observe that $\mathbf{x}\mathbf{y}^{\mathrm{T}}$ and $\mathbf{y}\mathbf{x}^{\mathrm{T}}$ are *different* (3×3) matrices (see Exercise 5, Section 1.5).

The transpose operation is used frequently in computational work, particularly in data-fitting problems and in other problems of approximation. In order to see why, let $\mathbf{x}$ and $\mathbf{y}$ be vectors in R^n,

$$\mathbf{x} = \begin{bmatrix} x_1 \\ x_2 \\ \vdots \\ x_n \end{bmatrix}, \quad \mathbf{y} = \begin{bmatrix} y_1 \\ y_2 \\ \vdots \\ y_n \end{bmatrix}$$

Then as we mentioned in Example 1.17, $\mathbf{x}^{\mathrm{T}}\mathbf{y}$ is the usual scalar or dot product of $\mathbf{x}$ and $\mathbf{y}$:

$$\mathbf{x}^{\mathrm{T}}\mathbf{y} = \sum_{i=1}^{n} x_i y_i. \tag{1.21}$$

In particular

$$\mathbf{x}^{\mathrm{T}}\mathbf{x} = \sum_{i=1}^{n} (x_i)^2$$

so that the number $\sqrt{\mathbf{x}^{\mathrm{T}}\mathbf{x}}$ represents the Euclidean length of the vector $\mathbf{x}$. Similarly $\sqrt{(\mathbf{x}-\mathbf{y})^{\mathrm{T}}(\mathbf{x}-\mathbf{y})}$ represents the Euclidean distance between $\mathbf{x}$ and $\mathbf{y}$ (or the size of the vector $\mathbf{x}-\mathbf{y}$) since

$$\sqrt{(\mathbf{x}-\mathbf{y})^{\mathrm{T}}(\mathbf{x}-\mathbf{y})} = \sqrt{(x_1-y_1)^2 + (x_2-y_2)^2 + \ldots + (x_n-y_n)^2}.$$

A convenient notation is often used to denote the length (or magnitude) of a vector $\mathbf{x}$. If $\mathbf{x}$ is an n-dimensional vector with real components, then

$$\|\mathbf{x}\| = \sqrt{\mathbf{x}^{\mathrm{T}}\mathbf{x}} \tag{1.22}$$

norm
measures the size of the vector X

and $\|\mathbf{x}\|$ is called the ***norm*** of $\mathbf{x}$. Thus $\|\mathbf{x}\|$ is a measure of the size of the vector $\mathbf{x}$. The double bar notation for the norm should remind the reader of the notation for the absolute value, or magnitude, of a real number. As we note later, there are a number of other ways to measure the size of a vector. The norm defined by (1.22) is usually called the Euclidean norm to distinguish it from other norms. The norm notation is employed in Section 2.6 when we consider linear systems $A\mathbf{x} = \mathbf{b}$ that are inconsistent, and develop a way to find a vector $\mathbf{x}^*$ that minimizes $\|A\mathbf{x}-\mathbf{b}\|$. In light of the discussion above, minimizing $\|A\mathbf{x}-\mathbf{b}\|$ means finding $\mathbf{x}^*$ so that the vector $A\mathbf{x}^* - \mathbf{b}$ is as "small" as possible among all choices for $\mathbf{x}$.

Three important properties of the transpose are given in Theorem 1.8.

Theorem 1.8

If A and B are $(m \times n)$ matrices, and C is an $(n \times p)$ matrix, then

1. $(A+B)^{\mathrm{T}} = A^{\mathrm{T}} + B^{\mathrm{T}}$
2. $(AC)^{\mathrm{T}} = C^{\mathrm{T}}A^{\mathrm{T}}$ *
3. $(A^{\mathrm{T}})^{\mathrm{T}} = A.$

Proof. We shall leave 1 and 3 to the reader and prove 2. Note first that $(AC)^{\mathrm{T}}$ and $C^{\mathrm{T}}A^{\mathrm{T}}$ are both $(p \times m)$ matrices; so we have only to show that their corresponding entries are equal. From Definition 1.6 the ijth entry of $(AC)^{\mathrm{T}}$ is the jith entry of AC.

Thus the ijth entry of $(AC)^{\mathrm{T}}$ is given by

$$\sum_{k=1}^{n} a_{jk}c_{ki}.$$

Next the ijth entry of $C^{\mathrm{T}}A^{\mathrm{T}}$ is the scalar product of the ith row of C^{T} with the jth column of A^{T}. In particular the ith row of C^{T} is $[c_{1i}, c_{2i}, \ldots, c_{ni}]$ (the ith column of C) while the jth column of A^{T} is

$$\begin{bmatrix} a_{j1} \\ a_{j2} \\ \vdots \\ a_{jn} \end{bmatrix}.$$

Therefore the ijth entry of $C^{\mathrm{T}}A^{T}$ is given by

$$c_{1i}a_{j1} + c_{2i}a_{j2} + \ldots + c_{ni}a_{jn} = \sum_{k=1}^{n} c_{ki}a_{jk}.$$

Finally since

$$\sum_{k=1}^{n} c_{ki}a_{jk} = \sum_{k=1}^{n} a_{jk}c_{ki},$$

we see that the ijth entries of $(AC)^{\mathrm{T}}$ and $C^{\mathrm{T}}A^{\mathrm{T}}$ agree. ■

The transpose operation is used also to define certain important types of matrices, such as positive-definite matrices, normal matrices, and symmetric matrices. We shall consider these in detail later and give the definition of only a symmetric matrix in this section.

DEFINITION 1.7.

A matrix A is called ***symmetric*** if

$$A = A^{\mathrm{T}}.$$

Note that this definition implies that a symmetric matrix must have the same number of rows and columns; for if A is $(m \times n)$, then A^{T} is $(n \times m)$; and we can have $A = A^{\mathrm{T}}$ only if $m = n$. It is common practice to refer to an $(n \times n)$ matrix as a ***square matrix***; so if a matrix is symmetric, then it must be a square matrix. Furthermore referring to Definition 1.6 if $A = (a_{ij})$ is an $(n \times n)$ symmetric matrix, then it follows that $a_{ij} = a_{ji}$ for all i and j, $1 \leqslant i, j \leqslant n$. Conversely if A is square, and if $a_{ij} = a_{ji}$ for all i and j, then A is symmetric.

EXAMPLE 1.18 Let A, B, and C be the matrices given below:

$$A = \begin{bmatrix} 1 & 2 \\ 2 & 3 \end{bmatrix}, \quad B = \begin{bmatrix} 1 & 2 \\ 1 & 2 \end{bmatrix}, \quad C = \begin{bmatrix} 1 & 6 \\ 3 & 1 \\ 2 & 0 \end{bmatrix}.$$

Then $A = A^T$ ∴ symmetric

$$A^{\mathrm{T}} = \begin{bmatrix} 1 & 2 \\ 2 & 3 \end{bmatrix}, \quad B^{\mathrm{T}} = \begin{bmatrix} 1 & 1 \\ 2 & 2 \end{bmatrix}, \quad C^{\mathrm{T}} = \begin{bmatrix} 1 & 3 & 2 \\ 6 & 1 & 0 \end{bmatrix}.$$

Therefore A is symmetric, but B and C are not symmetric. Note that even though B and C are not symmetric, the matrices $B^{\mathrm{T}}B$ and $C^{\mathrm{T}}C$ are symmetric, for

$$B^{\mathrm{T}}B = \begin{bmatrix} 2 & 4 \\ 4 & 8 \end{bmatrix} \quad \text{and} \quad C^{\mathrm{T}}C = \begin{bmatrix} 14 & 9 \\ 9 & 37 \end{bmatrix}.$$

In Exercise 13, Section 1.5, the reader is asked to show that $Q^{\mathrm{T}}Q$ is always a symmetric matrix regardless of whether Q is symmetric or not.

1.5 EXERCISES

1. Find the transpose of each matrix.

$$A = \begin{bmatrix} 3 & 1 \\ 4 & 7 \\ 2 & 6 \end{bmatrix} \quad B = \begin{bmatrix} 1 & 2 & 1 \\ 7 & 4 & 3 \\ 6 & 0 & 1 \end{bmatrix} \quad C = \begin{bmatrix} 2 & 1 & 4 & 0 \\ 6 & 1 & 3 & 5 \\ 2 & 1 & 1 & 0 \end{bmatrix}$$

$$D = \begin{bmatrix} 2 & 1 \\ 1 & 4 \end{bmatrix} \quad E = \begin{bmatrix} 3 & 6 \\ 2 & 3 \end{bmatrix} \quad F = \begin{bmatrix} 1 & 6 & 2 \\ 6 & 0 & 4 \\ 2 & 4 & 8 \end{bmatrix}$$

2. Use the matrices in Exercise 1.

a) Calculate $(DE)^{\mathrm{T}}$ and $E^{\mathrm{T}}D^{\mathrm{T}}$, and verify that they are equal.
b) Calculate $(AE)^{\mathrm{T}}$ and $E^{\mathrm{T}}A^{\mathrm{T}}$, and verify that they are equal.
c) Calculate $A(DE)$ and $(AD)E$, and verify that they are equal.
d) If R is $(p \times q)$ and S is $(q \times t)$, what is the size of the matrix $(RS)^{\mathrm{T}}$?

3. Which of the matrices in Exercise 1 are symmetric? In Exercise 1, is $A^{\mathrm{T}}A$ symmetric?

4. Find (2×2) matrices A and B such that A and B are each symmetric, but AB is not symmetric. [Hint: $(AB)^{\mathrm{T}} = B^{\mathrm{T}}A^{\mathrm{T}} = BA$.]

5. Given the vectors

$$\mathbf{x} = \begin{bmatrix} -1 \\ 2 \end{bmatrix} \quad \mathbf{y} = \begin{bmatrix} 3 \\ 4 \end{bmatrix} \quad \mathbf{u} = \begin{bmatrix} -2 \\ 1 \\ 3 \end{bmatrix} \quad \mathbf{b} = \begin{bmatrix} 1 \\ 1 \\ 2 \end{bmatrix}.$$

a) Calculate $\mathbf{y}^{\mathrm{T}}\mathbf{x}$, $\mathbf{x}^{\mathrm{T}}\mathbf{y}$, $\mathbf{x}^{\mathrm{T}}\mathbf{x}$, $\mathbf{x}\mathbf{y}^{\mathrm{T}}$, $\mathbf{y}\mathbf{x}^{\mathrm{T}}$, $\mathbf{u}^{\mathrm{T}}\mathbf{b}$.
b) Calculate $\|\mathbf{x}\|$, $\|\mathbf{y}\|$, $\|\mathbf{u}\|$, $\|\mathbf{b}\|$.
Use the matrices from Problem 1.
c) Calculate $\|A\mathbf{x}\|$, $\|A\mathbf{x} - \mathbf{b}\|$, $\|A\mathbf{y} - \mathbf{b}\|$.
d) Calculate $\mathbf{x}^{\mathrm{T}}E\mathbf{x}$, $\mathbf{x}^{\mathrm{T}}D\mathbf{y}$, $\mathbf{x}^{\mathrm{T}}D\mathbf{x}$.

e) Prove that $\mathbf{z}^T D\mathbf{z} > 0$ for any vector $\mathbf{z}$ of the form

$$\mathbf{z} = \begin{bmatrix} z_1 \\ z_2 \end{bmatrix} \quad \text{where } |z_1|+|z_2| > 0.$$

6. Let **a** and **b** be given by

$$\mathbf{a} = \begin{bmatrix} 1 \\ 2 \end{bmatrix} \quad \mathbf{b} = \begin{bmatrix} 3 \\ 4 \end{bmatrix}.$$

a) Find $\mathbf{x}$ in R^2 that satisfies both $\mathbf{x}^T\mathbf{a} = 6$ and $\mathbf{x}^T\mathbf{b} = 2$.
b) Find $\mathbf{x}$ in R^2 that satisfies both $\mathbf{x}^T(\mathbf{a}+\mathbf{b}) = 12$ and $\mathbf{x}^T\mathbf{a} = 2$.

7. Let A be a (2×2) matrix, and let B and C be given by

$$B = \begin{bmatrix} 1 & 3 \\ 1 & 4 \end{bmatrix}, \quad C = \begin{bmatrix} 2 & 3 \\ 4 & 5 \end{bmatrix}.$$

a) If $A^T + B = C$, what is A?
b) If $A^T B = C$, what is A?
c) Using the notation of (1.17a), calculate $B\mathbf{C}_1$, $\mathbf{B}_1^T C$, $(B\mathbf{C}_1)^T\mathbf{C}_2$, $\| C\mathbf{B}_2\|$.

8. Prove Theorem 1.5, parts (2), (3), and (4).

9. Prove Theorem 1.6.

10. Prove Theorem 1.7, parts (2), (3), and (4).

11. Prove Theorem 1.8, parts (1) and (3).

12. Let A be the (2×2) matrix

$$A = \begin{bmatrix} 1 & 2 \\ 3 & 6 \end{bmatrix}.$$

Choose some vector **b** in R^2 such that the equation $A\mathbf{x} = \mathbf{b}$ is inconsistent. Verify that the associated equation $A^T A\mathbf{x} = A^T\mathbf{b}$ *is* consistent for your choice of **b**. Let $\mathbf{x}^*$ be a solution to $A^T A\mathbf{x} = A^T\mathbf{b}$, and select some vectors $\mathbf{x}$ at random from R^2. Verify that $\| A\mathbf{x}^* - \mathbf{b}\| \leqslant \| A\mathbf{x} - \mathbf{b}\|$ for any of these random choices for $\mathbf{x}$. [In Chapter 2 we show that $A^T A\mathbf{x} = A^T\mathbf{b}$ is always consistent for any $(m \times n)$ matrix A regardless of whether $A\mathbf{x} = \mathbf{b}$ is consistent or not. We also show that any solution $\mathbf{x}^*$ of $A^T A\mathbf{x} = A^T\mathbf{b}$ satisfies $\| A\mathbf{x}^* - \mathbf{b}\| \leqslant \| A\mathbf{x} - \mathbf{b}\|$ for all x in R^n—that is, such a vector $\mathbf{x}^*$ minimizes the length of the residual vector $\mathbf{r} = A\mathbf{x} - \mathbf{b}$.]

13. Use Theorem 1.8 to prove the following.

a) If Q is any $(m \times n)$ matrix, then $Q^T Q$ and QQ^T are symmetric.
b) If A, B, and C are matrices such that the product ABC is defined, then $(ABC)^T = C^T B^T A^T$. (Hint: Set $BC = D$. Note: Both of these proofs can be done quickly without considering the entries of the matrices.)

14. Let Q be an $(m \times n)$ matrix, and let $\mathbf{x}$ be any vector in R^n. Prove that $\mathbf{x}^T Q^T Q\mathbf{x} \geqslant 0$. (Hint: Note that $Q\mathbf{x}$ is a vector in R^m.)

15. Let A and B be $(n \times n)$ symmetric matrices. Give a necessary and sufficient condition for AB to be symmetric.

1.6 LINEAR INDEPENDENCE

Section 1.4 illustrated how the general linear system (1.1)

$$\begin{array}{l} a_{11}x_1 + a_{12}x_2 + \ldots a_{1n}x_n = b_1 \\ a_{21}x_1 + a_{22}x_2 + \ldots a_{2n}x_n = b_2 \, . \\ \quad \vdots \qquad\qquad\qquad\quad \vdots \qquad \vdots \\ a_{m1}x_1 + a_{m2}x_2 + \ldots a_{mn}x_n = b_m \end{array}$$

can be expressed in matrix form as $A\mathbf{x} = \mathbf{b}$.

Alternatively with $A = [\mathbf{A}_1, \mathbf{A}_2, \ldots, \mathbf{A}_n]$ the equation $A\mathbf{x} = \mathbf{b}$ can be written in terms of the columns of A as

$$x_1\mathbf{A}_1 + x_2\mathbf{A}_2 + \ldots + x_n\mathbf{A}_n = \mathbf{b}. \tag{1.23}$$

(Again see Section 1.4.) From (1.23) it follows that (1.1) is consistent if and only if $\mathbf{b}$ can be written as a sum of scalar multiples of the column vectors of A. We call a sum such as $x_1\mathbf{A}_1 + x_2\mathbf{A}_2 + \ldots + x_n\mathbf{A}_n$ a ***linear combination*** of the vectors $\mathbf{A}_1, \mathbf{A}_2, \ldots, \mathbf{A}_n$. Thus $A\mathbf{x} = \mathbf{b}$ is consistent if and only if $\mathbf{b}$ is a linear combination of the columns of $\mathbf{A}$.

It is convenient at this point to introduce a special symbol, $\boldsymbol{\theta}$, to denote the m-dimensional ***zero vector***. Thus $\boldsymbol{\theta}$ is the vector in R^m all of whose components are zero:

$$\boldsymbol{\theta} = \begin{bmatrix} 0 \\ 0 \\ \vdots \\ 0 \end{bmatrix}.$$

zero vector.

We will use $\boldsymbol{\theta}$ throughout to designate zero vectors in order to avoid any possible confusion between a zero vector and the scalar zero. With this notation (1.23) can be easily rewritten as

$$x_1\mathbf{A}_1 + x_2\mathbf{A}_2 + \ldots + x_n\mathbf{A}_n - \mathbf{b} = \boldsymbol{\theta}. \tag{1.24a}$$

In particular if $A\mathbf{x} = \mathbf{b}$ is consistent, then there is a solution of

$$x_1\mathbf{A}_1 + x_2\mathbf{A}_2 + \ldots + x_n\mathbf{A}_n + x_{n+1}\mathbf{b} = \boldsymbol{\theta} \tag{1.24b}$$

for which $x_{n+1} = -1$. On the other hand if we can find scalars $a_1, a_2, \ldots, a_n, a_{n+1}$ such that

$$a_1\mathbf{A}_1 + a_2\mathbf{A}_2 + \ldots + a_n\mathbf{A}_n + a_{n+1}\mathbf{b} = \boldsymbol{\theta} \tag{1.24c}$$

and such that $a_{n+1} \neq 0$, then

$$\mathbf{b} = \frac{-1}{a_{n+1}}(a_1\mathbf{A}_1 + a_2\mathbf{A}_2 + \ldots + a_n\mathbf{A}_n).$$

In this case a solution to (1.23) and hence to $A\mathbf{x} = \mathbf{b}$ is provided by

$$x_1 = \frac{-a_1}{a_{n+1}}, \; x_2 = \frac{-a_2}{a_{n+1}}, \ldots, \; x_n = \frac{-a_n}{a_{n+1}}.$$

Thus $A\mathbf{x} = \mathbf{b}$ is consistent if and only if $\boldsymbol{\theta}$ can be expressed as a linear combination of the vectors $\mathbf{A}_1, \mathbf{A}_2, \ldots, \mathbf{A}_n, \mathbf{b}$, as in (1.24c) where $a_{n+1} \neq 0$. In general it is clear that if $\boldsymbol{v}_1, \boldsymbol{v}_2, \ldots, \boldsymbol{v}_p$ are any vectors in R^m, and if

$$a_1\boldsymbol{v}_1 + a_2\boldsymbol{v}_2 + \ldots + a_p\boldsymbol{v}_p = \boldsymbol{\theta},$$

then we can express $\boldsymbol{v}_p$ as a linear combination of the other vectors $\boldsymbol{v}_1, \boldsymbol{v}_2, \ldots, \boldsymbol{v}_{p-1}$ whenever $a_p \neq 0$.

This discussion leads us to the very important and fundamental concepts of linear independence and linear dependence. These concepts are defined as follows.

DEFINITION 1.8.

A set of m-dimensional vectors $\{\boldsymbol{v}_1, \boldsymbol{v}_2, \ldots, \boldsymbol{v}_p\}$ is said to be ***linearly dependent*** if there are scalars $a_1, a_2, \ldots, a_p$, not all of which are zero, such that

$$a_1\boldsymbol{v}_1 + a_2\boldsymbol{v}_2 + \ldots + a_p\boldsymbol{v}_p = \boldsymbol{\theta}.$$

A set $\{\boldsymbol{v}_1, \boldsymbol{v}_2, \ldots, \boldsymbol{v}_p\}$ of m-dimensional vectors is said to be ***linearly independent*** if it is not linearly dependent; that is, the only scalars for which $a_1\boldsymbol{v}_1 + a_2\boldsymbol{v}_2 + \ldots + a_p\boldsymbol{v}_p = \boldsymbol{\theta}$ are the scalars $a_1 = a_2 = \ldots = a_p = 0$.

A rephrasing of this definition is that a set of vectors $\{\boldsymbol{v}_1, \boldsymbol{v}_2, \ldots, \boldsymbol{v}_p\}$ is either linearly dependent or linearly independent; and the set is linearly independent if and only if the only solution to the vector equation

$$x_1\boldsymbol{v}_1 + x_2\boldsymbol{v}_2 + \ldots + x_p\boldsymbol{v}_p = \boldsymbol{\theta}$$

is the trivial solution $x_1 = x_2 = \ldots = x_p = 0$.

In practice we can employ the equation above to determine whether or not a set of vectors is linearly dependent. Suppose $\{\boldsymbol{v}_1, \boldsymbol{v}_2, \ldots, \boldsymbol{v}_p\}$ is a set of m-dimensional vectors; let $V = [\boldsymbol{v}_1, \boldsymbol{v}_2, \ldots, \boldsymbol{v}_p]$ be the $(m \times p)$ matrix whose jth column is $\boldsymbol{v}_j$. Recalling (1.23), we see that $\{\boldsymbol{v}_1, \boldsymbol{v}_2, \ldots, \boldsymbol{v}_p\}$ is linearly dependent if and only if the homogeneous equation

$$V\mathbf{x} = \boldsymbol{\theta} \tag{1.25}$$

has nontrivial solutions. This homogeneous $(m \times p)$ linear system can be solved using Gauss elimination as in Section 1.2. Thus to determine whether or not $\{\boldsymbol{v}_1, \boldsymbol{v}_2, \ldots, \boldsymbol{v}_p\}$ is a linearly dependent set, we solve $V\mathbf{x} = \boldsymbol{\theta}$ by forming the augmented matrix $[V \,\vdots\, \boldsymbol{\theta}]$ and reducing $[V \,\vdots\, \boldsymbol{\theta}]$ to echelon form. If $V\mathbf{x} = \boldsymbol{\theta}$ has nontrivial solutions, then $\{\boldsymbol{v}_1, \boldsymbol{v}_2, \ldots, \boldsymbol{v}_p\}$ is a linearly dependent set of vectors. If the only solution to $V\mathbf{x} = \boldsymbol{\theta}$ is $\mathbf{x} = \boldsymbol{\theta}$, then $\{\boldsymbol{v}_1, \boldsymbol{v}_2, \ldots, \boldsymbol{v}_p\}$ is a linearly independent set of vectors.

EXAMPLE 1.19 Consider the set of vectors $\{\boldsymbol{v}_1, \boldsymbol{v}_2, \boldsymbol{v}_3\}$ where

$$\boldsymbol{v}_1 = \begin{bmatrix} 1 \\ 2 \\ 3 \end{bmatrix}, \quad \boldsymbol{v}_2 = \begin{bmatrix} 2 \\ -1 \\ 4 \end{bmatrix}, \quad \boldsymbol{v}_3 = \begin{bmatrix} 0 \\ 5 \\ 2 \end{bmatrix}.$$

To determine whether this set is linearly dependent or not, we must determine whether there is a nontrivial solution of

$$x_1\boldsymbol{v}_1 + x_2\boldsymbol{v}_2 + x_3\boldsymbol{v}_3 = \boldsymbol{\theta}.$$

Equivalently according to the remarks above, we need to determine whether or not there is a nontrivial solution of $V\mathbf{x} = \boldsymbol{\theta}$ where $V = [\boldsymbol{v}_1, \boldsymbol{v}_2, \boldsymbol{v}_3]$.

As before, we solve $V\mathbf{x} = \boldsymbol{\theta}$ by forming the augmented matrix for the system

$$[V \,\vdots\, \boldsymbol{\theta}] = \begin{bmatrix} 1 & 2 & 0 & 0 \\ 2 & -1 & 5 & 0 \\ 3 & 4 & 2 & 0 \end{bmatrix} \tag{1.26}$$

and then reducing this matrix to a row-equivalent echelon matrix. In particular, the augmented matrix $[V \,\vdots\, \boldsymbol{\theta}]$ in (1.26) is row equivalent to

$$\begin{bmatrix} 1 & 2 & 0 & 0 \\ 0 & -5 & 5 & 0 \\ 0 & 0 & 0 & 0 \end{bmatrix}.$$

Backsolving the system

$$\begin{aligned} x_1 + 2x_2 \qquad &= 0 \\ -5x_2 + 5x_3 &= 0, \end{aligned}$$

we find the solution of $V\mathbf{x} = \boldsymbol{\theta}$, or of

$$x_1\boldsymbol{v}_1 + x_2\boldsymbol{v}_2 + x_3\boldsymbol{v}_3 = \boldsymbol{\theta},$$

is $x_2 = x_3$, $x_1 = -2x_3$ with x_3 arbitrary. Thus the equation $x_1\boldsymbol{v}_1 + x_2\boldsymbol{v}_2 + x_3\boldsymbol{v}_3 = \boldsymbol{\theta}$ has nontrivial solutions; for example with $x_3 = 1$

$$-2\boldsymbol{v}_1 + \boldsymbol{v}_2 + \boldsymbol{v}_3 = \boldsymbol{\theta}.$$

From this we conclude that $\{\boldsymbol{v}_1, \boldsymbol{v}_2, \boldsymbol{v}_3\}$ is a linearly dependent set of vectors. Note also from this equation that we can express $\boldsymbol{v}_3$ as a linear combination of $\boldsymbol{v}_1$ and $\boldsymbol{v}_2$:

$$\boldsymbol{v}_3 = 2\boldsymbol{v}_1 - \boldsymbol{v}_2.$$

Similarly of course, we can also express $\boldsymbol{v}_1$ as a linear combination of $\boldsymbol{v}_2$ and $\boldsymbol{v}_3$, and we can expression $\boldsymbol{v}_2$ as a linear combination of $\boldsymbol{v}_1$ and $\boldsymbol{v}_3$.

EXAMPLE 1.20 Let $\{\boldsymbol{v}_1, \boldsymbol{v}_2, \boldsymbol{v}_3\}$ be a set of vectors in R^3 where

$$\boldsymbol{v}_1 = \begin{bmatrix} 1 \\ 2 \\ -3 \end{bmatrix}, \quad \boldsymbol{v}_2 = \begin{bmatrix} -2 \\ 1 \\ 1 \end{bmatrix}, \quad \boldsymbol{v}_3 = \begin{bmatrix} 1 \\ -1 \\ -2 \end{bmatrix}.$$

To determine whether this set is linearly dependent or not, we must solve $V\mathbf{x} = \boldsymbol{\theta}$

where $V = [\boldsymbol{v}_1, \boldsymbol{v}_2, \boldsymbol{v}_3]$. The augmented matrix $[V \,\vdots\, \boldsymbol{\theta}]$ is row equivalent to

$$\begin{bmatrix} 1 & -2 & 1 & 0 \\ 0 & 5 & -3 & 0 \\ 0 & 0 & -2 & 0 \end{bmatrix}.$$

Thus the only solution of $x_1\boldsymbol{v}_1 + x_2\boldsymbol{v}_2 + x_3\boldsymbol{v}_3 = \boldsymbol{\theta}$ is the trivial solution $x_1 = x_2 = x_3 = 0$; so the set $\{\boldsymbol{v}_1, \boldsymbol{v}_2, \boldsymbol{v}_3\}$ is linearly independent. We observe, in contrast to the preceding example, that we cannot express $\boldsymbol{v}_3$ as a linear combination of $\boldsymbol{v}_1$ and $\boldsymbol{v}_2$. If there were scalars a_1 and a_2 such that

$$\boldsymbol{v}_3 = a_1\boldsymbol{v}_1 + a_2\boldsymbol{v}_2,$$

then there would be a nontrivial solution of $x_1\boldsymbol{v}_1 + x_2\boldsymbol{v}_2 + x_3\boldsymbol{v}_3 = \boldsymbol{\theta}$; namely, $x_1 = -a_1$, $x_2 = -a_2$, and $x_3 = 1$.

EXAMPLE 1.21 The unit vectors of Euclidean three-space

$$\mathbf{e}_1 = \begin{bmatrix} 1 \\ 0 \\ 0 \end{bmatrix}, \quad \mathbf{e}_2 = \begin{bmatrix} 0 \\ 1 \\ 0 \end{bmatrix}, \quad \mathbf{e}_3 = \begin{bmatrix} 0 \\ 0 \\ 1 \end{bmatrix}$$

are linearly independent. This linear independence is easily seen, for if $V = [\mathbf{e}_1, \mathbf{e}_2, \mathbf{e}_3]$, then

$$[V \,\vdots\, \boldsymbol{\theta}] = \begin{bmatrix} 1 & 0 & 0 & 0 \\ 0 & 1 & 0 & 0 \\ 0 & 0 & 1 & 0 \end{bmatrix};$$

so clearly the only solution of $V\mathbf{x} = \boldsymbol{\theta}$ (or of $x_1\mathbf{e}_1 + x_2\mathbf{e}_2 + x_3\mathbf{e}_3 = \boldsymbol{\theta}$) is the trivial solution, $x_1 = 0$, $x_2 = 0$, $x_3 = 0$.

The next example is a special case of Theorem 1.9, which we prove following Example 1.22. Theorem 1.9 is quite important as we shall see; it states that any set of m-dimensional vectors that contains more than m vectors must be a linearly dependent set. In Example 1.22 we see a specific set of three vectors in R^2; knowing Theorem 1.9, we could say immediately that the set is linearly dependent.

EXAMPLE 1.22 Let $\{\boldsymbol{v}_1, \boldsymbol{v}_2, \boldsymbol{v}_3\}$ be the set of vectors given by

$$\boldsymbol{v}_1 = \begin{bmatrix} 1 \\ 2 \end{bmatrix}, \quad \boldsymbol{v}_2 = \begin{bmatrix} 3 \\ 1 \end{bmatrix}, \quad \boldsymbol{v}_3 = \begin{bmatrix} 2 \\ 3 \end{bmatrix}.$$

The matrix $[V \,\vdots\, \boldsymbol{\theta}]$ where $V = [\boldsymbol{v}_1, \boldsymbol{v}_2, \boldsymbol{v}_3]$ is row equivalent to

$$\begin{bmatrix} 1 & 3 & 2 & 0 \\ 0 & -5 & -1 & 0 \end{bmatrix}.$$

Thus we see that the solution of $x_1\boldsymbol{v}_1 + x_2\boldsymbol{v}_2 + x_3\boldsymbol{v}_3 = \boldsymbol{\theta}$ is $x_3 = -5x_2$, $x_1 = 7x_2$ with x_2 arbitrary. In particular with $x_2 = 1$

$$7\boldsymbol{v}_1 + \boldsymbol{v}_2 - 5\boldsymbol{v}_3 = \boldsymbol{\theta};$$

so $\{\boldsymbol{v}_1, \boldsymbol{v}_2, \boldsymbol{v}_3\}$ is a linearly dependent set of vectors. The essential feature of this example, which we see in more generality in Theorem 1.9, is that $[V \,|\, \boldsymbol{\theta}]$ represents a (2×3) homogeneous system (a homogeneous system with fewer equations than unknowns), and hence $V\mathbf{x} = \boldsymbol{\theta}$ *must* have nontrivial solutions by Theorem 1.2.

The general result, of which Example 1.22 is a special case, is given in Theorem 1.9.

Theorem 1.9

Let $\{\boldsymbol{v}_1, \boldsymbol{v}_2, \ldots, \boldsymbol{v}_p\}$ be a set of vectors in R^m. If $p > m$, then this set is linearly dependent.

Proof. The set $\{\boldsymbol{v}_1, \boldsymbol{v}_2, \ldots, \boldsymbol{v}_p\}$ is linearly dependent if the equation $V\mathbf{x} = \boldsymbol{\theta}$ has a nontrivial solution, where $V = [\boldsymbol{v}_1, \boldsymbol{v}_2, \ldots \boldsymbol{v}_p]$; but $V\mathbf{x} = \boldsymbol{\theta}$ represents a homogeneous $(m \times p)$ system of linear equations with $p > m$; so by Theorem 1.2, $V\mathbf{x} = \boldsymbol{\theta}$ has nontrivial solutions. ■

Note that this theorem does *not* say that if $p \leqslant m$, then the set of vectors is linearly independent; and Example 1.19 provides an instance of three vectors in R^3 that are linearly dependent.

We conclude this section with some examples that are intended to illustrate further linear independence and linear dependence. The examples include some theoretical results, presented as challenging problems for the reader to try; but these results are not prerequisite for subsequent sections, and in fact the results will be repeated (using different terminology) in later chapters.

†**EXAMPLE 1.23** Let $B = [\mathbf{B}_1, \mathbf{B}_2, \ldots, \mathbf{B}_q]$ be a $(p \times q)$ matrix. In a later section we will define the ***rank*** of B to be the maximum number of linearly independent vectors in the set $\{\mathbf{B}_1, \mathbf{B}_2, \ldots, \mathbf{B}_q\}$. A result of fundamental theoretical importance is this: $A\mathbf{x} = \mathbf{b}$ is consistent if and only if the rank of A is equal to the rank of $[A \,|\, \mathbf{b}]$.

For example the matrix A given by

$$A = \begin{bmatrix} 1 & 1 & 2 \\ 1 & 2 & 1 \\ 1 & 1 & 3 \end{bmatrix} \qquad (1.27)$$

has rank 3. By Theorem 1.9 the rank of $[A \,|\, \mathbf{b}]$ must also be 3 for any $\mathbf{b}$; so $A\mathbf{x} = \mathbf{b}$ is

† Throughout, material designated with a dagger may be omitted without any loss of continuity. This material is not prerequisite for topics in later sections except occasionally for other material also designated by †.

always consistent (see Exercise 12, Section 1.6). On the other hand for

$$A = \begin{bmatrix} 1 & 2 & 0 \\ 2 & -1 & 5 \\ 3 & 4 & 2 \end{bmatrix} \tag{1.28}$$

we see that the rank of A is 2. A calculation shows that the rank of $[A \mid \mathbf{e}_1]$ is 3, and it is easy to show that $A\mathbf{x} = \mathbf{e}_1$ is inconsistent (also see Exercise 13, Section 1.6).

The next two examples use the concept of linear independence to characterize systems of linear equations that have unique solutions. Each example contains a theorem followed by a specific linear system that illustrates the theorem. The proof of each theorem is left as an exercise, but the special case that illustrates the theorem shows how the theorem might be proved.

† EXAMPLE 1.24 The theorem below gives a condition that ensures that a system with at least one solution has infinitely many solutions.

Theorem

Let $A\mathbf{x} = \mathbf{b}$ be an $(m \times n)$ system of linear equations with $A = [\mathbf{A}_1, \mathbf{A}_2, \ldots, \mathbf{A}_n]$. If $A\mathbf{x} = \mathbf{b}$ is consistent, and if $\{\mathbf{A}_1, \mathbf{A}_2, \ldots, \mathbf{A}_n\}$ is a linearly dependent set, then $A\mathbf{x} = \mathbf{b}$ has infinitely many solutions.

As an illustration of this theorem, we consider the system

$$\begin{aligned} x_1 + 2x_2 \phantom{{}+5x_3} &= 4 \\ 2x_1 - x_2 + 5x_3 &= -7 \\ 3x_1 + 4x_2 + 2x_3 &= 6, \end{aligned}$$

which we write as $A\mathbf{x} = \mathbf{b}$. The system $A\mathbf{x} = \mathbf{b}$ is consistent, and we can easily verify that the vector $\hat{\mathbf{x}}$ is a solution where

$$\hat{\mathbf{x}} = \begin{bmatrix} 2 \\ 1 \\ -2 \end{bmatrix}.$$

Writing A in column form as $A = [\mathbf{A}_1, \mathbf{A}_2, \mathbf{A}_3]$ where

$$\mathbf{A}_1 = \begin{bmatrix} 1 \\ 2 \\ 3 \end{bmatrix}, \quad \mathbf{A}_2 = \begin{bmatrix} 2 \\ -1 \\ 4 \end{bmatrix}, \quad \mathbf{A}_3 = \begin{bmatrix} 0 \\ 5 \\ 2 \end{bmatrix},$$

we also know from Example 1.19 that the columns of A are linearly dependent since

$$-2\mathbf{A}_1 + \mathbf{A}_2 + \mathbf{A}_3 = \boldsymbol{\theta}.$$

---

† Throughout, material designated with a dagger may be omitted without any loss of continuity. This material is not prerequisite for topics in later sections except occasionally for other material also designated by †.

Thus the system $A\mathbf{x} = \mathbf{b}$ fits the hypotheses of the theorem above; $A\mathbf{x} = \mathbf{b}$ has a solution, $\hat{\mathbf{x}}$; and the columns of the coefficient matrix, A, are linearly dependent. Now since $-2\mathbf{A}_1 + \mathbf{A}_2 + \mathbf{A}_3 = \boldsymbol{\theta}$, we know that $A\mathbf{w} = \boldsymbol{\theta}$ where

$$\mathbf{w} = \begin{bmatrix} -2 \\ 1 \\ 1 \end{bmatrix}.$$

Furthermore if a is any scalar, then

$$A(\hat{\mathbf{x}} + a\mathbf{w}) = A\hat{\mathbf{x}} + A(a\mathbf{w}) = A\hat{\mathbf{x}} + aA\mathbf{w} = \mathbf{b} + a\boldsymbol{\theta} = \mathbf{b}.$$

This equation means that any vector of the form $\hat{\mathbf{x}} + a\mathbf{w}$ is a solution of $A\mathbf{x} = \mathbf{b}$; and since a can be any scalar, $A\mathbf{x} = \mathbf{b}$ has infinitely many solutions. In many applications, $\hat{\mathbf{x}}$ is called a "particular solution" of $A\mathbf{x} = \mathbf{b}$. The example above illustrates that any solution of the homogeneous equation $A\mathbf{x} = \boldsymbol{\theta}$ can be added to a particular solution of $A\mathbf{x} = \mathbf{b}$ to produce other solutions of $A\mathbf{x} = \mathbf{b}$. Moreover it is easy to see that any solution of $A\mathbf{x} = \mathbf{b}$, say $\mathbf{x} = \mathbf{x}_s$, is the sum of $\hat{\mathbf{x}}$ and a solution of $A\mathbf{x} = \boldsymbol{\theta}$. To see this, suppose

$$A\hat{\mathbf{x}} = \mathbf{b} \qquad \text{and} \qquad A\mathbf{x}_s = \mathbf{b}.$$

Then

$$\boldsymbol{\theta} = A\mathbf{x}_s - A\hat{\mathbf{x}} = A(\mathbf{x}_s - \hat{\mathbf{x}}) = A\mathbf{x}_h;$$

and so $\mathbf{x}_s - \hat{\mathbf{x}} = \mathbf{x}_h$ is a solution of $A\mathbf{x} = \boldsymbol{\theta}$. Therefore

$$\mathbf{x}_s = \hat{\mathbf{x}} + \mathbf{x}_h;$$

and this equation shows that any solution of $A\mathbf{x} = \mathbf{b}$ is the sum of a fixed particular solution, $\hat{\mathbf{x}}$, and some solution, $\mathbf{x}_h$, of the homogeneous equation $A\mathbf{x} = \boldsymbol{\theta}$. As the reader may be aware, an analogous result holds for linear differential equations; and we note that the same result holds for any linear transformation defined between vector spaces (see Chapter 4).

†**EXAMPLE 1.25** The theorem below and the theorem in Example 1.24, completely characterize the uniqueness of solutions to a consistent linear system $A\mathbf{x} = \mathbf{b}$. Again we leave the proof to the reader and note only that the proof follows the lines of the example that illustrates the theorem.

Theorem

Let $A\mathbf{x} = \mathbf{b}$ be an $(m \times n)$ system of linear equations with $A = [\mathbf{A}_1, \mathbf{A}_2, \ldots, \mathbf{A}_n]$. If $A\mathbf{x} = \mathbf{b}$ is consistent, and if $\{\mathbf{A}_1, \mathbf{A}_2, \ldots, \mathbf{A}_n\}$ is a linearly independent set, then $A\mathbf{x} = \mathbf{b}$ has only one solution.

---

† Throughout, material designated with a dagger may be omitted without any loss of continuity. This material is not prerequisite for topics in later sections except occasionally for other material also designated by †.

We illustrate this theorem with the (3×3) system $A\mathbf{x} = \mathbf{b}$ where

$$A = \begin{bmatrix} 1 & 2 & 0 \\ 1 & 3 & 2 \\ 2 & 0 & 9 \end{bmatrix}, \qquad \mathbf{b} = \begin{bmatrix} 0 \\ 3 \\ 5 \end{bmatrix}.$$

We can immediately verify that

$$-2\mathbf{A}_1 + \mathbf{A}_2 + \mathbf{A}_3 = \mathbf{b};$$

so $A\mathbf{x} = \mathbf{b}$ is a consistent system with a solution $\hat{\mathbf{x}}$ where

$$\hat{\mathbf{x}} = \begin{bmatrix} -2 \\ 1 \\ 1 \end{bmatrix}.$$

As usual we can also verify that the columns of A are linearly independent by considering the augmented matrix for $A\mathbf{x} = \boldsymbol{\theta}$.
It is easy to show that $[A \,\vdots\, \boldsymbol{\theta}]$ is row equivalent to

$$\begin{bmatrix} 1 & 2 & 0 & 0 \\ 0 & 1 & 2 & 0 \\ 0 & 0 & 17 & 0 \end{bmatrix}$$

and therefore $\{\mathbf{A}_1, \mathbf{A}_2, \mathbf{A}_3\}$ is a linearly independent set.

From the foregoing, $A\mathbf{x} = \mathbf{b}$ is a consistent system and the columns of A are linearly independent; so $A\mathbf{x} = \mathbf{b}$ meets the hypotheses of the theorem. To show that $\hat{\mathbf{x}}$ is the only solution of $A\mathbf{x} = \mathbf{b}$, let us suppose that $\mathbf{x}_s$ is any solution of $A\mathbf{x} = \mathbf{b}$. Then $\mathbf{x}_h = \mathbf{x}_s - \hat{\mathbf{x}}$ is a solution of $A\mathbf{x} = \boldsymbol{\theta}$ since

$$A\mathbf{x}_h = A(\mathbf{x}_s - \hat{\mathbf{x}}) = A\mathbf{x}_s - A\hat{\mathbf{x}} = \mathbf{b} - \mathbf{b} = \boldsymbol{\theta}.$$

However since the columns of A are linearly independent, the only solution of $A\mathbf{x} = \boldsymbol{\theta}$ is $\mathbf{x} = \boldsymbol{\theta}$, and hence we must have $\mathbf{x}_h = \boldsymbol{\theta}$. Since $\mathbf{x}_h = \mathbf{x}_s - \hat{\mathbf{x}}$, we must conclude that $\mathbf{x}_s = \hat{\mathbf{x}}$ or that any solution of $A\mathbf{x} = \mathbf{b}$ is unique. In summary if $\hat{\mathbf{x}}$ is a solution of $A\mathbf{x} = \mathbf{b}$, and if the columns of the coefficient matrix A are linearly independent, then $\hat{\mathbf{x}}$ is the only solution of $A\mathbf{x} = \mathbf{b}$.

The theorems in Examples 1.24 and 1.25 can obviously be combined as:

Theorem

Solutions to a consistent system $A\mathbf{x} = \mathbf{b}$ are unique if and only if the column vectors of A are linearly independent.

The theorems above show as we stated in Section 1.1, that a system of linear equation either has

no solution,
a unique solution, or

infinitely many solutions.

In closing, we wish to stress that the observations and theorems given in Examples 1.23–1.25 are mainly qualitative rather than computational. Clearly questions about the existence and uniqueness of solutions to a *specific* system $A\mathbf{x} = \mathbf{b}$ can always be resolved quickly by Gauss elimination. That is, from a computational point of view we would expend as much effort in determining whether the columns of A are linearly independent as we would in actually solving $A\mathbf{x} = \mathbf{b}$ and *demonstrating* existence and/or uniqueness. Having said this, we must also say that there are many instances in which theoretical information about existence and uniqueness is extremely valuable to practical computations; a specific instance is given in the next section in Examples 1.29 and 1.30.

1.6 EXERCISES

1. Let

$$\mathbf{v}_1 = \begin{bmatrix} 1 \\ 2 \end{bmatrix}, \quad \mathbf{v}_2 = \begin{bmatrix} 2 \\ 3 \end{bmatrix}, \quad \mathbf{v}_3 = \begin{bmatrix} 2 \\ 4 \end{bmatrix}, \quad \mathbf{v}_4 = \begin{bmatrix} 1 \\ 1 \end{bmatrix}.$$

Use (1.25) to determine whether the following are linearly independent or linearly dependent sets of vectors. If the set is linearly dependent, express one vector in the set as a linear combination of the others.

a) $\{\mathbf{v}_1, \mathbf{v}_2\}$ b) $\{\mathbf{v}_1, \mathbf{v}_3\}$ c) $\{\mathbf{v}_2, \mathbf{v}_4\}$ d) $\{\mathbf{v}_1, \mathbf{v}_4\}$ e) $\{\mathbf{v}_1, \mathbf{v}_2, \mathbf{v}_4\}$
f) $\{\mathbf{v}_1, \mathbf{v}_2, \mathbf{v}_3\}$

2. Let

$$\mathbf{u}_1 = \begin{bmatrix} 1 \\ 2 \\ -1 \end{bmatrix}, \quad \mathbf{u}_2 = \begin{bmatrix} 2 \\ 1 \\ -3 \end{bmatrix}, \quad \mathbf{u}_3 = \begin{bmatrix} -1 \\ 4 \\ 3 \end{bmatrix}, \quad \mathbf{u}_4 = \begin{bmatrix} 0 \\ 3 \\ 1 \end{bmatrix}.$$

Repeat Exercise 1 for the sets

a) $\{\mathbf{u}_1, \mathbf{u}_2\}$ b) $\{\mathbf{u}_1, \mathbf{u}_2, \mathbf{u}_3\}$ c) $\{\mathbf{u}_1, \mathbf{u}_2, \mathbf{u}_4\}$

3. Let

$$\mathbf{w}_1 = \begin{bmatrix} 1 \\ 2 \\ -1 \end{bmatrix}, \quad \mathbf{w}_2 = \begin{bmatrix} 1 \\ 1 \\ 0 \end{bmatrix}, \quad \mathbf{w}_3 = \begin{bmatrix} 2 \\ 5 \\ 2 \end{bmatrix}, \quad \mathbf{w}_4 = \begin{bmatrix} -1 \\ 0 \\ 3 \end{bmatrix}, \quad \mathbf{w}_5 = \begin{bmatrix} 4 \\ 3 \\ -3 \end{bmatrix}.$$

Repeat Exercise 1 for the sets

a) $\{\mathbf{w}_1, \mathbf{w}_2\}$ b) $\{\mathbf{w}_1, \mathbf{w}_2, \mathbf{w}_3\}$ c) $\{\mathbf{w}_1, \mathbf{w}_2, \mathbf{w}_3, \mathbf{w}_4\}$ d) $\{\mathbf{w}_1, \mathbf{w}_2, \mathbf{w}_4\}$
e) $\{\mathbf{w}_2, \mathbf{w}_4, \mathbf{w}_5\}$ f) $\{\mathbf{w}_1, \mathbf{w}_2, \mathbf{w}_3, \mathbf{w}_5\}$

4. Let

$$\mathbf{v}_1 = \begin{bmatrix} 1 \\ 1 \\ 2 \\ 1 \end{bmatrix}, \quad \mathbf{v}_2 = \begin{bmatrix} 1 \\ 2 \\ 2 \\ 1 \end{bmatrix}, \quad \mathbf{v}_3 = \begin{bmatrix} 2 \\ 2 \\ 4 \\ a \end{bmatrix}.$$

For what values of a is the set $\{\mathbf{v}_1, \mathbf{v}_2, \mathbf{v}_3\}$ linearly independent, and for what values is the set linearly dependent?

5. Use the vectors in Exercise 3.

   a) Is $\mathbf{w}_3$ a linear combination of $\mathbf{w}_1$, $\mathbf{w}_2$, and $\mathbf{w}_4$?
   b) Is $\mathbf{w}_3$ a linear combination of $\mathbf{w}_2$, $\mathbf{w}_4$, and $\mathbf{w}_5$?
   (Note: Exercise 7 is a generalization of this problem.)

6. Using the vectors in Exercise 3, let A be the (3×3) matrix $A = [\mathbf{w}_1, \mathbf{w}_2, \mathbf{w}_3]$. Solve $A\mathbf{x} = \mathbf{w}_5$, and use your answer to write $\mathbf{w}_5$ as a linear combination of $\mathbf{w}_1$, $\mathbf{w}_2$, and $\mathbf{w}_3$.

7. Let $\{\mathbf{v}_1, \mathbf{v}_2, \mathbf{v}_3\}$ be a set of linearly independent vectors in R^3. Prove that if $\mathbf{u}$ is any vector in R^3, then $\mathbf{u}$ is a linear combination of $\mathbf{v}_1$, $\mathbf{v}_2$, and $\mathbf{v}_3$. [Hint: The set $\{\mathbf{v}_1, \mathbf{v}_2, \mathbf{v}_3, \mathbf{u}\}$ is linearly dependent. (Why?) Thus the equation $x_1\mathbf{v}_1 + x_2\mathbf{v}_2 + x_3\mathbf{v}_3 + x_4\mathbf{u} = \boldsymbol{\theta}$ has a nontrivial solution. Can this nontrivial solution have $x_4 = 0$?]

8. Let $\{\mathbf{v}_1, \mathbf{v}_2, \mathbf{v}_3\}$ be a set of linearly independent vectors in R^3, and let $\mathbf{u}$ be any vector in R^3. From Exercise 7, $\mathbf{u}$ is a linear combination of $\mathbf{v}_1$, $\mathbf{v}_2$, $\mathbf{v}_3$; so let us suppose that $\mathbf{u} = b_1\mathbf{v}_1 + b_2\mathbf{v}_2 + b_3\mathbf{v}_3$. Prove that this expression is unique—that is, if $\mathbf{u} = c_1\mathbf{v}_1 + c_2\mathbf{v}_2 + c_3\mathbf{v}_3$, then $c_1 = b_1$, $c_2 = b_2$, $c_3 = b_3$. (Hint: Subtract the two representations for $\mathbf{u}$.)

9. Let $\{\mathbf{v}_1, \mathbf{v}_2, \ldots, \mathbf{v}_p\}$ be a set of vectors in R^n, and suppose that $\mathbf{v}_1 = \boldsymbol{\theta}$. Prove that this set of vectors is linearly dependent.

10. Prove the theorem in Example 1.24.

11. Prove the theorem in Example 1.25.

12. Suppose $A = [\mathbf{A}_1, \mathbf{A}_2, \mathbf{A}_3]$ is a (3×3) matrix, and suppose $\{\mathbf{A}_1, \mathbf{A}_2, \mathbf{A}_3\}$ is linearly independent. Prove that the equation $A\mathbf{x} = \mathbf{b}$ is consistent for any $\mathbf{b}$ in R^3. [Hint: Recall (1.23) and Exercise 7.]

13. Let $A = [\mathbf{A}_1, \mathbf{A}_2, \mathbf{A}_3]$ be a (3×3) matrix, and suppose $\{\mathbf{A}_1, \mathbf{A}_2, \mathbf{A}_3\}$ contains just two linearly independent vectors. For definiteness, suppose that $\mathbf{A}_3 = c_1\mathbf{A}_1 + c_2\mathbf{A}_2$. If $\mathbf{b}$ is a vector in R^3 such that the set $\{\mathbf{A}_1, \mathbf{A}_2, \mathbf{b}\}$ contains three linearly independent vectors, show that $A\mathbf{x} = \mathbf{b}$ is not consistent. (Hint: Suppose that $A\mathbf{x} = \mathbf{b}$ is consistent and obtain a contradiction.)

14. Suppose that $\{\mathbf{v}_1, \mathbf{v}_2, \ldots, \mathbf{v}_p\}$ is a set of vectors in R^n such that

   a) $\mathbf{v}_i \neq \boldsymbol{\theta}$, $i = 1, 2, \ldots, p$
   b) $\mathbf{v}_i^T\mathbf{v}_j = 0$, $i \neq j$, $1 \leq i, j \leq p$.

   Prove that this set of vectors is linearly independent. [Hint: Use (b) in conjunction with the equation $x_1\mathbf{v}_1 + x_2\mathbf{v}_2 + \ldots + x_p\mathbf{v}_p = \boldsymbol{\theta}$.]

15. Let

$$\mathbf{v}_1 = \begin{bmatrix} 1 \\ 1 \\ 1 \end{bmatrix}, \quad \mathbf{v}_2 = \begin{bmatrix} 1 \\ -1 \\ 0 \end{bmatrix}, \quad \mathbf{v}_3 = \begin{bmatrix} 1 \\ 1 \\ -2 \end{bmatrix}$$

Verify that $\{\mathbf{v}_1, \mathbf{v}_2, \mathbf{v}_3\}$ satisfies conditions (a) and (b) of Exercise 14. From Exercises 7 and 14, it follows that any vector $\mathbf{u}$ in R^3 is a linear combination of $\mathbf{v}_1$, $\mathbf{v}_2$, and $\mathbf{v}_3$. Condition (b) of Exercise 14 can be used to find the coefficients a_1, a_2, a_3 in the expression

$$\mathbf{u} = a_1\mathbf{v}_1 + a_2\mathbf{v}_2 + a_3\mathbf{v}_3$$

without having to solve a (3×3) system. Determine how to do this and then find the coefficients in

$$\mathbf{u} = \begin{bmatrix} 5 \\ 13 \\ -9 \end{bmatrix} = a_1\mathbf{v}_1 + a_2\mathbf{v}_2 + a_3\mathbf{v}_3.$$

(Hint: Consider $\mathbf{v}_1^T\mathbf{u}$.)

1.7 NONSINGULAR MATRICES AND APPLICATIONS TO INTERPOLATING DATA

In this section we treat the important special case of an $(m \times n)$ system of linear equations, $A\mathbf{x} = \mathbf{b}$ where $m = n$. That is,

$$A = \begin{bmatrix} a_{11} & a_{12} & \dots & a_{1n} \\ a_{21} & a_{22} & \dots & a_{2n} \\ \vdots & & & \vdots \\ a_{n1} & a_{n2} & \dots & a_{nn} \end{bmatrix}, \quad \mathbf{x} = \begin{bmatrix} x_1 \\ x_2 \\ \vdots \\ x_n \end{bmatrix}, \quad \mathbf{b} = \begin{bmatrix} b_1 \\ b_2 \\ \vdots \\ b_n \end{bmatrix},$$

and A is a square matrix. For a square matrix A, we make the following definition.

DEFINITION 1.9.

An $(n \times n)$ matrix A is ***nonsingular*** if the only solution to $A\mathbf{x} = \boldsymbol{\theta}$ is $\mathbf{x} = \boldsymbol{\theta}$. Furthermore A is said to be ***singular*** if A is not nonsingular.

Note that for *any* $(n \times n)$ matrix A, $\mathbf{x} = \boldsymbol{\theta}$ is always a solution to $A\mathbf{x} = \boldsymbol{\theta}$; so Definition 1.9 says that A is nonsingular if and only if $\mathbf{x} = \boldsymbol{\theta}$ is the *only* solution of $A\mathbf{x} = \boldsymbol{\theta}$. Conversely A is singular if and only if there exist nontrivial solutions to $A\mathbf{x} = \boldsymbol{\theta}$. Also we wish to emphasize that the notions of singularity and nonsingularity are defined for *square* matrices only. Now since $A\mathbf{x} = \boldsymbol{\theta}$ can be written as

$$x_1\mathbf{A}_1 + x_2\mathbf{A}_2 + \dots + x_n\mathbf{A}_n = \boldsymbol{\theta},$$

we see immediately that A is nonsingular if and only if the column vectors of $A = [\mathbf{A}_1, \mathbf{A}_2, \dots, \mathbf{A}_n]$ are linearly independent. Although this observation is an immediate consequence of Definition 1.9, it is important enough to be stated as a theorem.

Theorem 1.10

The $(n \times n)$ matrix $A = [\mathbf{A}_1, \mathbf{A}_2, \dots, \mathbf{A}_n]$ is nonsingular if and only if $\{\mathbf{A}_1, \mathbf{A}_2, \dots, \mathbf{A}_n\}$ is a linearly independent set.

EXAMPLE 1.26 The (2×2) matrix

$$A = \begin{bmatrix} 1 & 3 \\ 2 & 2 \end{bmatrix}$$

is nonsingular since $[A \,|\, \boldsymbol{\theta}]$ is row equivalent to

$$\begin{bmatrix} 1 & 3 & 0 \\ 0 & -4 & 0 \end{bmatrix}.$$

On the other hand the (2×2) matrix

$$B = \begin{bmatrix} 1 & 2 \\ 2 & 4 \end{bmatrix}$$

is singular since the vector

$$\mathbf{x} = \begin{bmatrix} -2 \\ 1 \end{bmatrix}$$

is a nontrivial solution of $B\mathbf{x} = \boldsymbol{\theta}$. Equivalently the columns of B are linearly dependent since $2\mathbf{B}_1 - \mathbf{B}_2 = \boldsymbol{\theta}$. The importance of nonsingular matrices with respect to linear systems is demonstrated in the next theorem.

Theorem 1.11

Let A be an $(n \times n)$ matrix. The equation $A\mathbf{x} = \mathbf{b}$ has a unique solution for every $(n \times 1)$ column vector $\mathbf{b}$ if and only if A is nonsingular.

Proof. Suppose first that $A\mathbf{x} = \mathbf{b}$ has a unique solution no matter what choice we make for $\mathbf{b}$. Choosing $\mathbf{b} = \boldsymbol{\theta}$, we see that this assumption means that A is nonsingular (see Definition 1.9).

Conversely suppose $A = [\mathbf{A}_1, \mathbf{A}_2, \ldots, \mathbf{A}_n]$ is nonsingular, and let $\mathbf{b}$ be any $(n \times 1)$ column vector. To show $A\mathbf{x} = \mathbf{b}$ has a unique solution, we must show that the equation $A\mathbf{x} = \mathbf{b}$, or

$$x_1\mathbf{A}_1 + x_2\mathbf{A}_2 + \ldots + x_n\mathbf{A}_n = \mathbf{b},$$

is consistent and that the solution is unique. To show this, we observe first that

$$\{\mathbf{A}_1, \mathbf{A}_2, \ldots, \mathbf{A}_n, \mathbf{b}\}$$

is a set of $(n + 1)$ vectors in R^n; and so by Theorem 1.9 this set is linearly dependent. Thus there are scalars $a_1, a_2, \ldots, a_n, a_{n+1}$ such that

$$a_1\mathbf{A}_1 + a_2\mathbf{A}_2 + \ldots + a_n\mathbf{A}_n + a_{n+1}\mathbf{b} = \boldsymbol{\theta}; \tag{1.29a}$$

and moreover not all of these scalars are zero. In fact if $a_{n+1} = 0$ in (1.29a), then $\{\mathbf{A}_1, \mathbf{A}_2, \ldots, \mathbf{A}_n\}$ would be a linearly dependent set (contradicting the assumption that A is nonsingular); so we know that a_{n+1} is nonzero in (1.29a). Thus a solution to $A\mathbf{x} = \mathbf{b}$ is given by

$$\hat{\mathbf{x}} = \begin{bmatrix} \hat{x}_1 \\ \hat{x}_2 \\ \vdots \\ \hat{x}_n \end{bmatrix}$$

where

$$\hat{x}_1 = \frac{-a_1}{a_{n+1}}, \quad \hat{x}_2 = \frac{-a_2}{a_{n+1}}, \ldots, \quad \hat{x}_n = \frac{-a_n}{a_{n+1}}. \tag{1.29b}$$

Therefore $A\mathbf{x} = \mathbf{b}$ is always consistent when A is nonsingular.

The $(n \times 1)$ vector $\hat{\mathbf{x}}$ given by (1.29b) is one solution of $A\mathbf{x} = \mathbf{b}$. Now suppose that the $(n \times 1)$ vector $\mathbf{u}$ is any solution whatsoever to $A\mathbf{x} = \mathbf{b}$; that is, $A\mathbf{u} = \mathbf{b}$. Then $A\hat{\mathbf{x}} - A\mathbf{u} = \mathbf{b} - \mathbf{b}$ or

$$A(\hat{\mathbf{x}} - \mathbf{u}) = \boldsymbol{\theta},$$

and therefore $\mathbf{y} = \hat{\mathbf{x}} - \mathbf{u}$ is a solution of $A\mathbf{x} = \boldsymbol{\theta}$. But since A is nonsingular, we must have $\mathbf{y} = \boldsymbol{\theta}$ or $\hat{\mathbf{x}} = \mathbf{u}$; and thus $A\mathbf{x} = \mathbf{b}$ has one and only one solution. ∎

A system $A\mathbf{x} = \mathbf{b}$ where the coefficient matrix A is an $(n \times n)$ nonsingular matrix represents the simplest type of linear system to describe mathematically since the system always has one solution and that solution is the only one. By contrast we show later that if A is an $(m \times n)$ matrix, two possibilities exist.

1. If $m < n$, solutions (if any exist) to $A\mathbf{x} = \mathbf{b}$ are *never* unique.
2. If $m > n$, there are vectors $\mathbf{b}$ for which $A\mathbf{x} = \mathbf{b}$ has *no* solution.

If A is $(n \times n)$ and singular, then it can also be shown that both (1) and (2) hold; that is, there are $(n \times 1)$ vectors $\mathbf{b}$ for which $A\mathbf{x} = \mathbf{b}$ has no solution; and moreover if $A\mathbf{x} = \mathbf{b}$ is consistent, then solutions are not unique.

In order to discuss another important feature of nonsingular matrices, we introduce the concept of an identity matrix. The $(n \times n)$ ***identity matrix***, I_n, is the $(n \times n)$ matrix

$$I_n = \begin{bmatrix} 1 & 0 & 0 & \ldots & 0 \\ 0 & 1 & 0 & \ldots & 0 \\ 0 & 0 & 1 & \ldots & 0 \\ \vdots & & & & \\ 0 & 0 & 0 & \ldots & 1 \end{bmatrix}. \tag{1.30a}$$

Thus the ijth entry of I_n is 0 when $i \neq j$ and is 1 when $i = j$. It is easy to verify that if B is any $(m \times p)$ matrix, then $BI_p = B$ and $I_m B = B$. In particular $I_n\mathbf{x} = \mathbf{x}$ for all $\mathbf{x}$ in R^n. Usually the dimension of the identity matrix is clear from the context of the problem under consideration, and it is customary to drop the subscript n and denote the $(n \times n)$ identity matrix simply as I. [So for example if A is an $(n \times n)$ matrix, we will write $IA = AI = A$ instead of $I_nA = AI_n = A$.] The jth column of I is often denoted as $\mathbf{e}_j$; so if I is the $(n \times n)$ identity matrix, then

$$I = [\mathbf{e}_1, \mathbf{e}_2, \ldots, \mathbf{e}_n] \tag{1.30b}$$

where the vectors $\mathbf{e}_1, \mathbf{e}_2, \ldots, \mathbf{e}_n$ are the unit vectors in R^n:

$$\mathbf{e}_1 = \begin{bmatrix} 1 \\ 0 \\ 0 \\ \vdots \\ 0 \end{bmatrix}, \quad \mathbf{e}_2 = \begin{bmatrix} 0 \\ 1 \\ 0 \\ \vdots \\ 0 \end{bmatrix}, \quad \mathbf{e}_3 = \begin{bmatrix} 0 \\ 0 \\ 1 \\ \vdots \\ 0 \end{bmatrix}, \ldots, \quad \mathbf{e}_n = \begin{bmatrix} 0 \\ 0 \\ 0 \\ \vdots \\ 1 \end{bmatrix} \tag{1.30c}$$

EXAMPLE 1.27 The (3×3) identity matrix is

$$I = \begin{bmatrix} 1 & 0 & 0 \\ 0 & 1 & 0 \\ 0 & 0 & 1 \end{bmatrix}$$

Next let

$$A = \begin{bmatrix} 1 & 2 & 0 \\ -1 & 3 & 4 \\ 6 & 1 & 8 \end{bmatrix}, \quad B = \begin{bmatrix} 2 & 3 & 1 \\ 1 & 5 & 7 \end{bmatrix}, \quad C = \begin{bmatrix} -2 & 0 \\ 8 & 3 \\ 6 & 1 \end{bmatrix},$$

$$\mathbf{x} = \begin{bmatrix} 1 \\ 0 \\ 3 \end{bmatrix},$$

and note that

$$IA = AI = A$$
$$BI = B$$
$$IC = C$$
$$I\mathbf{x} = \mathbf{x}$$

whereas the products IB and CI are not defined.

Having defined the $(n \times n)$ identity matrix, we can establish the following important result.

Theorem 1.12

If A is an $(n \times n)$ nonsingular matrix, then there is a unique $(n \times n)$ matrix C such that $AC = I$.

Proof. Since A is nonsingular, we can apply Theorem 1.11 to assert that the equation $A\mathbf{x} = \mathbf{b}$ always has a unique solution. Thus there exist unique $(n \times 1)$ vectors $\mathbf{c}_1, \mathbf{c}_2, \ldots, \mathbf{c}_n$ such that

$$A\mathbf{c}_k = \mathbf{e}_k, \; k = 1, 2, \ldots, n.$$

Defining C by $C = [\mathbf{c}_1, \mathbf{c}_2, \ldots, \mathbf{c}_n]$ and appealing to Theorem 1.4, we see that

$$AC = [A\mathbf{c}_1, A\mathbf{c}_2, \ldots, A\mathbf{c}_n] = [\mathbf{e}_1, \mathbf{e}_2, \ldots, \mathbf{e}_n] = I,$$

which establishes the theorem. ∎

The matrix C defined by Theorem 1.12 is called the ***inverse*** of A and is denoted by A^{-1}. We show in the next section that if A is singular, then there is no matrix C such that $AC = I$. Thus A^{-1} exists if and only if A is nonsingular; and if A is nonsingular, then A^{-1} is unique. The next section also examines some of the properties of matrix inverses showing for example that A and A^{-1} commute (that is, $A^{-1}A = AA^{-1} = I$). One significance of A^{-1} is demonstrated by noting that if $AA^{-1} = I$, then solutions to $A\mathbf{x} = \mathbf{b}$ are given by $\mathbf{x} = A^{-1}\mathbf{b}$. To see this fact, we need only observe that

$$A(A^{-1}\mathbf{b}) = (AA^{-1})\mathbf{b} = I\mathbf{b} = \mathbf{b},$$

which verifies directly that $A^{-1}\mathbf{b}$ solves the equation $A\mathbf{x} = \mathbf{b}$. We shall comment in the next section on the practical aspects of calculating A^{-1} and on the role that A^{-1} plays in solving $(n \times n)$ nonsingular systems. For the moment we present only the following simple example.

EXAMPLE 1.28 Let the (2×2) matrix A be given by

$$A = \begin{bmatrix} 2 & 1 \\ 5 & 3 \end{bmatrix}.$$

Then it is easy to show that A is nonsingular; i.e., the only solution to $A\mathbf{x} = \boldsymbol{\theta}$ is $\mathbf{x} = \boldsymbol{\theta}$. To verify this statement, we note that the augmented matrix $[A \,\vdots\, \boldsymbol{\theta}]$ is row equivalent to

$$B = \begin{bmatrix} 2 & 1 & 0 \\ 0 & \frac{1}{2} & 0 \end{bmatrix}.$$

The matrix B represents a homogeneous system that has only the trivial solution; so A is nonsingular and Theorem 1.12 guarantees an inverse matrix for A. Furthermore it is easy to calculate A^{-1} by solving $A\mathbf{x} = \mathbf{e}_1$ and $A\mathbf{x} = \mathbf{e}_2$. In particular $A\mathbf{x} = \mathbf{e}_1$ is represented as $[A \,\vdots\, \mathbf{e}_1]$ where

$$[A \,\vdots\, \mathbf{e}_1] = \begin{bmatrix} 2 & 1 & 1 \\ 5 & 3 & 0 \end{bmatrix};$$

and this matrix is row equivalent to

$$\begin{bmatrix} 2 & 1 & 1 \\ 0 & \frac{1}{2} & -\frac{5}{2} \end{bmatrix}.$$

Therefore the solution of $A\mathbf{x} = \mathbf{e}_1$ is

$$\mathbf{c}_1 = \begin{bmatrix} 3 \\ -5 \end{bmatrix}.$$

Similarly it is easy to solve $A\mathbf{x} = \mathbf{e}_2$; and the solution is given by

$$\mathbf{c}_2 = \begin{bmatrix} -1 \\ 2 \end{bmatrix}.$$

Thus as in the proof of Theorem 1.12, we find that A^{-1} is the (2×2) matrix $A^{-1} = [\mathbf{c}_1, \mathbf{c}_2]$, or

$$A^{-1} = \begin{bmatrix} 3 & -1 \\ -5 & 2 \end{bmatrix}.$$

We can check these calculations by verifying that $AA^{-1} = I$:

$$\begin{bmatrix} 2 & 1 \\ 5 & 3 \end{bmatrix} \begin{bmatrix} 3 & -1 \\ -5 & 2 \end{bmatrix} = \begin{bmatrix} 1 & 0 \\ 0 & 1 \end{bmatrix}.$$

To demonstrate how A^{-1} can be used to solve a system, consider the (2×2) system

$$\begin{aligned} 2x_1 + x_2 &= 3 \\ 5x_1 + 3x_2 &= 4, \end{aligned}$$

which has A as the coefficient matrix. Writing this system in the form $A\mathbf{x} = \mathbf{b}$, we see that the solution vector $\mathbf{x}$ is given by $\mathbf{x} = A^{-1}\mathbf{b}$:

$$A^{-1}\mathbf{b} = \begin{bmatrix} 3 & -1 \\ -5 & 2 \end{bmatrix} \begin{bmatrix} 3 \\ 4 \end{bmatrix} = \begin{bmatrix} 5 \\ -7 \end{bmatrix}.$$

---

skip over

We conclude this section by giving several examples that show how matrix theory can be applied to the practical problem of interpolating data with polynomials. In particular Theorem 1.11 is used in Example 1.30 to establish a general existence and uniqueness result for polynomial interpolation.

†**EXAMPLE 1.29** As a simple example of polynomial interpolation, suppose we wish to find a quadratic polynomial, $q(t)$, such that the graph of $q(t)$ goes through the points (1, 2), (2, 3), (3, 6) in the ty-plane (see Fig. 1.3). Now any quadratic polynomial $q(t)$ has the form

$$q(t) = a + bt + ct^2; \tag{1.31a}$$

so our problem reduces to determining constants a, b, and c such that

$$\begin{aligned} q(1) &= 2 \\ q(2) &= 3 \\ q(3) &= 6. \end{aligned} \tag{1.31b}$$

† Throughout, material designated with a dagger may be omitted without any loss of continuity. This material is not prerequisite for topics in later sections except occasionally for other material also designated by †.

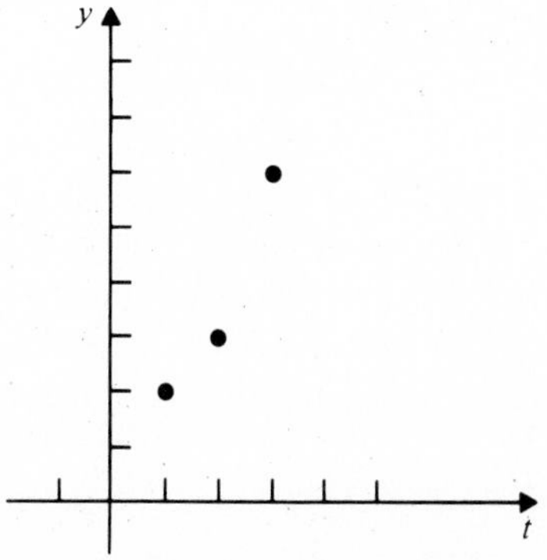

Fig. 1.3 Data points for $q(t)$.

The constraints in (1.31b) are by (1.31a) equivalent to

$$\begin{aligned} a+b+c &= 2 \\ a+2b+4c &= 3. \\ a+3b+9c &= 6. \end{aligned} \tag{1.31c}$$

Clearly (1.31c) is a system of three linear equations in the three unknowns a, b, and c; so solving (1.31c) will determine the polynomial $q(t)$ that we are searching for. Solving (1.31c), we find the (unique) solution $a = 3$, $b = -2$, $c = 1$; and therefore $q(t) = 3 - 2t + t^2$ is the unique quadratic polynomial satisfying the conditions in (1.31b). A portion of the graph of $q(t)$ is shown in Fig. 1.4.

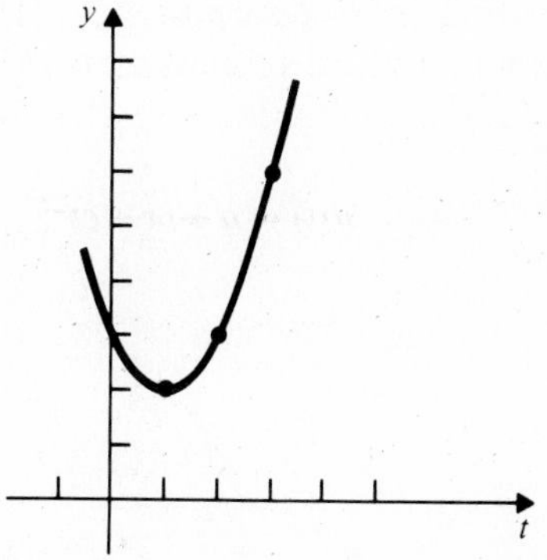

Fig. 1.4 Graph of $q(t) = 3 - 2t + t^2$.

Frequently polynomial interpolation is used when values of a function $f(t)$ are given in tabular form. For example if we are given a table of $n+1$ values of $f(t)$ (see

Table 1.1), it is easy to show (see Example 1.30) that there is a unique polynomial $p(t)$ of the form

$$p(t) = a_0 + a_1 t + a_2 t^2 + \ldots + a_n t^n$$

such that $p(t_i) = y_i$, $0 \leqslant i \leqslant n$. The polynomial $p(t)$ is called the ***interpolating polynomial*** for $f(t)$. Problems of interpolating in tables are quite common in scientific and engineering work; for example $y = f(t)$ might describe a temperature distribution as a function of time with $y_i = f(t_i)$ being observed (measured) temperatures. For a time $\hat{t}$ not listed in the table, an approximation to $f(\hat{t})$ is $f(\hat{t}) \simeq p(\hat{t})$ where $p(t)$ is the interpolating polynomial for $f(t)$.

TABLE 1.1

t	$f(t)$
t_0	y_0
t_1	y_1
t_2	y_2
$\vdots$	$\vdots$
t_n	y_n

As a specific example, consider a table of four observations (see Table 1.2). In this case the interpolating polynomial is a polynomial of degree three or less

$$p(t) = a_0 + a_1 t + a_2 t^2 + a_3 t^3$$

where $p(t)$ satisfies the four constraints $p(0) = 3$, $p(1) = 0$, $p(2) = -1$, $p(3) = 6$. As in (1.31), these constraints serve to determine the coefficients of $p(t)$ since a_0, a_1, a_2, a_3 must be chosen to satisfy

$$\begin{aligned} a_0 \qquad\qquad\qquad\quad &= 3 \\ a_0 + a_1 + a_2 + a_3 &= 0 \\ a_0 + 2a_1 + 4a_2 + 8a_3 &= -1 \\ a_0 + 3a_1 + 9a_2 + 27a_3 &= 6. \end{aligned}$$

TABLE 1.2

t	$f(t)$
0	3
1	0
2	−1
3	6

Solving this (4×4) system, we find $a_3 = 1, a_2 = -2, a_1 = -2$ and $a_0 = 3$. Hence the unique polynomial that interpolates the tabular data for $f(t)$ is $p(t) = 3 - 2t - 2t^2 + t^3$.

†**EXAMPLE 1.30** In this example we use Theorem 1.11 to prove the following general result of which two special cases were given in Example 1.29.

Theorem

Given $n+1$ distinct numbers $t_0, t_1, \ldots, t_n$ and any set of $n+1$ values $y_0, y_1, \ldots, y_n$, there is one and only one polynomial $p(t)$ of degree n or less, $p(t) = a_0 + a_1 t + \ldots + a_n t^n$, such that $p(t_i) = y_i$, $i = 0, 1, \ldots, n$.

To prove this theorem, we consider the constraints $p(t_i) = y_i$ as defining a linear system [just as in (1.31) above]:

$$\begin{aligned} a_0 + a_1 t_0 + a_2 (t_0)^2 + \ldots + a_n (t_0)^n &= y_0 \\ a_0 + a_1 t_1 + a_2 (t_1)^2 + \ldots + a_n (t_1)^n &= y_1 \\ \vdots \qquad\qquad\qquad & \vdots \quad \vdots \\ a_0 + a_1 t_n + a_2 (t_n)^2 + \ldots + a_n (t_n)^n &= y_n. \end{aligned} \tag{1.32}$$

We can write (1.32) in matrix form as $T\mathbf{a} = \mathbf{y}$.

By Theorem 1.11 we can guarantee that the equation $T\mathbf{a} = \mathbf{y}$ always has a unique solution for any $\mathbf{y}$ if we can show that T is nonsingular. To establish that T is nonsingular, consider the equation $T\mathbf{x} = \boldsymbol{\theta}$, and suppose that the vector $\mathbf{c}$ is a solution of $T\mathbf{x} = \boldsymbol{\theta}$ where

$$\mathbf{c} = \begin{bmatrix} c_0 \\ c_1 \\ \vdots \\ c_n \end{bmatrix}.$$

Using (1.32) and $T\mathbf{c} = \boldsymbol{\theta}$, we observe that the polynomial $q(t) = c_0 + c_1 t + \ldots + c_n t^n$ satisfies $q(t_i) = 0, 0 \leqslant i \leqslant n$. But if $q(t)$ is a polynomial of degree n with $n+1$ distinct real zeros, then by Exercise 15, Section 1.7, $q(t)$ must be identically zero. If $q(t)$ is identically zero, then we must have $c_0 = c_1 = \ldots c_n = 0$, or $\mathbf{c} = \boldsymbol{\theta}$. This equation proves that the matrix T (called the ***Vandermonde matrix***) is nonsingular when $t_0, t_1, \ldots, t_n$ are distinct. An important feature of this example is that we have used Theorem 1.11 to prove that there is always one and only one polynomial $p(t)$ of degree n or less that takes on $n+1$ prescribed values at $n+1$ distinct prescribed points.

---

† Throughout, material designated with a dagger may be omitted without any loss of continuity. This material is not prerequisite for topics in later sections except occasionally for other material also designated by †.

There are other equally easy ways to demonstrate the existence and uniqueness of interpolating polynomials, but the approach (of setting up a system of linear equations corresponding to the constraints) used in these two examples is applicable to more general problems in which we wish to choose the coefficients of a linear approximation to satisfy certain interpolatory constraints.

†**EXAMPLE 1.31** In this example we suppose that $0 \leqslant a < \pi$ and ask for constants A and B such that the function

$$y(t) = A \cos t + B \sin t$$

satisfies the constraints $y(a) = 0$, $y(\pi) = 1$. (Problems of this sort arise, for example, in solving two-point boundary-value problems in differential equations.) As before, we write out the constraints:

$$\begin{aligned} A \cos a + B \sin a &= 0 \\ A \cos \pi + B \sin \pi &= 1. \end{aligned}$$

Solving this system, we find that if $a = 0$, there is no solution while if $0 < a < \pi$, then $A = -1$, $B = \cot a$. Thus

$$y(t) = -\cos t + \cot a \sin t$$

defines a function of the prescribed form for which $y(a) = 0$, $y(\pi) = 1$ as long as $0 < a < \pi$.

1.7 EXERCISES

1. Determine whether the following matrices are singular or nonsingular. If a matrix is nonsingular, find its inverse.

$$A = \begin{bmatrix} 1 & 2 \\ 3 & 6 \end{bmatrix} \quad B = \begin{bmatrix} 3 & 2 \\ 4 & 3 \end{bmatrix} \quad C = \begin{bmatrix} 1 & 2 & -1 \\ 3 & 1 & 1 \\ 5 & 5 & -1 \end{bmatrix} \quad D = \begin{bmatrix} 1 & 1 & 1 \\ 0 & 1 & 3 \\ 5 & 5 & 4 \end{bmatrix}$$

2. Repeat Exercise 1 for these matrices.

$$A = \begin{bmatrix} 2 & 4 \\ 1 & 2 \end{bmatrix} \quad B = \begin{bmatrix} 5 & 3 \\ 3 & 2 \end{bmatrix} \quad C = \begin{bmatrix} 1 & 1 & 2 \\ 1 & 2 & 1 \\ 5 & 8 & 7 \end{bmatrix} \quad D = \begin{bmatrix} 1 & 2 & 2 \\ 0 & 1 & 1 \\ 3 & 6 & 7 \end{bmatrix}$$

3. a) Show that A is nonsingular, and find A^{-1} where

$$A = \begin{bmatrix} 1 & 0 & 1 \\ -1 & 1 & 2 \\ -4 & 1 & 0 \end{bmatrix}.$$

† Throughout, material designated with a dagger may be omitted without any loss of continuity. This material is not prerequisite for topics in later sections except occasionally for other material also designated by †.

b) Use A^{-1} as found in part (a) to solve

$$\begin{aligned} x_1 \qquad + x_3 &= 1 \\ -x_1 + x_2 + 2x_3 &= 3 \\ -4x_1 + x_2 \qquad &= 1. \end{aligned}$$

4. a) Verify that $AA^{-1} = I$ where

$$A = \begin{bmatrix} 1 & 0 & -1 \\ 0 & 1 & 1 \\ 3 & 0 & -2 \end{bmatrix} \qquad A^{-1} = \begin{bmatrix} -2 & 0 & 1 \\ 3 & 1 & -1 \\ -3 & 0 & 1 \end{bmatrix}.$$

b) Verify that $A^{-1}A = I$. (For this matrix, $AA^{-1} = A^{-1}A$.)

c) Verify that the inverse of A^{T} is the transpose of A^{-1}. [That is, for this matrix A check that $(A^{\mathrm{T}})^{-1} = (A^{-1})^{\mathrm{T}}$ by showing that $A^{\mathrm{T}}(A^{-1})^{\mathrm{T}} = I$.]

5. Let $A = (a_{ij})$ be a (2×2) matrix such that $\Delta \neq 0$ where $\Delta = a_{11}a_{22} - a_{12}a_{21}$. By multiplying A and the matrix below, verify that A has an inverse given by

$$A^{-1} = \frac{1}{\Delta} \begin{bmatrix} a_{22} & -a_{12} \\ -a_{21} & a_{11} \end{bmatrix}.$$

6. a) Use Exercise 5 to calculate the inverse of these matrices.

$$A = \begin{bmatrix} 2 & 5 \\ 1 & 3 \end{bmatrix} \qquad B = \begin{bmatrix} 1 & 2 \\ 3 & 5 \end{bmatrix} \qquad C = \begin{bmatrix} 2 & 4 \\ 2 & 5 \end{bmatrix}$$

Check your answer by multiplying the matrix and its inverse.

b) Use the results of part (a) to solve these systems.

$$\begin{aligned} 2x_1 + 5x_2 &= 3 \\ x_1 + 3x_2 &= 6 \end{aligned} \qquad \begin{aligned} x_1 + 2x_2 &= 0 \\ 3x_1 + 5x_2 &= 4 \end{aligned} \qquad \begin{aligned} 2x_1 + 4x_2 &= 1 \\ 2x_1 + 5x_2 &= 2 \end{aligned}$$

7. Let A be an $(n \times n)$ matrix, and prove the following: if A is nonsingular, then A^{T} is nonsingular. Carry out the proof by justifying the steps below.

a) Suppose $\mathbf{v}$ is any vector such that $A^{\mathrm{T}}\mathbf{v} = \boldsymbol{\theta}$. There is a vector $\mathbf{w}$ such that $\mathbf{v} = A\mathbf{w}$. (Why?)

b) By (a), $A^{\mathrm{T}}A\mathbf{w} = \boldsymbol{\theta}$; so $\mathbf{w}^{\mathrm{T}}A^{\mathrm{T}}A\mathbf{w} = \mathbf{w}^{\mathrm{T}}\boldsymbol{\theta} = 0$. Therefore $\|A\mathbf{w}\| = 0$. (Why?)

c) From (a) and (b) show that if $A^{\mathrm{T}}\mathbf{v} = \boldsymbol{\theta}$, then $\mathbf{v} = \boldsymbol{\theta}$—this will prove that A^{T} is nonsingular whenever A is.

8. Let A be an $(n \times n)$ matrix. Use Exercise 7 and Theorem 1.8(3) to give a quick proof of the following: A is nonsingular $\Leftrightarrow A^{\mathrm{T}}$ is nonsingular.

9. Let T be an $(n \times n)$ upper-triangular matrix.

$$T = \begin{bmatrix} t_{11} & t_{12} & t_{13} & \cdots & t_{1n} \\ 0 & t_{22} & t_{23} & \cdots & t_{2n} \\ 0 & 0 & t_{33} & \cdots & t_{3n} \\ \vdots & & & & \vdots \\ 0 & 0 & 0 & \cdots & t_{nn} \end{bmatrix}$$

Prove that if $t_{ii} = 0$ for some i, $1 \leqslant i \leqslant n$, then T is singular. (Hint: If $t_{11} = 0$, find a nonzero vector $\mathbf{v}$ such that $T\mathbf{v} = \boldsymbol{\theta}$. If $t_{rr} = 0$, but $t_{ii} \neq 0$ for $i = 1, 2, \ldots, r-1$, use Theorem 1.2 to show that the columns $\mathbf{T}_1, \mathbf{T}_2, \ldots, \mathbf{T}_r$ are linearly dependent; and then select a nonzero vector $\mathbf{v}$ such that $T\mathbf{v} = \boldsymbol{\theta}$.)

10. Let T be an $(n \times n)$ upper-triangular matrix as in Exercise 9. Prove that if $t_{ii} \neq 0$ for $i = 1, 2, \ldots, n$, then T is nonsingular. (Hint: Let $T = [\mathbf{T}_1, \mathbf{T}_2, \ldots, \mathbf{T}_n]$ and suppose $a_1\mathbf{T}_1 + a_2\mathbf{T}_2 + \ldots + a_n\mathbf{T}_n = \boldsymbol{\theta}$ for some scalars $a_1, a_2, \ldots, a_n$. Multiply both sides of this identity by $\mathbf{e}_n^T$, and deduce that $a_n = 0$. Next show $a_{n-1} = 0$, etc.) Note that Exercises 9 and 10 establish that an upper-triangular matrix is singular if and only if one of the diagonal entries $t_{11}, t_{22}, \ldots, t_{nn}$ is zero. By Exercise 8 this same result is true for lower-triangular matrices.

11. a) Find the inverse matrix for

$$T_1 = \begin{bmatrix} 1 & 2 \\ 0 & 1 \end{bmatrix} \qquad T_2 = \begin{bmatrix} 1 & 3 & 1 \\ 0 & 1 & 2 \\ 0 & 0 & 1 \end{bmatrix}.$$

b) In part (a) observe that the inverse of each upper-triangular matrix is again an upper-triangular matrix. Prove that the inverse of any nonsingular (4×4) upper-triangular matrix T is also an upper-triangular matrix. [Hint: Suppose $W = (w_{ij})$ is a (4×4) matrix such that $TW = I$. Use $t_{ii} \neq 0$ to show that $w_{41} = w_{42} = w_{43} = 0$; then move to the third row of W, etc.]

12. Suppose A is an $(m \times n)$ matrix. Prove that A^TA is singular $\Leftrightarrow$ the columns of A are linearly dependent.

13. Let $p(t) = a_0 + a_1t + a_2t^2$.

a) Find a_0, a_1, a_2 so that $p(-1) = 9$, $p(1) = 3$, $p(2) = 3$.
b) For $f(t) = 2t/(t^2 - 3)$, find a_0, a_1, a_2 so that $p(-2) = f(-2)$, $p(1) = f(1)$, $p(2) = f(2)$.
c) Find a_0, a_1, a_2 so that $p(1) = 0$, $p'(1) = 7$, $p(2) = 10$.

14. Let $q(t) = c_0 + c_1 \cos t + c_2 \sin t + c_3 \cos 2t$. Find c_0, c_1, c_2, c_3 so that $q(-\pi/2) = 4$, $q(0) = 2$, $q(\pi/4) = 2$, $q(\pi/2) = 6$.

15. In Example 1.30 the proof that the Vandermonde matrix is nonsingular depends on the fact that a real nth degree polynomial $q(t)$ with $n+1$ distinct real zeros must be identically zero. Use Rolle's theorem to prove this. [That is, if $q(t)$ has $n+1$ zeros, then $q'(t)$ has n zeros, and $q'(t)$ is of degree $n-1$, etc.]

1.8 MATRIX INVERSES AND THEIR PROPERTIES

In this section we will develop some properties of the matrix inverse, give an efficient scheme for calculating the inverse, and comment on the role inverses should play in solving systems of equations. Our first theoretical result is suggested by the following example.

EXAMPLE 1.32 From the arithmetic properties of the real number system we know that if $ab = 0$, then either $a = 0$ or $b = 0$ (or both). The set of $(n \times n)$ matrices does not have this cancellation property; and there are many pairs of matrices A and B such that $AB = \mathcal{O}$, but neither A nor B is the zero matrix. As an example, let

$$A = \begin{bmatrix} 1 & -1 \\ 0 & 0 \end{bmatrix}, \qquad B = \begin{bmatrix} 1 & 0 \\ 1 & 0 \end{bmatrix};$$

then

$$AB = \begin{bmatrix} 0 & 0 \\ 0 & 0 \end{bmatrix},$$

but $A \neq \mathcal{O}$, $B \neq \mathcal{O}$. Yet a further complication is introduced because matrix multiplication is not commutative. That is, even though $AB = \mathcal{O}$, BA is not the zero matrix, for

$$BA = \begin{bmatrix} 1 & -1 \\ 1 & -1 \end{bmatrix}.$$

We do observe from this example that A is singular. Moreover AB and BA are both singular matrices; the next theorem shows that this is a general result. That is, the matrix product AB is singular if and only if either A or B is singular. This result is thus analogous to the property of the real numbers that states that $ab = 0$ if and only if $a = 0$ or $b = 0$. Finally in Exercise 6, Section 1.8, the reader can show that if A is nonsingular, then $AB = \mathcal{O}$ if and only if $B = \mathcal{O}$.

Theorem 1.13

Let A and B be $(n \times n)$ matrices. The product AB is nonsingular if and only if both A and B are nonsingular.

Proof. We first suppose AB is nonsingular and show that this supposition implies that A and B are each nonsingular. To show this, suppose $\mathbf{x}_1$ is a solution of $B\mathbf{x} = \boldsymbol{\theta}$. But if $B\mathbf{x}_1 = \boldsymbol{\theta}$, then

$$(AB)\mathbf{x}_1 = A(B\mathbf{x}_1) = A\boldsymbol{\theta} = \boldsymbol{\theta};$$

so $\mathbf{x}_1$ is a solution of $(AB)\mathbf{x} = \boldsymbol{\theta}$. However AB is nonsingular; so $\mathbf{x}_1$ must be the zero vector, and thus the only solution of $B\mathbf{x} = \boldsymbol{\theta}$ is $\mathbf{x} = \boldsymbol{\theta}$. Having B nonsingular, we next suppose $\mathbf{x}_2$ is a solution of $A\mathbf{x} = \boldsymbol{\theta}$. Since B is nonsingular, there is a vector $\mathbf{y}_1$ such that $B\mathbf{y}_1 = \mathbf{x}_2$ (see Theorem 1.11). Thus $A\mathbf{x}_2 = \boldsymbol{\theta}$ means that

$$A\mathbf{x}_2 = A(B\mathbf{y}_1) = (AB)\mathbf{y}_1 = \boldsymbol{\theta};$$

and hence $\mathbf{y}_1$ is a solution of $(AB)\mathbf{x} = \boldsymbol{\theta}$. As before, the nonsingularity of AB implies that $\mathbf{y}_1 = \boldsymbol{\theta}$; so $B\mathbf{y}_1 = \mathbf{x}_2$ must also be the zero vector. Thus the only solution of $A\mathbf{x} = \boldsymbol{\theta}$ is $\mathbf{x} = \boldsymbol{\theta}$.

The proof that AB is nonsingular when both A and B are nonsingular is similar and is left to the reader. ■

Note that one implication of Theorem 1.13 is that a singular matrix cannot have an inverse. Specifically if A has an inverse, then the identity $AA^{-1} = I$ holds. Clearly I is nonsingular; so Theorem 1.13 asserts that A and A^{-1} are also nonsingular. In conjunction with Theorem 1.12 we thus have that A is nonsingular if and only if A^{-1} exists.

EXAMPLE 1.33 As Theorem 1.13 implies, the product of a singular matrix and a nonsingular matrix is always singular. For example let

$$A = \begin{bmatrix} 3 & 5 \\ 1 & 2 \end{bmatrix}, \qquad B = \begin{bmatrix} 1 & 2 \\ 2 & 4 \end{bmatrix}.$$

In this case A is nonsingular, B is singular, and the products AB and BA are each singular:

$$AB = \begin{bmatrix} 13 & 26 \\ 5 & 10 \end{bmatrix}, \qquad BA = \begin{bmatrix} 5 & 9 \\ 10 & 18 \end{bmatrix}.$$

The next theorem summarizes some useful matrix-inverse properties, including the fact that A and A^{-1} commute under multiplication. (Thus far we have established that if A is nonsingular, then there is a unique matrix, A^{-1}, such that $AA^{-1} = I$. We have not proved that there exists a unique matrix, R, such that $RA = I$ and that $R = A^{-1}$. The theorem below establishes this and a bit more.)

Theorem 1.14

Let A be an $(n \times n)$ nonsingular matrix. Then A has a unique inverse and

1. A^{-1} is nonsingular,
2. $(A^{-1})^{-1} = A$, and
3. $A^{-1}A = AA^{-1} = I$.
4. In addition if B is an $(n \times n)$ nonsingular matrix, then $(AB)^{-1} = B^{-1}A^{-1}$.

Proof. In Theorem 1.12 we have already established the existence and uniqueness of the inverse for a nonsingular matrix. This theorem, along with Theorem 1.13, makes parts (1) and (4) almost immediately apparent; and we leave their proofs to the reader. Assuming the proof of (1), that A^{-1} is nonsingular, we have from Theorem 1.12 that there is a unique matrix Q such that

$$A^{-1}Q = I.$$

[Note that Q is another name for the inverse of A^{-1}; that is, $(A^{-1})^{-1} = Q$. To prove (2), we need to show that $Q = A$.] Now since $AA^{-1} = I$, it follows that $(AA^{-1})Q = IQ = Q$. Using the basic associative property of matrix multiplication (Theorem 1.6), we have

$$Q = (AA^{-1})Q = A(A^{-1}Q) = AI = A.$$

Since $Q = A$, and since Q is the unique inverse of A^{-1}, we have shown $(A^{-1})^{-1} = A$. Our original definition of Q was from the equation $A^{-1}Q = I$; so we have also shown that

$$A^{-1}A = I.$$

But since $AA^{-1} = I$ as well, we know $A^{-1}A = AA^{-1}$; and this equation establishes (2) and (3). ■

EXAMPLE 1.34 As an illustration of Theorem 1.14, let A be given by

$$A = \begin{bmatrix} 3 & 5 \\ 1 & 2 \end{bmatrix}.$$

Then A^{-1} is

$$A^{-1} = \begin{bmatrix} 2 & -5 \\ -1 & 3 \end{bmatrix}.$$

A quick calculation shows that (as promised in Theorem 1.14)

$$I = AA^{-1} = \begin{bmatrix} 3 & 5 \\ 1 & 2 \end{bmatrix}\begin{bmatrix} 2 & -5 \\ -1 & 3 \end{bmatrix} = A^{-1}A = \begin{bmatrix} 2 & -5 \\ -1 & 3 \end{bmatrix}\begin{bmatrix} 3 & 5 \\ 1 & 2 \end{bmatrix}.$$

As an instance of $(AB)^{-1} = B^{-1}A^{-1}$, let B be the (2×2) matrix

$$B = \begin{bmatrix} 5 & 7 \\ 2 & 3 \end{bmatrix}.$$

In order to find B^{-1}, we must solve $B\mathbf{x} = \mathbf{e}_1$ and $B\mathbf{x} = \mathbf{e}_2$; and the augmented matrices, $[B \,\vdots\, \mathbf{e}_1]$ and $[B \,\vdots\, \mathbf{e}_2]$ are respectively row equivalent to

$$\begin{bmatrix} 5 & 7 & 1 \\ 0 & \frac{1}{5} & -\frac{2}{5} \end{bmatrix} \quad \text{and} \quad \begin{bmatrix} 5 & 7 & 0 \\ 0 & \frac{1}{5} & 1 \end{bmatrix}. \tag{1.33}$$

Therefore B^{-1} is given by

$$B^{-1} = \begin{bmatrix} 3 & -7 \\ -2 & 5 \end{bmatrix}.$$

Now AB is the matrix

$$AB = \begin{bmatrix} 25 & 36 \\ 9 & 13 \end{bmatrix}$$

while $B^{-1}A^{-1}$ is given by

$$B^{-1}A^{-1} = \begin{bmatrix} 13 & -36 \\ -9 & 25 \end{bmatrix}.$$

A direct multiplication shows that $(AB)(B^{-1}A^{-1}) = I$; so $(AB)^{-1} = B^{-1}A^{-1}$.

There is a simple and often useful formula for the inverse of a (2×2) matrix. (See Example 1.36 and Exercise 5, Section 1.7.) For now, we conclude our discussion of the properties of nonsingular matrices and inverses with Theorem 1.15, which summarizes some of the results already developed in the last section and in this section.

Theorem 1.15

Let A be an $(n \times n)$ matrix. Then the following are equivalent.

1. A is nonsingular.
2. The only solution of $A\mathbf{x} = \boldsymbol{\theta}$ is $\mathbf{x} = \boldsymbol{\theta}$.
3. The column vectors of A are linearly independent.
4. $A\mathbf{x} = \mathbf{b}$ always has a unique solution.
5. A has an inverse.

In order actually to compute A^{-1}, the procedure outlined in Example 1.35 below is a reasonably efficient way of organizing the calculations. This method essentially follows the steps in the proof of Theorem 1.12, in which we found the columns of A^{-1} by solving $A\mathbf{x} = \mathbf{e}_i$, $i = 1, 2, \ldots, n$.

EXAMPLE 1.35 Let A be the (3×3) matrix given by

$$A = \begin{bmatrix} 1 & -4 & 2 \\ 3 & 3 & 2 \\ 0 & 4 & -1 \end{bmatrix}.$$

To find A^{-1}, we have to solve the three systems

$$A\mathbf{x} = \mathbf{e}_1, \qquad A\mathbf{x} = \mathbf{e}_2, \qquad A\mathbf{x} = \mathbf{e}_3$$

and then use these solutions as the columns of A^{-1}. If we employ Gauss elimination to solve the three systems, we will have to reduce each of the matrices $[A \,\vdots\, \mathbf{e}_1]$, $[A \,\vdots\, \mathbf{e}_2]$, $[A \,\vdots\, \mathbf{e}_3]$ to echelon form. We note however that the reduction to echelon form for each of these three augmented matrices employs precisely the same elementary row operations in each case, namely the elementary row operations necessary to triangularize A. Hence the final echelon forms for each of these three augmented matrices will be identical except for the last column. [This similarity is exhibited, for instance, in Example 1.34, in which (1.33) shows the final reduced form of $[B \,\vdots\, \mathbf{e}_1]$ and $[B \,\vdots\, \mathbf{e}_2]$ for the given (2×2) matrix B.]

Therefore the computation of A^{-1} can be organized so that triangularizing A need be performed only once instead of three times. To organize the computation, we form the (3×6) matrix

$$[A \,\vdots\, \mathbf{e}_1, \mathbf{e}_2, \mathbf{e}_3], \tag{1.34a}$$

which we will write as $[A \,\vdots\, I]$ since $I = [\mathbf{e}_1, \mathbf{e}_2, \mathbf{e}_3]$. We now use elementary row operations to transform $[A \,\vdots\, I]$ to the echelon form, $[T \,\vdots\, \mathbf{v}_1, \mathbf{v}_2, \mathbf{v}_3]$; and we can immediately backsolve this final echelon form to find the solutions of $T\mathbf{x} = \mathbf{v}_1$, $T\mathbf{x} = \mathbf{v}_2$, and $T\mathbf{x} = \mathbf{v}_3$. These solutions are the solutions of $A\mathbf{x} = \mathbf{e}_1$, $A\mathbf{x} = \mathbf{e}_2$, and $A\mathbf{x}$

$= \mathbf{e}_3$, respectively, and hence are the columns of A^{-1}. For the matrix A above, $[A \vdots I]$ is row equivalent to

$$\begin{bmatrix} 1 & -4 & 2 & 1 & 0 & 0 \\ 0 & 15 & -4 & -3 & 1 & 0 \\ 0 & 4 & -1 & 0 & 0 & 1 \end{bmatrix},$$

which in turn is row equivalent to

$$\begin{bmatrix} 1 & -4 & 2 & 1 & 0 & 0 \\ 0 & 15 & -4 & -3 & 1 & 0 \\ 0 & 0 & \frac{1}{15} & \frac{12}{15} & \frac{-4}{15} & 1 \end{bmatrix} = [T \vdots \mathbf{v}_1, \mathbf{v}_2, \mathbf{v}_3]. \tag{1.34b}$$

The first four columns of (1.34b) give the augmented matrix for $T\mathbf{x} = \mathbf{v}_1$; columns 1, 2, 3, 5 give the augmented matrix for $T\mathbf{x} = \mathbf{v}_2$; and columns 1, 2, 3, 6 give the augmented matrix for $T\mathbf{x} = \mathbf{v}_3$. These three systems can be backsolved immediately to obtain the three columns of A^{-1}. For example the third system is

$$\begin{aligned} x_1 - 4x_2 + 2x_3 &= 0 \\ 15x_2 - 4x_3 &= 0 \\ \tfrac{1}{15}x_3 &= 1, \end{aligned}$$

which yields the third column of A^{-1}.

Customarily, especially when doing calculations by hand, this procedure does not stop with the form of (1.34b). Instead we continue the elimination process on (1.34b) by using elementary row operations to place zeros in the (1, 2), (1, 3), and (2, 3) positions, and to place ones in the (1, 1), (2, 2), and (3, 3) positions. That is, we use row operations to transform $[A \vdots I]$ to a (3×6) matrix of the form $[I \vdots C]$; and we shall see that $A^{-1} = C$. To reduce $[A \vdots I]$ to $[I \vdots C]$, we first multiply the rows of (1.34b) by the appropriate scalars to obtain a row-equivalent matrix with ones on the diagonal:

$$\begin{bmatrix} 1 & -4 & 2 & 1 & 0 & 0 \\ 0 & 1 & \frac{-4}{15} & \frac{-3}{15} & \frac{1}{15} & 0 \\ 0 & 0 & 1 & 12 & -4 & 15 \end{bmatrix}.$$

To introduce zeros into the (1, 3) and (2, 3) positions, we multiply row three by -2 and add the result to row one, and multiply row three by 4/15 and add the result to row two; the result is

$$\begin{bmatrix} 1 & -4 & 0 & -23 & 8 & -30 \\ 0 & 1 & 0 & 3 & -1 & 4 \\ 0 & 0 & 1 & 12 & -4 & 15 \end{bmatrix}.$$

Completing the reduction, we multiply row two by 4 and add the result to row one:

$$[I \vdots C] = \begin{bmatrix} 1 & 0 & 0 & -11 & 4 & -14 \\ 0 & 1 & 0 & 3 & -1 & 4 \\ 0 & 0 & 1 & 12 & -4 & 15 \end{bmatrix}. \tag{1.35}$$

Interpreting (1.35) in terms of the three systems $A\mathbf{x} = \mathbf{e}_1$, $A\mathbf{x} = \mathbf{e}_2$, and $A\mathbf{x} = \mathbf{e}_3$, we see that the system $A\mathbf{x} = \mathbf{e}_1$ is equivalent to

$$\begin{aligned} x_1 \qquad &= -11 \\ x_2 \quad &= 3 \\ x_3 &= 12 \end{aligned}$$

while the systems $A\mathbf{x} = \mathbf{e}_2$ and $A\mathbf{x} = \mathbf{e}_3$ are equivalent to

$$\begin{aligned} x_1 \qquad &= 4 \\ x_2 \quad &= -1 \\ x_3 &= -4 \end{aligned} \qquad \text{and} \qquad \begin{aligned} x_1 \qquad &= -14 \\ x_2 \quad &= 4. \\ x_3 &= 15 \end{aligned}$$

Therefore the matrix C is A^{-1} and we have

$$A^{-1} = \begin{bmatrix} -11 & 4 & -14 \\ 3 & -1 & 4 \\ 12 & -4 & 15 \end{bmatrix}.$$

In general, given a nonsingular $(n \times n)$ matrix A, we can calculate A^{-1} by transforming the $(n \times 2n)$ matrix $[A \vdots I]$ by elementary row operations to the form $[I \vdots C]$; and in this final form, $C = A^{-1}$. This scheme for organizing the computation of A^{-1} can be used in other situations also. Suppose we are given a number of $(m \times n)$ linear systems, all of which have the same coefficient matrix: for example

$$A\mathbf{x} = \mathbf{b}_i,\ i = 1, 2, \ldots, p.$$

We can solve these p systems by defining the $(m \times p)$ matrix B by

$$B = [\mathbf{b}_1, \mathbf{b}_2, \ldots, \mathbf{b}_p]$$

and then forming the $(m \times (n+p))$ matrix $[A \vdots B]$. Although the bookkeeping details vary slightly from those in the construction of A^{-1}, clearly we can use row operations to transfer $[A \vdots B]$ into a row-equivalent matrix $[P \vdots Q]$ where P is in echelon form and is row equivalent to A, and where Q is row equivalent to B corresponding to the elementary operations required to transform A into echelon form. If we let $Q = [\mathbf{q}_1, \mathbf{q}_2, \ldots, \mathbf{q}_p]$, then each system, $P\mathbf{x} = \mathbf{q}_i$, $1 \leqslant i \leqslant p$, is equivalent to $A\mathbf{x} = \mathbf{b}_i$, $1 \leqslant i \leqslant p$. Inspecting each system, $P\mathbf{x} = \mathbf{q}_i$, we can see if the system is consistent; and if so, we can backsolve to find the solution(s).

EXAMPLE 1.36 This example provides a formula for the inverse of a (2×2) matrix, together with a characterization of nonsingularity for (2×2) matrices.

Theorem

Let A be the (2×2) matrix

$$A = \begin{bmatrix} a & b \\ c & d \end{bmatrix}.$$

Then A is nonsingular if and only if $ad-bc \neq 0$. Furthermore if A is nonsingular, and if $\Delta = ad-bc$, then A^{-1} is given by

$$A^{-1} = \frac{1}{\Delta}\begin{bmatrix} d & -b \\ -c & a \end{bmatrix}.$$

(The reader who is familiar with determinants will recognize the number Δ as the determinant of A.) We note that similar formulas (see Chapter 5) for the inverse of an $(n \times n)$ nonsingular matrix also involve determinants. However even for relatively small values of n, it is computationally much more efficient to calculate A^{-1} by reducing $[A \mid I]$ to $[I \mid A^{-1}]$ as in Example 1.35.

We leave the proof of the theorem above to Exercise 7, Section 1.8, and merely illustrate the theorem here. Let A and B be given by

$$A = \begin{bmatrix} 6 & 8 \\ 3 & 4 \end{bmatrix}, \qquad B = \begin{bmatrix} 1 & 7 \\ 3 & 5 \end{bmatrix}.$$

For A we see that the number Δ is $\Delta = 6\cdot4-8\cdot3 = 0$; so A is a singular matrix and cannot have an inverse. On the other hand for B the number Δ is $\Delta = 5-21 = -16$; so B is nonsingular; and according to the theorem above,

$$B^{-1} = \frac{-1}{16}\begin{bmatrix} 5 & -7 \\ -3 & 1 \end{bmatrix}.$$

As we have shown, A^{-1} can be used (in principle) to solve $A\mathbf{x} = \mathbf{b}$ whenever A is nonsingular—if A is nonsingular, the solution of $A\mathbf{x} = \mathbf{b}$ is $\mathbf{x} = A^{-1}\mathbf{b}$. For example if we write the system

$$\begin{aligned} x_1 - 4x_2 + 2x_3 &= 1 \\ 3x_1 + 3x_2 + 2x_3 &= 2 \\ 4x_2 - x_3 &= 1 \end{aligned}$$

as $A\mathbf{x} = \mathbf{b}$, then (see Example 1.35)

$$A^{-1} = \begin{bmatrix} -11 & 4 & -14 \\ 3 & -1 & 4 \\ 12 & -4 & 15 \end{bmatrix}.$$

Therefore the solution of the system $A\mathbf{x} = \mathbf{b}$ is $\mathbf{x} = A^{-1}\mathbf{b}$, or

$$\begin{bmatrix} x_1 \\ x_2 \\ x_3 \end{bmatrix} = \begin{bmatrix} -11 & 4 & -14 \\ 3 & -1 & 4 \\ 12 & -4 & 15 \end{bmatrix}\begin{bmatrix} 1 \\ 2 \\ 1 \end{bmatrix} = \begin{bmatrix} -17 \\ 5 \\ 19 \end{bmatrix}.$$

Having the example above, we now pause to make some comments about the use of the inverse when actually solving an $(n \times n)$ system $A\mathbf{x} = \mathbf{b}$. Most systems of any appreciable size are solved on a computer, and what we are going to say has relevance mostly to computer solutions of $A\mathbf{x} = \mathbf{b}$. If we are to solve a system $A\mathbf{x} = \mathbf{b}$

where A is nonsingular, then we might consider calculating A^{-1} and finding the solution from $\mathbf{x} = A^{-1}\mathbf{b}$. Computationally this procedure is unattractive for two reasons, both of which are discussed in more detail in Chapter 6. First of all, finding A^{-1} and then calculating $\mathbf{x} = A^{-1}\mathbf{b}$ require a minimum of twice as much effort and machine time as solving $A\mathbf{x} = \mathbf{b}$ directly by Gauss elimination. More important, the roundoff errors inherent in most computer calculations may well make the solution $\mathbf{x} = A^{-1}\mathbf{b}$ more inaccurate than the solution found directly by Gauss elimination. In addition we normally do not know in advance whether A has an inverse or not whereas even if A is singular, Gauss elimination will find all solutions to $A\mathbf{x} = \mathbf{b}$ or will indicate that there is no solution.

Finally we might assume that if we are given a number of systems

$$A\mathbf{x} = \mathbf{b}_i, \ i = 1, 2, \ldots, p,$$

each of which has the same coefficient matrix, then it is efficient to calculate and use A^{-1} in order to solve each of these systems with one matrix multiplication, namely $\mathbf{x} = A^{-1}\mathbf{b}_i$, $i = 1, 2, \ldots, p$. This assumption is also incorrect as we point out in Chapter 6. In summary while we may *represent* solutions to $A\mathbf{x} = \mathbf{b}$ as $\mathbf{x} = A^{-1}\mathbf{b}$, $A\mathbf{x} = \mathbf{b}$ should seldom actually be solved by finding A^{-1}.

Having said this, we must say also that there are instances in which the matrix A^{-1} is needed, particularly in problems of statistical analysis. Other applications of the inverse matrix are found in certain coordinate transformations and in the eigenvalue problem. Frequently these matrices are of a special type, and the inverse is readily available. We shall encounter some of these special matrices in Chapter 3 and will have occasion to use their inverses repeatedly.

1.8 EXERCISES

1. Find A^{-1} by reducing $[A \vdots I]$ to $[I \vdots C]$ for

$$A = \begin{bmatrix} -1 & -2 & 11 \\ 1 & 3 & -15 \\ 0 & -1 & 5 \end{bmatrix} \quad A = \begin{bmatrix} 1 & 2 & 3 & 1 \\ -1 & 0 & 2 & 1 \\ 2 & 1 & -3 & 0 \\ 1 & 1 & 2 & 1 \end{bmatrix} \quad A = \begin{bmatrix} 1 & 0 & 0 \\ 2 & 1 & 0 \\ 3 & 4 & 1 \end{bmatrix}.$$

2. Repeat Exercise 1 for the following and check your result by verifying that $AA^{-1} = I$:

$$A = \begin{bmatrix} 1 & 3 & 5 \\ 0 & 1 & 4 \\ 0 & 2 & 7 \end{bmatrix} \quad A = \begin{bmatrix} 2 & 3 \\ 6 & 7 \end{bmatrix} \quad A = \begin{bmatrix} 1 & 4 & 2 \\ 0 & 2 & 1 \\ 3 & 5 & 3 \end{bmatrix}.$$

3. Use an inverse found in Exercise 1 to solve this system for the values listed:

$$\begin{aligned} -x_1 - 2x_2 + 11x_3 &= b_1 \\ x_1 + 3x_2 - 15x_3 &= b_2 \\ -x_2 + 5x_3 &= b_3. \end{aligned}$$

a) For $b_1 = 8$, $b_2 = -11$, $b_3 = 4$
b) For $b_1 = 7$, $b_2 = -9$, $b_3 = 3$
c) For $b_1 = -6$, $b_2 = 8$, $b_3 = -2$

4. Let B and D denote the (3×3) matrices in Exercise 1. Calculate BD and $D^{-1}B^{-1}$. Verify that $(BD)(D^{-1}B^{-1}) = I$.

5. Solve the systems $A\mathbf{x} = \mathbf{b}_k$, $k = 1, 2, 3, 4$ by row reduction of $[A \,\vdots\, \mathbf{b}_1, \mathbf{b}_2, \mathbf{b}_3, \mathbf{b}_4]$ where

$$A = \begin{bmatrix} 1 & 3 & -4 \\ 2 & -1 & 3 \\ 8 & 3 & 1 \end{bmatrix}, \quad \mathbf{b}_1 = \begin{bmatrix} 1 \\ 1 \\ 5 \end{bmatrix}, \quad \mathbf{b}_2 = \begin{bmatrix} 2 \\ 3 \\ -4 \end{bmatrix}, \quad \mathbf{b}_3 = \begin{bmatrix} 4 \\ -1 \\ 5 \end{bmatrix},$$

$$\mathbf{b}_4 = \begin{bmatrix} 0 \\ 0 \\ 0 \end{bmatrix}.$$

6. a) Suppose $AB = \mathcal{O}$ where A is nonsingular. Use Theorem 1.15(5) and Theorem 1.14(3) to prove that $B = \mathcal{O}$.
   b) Find a (2×2) matrix B such that $AB = \mathcal{O}$ where B has nonzero entries and where A is the matrix

$$A = \begin{bmatrix} 1 & 1 \\ 1 & 1 \end{bmatrix}.$$

   Why does this example not contradict part (a)?

7. Let A be the (2×2) matrix given in Example 1.36, with $\Delta = ad - bc$. Prove that if $\Delta = 0$, then A is singular. (Hint: Consider the vector

$$\mathbf{v} = \begin{bmatrix} d \\ -c \end{bmatrix}$$

   and also treat the special case that $\mathbf{v} = \boldsymbol{\theta}$.)

8. a) Find two values of λ for which $(A - \lambda I)$ is singular where

$$A = \begin{bmatrix} 1 & -1 \\ 2 & 4 \end{bmatrix} \qquad \lambda I = \begin{bmatrix} \lambda & 0 \\ 0 & \lambda \end{bmatrix}$$

   b) Let λ_1 and λ_2 be the values found in part (a). Find nonzero vectors $\mathbf{u}_1$ and $\mathbf{u}_2$ such that $A\mathbf{u}_1 = \lambda_1\mathbf{u}_1$ and $A\mathbf{u}_2 = \lambda_2\mathbf{u}_2$.
   c) Show that $\{\mathbf{u}_1, \mathbf{u}_2\}$ is linearly independent.
   d) Let $U = [\mathbf{u}_1, \mathbf{u}_2]$. Verify that

$$U^{-1}AU = \begin{bmatrix} \lambda_1 & 0 \\ 0 & \lambda_2 \end{bmatrix}.$$

9. Complete the proof of Theorem 1.13 by showing that if A and B are nonsingular $(n \times n)$ matrices, then AB is nonsingular. [Hint: Suppose $\mathbf{v}$ is any vector such that $(AB)\mathbf{v} = \boldsymbol{\theta}$, and write $(AB)\mathbf{v}$ as $A(B\mathbf{v})$.]

10. Let A be an $(n \times n)$ singular matrix. Argue that at least one of the systems $A\mathbf{x} = \mathbf{e}_k$, $k = 1, 2, \ldots, n$ must be inconsistent where $\mathbf{e}_k$ is the kth column of I.

11. If A is a square matrix, we define A^2 by $A^2 = AA$, A^3 by $A^3 = AA^2$, etc. Suppose A is $(n \times n)$, and suppose

$$A^3 - 2A^2 + 3A - I = \mathcal{O}.$$

   Show that $AB = I$ where $B = A^2 - 2A + 3I$.

12. Suppose A is $(n \times n)$, and suppose

$$A^3 + b_2 A^2 + b_1 A + b_0 I = \mathcal{O}$$

where $b_0 \neq 0$. Show that A is nonsingular by giving a formula for A^{-1}. (Hint: If A^{-1} does exist, what relationship must it satisfy?)

13. Prove Theorem 1.14(4).

14. Prove that if A is $(n \times n)$ and nonsingular, then $(A^{\mathrm{T}})^{-1} = (A^{-1})^{\mathrm{T}}$, [Hint: First use Exercise 8, Section 1.7, to observe that A^{T} has an inverse. Then consider the product $(A^{-1})^{\mathrm{T}} A^{\mathrm{T}}$.]

2

THE VECTOR SPACE R^n

2.1 INTRODUCTION

This chapter is concerned with the algebraic and geometric structure of R^n and subsets of R^n. Some of the concepts fundamental to describing this structure are subspace, basis, and dimension, which are discussed in the first few sections. While these ideas are relatively abstract, they are easy to understand in R^n and they also have application to concrete problems. In addition R^n serves as an example and as a model for a general vector space (see Chapter 4). In this regard, the insight provided by a study of R^n is valuable as an introduction to the important theory of vector spaces.

2.2 SUBSPACES IN R^n

To begin our discussion of the algebraic and geometric structure of R^n, we recall that R^n is the set of all n-dimensional vectors with real components:

$$R^n = \left\{\mathbf{x}:\mathbf{x} = \begin{bmatrix} x_1 \\ x_2 \\ \vdots \\ x_n \end{bmatrix}, x_1, x_2, \ldots, x_n \text{ real numbers}\right\}.$$

We next observe that certain subsets of R^n behave like R^p where $p \leqslant n$. For example if we think of R^3 as representing three-dimensional Euclidean space, then we can think of any plane in R^3 through the origin as being represented by R^2 (see Fig. 2.1). This concept can be made more precise by the following definition of a subspace of R^n.

DEFINITION 2.1.

Let W be a nonempty subset of R^n. Then W is a ***subspace*** of R^n if W satisfies two conditions.

1. Whenever $\mathbf{x}$ and $\mathbf{y}$ are vectors in W, then $\mathbf{x}+\mathbf{y}$ is a vector in W.

2. Whenever $\mathbf{x}$ is a vector in W, then $a\mathbf{x}$ is a vector in W for any scalar a.

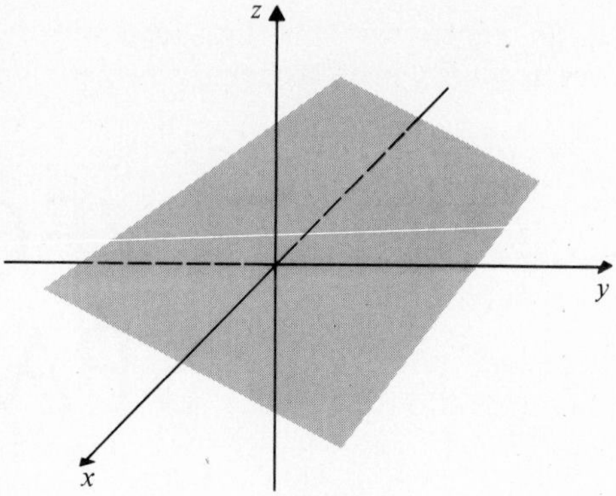

Fig. 2.1 A plane in R^3 through the origin.

EXAMPLE 2.1 Let W be the subset of R^3 defined by

$$W = \{\mathbf{x}: \mathbf{x} = \begin{bmatrix} x_1 \\ x_2 \\ 0 \end{bmatrix}, \ x_1 \text{ and } x_2 \text{ any real numbers}\}. \tag{2.1}$$

Geometrically W can be interpreted as the xy-plane in three-space (see Fig. 2.2). Furthermore this subset W is specified in (2.1) by an algebraic condition: "A three-dimensional vector $\mathbf{x}$ is in W if and only if the third component of $\mathbf{x}$ is 0."

To verify that W is a subspace of R^3, we must check conditions (1) and (2) of Definition 2.1, and show that sums and scalar multiples of vectors in W remain in W. In general such a verification normally proceeds along the following lines.

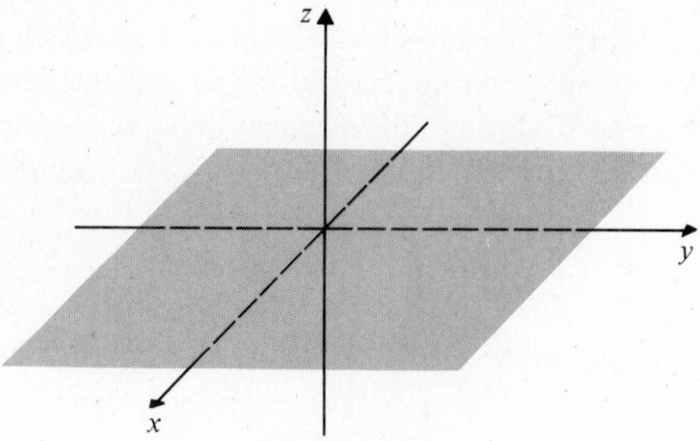

Fig. 2.2 W as a subset iof R^3.

1. We are given an algebraic specification for a subset W as in (2.1), and this specification serves as a test for determining whether a vector is in W or not.
2. We choose arbitrary vectors $\mathbf{x}$ and $\mathbf{y}$ from W (that is, we choose vectors $\mathbf{x}$ and $\mathbf{y}$ that satisfy the algebraic specification of W); and we then test the sum, $\mathbf{x}+\mathbf{y}$, to see if it meets the specification of W.
3. As in (2), we choose an arbitrary vector $\mathbf{x}$ from W and an arbitrary scalar a; and we test the scalar multiple $a\mathbf{x}$ to see if it is in W.

The program outlined above is usually quite easy to follow in a specific case such as (2.1). For example to show that W as defined by (2.1) is a subspace of R^3, we choose two arbitrary vectors, $\mathbf{x}$ and $\mathbf{y}$, from W. Now according to the specification of W, the third component of both $\mathbf{x}$ and $\mathbf{y}$ must be zero. Thus if $\mathbf{x}$ and $\mathbf{y}$ are in W, they must be of the form

$$\mathbf{x} = \begin{bmatrix} x_1 \\ x_2 \\ 0 \end{bmatrix}, \qquad \mathbf{y} = \begin{bmatrix} y_1 \\ y_2 \\ 0 \end{bmatrix}$$

Clearly then the sum $\mathbf{x}+\mathbf{y}$ and the multiple $a\mathbf{x}$ are given by

$$\mathbf{x}+\mathbf{y} = \begin{bmatrix} x_1+y_1 \\ x_2+y_2 \\ 0 \end{bmatrix}, \qquad a\mathbf{x} = \begin{bmatrix} ax_1 \\ ax_2 \\ 0 \end{bmatrix}.$$

Since the third component of $\mathbf{x}+\mathbf{y}$ is zero, we know that $\mathbf{x}+\mathbf{y}$ is in W. Similarly the scalar multiple $a\mathbf{x}$ has the form of a vector in W. Since $\mathbf{x}+\mathbf{y}$ and $a\mathbf{x}$ are in W whenever $\mathbf{x}$ and $\mathbf{y}$ are in W, we know that W is a subspace of R^3.

EXAMPLE 2.2 Let W be the subset of R^3 defined by

$$W = \left\{\mathbf{x} : \mathbf{x} = \begin{bmatrix} x_1 \\ x_2 \\ x_3 \end{bmatrix},\ x_1 = x_2 - x_3,\ x_2 \text{ and } x_3 \text{ any real numbers}\right\}.$$

In this case the algebraic specification of the subset W is this: a three-dimensional vector $\mathbf{x}$ is in W if and only if $x_1 = x_2 - x_3$. So for example the vectors $\mathbf{u}$ and $\mathbf{v}$ are in W while $\mathbf{r}$ and $\mathbf{s}$ are not:

$$\mathbf{u} = \begin{bmatrix} 1 \\ 2 \\ 1 \end{bmatrix}, \quad \mathbf{v} = \begin{bmatrix} 3 \\ 5 \\ 2 \end{bmatrix}, \quad \mathbf{r} = \begin{bmatrix} 1 \\ 2 \\ 2 \end{bmatrix}, \quad \mathbf{s} = \begin{bmatrix} 3 \\ 1 \\ 4 \end{bmatrix}.$$

To determine whether W is a subspace of R^3 or not, we choose two arbitrary vectors from W

$$\mathbf{x} = \begin{bmatrix} x_1 \\ x_2 \\ x_3 \end{bmatrix}, \qquad \mathbf{y} = \begin{bmatrix} y_1 \\ y_2 \\ y_3 \end{bmatrix}$$

where since $\mathbf{x}$ and $\mathbf{y}$ are in W, we know that

$$x_1 = x_2 - x_3 \qquad \text{and} \qquad y_1 = y_2 - y_3. \tag{2.2}$$

Forming the sum $\mathbf{x}+\mathbf{y}$ and the scalar multiple $a\mathbf{x}$, we obtain

$$\mathbf{x}+\mathbf{y} = \begin{bmatrix} x_1+y_1 \\ x_2+y_2 \\ x_3+y_3 \end{bmatrix}, \qquad a\mathbf{x} = \begin{bmatrix} ax_1 \\ ax_2 \\ ax_3 \end{bmatrix}.$$

In order to show that the vector $\mathbf{x}+\mathbf{y}$ is in W, we must show that $x_1+y_1 = (x_2+y_2) - (x_3+y_3)$. From (2.2) we know that x_1+y_1 is given by

$$x_1+y_1 = (x_2-x_3)+(y_2-y_3) = (x_2+y_2)-(x_3+y_3);$$

so it follows that sums of vectors from W are again in W. Checking that scalar multiples of vectors in W are in W is equally easy, for if $a\mathbf{x}$ is any scalar multiple of a vector $\mathbf{x}$ in W, then by (2.2) it is clear that $ax_1 = ax_2 - ax_3$. Thus $a\mathbf{x}$ is in W, and the verification that W is a subspace of R^3 is complete.

EXAMPLE 2.3 Not every subset of R^3 that looks like a plane geometrically is a subspace of R^3. For example let W be the subset

$$W = \left\{\mathbf{x}: \mathbf{x} = \begin{bmatrix} x_1 \\ x_2 \\ 1 \end{bmatrix},\ x_1 \text{ and } x_2 \text{ any real numbers}\right\}.$$

W must contain the {0} vector

Geometrically W can be interpreted as the plane $z = 1$ in R^3. However W is not a subspace. To see that W is not a subspace, suppose $\mathbf{x}$ and $\mathbf{y}$ are in W. Then $\mathbf{x}$ and $\mathbf{y}$ have the form

$$\mathbf{x} = \begin{bmatrix} x_1 \\ x_2 \\ 1 \end{bmatrix} \quad \text{and} \quad \mathbf{y} = \begin{bmatrix} y_1 \\ y_2 \\ 1 \end{bmatrix};$$

and so $\mathbf{x}+\mathbf{y}$ has the form

$$\mathbf{x}+\mathbf{y} = \begin{bmatrix} x_1+y_1 \\ x_2+y_2 \\ 2 \end{bmatrix}.$$

Since the third component of $\mathbf{x}+\mathbf{y}$ is 2, $\mathbf{x}+\mathbf{y}$ cannot be in W; hence W is not a subspace of R^3. (In Exercise 8, Section 2.2, the reader is asked to show that if W is a subspace of R^n, then W must contain the zero vector, $\boldsymbol{\theta}$. Knowing this result, we could have concluded immediately that W was not a subspace.)

We note that Definition 2.1 does not exclude the special case that W is equal to R^n. Hence R^n itself may be regarded as a subspace. At the other extreme, it is not hard to show that the subset $W = \{\boldsymbol{\theta}\}$, consisting solely of the zero vector from R^n, is a subspace of R^n; this trivial subspace is called the ***zero subspace***. We now introduce two subspaces that have particular relevance to the linear system of equations $A\mathbf{x} = \mathbf{b}$ where A is an $(m \times n)$ matrix. The first of these subspaces is called the ***null***

space of A (or the ***kernel*** of A) and consists of all the solutions of $A\mathbf{x} = \boldsymbol{\theta}$.

DEFINITION 2.2.

Let A be an $(m \times n)$ matrix. The ***null space*** of A [denoted $\mathcal{N}(A)$] is the set of vectors in R^n:

$$\mathcal{N}(A) = \{\mathbf{x}: A\mathbf{x} = \boldsymbol{\theta},\ \mathbf{x} \text{ in } R^n\}.$$

We will now show that the null space of an $(m \times n)$ matrix is a subspace of R^n.

Theorem 2.1

If A is an $(m \times n)$ matrix, and if $\mathcal{N}(A)$ is the null space of A, then $\mathcal{N}(A)$ is a subspace of R^n.

Proof. If $\mathbf{u}$ and $\boldsymbol{v}$ are vectors in $\mathcal{N}(A)$, then we want to show that $\mathbf{u} + \boldsymbol{v}$ and $a\mathbf{u}$ are also in $\mathcal{N}(A)$. Given that $\mathbf{u}$ and $\boldsymbol{v}$ are in $\mathcal{N}(A)$, we know that $A\mathbf{u} = \boldsymbol{\theta}$ and $A\boldsymbol{v} = \boldsymbol{\theta}$. Therefore testing $\mathbf{u} + \boldsymbol{v}$ against the algebraic specification of $\mathcal{N}(A)$, we check to see if $A(\mathbf{u} + \boldsymbol{v}) = \boldsymbol{\theta}$ and find that

$$A(\mathbf{u} + \boldsymbol{v}) = A\mathbf{u} + A\boldsymbol{v} = \boldsymbol{\theta} + \boldsymbol{\theta} = \boldsymbol{\theta}.$$

From this equality it follows that $\mathbf{u} + \boldsymbol{v}$ is in $\mathcal{N}(A)$ whenever $\mathbf{u}$ and $\boldsymbol{v}$ are.
Similarly if $\mathbf{u}$ is in $\mathcal{N}(A)$, then

$$A(a\mathbf{u}) = a(A\mathbf{u}) = a\boldsymbol{\theta} = \boldsymbol{\theta};$$

so $a\mathbf{u}$ is in $\mathcal{N}(A)$ for any scalar a. ∎

Another important subspace associated with an $(m \times n)$ matrix A is the ***range space*** of A.

DEFINITION 2.3.

Let A be an $(m \times n)$ matrix. The ***range space*** of A [denoted $\mathcal{R}(A)$] is the set of vectors in R^m:

$$\mathcal{R}(A) = \{\mathbf{y}: \mathbf{y} = A\mathbf{x} \text{ for some } \mathbf{x} \text{ in } R^n\}.$$

Thus $\mathcal{R}(A)$ is the set of all vectors $\mathbf{y}$ in R^m such that the linear system $A\mathbf{x} = \mathbf{y}$ is consistent. As another way to view $\mathcal{R}(A)$, suppose A is an $(m \times n)$ matrix. We can regard A as defining a function from R^n to R^m:

$$\mathbf{y} = A\mathbf{x}.$$

In this sense the set of all vectors $A\mathbf{x}$ that are produced in R^m as $\mathbf{x}$ varies through R^n constitutes the "range" of the function and hence suggests the name range space. As a companion result to Theorem 2.1, we have Theorem 2.2.

Theorem 2.2

If A is an $(m \times n)$ matrix, and if $\mathscr{R}(A)$ is the range space of A, then $\mathscr{R}(A)$ is a subspace of R^m.

Proof. According to Definition 2.1, we have to show that sums and scalar multiples of vectors in $\mathscr{R}(A)$ are again in $\mathscr{R}(A)$ where the algebraic specification of $\mathscr{R}(A)$ is this: a vector $\mathbf{y}$ in R^m is in $\mathscr{R}(A)$ if and only if there is at least one vector $\mathbf{x}$ in R^n such that $\mathbf{y} = A\mathbf{x}$. So we let $\mathbf{y}_1$ and $\mathbf{y}_2$ be *any* two vectors in $\mathscr{R}(A)$. Then we know there exist vectors $\mathbf{x}_1$ and $\mathbf{x}_2$ in R^n such that $\mathbf{y}_1 = A\mathbf{x}_1$ and $\mathbf{y}_2 = A\mathbf{x}_2$. Forming the sum $\mathbf{y}_1 + \mathbf{y}_2$, we find

$$\mathbf{y}_1 + \mathbf{y}_2 = A\mathbf{x}_1 + A\mathbf{x}_2 = A(\mathbf{x}_1 + \mathbf{x}_2);$$

and thus the vector $\mathbf{y}_1 + \mathbf{y}_2$ is in $\mathscr{R}(A)$ since there is a vector $\mathbf{x}$ in R^n, namely $\mathbf{x} = \mathbf{x}_1 + \mathbf{x}_2$, that is a solution to the equation $\mathbf{y}_1 + \mathbf{y}_2 = A\mathbf{x}$.

Next to show that $a\mathbf{y}_1$ is in $\mathscr{R}(A)$, we merely observe that

$$a\mathbf{y}_1 = a(A\mathbf{x}_1) = A(a\mathbf{x}_1);$$

and thus the equation $a\mathbf{y}_1 = A\mathbf{x}$ has a solution—namely, $\mathbf{x} = a\mathbf{x}_1$. ∎

The next example illustrates how we can represent vectors in the subspaces $\mathscr{N}(A)$ and $\mathscr{R}(A)$ as linear combinations of certain fixed vectors. We will return to this idea in the next section when we discuss spanning sets and bases for subspaces.

EXAMPLE 2.4 Let A be the (3×4) matrix

$$A = \begin{bmatrix} 1 & 1 & 3 & 1 \\ 2 & 1 & 5 & 4 \\ 1 & 2 & 4 & -1 \end{bmatrix}.$$

Our first objective is to determine $\mathscr{N}(A)$, which we can do by reducing the augmented matrix $[A \,\vdots\, \boldsymbol{\theta}]$ to echelon form. It is easy to verify that $[A \,\vdots\, \boldsymbol{\theta}]$ is row equivalent to

$$\begin{bmatrix} 1 & 1 & 3 & 1 & 0 \\ 0 & -1 & -1 & 2 & 0 \\ 0 & 0 & 0 & 0 & 0 \end{bmatrix}.$$

Therefore a vector $\boldsymbol{v}$ in R^4,

$$\boldsymbol{v} = \begin{bmatrix} v_1 \\ v_2 \\ v_3 \\ v_4 \end{bmatrix},$$

is in $\mathscr{N}(A)$ if and only if

$$\begin{aligned} v_1 &= -2v_3 - 3v_4 \\ v_2 &= -v_3 + 2v_4 \end{aligned} \tag{2.3}$$

where v_3 and v_4 are arbitrary.

Equivalently $\mathcal{N}(A)$ consists of all vectors $\boldsymbol{v}$ of the form

$$\boldsymbol{v} = \begin{bmatrix} -2v_3 - 3v_4 \\ -\ v_3 + 2v_4 \\ v_3 \\ v_4 \end{bmatrix}, \tag{2.4}$$

and (2.3) constitutes an algebraic specification for $\mathcal{N}(A)$. To illustrate the structure of $\mathcal{N}(A)$, it is convenient to separate the parameters v_3 and v_4; and we decompose $\boldsymbol{v}$ in (2.4) as

$$\boldsymbol{v} = v_3 \begin{bmatrix} -2 \\ -1 \\ 1 \\ 0 \end{bmatrix} + v_4 \begin{bmatrix} -3 \\ 2 \\ 0 \\ 1 \end{bmatrix}.$$

If we set

$$\mathbf{u}_1 = \begin{bmatrix} -2 \\ -1 \\ 1 \\ 0 \end{bmatrix} \quad \text{and} \quad \mathbf{u}_2 = \begin{bmatrix} -3 \\ 2 \\ 0 \\ 1 \end{bmatrix}, \tag{2.5}$$

then we have that $\boldsymbol{v}$ is in $\mathcal{N}(A)$ if and only if $\boldsymbol{v}$ is a linear combination of $\mathbf{u}_1$ and $\mathbf{u}_2$:

$$\boldsymbol{v} = c_1\mathbf{u}_1 + c_2\mathbf{u}_2.$$

We can characterize $\mathcal{N}(A)$ as being the set of all linear combinations of $\mathbf{u}_1$ and $\mathbf{u}_2$.

Similarly it is easy to represent $\mathcal{R}(A)$ as a set of linear combinations of vectors in R^3. To do this, we write A in column form as $A = [\mathbf{A}_1, \mathbf{A}_2, \mathbf{A}_3, \mathbf{A}_4]$. As we know, if $\mathbf{x}$ is given by

$$\mathbf{x} = \begin{bmatrix} x_1 \\ x_2 \\ x_3 \\ x_4 \end{bmatrix},$$

then $A\mathbf{x} = x_1\mathbf{A}_1 + x_2\mathbf{A}_2 + x_3\mathbf{A}_3 + x_4\mathbf{A}_4$. Thus $\mathcal{R}(A)$ is precisely the set of all linear combinations of the columns of A. [In fact because of this characterization, $\mathcal{R}(A)$ is frequently called the column space of A.]

An alternate way of describing $\mathcal{R}(A)$ is provided by the following discussion, which determines conditions that $\mathbf{b}$ must satisfy if $A\mathbf{x} = \mathbf{b}$ is to be consistent. In particular let

$$\mathbf{b} = \begin{bmatrix} b_1 \\ b_2 \\ b_3 \end{bmatrix}$$

be an arbitrary vector in R^3. We now reduce the augmented matrix $[A \mid \mathbf{b}]$ to echelon form where

$$[A \mid \mathbf{b}] = \begin{bmatrix} 1 & 1 & 3 & 1 & b_1 \\ 2 & 1 & 5 & 4 & b_2 \\ 1 & 2 & 4 & -1 & b_3 \end{bmatrix}.$$

We find that $[A \,\vdots\, \mathbf{b}]$ is row equivalent to

$$\begin{bmatrix} 1 & 1 & 3 & 1 & b_1 \\ 0 & -1 & -1 & 2 & b_2 - 2b_1 \\ 0 & 1 & 1 & -2 & b_3 - b_1 \end{bmatrix},$$

which is in turn equivalent to

$$\begin{bmatrix} 1 & 1 & 3 & 1 & b_1 \\ 0 & -1 & -1 & 2 & b_2 - 2b_1 \\ 0 & 0 & 0 & 0 & -3b_1 + b_2 + b_3 \end{bmatrix}.$$

From this reduction we see that $A\mathbf{x} = \mathbf{b}$ has a solution [that is, $\mathbf{b}$ is in $\mathscr{R}(A)$] if and only if $-3b_1 + b_2 + b_3 = 0$, or $b_3 = 3b_1 - b_2$, where b_1 and b_2 are arbitrary. Hence if $\mathbf{b}$ is in $\mathscr{R}(A)$, then $\mathbf{b}$ has the form

$$\mathbf{b} = \begin{bmatrix} b_1 \\ b_2 \\ b_3 \end{bmatrix} = \begin{bmatrix} b_1 \\ b_2 \\ 3b_1 - b_2 \end{bmatrix} = b_1 \begin{bmatrix} 1 \\ 0 \\ 3 \end{bmatrix} + b_2 \begin{bmatrix} 0 \\ 1 \\ -1 \end{bmatrix} = b_1\mathbf{w}_1 + b_2\mathbf{w}_2.$$

From these two analyses, we are led to two different descriptions of $\mathscr{R}(A)$. First, $\mathbf{b}$ is in $\mathscr{R}(A)$ if and only if $\mathbf{b}$ is a linear combination of $\mathbf{A}_1$, $\mathbf{A}_2$, $\mathbf{A}_3$, and $\mathbf{A}_4$. Second, $\mathbf{b}$ is in $\mathscr{R}(A)$ if and only if $\mathbf{b}$ is a linear combination of $\mathbf{w}_1$ and $\mathbf{w}_2$. It is not hard to show that these two characterizations of $\mathscr{R}(A)$ are consistent.

Finally we can show that any plane through the origin in R^3 is indeed a subspace of R^3, and that planes containing the origin in R^3 as they are defined in analytic geometry correspond to subspaces as defined in Definition 2.1. To show this, recall that the equation of a plane in three-space through the origin is

$$ax_1 + bx_2 + cx_3 = 0 \tag{2.6}$$

where a, b, and c are specified constants, not all of which are zero.

To see that the set of vectors in R^3 whose coordinates satisfy (2.6) forms a subspace of R^3, consider the (1×3) matrix A given by

$$A = [a \quad b \quad c].$$

Then for $\mathbf{x}$ in R^3, we see that $A\mathbf{x} = \boldsymbol{\theta}$ if and only if the components of $\mathbf{x}$ satisfy (2.6). Thus $\mathbf{x}$ in R^3 is on the plane defined by (2.6) if and only if $\mathbf{x}$ is in the null space of the (1×3) matrix A. Since $\mathscr{N}(A)$ is a subspace by Theorem 2.1, any plane through the origin is a subspace of R^3 in the sense of Definition 2.1. A similar argument (see Exercise 18, Section 2.2) will show that every line through the origin in R^3 is also a subspace of R^3. However since lines and planes are intrinsically different geometrically, there should be some way to distinguish them algebraically. This in part is the topic of Section 2.4.

2.2 EXERCISES

1. Let V be a subset of R^3 consisting of vectors of the form

$$\mathbf{x} = \begin{bmatrix} x_1 \\ x_2 \\ x_3 \end{bmatrix}.$$

For which conditions on the components of $\mathbf{x}$ is V a subspace of R^3?

a) $x_3 = 2x_1 - x_2$ b) $x_2 = x_3 + x_1$ c) $x_1x_2 = x_3$ d) $x_1 = 2x_3$
e) $x_1^2 = x_1 + x_2$ f) $x_2 = 0$

2. Repeat Exercise 1 for the following conditions.

a) $x_3 = x_1 + 4x_2$ b) $2x_1 - x_2 - x_3 = 0$ c) $x_1^2 + x_2 = 1$ d) $x_2 = x_1$
e) $x_1 = 0$ f) $x_1x_2 = 0$

3. For each subset V in Exercise 1 that is a subspace, find vectors $\mathbf{u}_1$ and $\mathbf{u}_2$ such that every vector in V is a linear combination of $\mathbf{u}_1$ and $\mathbf{u}_2$. [Hint: Recall Example 2.4 and consider how the vectors in (2.5) were formed.]

4. Repeat Exercise 3 for the subsets V in Exercise 2.

5. Let V be a subset of R^3 as in Exercise 1 where

$$\begin{aligned} x_1 - x_2 + x_3 &= 0 \\ x_1 \qquad - x_3 &= 0. \end{aligned}$$

a) Verify that V is a subspace of R^3.
b) Find a vector $\mathbf{u}_1$ such that every vector in V is a scalar multiple of $\mathbf{u}_1$.
c) Find a (2×3) matrix A such that $V = \mathcal{N}(A)$.

6. Repeat Exercise 5 for the conditions

$$\begin{aligned} x_1 - 2x_2 + \ x_3 &= 0 \\ x_1 \qquad - 3x_3 &= 0. \end{aligned}$$

7. For each matrix, proceed as in Example 2.4 and find a set of vectors Q such that every vector in the null space is a linear combination of vectors from Q.

a) $\begin{bmatrix} 1 & -2 & 4 & 1 \\ 2 & 1 & 3 & 7 \\ 1 & -1 & 3 & 2 \end{bmatrix}$ b) $\begin{bmatrix} 1 & 2 & 3 & 4 & 5 \\ 2 & 3 & 4 & 5 & 1 \\ 1 & 1 & 1 & 1 & 2 \\ 2 & 3 & 4 & 5 & 4 \end{bmatrix}$ c) $\begin{bmatrix} 1 & 2 & 7 \\ 2 & -2 & 2 \\ -1 & 3 & 3 \end{bmatrix}$

8. Let V be a subspace of R^n. Prove that $\boldsymbol{\theta}$ is in V. [Hint: A very short proof can be given using condition (2) of Definition 2.1.]

9. Let A be the (3×4) matrix in part (a) of Exercise 7. As in Example 2.4, reduce $[A \,\vdots\, \mathbf{b}]$ to echelon form to determine the conditions that $\mathbf{b}$ must satisfy if $A\mathbf{x} = \mathbf{b}$ is consistent. Use this form to find a set of vectors Q such that every vector in $\mathcal{R}(A)$ is a linear combination of vectors from Q. Repeat this procedure for parts (b) and (c) of Exercise 7.

10. Let $\mathbf{a}$ be a fixed vector in R^3, and let V be the subset of R^3 defined thus: $\mathbf{x}$ is in V if and only if $\mathbf{a}^T\mathbf{x} = 0$. Verify that V is a subspace of R^3.

11. Let V be the subspace defined in Exercise 10. For each vector $\mathbf{a}$ below, find vectors $\mathbf{u}_1$ and

$\mathbf{u}_2$ such that every vector $\mathbf{x}$ in V is a linear combination of $\mathbf{u}_1$ and $\mathbf{u}_2$.

a) $\mathbf{a} = \begin{bmatrix} 1 \\ 1 \\ 1 \end{bmatrix}$ b) $\mathbf{a} = \begin{bmatrix} -1 \\ 0 \\ 1 \end{bmatrix}$ c) $\mathbf{a} = \begin{bmatrix} 0 \\ 3 \\ 0 \end{bmatrix}$ d) $\mathbf{a} = \begin{bmatrix} 2 \\ -1 \\ 3 \end{bmatrix}$

12. Let $\mathbf{a}$ and $\mathbf{b}$ be fixed vectors in R^4, and let V be the subset of R^4 defined thus: $\mathbf{x}$ is in V if and only if $\mathbf{a}^T\mathbf{x} = 0$ and $\mathbf{b}^T\mathbf{x} = 0$. Verify that V is a subspace of R^4.

13. For V as in Exercise 12, find a set of vectors Q such that every vector $\mathbf{x}$ in V is a linear combination of vectors from Q.

a) $\mathbf{a} = \begin{bmatrix} 1 \\ 1 \\ 0 \\ -1 \end{bmatrix}$, $\mathbf{b} = \begin{bmatrix} 0 \\ 1 \\ 1 \\ 0 \end{bmatrix}$ b) $\mathbf{a} = \begin{bmatrix} 1 \\ 1 \\ 1 \\ -1 \end{bmatrix}$, $\mathbf{b} = \begin{bmatrix} 1 \\ 2 \\ 1 \\ 1 \end{bmatrix}$

c) $\mathbf{a} = \begin{bmatrix} 1 \\ 2 \\ 1 \\ 0 \end{bmatrix}$, $\mathbf{b} = \begin{bmatrix} 3 \\ 6 \\ 3 \\ 0 \end{bmatrix}$ d) $\mathbf{a} = \begin{bmatrix} 1 \\ -1 \\ 0 \\ 0 \end{bmatrix}$, $\mathbf{b} = \begin{bmatrix} 0 \\ 0 \\ 0 \\ 0 \end{bmatrix}$

e) $\mathbf{a} = \begin{bmatrix} 1 \\ -1 \\ 0 \\ 0 \end{bmatrix}$ $\mathbf{b} = \begin{bmatrix} 1 \\ 1 \\ 0 \\ 0 \end{bmatrix}$

14. Identify the range space and the null space for the $(n \times n)$ identity matrix and the $(n \times n)$ zero matrix. Also identify the range space and the null space for any $(n \times n)$ nonsingular matrix A.

15. If U and V are subsets of R^n, the set $U + V$ is defined to be

$$U + V = \{\mathbf{x} : \mathbf{x} = \mathbf{u} + \mathbf{v},\ \mathbf{u} \text{ in } U,\ \mathbf{v} \text{ in } V\}.$$

Prove that if U and V are subspaces of R^n, then $U + V$ is a subspace of R^n.

16. Let U and V be subspaces of R^3. Prove or find a counterexample to these statements.

a) The intersection of U and V is a subspace of R^3.
b) The union of U and V is a subspace of R^3.

17. Let U be a subspace of R^n, and let A be an $(m \times n)$ matrix. Let V be the subset of R^m defined by

$$V = \{\mathbf{y} : \mathbf{y} = A\mathbf{x} \text{ for some } \mathbf{x} \text{ in } U\}.$$

Prove that V is a subspace of R^m.

18. In R^3 a line through the origin is the set of all points in R^3 whose coordinates satisfy $x_1 = at$, $x_2 = bt$, $x_3 = ct$ where t is a variable and the constants a, b, and c are not all zero. Show that a line through the origin is a subspace of R^3.

19. Let A and B be $(n \times n)$ matrices. Verify that $\mathcal{N}(A) \cap \mathcal{N}(B) \subseteq \mathcal{N}(A + B)$.

20. Let A and B be (2×2) matrices. Give an example to show that $\mathcal{N}(AB)$ and $\mathcal{N}(BA)$ can be different subspaces. Show that if A and B are nonsingular, then $\mathcal{N}(AB) = \mathcal{N}(BA)$.

2.3 BASES FOR SUBSPACES

The topic of this section is suggested by Example 2.4, Section 2.2, which showed that we could express any vector in the subspace $\mathscr{R}(A)$ as a linear combination of the four vectors $\mathbf{A}_1$, $\mathbf{A}_2$, $\mathbf{A}_3$, and $\mathbf{A}_4$, or alternatively as a linear combination of the two vectors $\mathbf{w}_1$ and $\mathbf{w}_2$. Thus a question arises as to what is the fewest number of vectors necessary to characterize a given subspace. In order to answer this question and to introduce some related and useful ideas, we begin with the following definition.

DEFINITION 2.4

Let W be a subspace of R^n, and let $S = \{\mathbf{s}_1, \mathbf{s}_2, \ldots, \mathbf{s}_m\}$ be a set of vectors in W. We say that S is a ***spanning set*** for W if every vector $\mathbf{w}$ in W can be expressed as a linear combination of vectors from S;

$$\mathbf{w} = a_1\mathbf{s}_1 + a_2\mathbf{s}_2 + \ldots + a_m\mathbf{s}_m.$$

In Example 2.4 for instance, the set $\{\mathbf{u}_1, \mathbf{u}_2\}$ is a spanning set for $\mathscr{N}(A)$ while $\{\mathbf{A}_1, \mathbf{A}_2, \mathbf{A}_3, \mathbf{A}_4\}$ and $\{\mathbf{w}_1, \mathbf{w}_2\}$ are examples of two different spanning sets for the same subspace, $\mathscr{R}(A)$. Frequently it is quite easy to find a spanning set for a subspace as the next two examples illustrate.

EXAMPLE 2.5 As in Example 2.1, let W be the subspace of R^3 defined by

$$W = \{\mathbf{x}: \mathbf{x} = \begin{bmatrix} x_1 \\ x_2 \\ 0 \end{bmatrix}, x_1 \text{ and } x_2 \text{ any real numbers}\}.$$

It is almost obvious that a subset S of W that meets the conditions of Definition 2.4 is the set $S = \{\mathbf{s}_1, \mathbf{s}_2\}$ where

$$\mathbf{s}_1 = \begin{bmatrix} 1 \\ 0 \\ 0 \end{bmatrix} \qquad \mathbf{s}_2 = \begin{bmatrix} 0 \\ 1 \\ 0 \end{bmatrix}.$$

Clearly $\mathbf{s}_1$ and $\mathbf{s}_2$ are in W; and if $\mathbf{x}$ is any vector in W, then $\mathbf{x} = x_1\mathbf{s}_1 + x_2\mathbf{s}_2$. Since every vector in W is a linear combination of vectors in S, we see that S is a spanning set for W.

EXAMPLE 2.6 As in Example 2.2, let W be the subspace of R^3 defined by

$$W = \{\mathbf{x}: \mathbf{x} = \begin{bmatrix} x_1 \\ x_2 \\ x_3 \end{bmatrix}, x_1 = x_2 - x_3\}.$$

From the specification above, every vector $\mathbf{x}$ in W has the form

$$\mathbf{x} = \begin{bmatrix} x_1 \\ x_2 \\ x_3 \end{bmatrix} = \begin{bmatrix} x_2 - x_3 \\ x_2 \\ x_3 \end{bmatrix}. \tag{2.7}$$

Given (2.7), it is natural to decompose $\mathbf{x}$ as

$$\mathbf{x} = \begin{bmatrix} x_2 \\ x_2 \\ 0 \end{bmatrix} + \begin{bmatrix} -x_3 \\ 0 \\ x_3 \end{bmatrix} = x_2 \begin{bmatrix} 1 \\ 1 \\ 0 \end{bmatrix} + x_3 \begin{bmatrix} -1 \\ 0 \\ 1 \end{bmatrix}.$$

Thus with $\mathbf{s}_1$ and $\mathbf{s}_2$ given by

$$\mathbf{s}_1 = \begin{bmatrix} 1 \\ 1 \\ 0 \end{bmatrix} \qquad \mathbf{s}_2 = \begin{bmatrix} -1 \\ 0 \\ 1 \end{bmatrix}$$

we see that any vector $\mathbf{x}$ in W can be expressed as a linear combination of $\mathbf{s}_1$ and $\mathbf{s}_2$. So $S = \{\mathbf{s}_1, \mathbf{s}_2\}$ is a spanning set for W.

We return now to Example 2.4, in which we exhibited two spanning sets for the subspace $\mathcal{R}(A)$. These are the sets

$$S_1 = \{\mathbf{A}_1, \mathbf{A}_2, \mathbf{A}_3, \mathbf{A}_4\} \quad \text{and} \quad S_2 = \{\mathbf{w}_1, \mathbf{w}_2\}$$

where

$$\mathbf{A}_1 = \begin{bmatrix} 1 \\ 2 \\ 1 \end{bmatrix}, \mathbf{A}_2 = \begin{bmatrix} 1 \\ 1 \\ 2 \end{bmatrix}, \mathbf{A}_3 = \begin{bmatrix} 3 \\ 5 \\ 4 \end{bmatrix}, \mathbf{A}_4 = \begin{bmatrix} 1 \\ 4 \\ -1 \end{bmatrix}, \mathbf{w}_1 = \begin{bmatrix} 1 \\ 0 \\ 3 \end{bmatrix}, \mathbf{w}_2 = \begin{bmatrix} 0 \\ 1 \\ -1 \end{bmatrix}.$$

As we saw in Example 2.4, $\mathbf{y}$ is in $\mathcal{R}(A)$ if and only if $\mathbf{y}$ is a linear combination of $\mathbf{A}_1, \mathbf{A}_2, \mathbf{A}_3, \mathbf{A}_4$:

$$\mathbf{y} = c_1\mathbf{A}_1 + c_2\mathbf{A}_2 + c_3\mathbf{A}_3 + c_4\mathbf{A}_4. \tag{2.8}$$

However some reflection reveals that the spanning set $S_1 = \{\mathbf{A}_1, \mathbf{A}_2, \mathbf{A}_3, \mathbf{A}_4\}$ contains information that is redundant. In particular since a set of four vectors from R^3 must be linearly dependent (see Theorem 1.9), we know that at least one of the vectors in S_1 can be expressed as a linear combination of the other three. In fact a calculation shows that $\mathbf{A}_3 = 2\mathbf{A}_1 + \mathbf{A}_2$ and $\mathbf{A}_4 = 3\mathbf{A}_1 - 2\mathbf{A}_2$. Thus the expression for $\mathbf{y}$ in (2.8) can be replaced by

$$\mathbf{y} = c_1\mathbf{A}_1 + c_2\mathbf{A}_2 + c_3(2\mathbf{A}_1 + \mathbf{A}_2) + c_4(3\mathbf{A}_1 - 2\mathbf{A}_2),$$

or regrouping, we can express $\mathbf{y}$ as

$$\mathbf{y} = (c_1 + 2c_3 + 3c_4)\mathbf{A}_1 + (c_2 + c_3 - 2c_4)\mathbf{A}_2. \tag{2.9}$$

From (2.9) it follows that $\mathbf{y}$ is a linear combination of $\mathbf{A}_1$ and $\mathbf{A}_2$ whenever $\mathbf{y}$ is in $\mathcal{R}(A)$. We conclude that $\mathbf{A}_3$ and $\mathbf{A}_4$ can be discarded from S_1, and the set $S_3 = \{\mathbf{A}_1, \mathbf{A}_2\}$ is still a spanning set for $\mathcal{R}(A)$.

The lesson to be drawn from this example is that a linearly dependent spanning set contains redundant information. That is, if $S = \{\mathbf{w}_1, \mathbf{w}_2, \ldots, \mathbf{w}_p\}$ is a linearly dependent spanning set for a subspace W, then at least one vector from S is a linear combination of the other $p-1$ vectors and can be discarded from S to produce a smaller spanning set. On the other hand if $B = \{v_1, v_2, \ldots, v_m\}$ is a linearly independent spanning set for W, then no vector in B is a linear combination of the other $m-1$ vectors in B. Hence if a vector is removed from B, this smaller set cannot be a spanning set for W (in particular, the vector removed from B is in W but cannot be expressed as a linear combination of the vectors retained). In this sense a linearly independent spanning set is a minimal spanning set and hence represents the most efficient way of characterizing the subspace. This idea leads to the following definition.

DEFINITION 2.5.

Let W be a nonzero subspace of R^n. A ***basis*** for W is a linearly independent spanning set for W.
(Note: The concept of a basis is not meaningful for the zero subspace.)

EXAMPLE 2.7 In Example 2.4 the set $\{\mathbf{w}_1, \mathbf{w}_2\}$ is a spanning set for $\mathcal{R}(A)$; and since $\{\mathbf{w}_1, \mathbf{w}_2\}$ is a linearly independent set, it is a basis for $\mathcal{R}(A)$. In (2.9) we also saw that $\{\mathbf{A}_1, \mathbf{A}_2\}$ is a spanning set for $\mathcal{R}(A)$, and it is easy to check that this set is also linearly independent. Thus $\{\mathbf{A}_1, \mathbf{A}_2\}$ is another basis for $\mathcal{R}(A)$. On the other hand the spanning set $\{\mathbf{A}_1, \mathbf{A}_2, \mathbf{A}_3, \mathbf{A}_4\}$ is not a basis for $\mathcal{R}(A)$ since it is a linearly dependent set.

This example not only illustrates that bases for subspaces are not unique, but also illustrates a fact that we will establish in the next section: if W is a subspace of R^n, then every basis for W contains the same number of vectors.

EXAMPLE 2.8 The spanning set $\{\mathbf{s}_1, \mathbf{s}_2\}$ given in Example 2.5 is a basis since it is linearly independent. Similarly the spanning set shown in Example 2.6 is a basis for the subspace in Example 2.6.

EXAMPLE 2.9 The unit vectors

$$\mathbf{e}_1 = \begin{bmatrix} 1 \\ 0 \\ 0 \end{bmatrix}, \quad \mathbf{e}_2 = \begin{bmatrix} 0 \\ 1 \\ 0 \end{bmatrix}, \quad \mathbf{e}_3 = \begin{bmatrix} 0 \\ 0 \\ 1 \end{bmatrix}$$

constitute a basis for R^3. In general the n-dimensional unit vectors $\mathbf{e}_1, \mathbf{e}_2, \ldots, \mathbf{e}_n$ form a basis for R^n, frequently called the natural basis.

Another basis for R^3 is provided by the vectors

$$v_1 = \begin{bmatrix} 1 \\ 0 \\ 0 \end{bmatrix}, \quad v_2 = \begin{bmatrix} 1 \\ 1 \\ 0 \end{bmatrix}, \quad v_3 = \begin{bmatrix} 1 \\ 1 \\ 1 \end{bmatrix}.$$

It is easy to see that $\{v_1, v_2, v_3\}$ is a linearly independent set. In Exercise 12, Section 2.3, we ask the reader to use Theorem 1.11 to show that any set of three linearly independent vectors in R^3 is a spanning set for R^3. Thus it follows that $\{v_1, v_2, v_3\}$ is a basis for R^3.

EXAMPLE 2.10 Let W be the subset of R^5 defined by

$$W = \{\mathbf{x} : \mathbf{x} = \begin{bmatrix} x_1 \\ x_2 \\ x_3 \\ x_4 \\ x_5 \end{bmatrix},\ x_1 = x_3 - x_4 + 2x_5,\ x_2 = 3x_3 + x_4 + x_5\}.$$

It is easy to verify that W is a subspace, and the specification of W shows that $\mathbf{x}$ is in W if and only if $\mathbf{x}$ has the form

$$\mathbf{x} = \begin{bmatrix} x_3 - x_4 + 2x_5 \\ 3x_3 + x_4 + \ x_5 \\ x_3 \\ x_4 \\ x_5 \end{bmatrix} = x_3 \begin{bmatrix} 1 \\ 3 \\ 1 \\ 0 \\ 0 \end{bmatrix} + x_4 \begin{bmatrix} -1 \\ 1 \\ 0 \\ 1 \\ 0 \end{bmatrix} + x_5 \begin{bmatrix} 2 \\ 1 \\ 0 \\ 0 \\ 1 \end{bmatrix}.$$

The three vectors

$$\boldsymbol{v}_1 = \begin{bmatrix} 1 \\ 3 \\ 1 \\ 0 \\ 0 \end{bmatrix}, \quad \boldsymbol{v}_2 = \begin{bmatrix} -1 \\ 1 \\ 0 \\ 1 \\ 0 \end{bmatrix}, \quad \boldsymbol{v}_3 = \begin{bmatrix} 2 \\ 1 \\ 0 \\ 0 \\ 1 \end{bmatrix}$$

form a linearly independent spanning set for W and thus give a basis for W.

We conclude this section by describing a procedure that can be used to construct a basis from a spanning set. Let W be a subspace of R^m, and let $\{\mathbf{s}_1, \mathbf{s}_2, \ldots, \mathbf{s}_n\}$ be a spanning set for W. If we form the $(m \times n)$ matrix $A = [\mathbf{s}_1, \mathbf{s}_2, \ldots, \mathbf{s}_n]$, then it is easy to see that $W = \mathcal{R}(A)$ (since $\mathbf{w}$ is in W if and only if $\mathbf{w}$ is a linear combination of $\mathbf{s}_1, \mathbf{s}_2, \ldots, \mathbf{s}_n$). Thus if we can construct a basis for $\mathcal{R}(A)$, we have one for W also. A rather mechanical procedure for constructing a basis for $\mathcal{R}(A)$ consists of the following steps.

1. Form A^T and reduce A^T to echelon form. That is, find a matrix B where B is row equivalent to A^T and B is in echelon form.
2. The nonzero columns of B^T form a basis for $\mathcal{R}(A)$.

EXAMPLE 2.11 By way of illustration we use the method given above to find a basis for $\mathcal{R}(A)$ where A is the (3×4) matrix

$$A = \begin{bmatrix} 1 & 2 & 3 & 0 \\ 3 & 5 & 8 & 1 \\ 4 & 1 & 5 & 7 \end{bmatrix}.$$

Carrying out the steps described above, we find that A^{T} is row equivalent to C and C is row equivalent to B where

$$A^{\mathrm{T}} = \begin{bmatrix} 1 & 3 & 4 \\ 2 & 5 & 1 \\ 3 & 8 & 5 \\ 0 & 1 & 7 \end{bmatrix}, \quad C = \begin{bmatrix} 1 & 3 & 4 \\ 0 & -1 & -7 \\ 0 & -1 & -7 \\ 0 & 1 & 7 \end{bmatrix}, \quad B = \begin{bmatrix} 1 & 3 & 4 \\ 0 & -1 & -7 \\ 0 & 0 & 0 \\ 0 & 0 & 0 \end{bmatrix}.$$

Forming B^{T}, we obtain

$$B^{\mathrm{T}} = \begin{bmatrix} 1 & 0 & 0 & 0 \\ 3 & -1 & 0 & 0 \\ 4 & -7 & 0 & 0 \end{bmatrix}.$$

According to step (2), a basis for $\mathscr{R}(A)$ is $\{\boldsymbol{v}_1, \boldsymbol{v}_2\}$ where

$$\boldsymbol{v}_1 = \begin{bmatrix} 1 \\ 3 \\ 4 \end{bmatrix}, \quad \boldsymbol{v}_2 = \begin{bmatrix} 0 \\ -1 \\ -7 \end{bmatrix}. \tag{2.10}$$

To sketch the reasons that this procedure is effective, we observe that because of the way step (1) is carried out, the row vectors of B are linear combinations of the row vectors of A^{T}. Equivalently each column vector of B^{T} is a linear combination of the column vectors of A [hence the columns of B^{T} are in $\mathscr{R}(A)$]. Furthermore the nonzero column vectors of B^{T} are easily seen to be linearly independent. [Since B is in echelon form, the ith column of B^{T} has its first nonzero entry above the first nonzero entry of the $(i+1)$st column of B^{T}, See Example 2.11 for an illustration.] Thus the nonzero columns of B^{T} are a linearly independent set in $\mathscr{R}(A)$. Next the row operations used to produce B starting with A^{T} are reversible. That is, we can begin with B and use a series of elementary row operations to produce A^{T}; and therefore the row vectors of A^{T} are linear combinations of the row vectors of B. Equivalently the columns of A are linear combinations of the nonzero columns of B^{T}; in other words, the nonzero columns of B^{T} form a spanning set for $\mathscr{R}(A)$ that is linearly independent. (For instance in Example 2.11 we can produce C from B by elementary row operations. Next we can produce A^{T} from C by elementary row operations; and therefore the rows of A^{T} are linear combinations of the rows of B.)

The process described above can always be used to find a basis for $\mathscr{R}(A)$. We can always find a basis for $\mathscr{N}(A)$ by reducing $[A \,\vdots\, \boldsymbol{\theta}]$ to echelon form as is illustrated below.

EXAMPLE 2.12 In this example we find a basis for $\mathscr{N}(A)$ where A is the (3×4) matrix in Example 2.11. In order to find this basis, we need some sort of specification for the subspace $\mathscr{N}(A)$; we can always get this by solving $A\mathbf{x} = \boldsymbol{\theta}$. To begin, the augmented matrix $[A \mid \boldsymbol{\theta}]$ is row equivalent to

$$\begin{bmatrix} 1 & 2 & 3 & 0 & 0 \\ 0. & -1 & -1 & 1 & 0 \\ 0 & 0 & 0 & 0 & 0 \end{bmatrix};$$

so a vector $\mathbf{x}$ in R^4 is in $\mathcal{N}(A)$ if and only if

$$\begin{aligned} x_1 + 2x_2 &= -3x_3 \\ -\ x_2 &= \quad x_3 - x_4. \end{aligned}$$

Expressing x_1 and x_2 in terms of x_3 and x_4, we have the algebraic condition: a vector $\mathbf{x}$ in R^4 is in $\mathcal{N}(A)$ if and only if $\mathbf{x}$ is of the form

$$\mathbf{x} = \begin{bmatrix} -x_3 - 2x_4 \\ -x_3 + x_4 \\ x_3 \\ x_4 \end{bmatrix} = x_3 \begin{bmatrix} -1 \\ -1 \\ 1 \\ 0 \end{bmatrix} + x_4 \begin{bmatrix} -2 \\ 1 \\ 0 \\ 1 \end{bmatrix} = x_3\mathbf{u}_1 + x_4\mathbf{u}_2.$$

Since the vectors $\mathbf{u}_1$ and $\mathbf{u}_2$ are also linearly independent, $\{\mathbf{u}_1, \mathbf{u}_2\}$ is a basis for $\mathcal{N}(A)$.

2.3 EXERCISES

1. Let V be a subspace of R^4 consisting of vectors of the form

$$\mathbf{x} = \begin{bmatrix} x_1 \\ x_2 \\ x_3 \\ x_4 \end{bmatrix}.$$

As in Examples 2.6 and 2.10, find a spanning set for V when the components of $\mathbf{x}$ satisfy these conditions.

a) $\begin{aligned} x_1 + x_2 - x_3 \quad &= 0 \\ x_2 \quad - x_4 &= 0 \end{aligned}$ b) $\begin{aligned} x_1 + x_2 - x_3 + x_4 &= 0 \\ x_2 - 2x_3 - x_4 &= 0 \end{aligned}$

c) $x_1 - x_2 + x_3 - 3x_4 = 0$ d) $\begin{aligned} x_1 - x_2 \quad &= 0 \\ x_2 - 2x_3 \quad &= 0 \\ x_3 - x_4 &= 0 \end{aligned}$

2. Repeat Exercise 1 for these conditions.

a) $\begin{aligned} -x_1 + 2x_2 \quad - x_4 &= 0 \\ x_2 + x_3 \quad &= 0 \end{aligned}$ b) $\begin{aligned} x_1 - x_2 - x_3 + x_4 &= 0 \\ x_2 + x_3 \quad &= 0 \end{aligned}$

c) $x_1 + x_2 = 0$ d) $x_1 - x_2 + x_3 = 0$

3. As in Example 2.11, find a basis for the range space of the following matrices.

a) $\begin{bmatrix} 1 & 2 & 3 & -1 \\ 3 & 5 & 8 & -2 \\ 1 & 1 & 2 & 0 \end{bmatrix}$ b) $\begin{bmatrix} 1 & 1 & 2 \\ 1 & 1 & 2 \\ 2 & 3 & 5 \end{bmatrix}$ c) $\begin{bmatrix} 1 & 2 & 1 & 0 \\ 2 & 5 & 3 & -1 \\ 2 & 2 & 0 & 2 \\ 0 & 1 & 1 & -1 \end{bmatrix}$

4. As in Example 2.11, find a basis for the range space of the following matrices.

a) $\begin{bmatrix} 2 & 2 & 0 \\ 2 & 1 & 1 \\ 4 & 3 & 1 \end{bmatrix}$ b) $\begin{bmatrix} 1 & 2 & 1 \\ 2 & 4 & 1 \\ 3 & 6 & 2 \end{bmatrix}$ c) $\begin{bmatrix} 2 & 1 & 2 \\ 2 & 2 & 1 \\ 2 & 3 & 0 \end{bmatrix}$

5. Find a basis for the null space of each matrix in Exercise 3.

6. Find a basis for the null space of each matrix in Exercise 4.

7. Let V be a subspace, and let S be a spanning set for V. Find a basis for V when S consists of the following vectors. [Recall that if A is a matrix whose columns are the vectors in S, then a basis for $\mathscr{R}(A)$ is a basis for V.]

a) $\begin{bmatrix} 1 \\ 2 \end{bmatrix}, \begin{bmatrix} 2 \\ 4 \end{bmatrix}$ b) $\begin{bmatrix} 1 \\ 2 \end{bmatrix}, \begin{bmatrix} 2 \\ 1 \end{bmatrix}, \begin{bmatrix} 3 \\ 2 \end{bmatrix}$ 3 vectors, 2 equations → not independent ∴ cannot be a basis (need to reduce)

c) $\begin{bmatrix} 1 \\ 2 \\ -1 \\ 3 \end{bmatrix}, \begin{bmatrix} -2 \\ 1 \\ 2 \\ -1 \end{bmatrix}, \begin{bmatrix} -1 \\ -1 \\ 1 \\ -3 \end{bmatrix}, \begin{bmatrix} -2 \\ 2 \\ 2 \\ 0 \end{bmatrix}$ d) $\begin{bmatrix} 1 \\ 2 \\ 1 \end{bmatrix}, \begin{bmatrix} 2 \\ 5 \\ 0 \end{bmatrix}, \begin{bmatrix} 3 \\ 7 \\ 1 \end{bmatrix}, \begin{bmatrix} 1 \\ 1 \\ 3 \end{bmatrix}$

8. Find a basis for the null space of the following matrices.

a) $\begin{bmatrix} 1 & 0 & 0 \\ 1 & 0 & 1 \end{bmatrix}$ b) $\begin{bmatrix} 1 & 1 & 0 \\ 1 & 1 & 0 \end{bmatrix}$ c) $\begin{bmatrix} 1 & 1 & 0 \\ 1 & 1 & 1 \end{bmatrix}$

9. Find a basis for the range space of each matrix in Exercise 8.

10. Let $A = [\mathbf{A}_1, \mathbf{A}_2, \mathbf{A}_3]$ be the (3×3) matrix in part (a) of Exercise 4. Verify that $\{\mathbf{A}_1, \mathbf{A}_2\}$ is a spanning set for $\mathscr{R}(A)$ by showing that $\mathbf{A}_3$ is a linear combination of $\mathbf{A}_1$ and $\mathbf{A}_2$. Verify that $\{\mathbf{A}_1, \mathbf{A}_2\}$ is linearly independent [and thus is a basis for $\mathscr{R}(A)$].

11. With A as in Exercise 10, determine whether $\mathbf{b}$ is a linear combination of $\{\mathbf{A}_1, \mathbf{A}_2\}$—that is, is $\mathbf{b}$ in $\mathscr{R}(A)$? For $\mathbf{b}$ in $\mathscr{R}(\mathbf{A})$, find $\mathbf{x}_p$ in R^3 such that $A\mathbf{x}_p = \mathbf{b}$ and such that $\mathbf{e}_3^T\mathbf{x}_p = 0$. Give the "general" solution of $A\mathbf{x} = \mathbf{b}$ in the form $\mathbf{x} = \mathbf{x}_p + c\mathbf{u}$ where $\mathbf{u}$ is in $\mathscr{N}(A)$.

a) $\mathbf{b} = \begin{bmatrix} 6 \\ 5 \\ 11 \end{bmatrix}$ b) $\mathbf{b} = \begin{bmatrix} 0 \\ 1 \\ 1 \end{bmatrix}$ c) $\mathbf{b} = \begin{bmatrix} 1 \\ 1 \\ 1 \end{bmatrix}$ d) $\mathbf{b} = \begin{bmatrix} 1 \\ 2 \\ 2 \end{bmatrix}$

e) $\mathbf{b} = \begin{bmatrix} -2 \\ 2 \\ 0 \end{bmatrix}$

12. Let $B = \{\mathbf{v}_1, \mathbf{v}_2, \mathbf{v}_3\}$ be a set of linearly independent vectors in R^3. Prove that B is a basis for R^3. (Hint: Use Theorem 1.11 to show that B is a spanning set for R^3.)

13. Prove that every basis for R^2 contains exactly two vectors. Proceed by showing the following.

a) A basis for R^2 cannot have more than two vectors.

b) A basis for R^2 cannot have one vector.

(Hint: Suppose that a basis for R^2 could contain one vector. Represent $\mathbf{e}_1$ and $\mathbf{e}_2$ in terms of the basis, and obtain a contradiction.)

14. Find a vector $\mathbf{w}$ in R^3 such that $\mathbf{w}$ is not a linear combination of $\mathbf{v}_1$ and $\mathbf{v}_2$:

$$\mathbf{v}_1 = \begin{bmatrix} 1 \\ 2 \\ -1 \end{bmatrix}, \quad \mathbf{v}_2 = \begin{bmatrix} 2 \\ -1 \\ -2 \end{bmatrix}.$$

15. Show that any spanning set for R^n must contain at least n vectors. Proceed by showing that if $\mathbf{u}_1, \mathbf{u}_2, \ldots, \mathbf{u}_p$ are vectors in R^n, and if $p < n$, then there is a nonzero vector $\mathbf{v}$ such that $\mathbf{v}^T\mathbf{u}_i = 0$, $1 \leqslant i \leqslant p$. [Hint: Write the constraints as a $(p \times n)$ system and use Theorem 1.2.] Given $\mathbf{v}$ as above, can $\mathbf{v}$ be a linear combination of $\mathbf{u}_1, \mathbf{u}_2, \ldots, \mathbf{u}_p$?

16. Recalling Exercise 15, prove that every basis for R^n contains exactly n vectors.

2.4 DIMENSION

We have seen several examples showing that a given subspace W may have many different bases. In fact Exercise 12, Section 2.3, shows that *any* set of three linearly independent vectors in R^3 is a basis for R^3. In this section we will show that all bases for a subspace W do have one property in common: if W is a subspace, then every basis for W contains exactly the same number of vectors. We will single out this integer—the number of vectors in any basis for W—and call it the dimension of W. But first to put the concept of dimension on a firm mathematical foundation, we prove the following theorem.

Theorem 2.3

Let W be a subspace of R^n, and let $B = \{\mathbf{w}_1, \mathbf{w}_2, \ldots, \mathbf{w}_p\}$ be a basis for W containing p vectors. Then any set of $p+1$ or more vectors in W is linearly dependent.

Proof. Let $\{\mathbf{s}_1, \mathbf{s}_2, \ldots, \mathbf{s}_m\}$ be any set of m vectors in W where $m > p$. To show this set is linearly dependent, we first express each $\mathbf{s}_i$ in terms of the basis B:

$$\begin{aligned}
\mathbf{s}_1 &= a_{11}\mathbf{w}_1 + a_{21}\mathbf{w}_2 + \ldots + a_{p1}\mathbf{w}_p \\
\mathbf{s}_2 &= a_{12}\mathbf{w}_1 + a_{22}\mathbf{w}_2 + \ldots + a_{p2}\mathbf{w}_p \\
&\vdots \\
\mathbf{s}_m &= a_{1m}\mathbf{w}_1 + a_{2m}\mathbf{w}_2 + \ldots + a_{pm}\mathbf{w}_p.
\end{aligned} \tag{2.11}$$

To show that $\{\mathbf{s}_1, \mathbf{s}_2, \ldots, \mathbf{s}_m\}$ is linearly dependent, we must show that there is a nontrivial solution of

$$c_1\mathbf{s}_1 + c_2\mathbf{s}_2 + \ldots + c_m\mathbf{s}_m = \boldsymbol{\theta}. \tag{2.12}$$

Now using (2.11), we can rewrite (2.12) in terms of the basis vectors as

$$\begin{aligned}
&c_1(a_{11}\mathbf{w}_1 + a_{21}\mathbf{w}_2 + \ldots + a_{p1}\mathbf{w}_p) + \\
&c_2(a_{12}\mathbf{w}_1 + a_{22}\mathbf{w}_2 + \ldots + a_{p2}\mathbf{w}_p) + \\
&\ldots + c_m(a_{1m}\mathbf{w}_1 + a_{2m}\mathbf{w}_2 + \ldots + a_{pm}\mathbf{w}_p) = \boldsymbol{\theta}.
\end{aligned} \tag{2.13a}$$

Equation (2.13a) can be regrouped as

$$\begin{aligned}&(c_1a_{11}+c_2a_{12}+\ldots+c_ma_{1m})\mathbf{w}_1+\\&(c_1a_{21}+c_2a_{22}+\ldots+c_ma_{2m})\mathbf{w}_2+\\&\ldots+(c_1a_{p1}+c_2a_{p2}+\ldots+c_ma_{pm})\mathbf{w}_p=\boldsymbol{\theta}.\end{aligned} \qquad (2.13b)$$

Now finding $c_1, c_2, \ldots, c_m$ to satisfy (2.12) is the same as finding $c_1, c_2, \ldots, c_m$ to satisfy (2.13b). Furthermore we can clearly satisfy (2.13b) if we can choose each coefficient of each $\mathbf{w}_i$ to be 0. [In fact since $\mathbf{w}_1, \mathbf{w}_2, \ldots, \mathbf{w}_p$ are linearly independent, this is the only way we can satisfy (2.13b).] Therefore equation (2.13b) is equivalent to the system

$$\begin{aligned}a_{11}c_1+a_{12}c_2+\ldots+a_{1m}c_m&=0\\a_{21}c_1+a_{22}c_2+\ldots+a_{2m}c_m&=0\\&\vdots\\a_{p1}c_1+a_{p2}c_2+\ldots+a_{pm}c_m&=0.\end{aligned} \qquad (2.14)$$

[Recall that each a_{ij} is a specified constant determined by (2.11) whereas each c_i is an unknown parameter of (2.12).] The homogeneous system in (2.14) has more unknowns than equations; so by Theorem 1.2 there is a nontrivial solution to (2.14). Since (2.14) and (2.12) are equivalent systems, (2.12) has a nontrivial solution, and the theorem is established. ■

As an immediate corollary of Theorem 2.3, we can show that all bases for a subspace contain the same number of vectors.

Corollary

Let W be a subspace of R^n, and let $B = \{\mathbf{w}_1, \mathbf{w}_2, \ldots, \mathbf{w}_p\}$ be a basis for W containing p vectors. Then every basis for W contains p vectors.

Proof. Let $Q = \{\mathbf{u}_1, \mathbf{u}_2, \ldots, \mathbf{u}_r\}$ be any basis for W. By Theorem 2.3 any set of $r+1$ or more vectors in W is linearly dependent. Since B is a linearly independent set of p vectors in W, we know that $p \leqslant r$. Similarly since B is a basis of p vectors for W, any set of $p+1$ or more vectors in W is linearly dependent. By assumption, Q is a set of r linearly independent vectors in W; so $r \leqslant p$. Now since we have $p \leqslant r$ and $r \leqslant p$, it must be that $r = p$. ■

Given that every basis for a subspace contains the same number of vectors, we can make the following definition without any possibility of ambiguity.

DEFINITION 2.6.

Let W be a subspace of R^n. If W has a basis $B = \{\mathbf{w}_1, \mathbf{w}_2, \ldots, \mathbf{w}_p\}$ of p vectors, then we say that W is a subspace of ***dimension*** p, and we write $\dim(W) = p$.

In Exercise 7, Section 2.4, the reader is asked to show that every nonzero subspace of R^n does have a basis. Thus a value for dimension can be assigned to any subspace of R^n where for completeness we define $\dim(W) = 0$ if W is the zero subspace.

EXAMPLE 2.13 Since R^3 has a basis of three vectors, $\{\mathbf{e}_1, \mathbf{e}_2, \mathbf{e}_3\}$, we see that dim $(R^3) = 3$. In general, R^n has a basis of n vectors; so dim $(R^n) = n$. Thus the definition of dimension—the number of vectors in a basis—agrees with the usual terminology; R^3 is three-dimensional and, in general, R^n is n-dimensional.

EXAMPLE 2.14 In Example 2.8 we saw that the subspace

$$W = \{\mathbf{x}:\mathbf{x} = \begin{bmatrix} x_1 \\ x_2 \\ 0 \end{bmatrix}, x_1 \text{ and } x_2 \text{ any real numbers}\}$$

has a basis of two vectors:

$$\mathbf{w}_1 = \begin{bmatrix} 1 \\ 0 \\ 0 \end{bmatrix}, \quad \mathbf{w}_2 = \begin{bmatrix} 0 \\ 1 \\ 0 \end{bmatrix}.$$

Therefore dim $(W) = 2$; this accords with our geometric intuition since we can visualize W as representing the xy-plane in three-dimensional space.

Similarly the subspace W in Example 2.6 can be visualized as the plane $x = y - z$ in three-dimensional space. In Example 2.8 we noted that this subspace had a basis of two vectors; so it is two-dimensional as our intuition would suggest.

An important feature of dimension is that a p-dimensional subspace W has many of the same properties as R^p. For example in R^p any set of $p+1$ or more vectors is linearly dependent, and this same property holds in W when dim $(W) = p$.

Theorem 2.4

Let W be a subspace of R^n with dim $(W) = p$.

1. Any set of $p+1$ or more vectors in W is linearly dependent.
2. Any set of p linearly independent vectors in W is a basis for W.
3. Any set of p vectors that spans W is a basis for W.

Proof. Part (1) follows immediately from Theorem 2.3 since dim $(W) = p$ means that W has a basis of p vectors.

The proof of part (2) is given below while part (3) is left as an exercise. To establish part (2), let $\{\mathbf{u}_1, \mathbf{u}_2, \ldots, \mathbf{u}_p\}$ be a set of p linearly independent vectors in W; and let $\mathbf{v}$ be any vector in W. Then by part (1) the set $\{\mathbf{v}, \mathbf{u}_1, \mathbf{u}_2, \ldots, \mathbf{u}_p\}$ is a linearly dependent set of vectors since the set contains $p+1$ vectors. Thus there are scalars $a_0, a_1, \ldots, a_p$ (not all of which are zero) such that

$$a_0\mathbf{v} + a_1\mathbf{u}_1 + a_2\mathbf{u}_2 + \ldots + a_p\mathbf{u}_p = \boldsymbol{\theta}. \qquad (2.15)$$

In addition in (2.15), a_0 cannot be zero since $\{\mathbf{u}_1, \mathbf{u}_2, \ldots, \mathbf{u}_p\}$ is a linearly independent set. Therefore rewriting (2.15), we find

$$\mathbf{v} = -\frac{1}{a_0}(a_1\mathbf{u}_1 + a_2\mathbf{u}_2 + \ldots + a_p\mathbf{u}_p). \tag{2.16}$$

Thus from (2.16) we see that any vector can be expressed as a linear combination of $\mathbf{u}_1, \mathbf{u}_2, \ldots, \mathbf{u}_p$. Since it is a spanning set for W, this linearly independent set is a basis. ■

2.4 EXERCISES

1. Find the dimension of the range space and the dimension of the null space. Also give a basis for $\mathcal{N}(A)$, $\mathcal{R}(B)$, $\mathcal{N}(B)$, and $\mathcal{R}(D)$.

$$A = \begin{bmatrix} 1 & 3 & 4 & 5 \\ 3 & 2 & 5 & 1 \end{bmatrix} \qquad B = \begin{bmatrix} 1 & 1 & 1 & 2 \\ 2 & 3 & 4 & 5 \\ 3 & 4 & 5 & 1 \end{bmatrix}$$

$$C = \begin{bmatrix} 1 & 0 & 0 \\ 2 & 4 & 0 \\ 1 & 1 & 5 \end{bmatrix} \qquad D = \begin{bmatrix} 1 & 2 & 0 \\ 3 & 7 & 4 \\ 1 & 3 & 4 \end{bmatrix}$$

2. If A is an $(n \times n)$ nonsingular matrix, what is the dimension of $\mathcal{R}(A)$ and $\mathcal{N}(A)$?

3. Let W be a subspace and let S be a spanning set for W. Find a basis for W, and calculate $\dim(W)$ when S contains the following vectors.

a) $\begin{bmatrix} 1 \\ 1 \\ -2 \end{bmatrix}, \begin{bmatrix} -1 \\ -2 \\ 3 \end{bmatrix}, \begin{bmatrix} 1 \\ 0 \\ -1 \end{bmatrix}, \begin{bmatrix} 2 \\ -1 \\ 0 \end{bmatrix}$ b) $\begin{bmatrix} 1 \\ 2 \\ -1 \\ 1 \end{bmatrix}, \begin{bmatrix} 3 \\ 1 \\ 1 \\ 2 \end{bmatrix}, \begin{bmatrix} -1 \\ 1 \\ -2 \\ 2 \end{bmatrix}, \begin{bmatrix} 0 \\ -2 \\ 1 \\ 2 \end{bmatrix}$

4. Even though $\{\mathbf{e}_1, \mathbf{e}_2, \mathbf{e}_3\}$ is a spanning set for R^3, these vectors may not be in a subspace of R^3. Give an example of a subspace W of R^3 where $\dim(W) = 2$ and W does not contain any of $\mathbf{e}_1$, $\mathbf{e}_2$, or $\mathbf{e}_3$.

5. a) Let $\mathbf{u}_1$ and $\mathbf{u}_2$ be vectors in R^3, and let W be the set of all linear combinations of $\mathbf{u}_1$ and $\mathbf{u}_2$:

$$W = \{\mathbf{x}: \mathbf{x} = a_1\mathbf{u}_1 + a_2\mathbf{u}_2,\ a_1 \text{ and } a_2 \text{ any real numbers}\}.$$

Prove that W is a subspace of R^3 by showing that $\mathbf{x} + \mathbf{y}$ and $a\mathbf{x}$ are in W whenever $\mathbf{x}$ and $\mathbf{y}$ are in W and a is any scalar.

b) Let $\mathbf{u}_1$ and $\mathbf{u}_2$ be the vectors

$$\mathbf{u}_1 = \begin{bmatrix} 1 \\ 2 \\ -1 \end{bmatrix} \qquad \mathbf{u}_2 = \begin{bmatrix} 3 \\ 1 \\ 2 \end{bmatrix},$$

and let W be defined as above. By the definition of W, $\{\mathbf{u}_1, \mathbf{u}_2\}$ is spanning set for W. Show that $\dim(W) = 2$.

c) With W as in part (b), show that the set W can be visualized as a plane in R^3 by showing that the components of every vector in W satisfy a relationship $ax_1 + bx_2 + x_3 = 0$. What are the numerical values of a and b?

6. Let W be the set of vectors in R^3 whose components satisfy $ax_1 + bx_2 + cx_3 = 0$ where not all of a, b, and c are zero (that is, W represents a plane in R^3). Prove that $\dim(W) = 2$; for simplicity assume that $c = -1$. (Hint: W has a natural spanning set of two vectors.)
7. Let W be a nonzero subspace of R^n. Show that W has a basis. [Hint: Let $\mathbf{w}_1$ be any nonzero vector in W. If $\{\mathbf{w}_1\}$ is a spanning set for W, then we are done. If not, there is a vector $\mathbf{w}_2$ in W such that $\{\mathbf{w}_1, \mathbf{w}_2\}$ is linearly independent. (Why?) Continue by asking if this is a spanning set for W. Why must this process eventually stop?]
8. Prove part (3) of Theorem 2.4.
9. Let W be the subspace of R^4 defined by $W = \{\mathbf{x}: \mathbf{v}^T\mathbf{x} = 0\}$, and calculate $\dim(W)$ where

$$\mathbf{v} = \begin{bmatrix} 1 \\ 2 \\ -3 \\ -1 \end{bmatrix}.$$

10. Let W be the subspace of R^4 defined by $W = \{\mathbf{x} \text{ in } R^4: \mathbf{a}^T\mathbf{x} = 0 \text{ and } \mathbf{b}^T\mathbf{x} = 0 \text{ and } \mathbf{c}^T\mathbf{x} = 0\}$. Calculate $\dim(W)$ for the following vectors.

a) $\mathbf{a} = \begin{bmatrix} 1 \\ -1 \\ 0 \\ 0 \end{bmatrix}, \mathbf{b} = \begin{bmatrix} 1 \\ 0 \\ -1 \\ 0 \end{bmatrix}, \mathbf{c} = \begin{bmatrix} 0 \\ 1 \\ -1 \\ 0 \end{bmatrix}$ b) $\mathbf{a} = \begin{bmatrix} 1 \\ 0 \\ 0 \\ 0 \end{bmatrix}, \mathbf{b} = \begin{bmatrix} 1 \\ 1 \\ 0 \\ 0 \end{bmatrix}, \mathbf{c} = \begin{bmatrix} 1 \\ 0 \\ 0 \\ 1 \end{bmatrix}$

c) $\mathbf{a} = \begin{bmatrix} 0 \\ 0 \\ 0 \\ 1 \end{bmatrix}, \mathbf{b} = \begin{bmatrix} 0 \\ 0 \\ 0 \\ 0 \end{bmatrix}, \mathbf{c} = \begin{bmatrix} 1 \\ 0 \\ 0 \\ -1 \end{bmatrix}$ d) $\mathbf{a} = \begin{bmatrix} 1 \\ 0 \\ 0 \\ 0 \end{bmatrix}, \mathbf{b} = \begin{bmatrix} 1 \\ 1 \\ 0 \\ 0 \end{bmatrix}, \mathbf{c} = \begin{bmatrix} 2 \\ 3 \\ 0 \\ 0 \end{bmatrix}$

11. Repeat Exercise 10 for the following vectors.

a) $\mathbf{a} = \begin{bmatrix} 1 \\ 1 \\ 0 \\ 0 \end{bmatrix}, \mathbf{b} = \begin{bmatrix} 1 \\ 0 \\ 0 \\ -1 \end{bmatrix}, \mathbf{c} = \begin{bmatrix} 0 \\ 0 \\ 1 \\ -1 \end{bmatrix}$ b) $\mathbf{a} = \begin{bmatrix} 1 \\ -1 \\ 0 \\ 0 \end{bmatrix}, \mathbf{b} = \begin{bmatrix} 0 \\ 1 \\ -1 \\ 0 \end{bmatrix},$

$\mathbf{c} = \begin{bmatrix} -1 \\ 3 \\ -2 \\ 0 \end{bmatrix}$ c) $\mathbf{a} = \begin{bmatrix} 1 \\ -1 \\ 0 \\ 0 \end{bmatrix}. \mathbf{b} = \begin{bmatrix} 0 \\ 0 \\ 1 \\ -1 \end{bmatrix}, \mathbf{c} = \begin{bmatrix} 0 \\ 0 \\ 0 \\ 0 \end{bmatrix}$ d) $\mathbf{a} = \begin{bmatrix} 1 \\ -1 \\ 1 \\ -1 \end{bmatrix},$

$\mathbf{b} = \begin{bmatrix} 2 \\ -2 \\ 2 \\ -2 \end{bmatrix}, \mathbf{c} = \begin{bmatrix} 3 \\ -3 \\ 3 \\ -3 \end{bmatrix}$

12. Suppose $\{\mathbf{u}_1, \mathbf{u}_2, \ldots, \mathbf{u}_p\}$ is a basis for a subspace W, and suppose $\mathbf{x}$ is in W with $\mathbf{x} = a_1\mathbf{u}_1 + a_2\mathbf{u}_2 + \ldots + a_p\mathbf{u}_p$. Show that this representation for $\mathbf{x}$ in terms of the basis is unique—that is, if $\mathbf{x} = b_1\mathbf{u}_1 + b_2\mathbf{u}_2 + \ldots + b_p\mathbf{u}_p$, then $b_1 = a_1, b_2 = a_2, \ldots, b_p = a_p$.
13. Let U and V be subspaces of R^n, and suppose that U is a subset of V. Prove that $\dim(U) \leqslant \dim(V)$. If $\dim(U) = \dim(V)$, prove that V is contained in U, and thus conclude that $U = V$.

14. Suppose $S = \{\mathbf{u}_1, \mathbf{u}_2, \ldots, \mathbf{u}_p\}$ is a set of linearly independent vectors in a subspace W where dim $(W) = m$ and $m > p$. Prove that there is a vector $\mathbf{u}_{p+1}$ in W such that $\{\mathbf{u}_1, \mathbf{u}_2, \ldots, \mathbf{u}_p, \mathbf{u}_{p+1}\}$ is linearly independent. Use this proof to prove that a basis that contains all the vectors in S can be constructed for W.

†2.5 ORTHOGONAL BASES FOR SUBSPACES

If W is a p-dimensional subspace of R^n, then W has a basis of p vectors, say $B = \{\mathbf{w}_1, \mathbf{w}_2, \ldots, \mathbf{w}_p\}$. Consequently if $\boldsymbol{v}$ is any vector in W, we know that we can express $\boldsymbol{v}$ uniquely as the linear combination

$$\boldsymbol{v} = a_1\mathbf{w}_1 + a_2\mathbf{w}_2 + \ldots + a_p\mathbf{w}_p. \tag{2.17}$$

[Uniqueness in (2.17) follows from Exercise 12, Section 2.4.] In (2.17) the coefficients $a_1, a_2, \ldots, a_p$ are called the ***coordinates*** of $\boldsymbol{v}$ with respect to the basis B. As it stands, this is purely a theoretical statement of existence—that is, we know that a given vector $\boldsymbol{v}$ in W can be expressed as in (2.17), but we do not know the coordinates $a_1, a_2, \ldots, a_p$ without solving the $(n \times p)$ system represented by (2.17). In many applications it is necessary to express $\boldsymbol{v}$ in terms of a basis B for W, but it is irrelevant (to the application) what particular basis we choose for W. Given such a problem, one logically chooses a basis for W that facilitates the solution of (2.17) for the coordinates of $\boldsymbol{v}$. As defined below, an orthogonal basis for W is just such a basis.

To be precise, we say that two vectors $\mathbf{u}$ and $\boldsymbol{v}$ in R^n are ***orthogonal*** if their scalar product is zero. That is, $\mathbf{u}$ and $\boldsymbol{v}$ are orthogonal if

$$\mathbf{u}^{\mathrm{T}}\boldsymbol{v} = 0. \tag{2.18}$$

(Note that the order of the scalar product in(2.18) is immaterial since $\mathbf{u}^{\mathrm{T}}\boldsymbol{v} = \boldsymbol{v}^{\mathrm{T}}\mathbf{u}$.) In R^n we can visualize orthogonal vectors as being "perpendicular." For instance in R^2 the vectors

$$\mathbf{u} = \begin{bmatrix} 1 \\ -2 \end{bmatrix} \qquad \boldsymbol{v} = \begin{bmatrix} 6 \\ 3 \end{bmatrix}$$

are orthogonal since $\mathbf{u}^{\mathrm{T}}\boldsymbol{v} = 0$. Moreover the vectors $\mathbf{u}$ and $\boldsymbol{v}$ are perpendicular in the usual vector geometry sense. (See Fig. 2.3.)

Extending the idea of orthogonality to sets of vectors, we say that $\{\mathbf{u}_1, \mathbf{u}_2, \ldots, \mathbf{u}_p\}$ is an ***orthogonal set*** of vectors if

$$\mathbf{u}_i^{\mathrm{T}}\mathbf{u}_j = 0, \qquad i \neq j, \qquad 1 \leqslant i, j \leqslant p.$$

Thus a set of vectors is orthogonal if all the vectors in the set are "mutually perpendicular." Finally if W is a p-dimensional subspace of R^n, and if $B = \{\mathbf{u}_1, \mathbf{u}_2, \ldots, \mathbf{u}_p\}$ is a basis for W, we say that B is an ***orthogonal basis*** for W if B is an orthogonal set of vectors.

---

† Sections 2.5 and 2.6 form a self-contained unit. In later chapters the only section that refers to 2.5 and 2.6 is 4.6. These three sections may be omitted without loss of continuity.

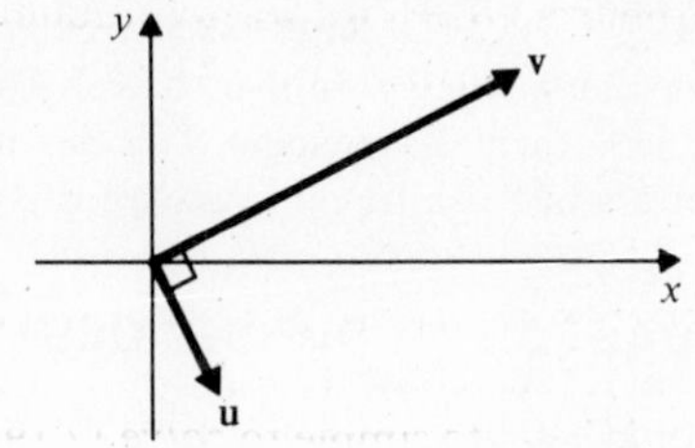

Fig. 2.3 Orthogonal vectors, **u** and **v**, in R^2.

EXAMPLE 2.15 The unit vectors in R^3, $\mathbf{e}_1$, $\mathbf{e}_2$, and $\mathbf{e}_3$ are orthogonal (since $\mathbf{e}_1^T\mathbf{e}_2 = 0$, $\mathbf{e}_1^T\mathbf{e}_3 = 0$, $\mathbf{e}_2^T\mathbf{e}_3 = 0$). Therefore the natural basis $\{\mathbf{e}_1, \mathbf{e}_2, \mathbf{e}_3\}$ for R^3 is an orthogonal basis for R^3. In general the natural basis $\{\mathbf{e}_1, \mathbf{e}_2, \ldots, \mathbf{e}_n\}$ for R^n is an orthogonal basis for R^n.

EXAMPLE 2.16 In R^3 the set of vectors $\{\mathbf{u}_1, \mathbf{u}_2, \mathbf{u}_3\}$ where

$$\mathbf{u}_1 = \begin{bmatrix} 1 \\ 2 \\ 1 \end{bmatrix}, \quad \mathbf{u}_2 = \begin{bmatrix} 3 \\ -1 \\ -1 \end{bmatrix}, \quad \mathbf{u}_3 = \begin{bmatrix} 1 \\ -4 \\ 7 \end{bmatrix}$$

is an orthogonal set since

$$\mathbf{u}_1^T\mathbf{u}_2 = 0, \qquad \mathbf{u}_1^T\mathbf{u}_3 = 0, \qquad \mathbf{u}_2^T\mathbf{u}_3 = 0.$$

Also it is not hard to show that $\{\mathbf{u}_1, \mathbf{u}_2, \mathbf{u}_3\}$ is a linearly independent set of vectors; so by Theorem 2.4, $\{\mathbf{u}_1, \mathbf{u}_2, \mathbf{u}_3\}$ is an orthogonal basis for R^3. Finally $\{\mathbf{e}_1, \mathbf{u}_2, \mathbf{u}_3\}$ is a basis for R^3 since this is a linearly independent set of three vectors in R^3. However this set is not an orthogonal basis since $\mathbf{e}_1^T\mathbf{u}_2 \neq 0$.

In Exercise 4, Section 2.5, the reader is asked to prove a general result illustrated in Example 2.16: any orthogonal set of nonzero vectors is linearly independent. Therefore in view of Theorem 2.4 any orthogonal set of p *nonzero* vectors in a p-dimensional subspace is an orthogonal basis for the subspace. The requirement that the vectors be nonzero is necessary since, for example, the set $\{\boldsymbol{\theta}, \mathbf{e}_1, \mathbf{e}_2\}$ is an orthogonal set of three vectors in R^3 (clearly, $\boldsymbol{\theta}^T\mathbf{e}_1 = 0$, $\boldsymbol{\theta}^T\mathbf{e}_2 = 0$, $\mathbf{e}_1^T\mathbf{e}_2 = 0$); but this set is not a basis for R^3. (The set is not linearly independent.) Finally we mention that the vectors in an orthogonal basis can be "normalized" by dividing each by its magnitude to produce an ***orthonormal basis*** (see Exercise 8, Section 2.5).

Our next example shows how we can use the property of orthogonality to find the coordinates of a vector.

EXAMPLE 2.17 Since the set $B = \{\mathbf{u}_1, \mathbf{u}_2, \mathbf{u}_3\}$ defined in Example 2.16 is an orthogonal basis for R^3, we know that any vector $\boldsymbol{v}$ in R^3 can be expressed in the form

$$\boldsymbol{v} = a_1\mathbf{u}_1 + a_2\mathbf{u}_2 + a_3\mathbf{u}_3. \tag{2.19}$$

If we want the coefficients a_1, a_2, a_3 in (2.19) for a given vector $\boldsymbol{v}$ in R^3, we can use Gauss elimination to solve the (3×3) system of linear equations represented by (2.19). However it is quite simple to solve (2.19) when we utilize the fact that $\{\mathbf{u}_1, \mathbf{u}_2, \mathbf{u}_3\}$ is an orthogonal set. Before stating the general procedure in the next theorem, we give an illustration.

As a particular instance, let $\boldsymbol{v}$ be the vector

$$\boldsymbol{v} = \begin{bmatrix} 12 \\ -3 \\ 6 \end{bmatrix}.$$

Now given that $\boldsymbol{v}$ can be expressed as in (2.19), we can obtain the three constraints

$$\begin{aligned} \mathbf{u}_1^T\boldsymbol{v} &= \mathbf{u}_1^T(a_1\mathbf{u}_1 + a_2\mathbf{u}_2 + a_3\mathbf{u}_3) = a_1\mathbf{u}_1^T\mathbf{u}_1 + a_2\mathbf{u}_1^T\mathbf{u}_2 + a_3\mathbf{u}_1^T\mathbf{u}_3 \\ \mathbf{u}_2^T\boldsymbol{v} &= \mathbf{u}_2^T(a_1\mathbf{u}_1 + a_2\mathbf{u}_2 + a_3\mathbf{u}_3) = a_1\mathbf{u}_2^T\mathbf{u}_1 + a_2\mathbf{u}_2^T\mathbf{u}_2 + a_3\mathbf{u}_2^T\mathbf{u}_3 \\ \mathbf{u}_3^T\boldsymbol{v} &= \mathbf{u}_3^T(a_1\mathbf{u}_1 + a_2\mathbf{u}_2 + a_3\mathbf{u}_3) = a_1\mathbf{u}_3^T\mathbf{u}_1 + a_2\mathbf{u}_3^T\mathbf{u}_2 + a_3\mathbf{u}_3^T\mathbf{u}_3. \end{aligned} \tag{2.20a}$$

Next since $\mathbf{u}_1^T\mathbf{u}_2 = 0$, $\mathbf{u}_1^T\mathbf{u}_3 = 0$, and $\mathbf{u}_2^T\mathbf{u}_3 = 0$, we see that (2.20a) reduces dramatically to a diagonal system

$$\begin{aligned} \mathbf{u}_1^T\boldsymbol{v} &= a_1\mathbf{u}_1^T\mathbf{u}_1 \\ \mathbf{u}_2^T\boldsymbol{v} &= a_2\mathbf{u}_2^T\mathbf{u}_2 \\ \mathbf{u}_3^T\boldsymbol{v} &= a_3\mathbf{u}_3^T\mathbf{u}_3. \end{aligned} \tag{2.20b}$$

Thus (2.20b) constitutes a set of formulas for the coordinates of $\boldsymbol{v}$ with respect to B:

$$a_1 = \frac{\mathbf{u}_1^T\boldsymbol{v}}{\mathbf{u}_1^T\mathbf{u}_1}, \qquad a_2 = \frac{\mathbf{u}_2^T\boldsymbol{v}}{\mathbf{u}_2^T\mathbf{u}_2}, \qquad a_3 = \frac{\mathbf{u}_3^T\boldsymbol{v}}{\mathbf{u}_3^T\mathbf{u}_3}.$$

For the particular vector $\boldsymbol{v}$ above,

$$\begin{aligned} a_1 &= 12/6 = 2 \\ a_2 &= 33/11 = 3 \\ a_3 &= 66/66 = 1; \end{aligned}$$

and it is easy to verify that $\boldsymbol{v} = 2\mathbf{u}_1 + 3\mathbf{u}_2 + \mathbf{u}_3$. The general result, of which Example 2.17 gives a special case, is Theorem 2.5.

Theorem 2.5

Let W be a p-dimensional subspace of R^n, and let $B = \{\mathbf{u}_1, \mathbf{u}_2, \ldots, \mathbf{u}_p\}$ be an orthogonal basis for W. If $\mathbf{v}$ is in W, we can express $\mathbf{v}$ as

$$\mathbf{v} = \frac{\mathbf{u}_1^T\mathbf{v}}{\mathbf{u}_1^T\mathbf{u}_1}\mathbf{u}_1 + \frac{\mathbf{u}_2^T\mathbf{v}}{\mathbf{u}_2^T\mathbf{u}_2}\mathbf{u}_2 + \ldots + \frac{\mathbf{u}_p^T\mathbf{v}}{\mathbf{u}_p^T\mathbf{u}_p}\mathbf{u}_p.$$

Proof. First of all, we note that since B is a basis, none of the vectors $\mathbf{u}_1, \mathbf{u}_2, \ldots, \mathbf{u}_p$ is the zero vector, and therefore $\mathbf{u}_i^T\mathbf{u}_i > 0$ for $i = 1, 2, \ldots, p$. Next B is a basis for W; so if $\mathbf{v}$ is in W, then $\mathbf{v}$ must be expressible as

$$\mathbf{v} = a_1\mathbf{u}_1 + a_2\mathbf{u}_2 + \ldots + a_i\mathbf{u}_i + \ldots + a_p\mathbf{u}_p. \tag{2.21}$$

Therefore as in Example 2.17,

$$\mathbf{u}_i^T\mathbf{v} = a_i\mathbf{u}_i^T\mathbf{u}_i, \; i = 1, 2, \ldots, p;$$

and since $\mathbf{u}_i^T\mathbf{u}_i \neq 0$, we obtain

$$a_i = \mathbf{u}_i^T\mathbf{v}/\mathbf{u}_i^T\mathbf{u}_i, \; i = 1, 2, \ldots, p. \tag{2.22}$$

In view of (2.21), the theorem is proved; and (2.22) gives a set of formulas for the coordinates of $\mathbf{v}$ with respect to B. ■

The next theorem gives a procedure that can be used to generate an orthogonal basis from any given basis. This procedure, called the Gram–Schmidt process, is quite practical from a computational standpoint (although some care must be exercised when programming the procedure for the computer). Generating an orthogonal basis is often the first step in solving problems in least-squares approximation; so Gram–Schmidt orthogonalization is of more than theoretical interest.

Theorem 2.6 (Gram–Schmidt)

Let W be a p-dimensional subspace of R^n, and let $\{\mathbf{w}_1, \mathbf{w}_2, \ldots, \mathbf{w}_p\}$ be any basis for W. Then the set of vectors $\{\mathbf{u}_1, \mathbf{u}_2, \ldots, \mathbf{u}_p\}$ is an orthogonal basis for W where

$$\mathbf{u}_1 = \mathbf{w}_1$$

$$\mathbf{u}_2 = \mathbf{w}_2 - \frac{\mathbf{u}_1^T\mathbf{w}_2}{\mathbf{u}_1^T\mathbf{u}_1}\mathbf{u}_1$$

$$\mathbf{u}_3 = \mathbf{w}_3 - \frac{\mathbf{u}_1^T\mathbf{w}_3}{\mathbf{u}_1^T\mathbf{u}_1}\mathbf{u}_1 - \frac{\mathbf{u}_2^T\mathbf{w}_3}{\mathbf{u}_2^T\mathbf{u}_2}\mathbf{u}_2,$$

and where in general

$$\mathbf{u}_i = \mathbf{w}_i - \sum_{k=1}^{i-1} \frac{\mathbf{u}_k^T\mathbf{w}_i}{\mathbf{u}_k^T\mathbf{u}_k}\mathbf{u}_k, \qquad 2 \leqslant i \leqslant p. \tag{2.23}$$

Proof. We first show that the expression given in (2.23) is always defined and that the vectors $\mathbf{u}_1, \mathbf{u}_2, \ldots, \mathbf{u}_p$ are all nonzero. To begin, $\mathbf{u}_1$ is a nonzero vector since $\mathbf{u}_1 = \mathbf{w}_1$. Thus $\mathbf{u}_1^T\mathbf{u}_1 > 0$, and so we can define $\mathbf{u}_2$. Furthermore we observe that $\mathbf{u}_2$ has the form $\mathbf{u}_2 = \mathbf{w}_2 - b_1\mathbf{u}_1 = \mathbf{w}_2 - b_1\mathbf{w}_1$; so $\mathbf{u}_2$ is nonzero since it is a nontrivial linear combination of $\mathbf{w}_1$ and $\mathbf{w}_2$. Proceeding inductively, suppose $\mathbf{u}_1, \mathbf{u}_2, \ldots, \mathbf{u}_{i-1}$ have been generated by (2.23); and suppose each $\mathbf{u}_k$ has the form

$$\mathbf{u}_k = \mathbf{w}_k - c_1\mathbf{w}_1 - c_2\mathbf{w}_2 - \cdots - c_{k-1}\mathbf{w}_{k-1};\ 1 \leqslant k \leqslant i-1.$$

From this equation, each $\mathbf{u}_k$ is nonzero; and it follows that (2.23) is a well-defined expression [since $\mathbf{u}_k^T\mathbf{u}_k > 0$ for $1 \leqslant k \leqslant (i-1)$]. Finally since each $\mathbf{u}_k$ in (2.23) is a linear combination of $\mathbf{w}_1, \mathbf{w}_2, \ldots, \mathbf{w}_k$, we see that $\mathbf{u}_i$ is a nontrivial linear combination of $\mathbf{w}_1, \mathbf{w}_2, \ldots, \mathbf{w}_i$; and therefore $\mathbf{u}_i$ is nonzero.

All that remains to be proved is that the vectors generated by (2.23) are orthogonal. Clearly $\mathbf{u}_1^T\mathbf{u}_2 = 0$. Proceeding inductively again, suppose $\mathbf{u}_j^T\mathbf{u}_k = 0$ for any j and k where $j \neq k$ and $1 \leqslant j, k \leqslant i-1$. From (2.23) we have

$$\mathbf{u}_j^T\mathbf{u}_i = \mathbf{u}_j^T\left(\mathbf{w}_i - \sum_{k=1}^{i-1}\frac{\mathbf{u}_k^T\mathbf{w}_i}{\mathbf{u}_k^T\mathbf{u}_k}\mathbf{u}_k\right) = \mathbf{u}_j^T\mathbf{w}_i - \sum_{k=1}^{i-1}\left(\frac{\mathbf{u}_k^T\mathbf{w}_i}{\mathbf{u}_k^T\mathbf{u}_k}\right)(\mathbf{u}_j^T\mathbf{u}_k)$$

$$= \mathbf{u}_j^T\mathbf{w}_i - \left(\frac{\mathbf{u}_j^T\mathbf{w}_i}{\mathbf{u}_j^T\mathbf{u}_j}\right)(\mathbf{u}_j^T\mathbf{u}_j) = 0.$$

Thus $\mathbf{u}_i$ is orthogonal to $\mathbf{u}_j$ for $1 \leqslant j \leqslant i-1$. Having this result, we have shown that $\{\mathbf{u}_1, \mathbf{u}_2, \ldots, \mathbf{u}_p\}$ is an orthogonal set of p nonzero vectors. So by Theorem 2.4 the vectors $\mathbf{u}_1, \mathbf{u}_2, \ldots, \mathbf{u}_p$ generated by (2.23) are an orthogonal basis for W. ∎

In order to use the Gram–Schmidt orthogonalization process to find an orthogonal basis for W, we need some basis for W as a starting point. In many of the applications that require an orthogonal basis for a subspace W, it is relatively easy to produce this initial basis—we shall give some examples in the next section. Then given a basis for W, the Gram–Schmidt process proceeds in a mechanical fashion according to (2.23). Finally in conjunction with Exercise 7, Section 2.4, Theorem 2.6 guarantees us that every subspace of R^n has an orthogonal basis.

EXAMPLE 2.18 The subspace W in Example 2.6 had a basis $\{\mathbf{w}_1, \mathbf{w}_2\}$ where

$$\mathbf{w}_1 = \begin{bmatrix} 1 \\ 1 \\ 0 \end{bmatrix}, \quad \mathbf{w}_2 = \begin{bmatrix} -1 \\ 0 \\ 1 \end{bmatrix}.$$

using Gram-Schmidt

We can generate an orthogonal basis of two vectors for W using $\{\mathbf{w}_1, \mathbf{w}_2\}$ as a starting point. In particular we set $\mathbf{u}_1 = \mathbf{w}_1$ and use (2.23) to define $\mathbf{u}_2$:

$$\mathbf{u}_2 = \mathbf{w}_2 - \frac{\mathbf{u}_1^T\mathbf{w}_2}{\mathbf{u}_1^T\mathbf{u}_1}\mathbf{u}_1.$$

From this equation we obtain

$$\mathbf{u}_2 = \begin{bmatrix} -1 \\ 0 \\ 1 \end{bmatrix} - \frac{-1}{2}\begin{bmatrix} 1 \\ 1 \\ 0 \end{bmatrix} = \begin{bmatrix} -1 \\ 0 \\ 1 \end{bmatrix} + \begin{bmatrix} \frac{1}{2} \\ \frac{1}{2} \\ 0 \end{bmatrix} = \begin{bmatrix} -\frac{1}{2} \\ \frac{1}{2} \\ 1 \end{bmatrix}.$$

Clearly $\mathbf{u}_1^{\mathrm{T}}\mathbf{u}_2 = 0$, and thus $\{\mathbf{u}_1, \mathbf{u}_2\}$ is an orthogonal basis for W.

For convenience in hand calculations, we can always eliminate fractional components in a set of orthogonal vectors. Specifically if $\mathbf{x}$ and $\mathbf{y}$ are orthogonal, then so are $a\mathbf{x}$ and $\mathbf{y}$ for any scalar a:

$$\text{if } \mathbf{x}^{\mathrm{T}}\mathbf{y} = 0, \text{ then } (a\mathbf{x})^{\mathrm{T}}\mathbf{y} = a(\mathbf{x}^{\mathrm{T}}\mathbf{y}) = 0.$$

Having made this observation, we see that $\{\boldsymbol{v}_1, \boldsymbol{v}_2\}$ is also an orthogonal basis for W where $\boldsymbol{v}_1 = \mathbf{u}_1$, $\boldsymbol{v}_2 = 2\mathbf{u}_2$:

$$\boldsymbol{v}_1 = \begin{bmatrix} 1 \\ 1 \\ 0 \end{bmatrix}, \qquad \boldsymbol{v}_2 = \begin{bmatrix} -1 \\ 1 \\ 2 \end{bmatrix}.$$

EXAMPLE 2.19 As an example to illustrate further the mechanics of Gram–Schmidt orthogonalization, consider the set $\{\mathbf{w}_1, \mathbf{w}_2, \mathbf{w}_3\}$, which is a linearly independent set in R^4 where

$$\mathbf{w}_1 = \begin{bmatrix} 0 \\ 1 \\ 2 \\ 1 \end{bmatrix}, \qquad \mathbf{w}_2 = \begin{bmatrix} 0 \\ 1 \\ 3 \\ 1 \end{bmatrix}, \qquad \mathbf{w}_3 = \begin{bmatrix} 1 \\ 1 \\ 1 \\ 0 \end{bmatrix}.$$

To generate an orthogonal basis $\{\mathbf{u}_1, \mathbf{u}_2, \mathbf{u}_3\}$ from $\{\mathbf{w}_1, \mathbf{w}_2, \mathbf{w}_3\}$, we first set $\mathbf{u}_1 = \mathbf{w}_1$ and then calculate $\mathbf{u}_2$ from

$$\mathbf{u}_2 = \mathbf{w}_2 - \frac{\mathbf{u}_1^{\mathrm{T}}\mathbf{w}_2}{\mathbf{u}_1^{\mathrm{T}}\mathbf{u}_1}\mathbf{u}_1.$$

Thus

$$\mathbf{u}_2 = \begin{bmatrix} 0 \\ 1 \\ 3 \\ 1 \end{bmatrix} - \frac{8}{6}\begin{bmatrix} 0 \\ 1 \\ 2 \\ 1 \end{bmatrix} = \begin{bmatrix} 0 \\ -\frac{1}{3} \\ \frac{1}{3} \\ -\frac{1}{3} \end{bmatrix}.$$

Next

$$\mathbf{u}_3 = \mathbf{w}_3 - \frac{\mathbf{u}_1^{\mathrm{T}}\mathbf{w}_3}{\mathbf{u}_1^{\mathrm{T}}\mathbf{u}_1}\mathbf{u}_1 - \frac{\mathbf{u}_2^{\mathrm{T}}\mathbf{w}_3}{\mathbf{u}_2^{\mathrm{T}}\mathbf{u}_2}\mathbf{u}_2,$$

or

$$\mathbf{u}_3 = \begin{bmatrix} 1 \\ 1 \\ 1 \\ 0 \end{bmatrix} - \frac{3}{6}\begin{bmatrix} 0 \\ 1 \\ 2 \\ 1 \end{bmatrix} - \frac{0}{1/3}\begin{bmatrix} 0 \\ -\frac{1}{3} \\ \frac{1}{3} \\ -\frac{1}{3} \end{bmatrix} = \begin{bmatrix} 1 \\ \frac{1}{2} \\ 0 \\ -\frac{1}{2} \end{bmatrix}.$$

For convenience we can eliminate the fractional components in $\mathbf{u}_2$ and $\mathbf{u}_3$ as in Example 2.18 and obtain an orthogonal basis $\{\boldsymbol{v}_1, \boldsymbol{v}_2, \boldsymbol{v}_3\}$ where

$$\boldsymbol{v}_1 = \begin{bmatrix} 0 \\ 1 \\ 2 \\ 1 \end{bmatrix}, \quad \boldsymbol{v}_2 = \begin{bmatrix} 0 \\ -1 \\ 1 \\ -1 \end{bmatrix}, \quad \boldsymbol{v}_3 = \begin{bmatrix} 2 \\ 1 \\ 0 \\ -1 \end{bmatrix}.$$

2.5 EXERCISES

1. Verify that the set $S = \{\mathbf{u}_1, \mathbf{u}_2, \mathbf{u}_3\}$ is an orthogonal set of vectors where

$$\mathbf{u}_1 = \begin{bmatrix} 1 \\ 1 \\ 1 \end{bmatrix} \quad \mathbf{u}_2 = \begin{bmatrix} -1 \\ 0 \\ 1 \end{bmatrix} \quad \mathbf{u}_3 = \begin{bmatrix} -1 \\ 2 \\ -1 \end{bmatrix}.$$

By Exercise 3, S is an orthogonal basis for R^3. Use (2.22) in Theorem 2.5 to express the vectors below in terms of S.

$$\mathbf{a} = \begin{bmatrix} 1 \\ 1 \\ 0 \end{bmatrix} \quad \mathbf{b} = \begin{bmatrix} 0 \\ 1 \\ 2 \end{bmatrix} \quad \mathbf{c} = \begin{bmatrix} 3 \\ 3 \\ 3 \end{bmatrix} \quad \mathbf{d} = \begin{bmatrix} 1 \\ 2 \\ 1 \end{bmatrix}$$

2. Show that the set of vectors $\{\mathbf{u}_1, \mathbf{u}_2, \mathbf{u}_3\}$ given in Example 2.16 is a basis for R^3. Express the vectors **a**, **b**, **c**, **d** of Exercise 1 in terms of this basis, using (2.22) in Theorem 2.5.

3. Let $S = \{\boldsymbol{v}_1, \boldsymbol{v}_2, \boldsymbol{v}_3\}$ be any orthogonal set of nonzero vectors in R^3. Prove that S is a linearly independent set of vectors. (Hint: Consider $a_1\boldsymbol{v}_1 + a_2\boldsymbol{v}_2 + a_3\boldsymbol{v}_3 = \boldsymbol{\theta}$, and multiply both sides by $\boldsymbol{v}_1^T$.) Knowing that S is linearly independent, cite a theorem that shows that S is a basis for R^3.

4. Let $S = \{\mathbf{u}_1, \mathbf{u}_2, \ldots, \mathbf{u}_p\}$ be an orthogonal set of nonzero vectors in a p-dimensional subspace W. Prove that S is a basis for W.

5. Use the Gram–Schmidt process (Theorem 2.6) to generate an orthogonal set from $S = \{\mathbf{w}_1, \mathbf{w}_2, \mathbf{w}_3\}$ when S consists of the following.

$$\text{a)} \begin{bmatrix} 0 \\ 0 \\ 1 \\ 0 \end{bmatrix}, \begin{bmatrix} 1 \\ 1 \\ 2 \\ 1 \end{bmatrix}, \begin{bmatrix} 1 \\ 0 \\ 1 \\ 1 \end{bmatrix} \qquad \text{(b)} \begin{bmatrix} 1 \\ 0 \\ 1 \\ 2 \end{bmatrix}, \begin{bmatrix} 2 \\ 1 \\ 0 \\ 2 \end{bmatrix}, \begin{bmatrix} 1 \\ -1 \\ 0 \\ 1 \end{bmatrix}$$

6. Repeat Exercise 5 for these vectors.

$$\text{a)}\ \begin{bmatrix}0\\1\\0\\1\end{bmatrix}, \begin{bmatrix}1\\2\\0\\0\end{bmatrix}, \begin{bmatrix}0\\2\\1\\0\end{bmatrix} \qquad \text{b)}\ \begin{bmatrix}1\\1\\0\\2\end{bmatrix}, \begin{bmatrix}0\\2\\1\\2\end{bmatrix}, \begin{bmatrix}0\\1\\0\\2\end{bmatrix}.$$

7. Find a basis for the null space and the range of the matrices below. Then use the Gram–Schmidt process to obtain an orthogonal basis.

$$A = \begin{bmatrix}1 & -2 & 1 & -5\\ 2 & 1 & 7 & 5\\ 1 & -1 & 2 & -2\end{bmatrix} \qquad B = \begin{bmatrix}1 & 3 & 10 & 11 & 9\\ -1 & 2 & 5 & 4 & 1\\ 2 & -1 & -1 & 1 & 4\end{bmatrix}$$

8. Let $S = \{\mathbf{u}_1, \mathbf{u}_2, \ldots, \mathbf{u}_p\}$ be an orthogonal basis for a p-dimensional subspace W. Let $a_i = 1/\|\mathbf{u}_i\|$ for $1 \leqslant i \leqslant p$; and let $B = \{\boldsymbol{v}_1, \boldsymbol{v}_2, \ldots, \boldsymbol{v}_p\}$ where $\boldsymbol{v}_i = a_i\mathbf{u}_i$, $1 \leqslant i \leqslant p$. Show that B is an orthogonal basis for W; and also show that $\boldsymbol{v}_i^{\mathrm{T}}\boldsymbol{v}_i = 1$, $1 \leqslant i \leqslant p$. (Such a basis is called an ***orthonormal*** basis.) How does Theorem 2.5 simplify when B is an orthonormal basis?

9. Use Exercise 8 to convert the orthogonal bases in Exercises 1 and 2 to orthonormal bases.

10. Let W be a p-dimensional subspace of R^n where $p \geqslant 1$. If $\boldsymbol{v}$ is a vector in W such that $\boldsymbol{v}^{\mathrm{T}}\mathbf{w} = 0$ for every $\mathbf{w}$ in W, show that $\boldsymbol{v}$ must be the zero vector. (Hint: Begin with Exercise 7, Section 2.4, and Theorem 2.6 to note that W has an orthogonal basis.)

11. (The Cauchy–Schwarz inequality) Let $\mathbf{x}$ and $\mathbf{y}$ be vectors in R^n. Prove that $|\mathbf{x}^{\mathrm{T}}\mathbf{y}| \leqslant \|\mathbf{x}\|\,\|\mathbf{y}\|$. (Hint: Observe that $\|\mathbf{x} - c\mathbf{y}\|^2 \geqslant \theta$ for any scalar c. If $\mathbf{y} \neq \boldsymbol{\theta}$, let $c = \mathbf{x}^{\mathrm{T}}\mathbf{y}/\mathbf{y}^{\mathrm{T}}\mathbf{y}$; and also treat the case that $\mathbf{y} = \boldsymbol{\theta}$.)

12. (The triangle inequality) Let $\mathbf{x}$ and $\mathbf{y}$ be vectors in R^n. Prove that $\|\mathbf{x}+\mathbf{y}\| \leqslant \|\mathbf{x}\| + \|\mathbf{y}\|$. (Hint: Expand $\|\mathbf{x}+\mathbf{y}\|^2$ and use Exercise 11.)

13. Let $\mathbf{x}$ and $\mathbf{y}$ be vectors in R^n. Prove that $\big|\,\|\mathbf{x}\| - \|\mathbf{y}\|\,\big| \leqslant \|\mathbf{x}-\mathbf{y}\|$. (Hint: For one part consider $\|\mathbf{x} + (\mathbf{y}-\mathbf{x})\|$ and Exercise 12.)

14. If the hypotheses for Theorem 2.6 were altered so that $\{\mathbf{w}_i\}_{i=1}^{p-1}$ is linearly independent and $\{\mathbf{w}\}_{i=1}^{p}$ is linearly dependent, use Exercise 10 to show that (2.23) yields $\mathbf{u}_p = \boldsymbol{\theta}$.

†2.6 PROJECTIONS, LEAST-SQUARES SOLUTIONS OF INCONSISTENT SYSTEMS, AND APPLICATIONS TO DATA FITTING

Orthogonal bases have quite a number of applications to both theoretical and computational problems. In this section we would like to focus on one such application, the "nearest point" problem of finding in a subspace W a vector $\mathbf{w}^*$ that is nearest to a vector $\boldsymbol{v}$ that is not in W. As a very simple example, let W be the subspace of R^3 defined by

$$W = \{\mathbf{x} : \mathbf{x} = \begin{bmatrix}x_1\\x_2\\0\end{bmatrix},\ x_1 \text{ and } x_2 \text{ any real numbers}\};$$

and let $\boldsymbol{v}$ be the vector

$$\boldsymbol{v} = \begin{bmatrix} 2 \\ 3 \\ 2 \end{bmatrix}.$$

We can visualize W as the xy-plane in three-space as is depicted in Fig. 2.4; and we can argue from geometric principles that the vector $\mathbf{w}^*$, given by

$$\mathbf{w}^* = \begin{bmatrix} 2 \\ 3 \\ 0 \end{bmatrix},$$

is the nearest vector in W to $\boldsymbol{v}$. [From geometry, we know that $\mathbf{w}^*$ is the nearest vector to $\boldsymbol{v}$ if the vector $\boldsymbol{v} - \mathbf{w}^*$ is normal (perpendicular) to the subspace W. We show below that this principle generalizes to the situation in which W is any subspace of R^n.]

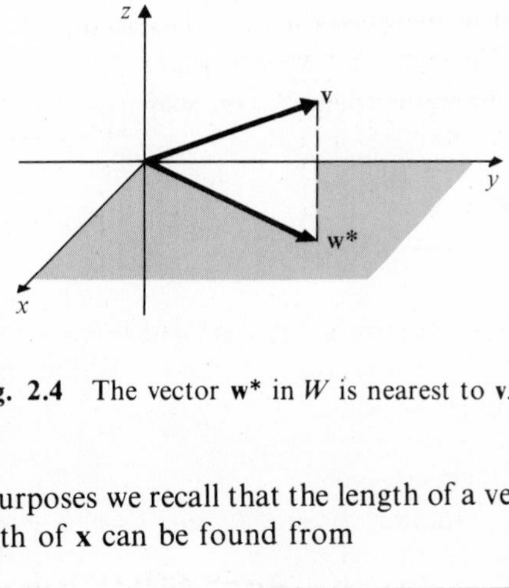

Fig. 2.4 The vector $\mathbf{w}^*$ in W is nearest to $\mathbf{v}$.

For notational purposes we recall that the length of a vector $\mathbf{x}$ in R^n is denoted by $\|\mathbf{x}\|$ and the length of $\mathbf{x}$ can be found from

$$\|\mathbf{x}\| = \sqrt{\mathbf{x}^{\mathrm{T}}\mathbf{x}} = \sqrt{x_1^2 + x_2^2 + \ldots + x_n^2}.$$

Furthermore the distance between two vectors $\mathbf{x}$ and $\mathbf{y}$ is defined to be $\|\mathbf{x} - \mathbf{y}\|$:

$$\|\mathbf{x} - \mathbf{y}\| = \sqrt{(\mathbf{x} - \mathbf{y})^{\mathrm{T}}(\mathbf{x} - \mathbf{y})}$$

(the distance between $\mathbf{x}$ and $\mathbf{y}$ is the length of the vector $\mathbf{x} - \mathbf{y}$). The problem we wish to consider is the following:

Given a subspace W of R^n, and given a vector $\boldsymbol{v}$ in R^n, find a vector $\mathbf{w}^*$ in W such that

$$\|\boldsymbol{v} - \mathbf{w}^*\| \leqslant \|\boldsymbol{v} - \mathbf{w}\| \text{ for all } \mathbf{w} \text{ in } W. \tag{2.24}$$

Thus (2.24) is the mathematical statement of the problem of finding a vector in W that is nearest to a given vector $\mathbf{v}$. A number of problems can be put in the form of (2.24), and a simple solution can be found using an orthogonal basis for W. The next theorem provides the means for solving this problem and is suggested by Fig. 2.4: if $\mathbf{w}^*$ is the nearest vector to $\mathbf{v}$, we expect that $\mathbf{v}-\mathbf{w}^*$ is "normal" to the subspace W. Before stating the next theorem, we note that $\mathbf{w}^*$ is usually called the ***projection*** of $\mathbf{v}$ onto W. (We note also that typically $\mathbf{v}$ is not in W although Theorem 2.7 below does not insist upon this. If $\mathbf{v}$ is in W, then obviously $\mathbf{w}^* = \mathbf{v}$ is the nearest vector in W to $\mathbf{v}$; the projection of $\mathbf{v}$ is $\mathbf{v}$.)

The observation above suggests a way to calculate $\mathbf{w}^*$. That is, if $\mathbf{v}-\mathbf{w}^*$ is orthogonal to every vector in W, then $\mathbf{v}-\mathbf{w}^*$ must be orthogonal to every vector in a basis for W. In particular if $\{\mathbf{u}_1, \mathbf{u}_2, \ldots, \mathbf{u}_p\}$ is a basis for W, then the p constraints

$$\begin{aligned} (\mathbf{v}-\mathbf{w}^*)^{\mathrm{T}}\mathbf{u}_1 &= 0 \\ (\mathbf{v}-\mathbf{w}^*)^{\mathrm{T}}\mathbf{u}_2 &= 0 \\ &\vdots \\ (\mathbf{v}-\mathbf{w}^*)^{\mathrm{T}}\mathbf{u}_p &= 0 \end{aligned} \tag{2.25}$$

should be sufficient to determine $\mathbf{w}^*$. Next we can obtain a considerable simplification of (2.25) if we assume $\{\mathbf{u}_1, \mathbf{u}_2, \ldots, \mathbf{u}_p\}$ is an orthogonal basis for W. That is, we can express $\mathbf{w}^*$ in terms of an orthogonal basis as

$$\mathbf{w}^* = a_1\mathbf{u}_1 + a_2\mathbf{u}_2 + \ldots + a_p\mathbf{u}_p$$

where $a_i = \mathbf{u}_i^{\mathrm{T}}\mathbf{w}^*/\mathbf{u}_i^{\mathrm{T}}\mathbf{u}_i$. Now the ith equation of (2.25) is equivalent to

$$\mathbf{u}_i^{\mathrm{T}}\mathbf{w}^* = \mathbf{u}_i^{\mathrm{T}}\mathbf{v};$$

so if we choose $a_i = \mathbf{u}_i^{\mathrm{T}}\mathbf{v}/\mathbf{u}_i^{\mathrm{T}}\mathbf{u}_i$, then $(\mathbf{v}-\mathbf{w}^*)^{\mathrm{T}}\mathbf{u}_i = 0$, $1 \leqslant i \leqslant p$. This result means we can always calculate a vector $\mathbf{w}^*$ that satisfies (2.25), and all that remains to be shown is that $\mathbf{w}^*$ is the nearest vector in W to $\mathbf{v}$.

Theorem 2.7

Let W be a p-dimensional subspace of R^n, and let $\mathbf{v}$ be a vector in R^n. Then there is a unique vector $\mathbf{w}^*$ in W such that

$$\|\mathbf{v}-\mathbf{w}^*\| \leqslant \|\mathbf{v}-\mathbf{w}\| \text{ for all } \mathbf{w} \text{ in } W.$$

Furthermore if $\{\mathbf{u}_1, \mathbf{u}_2, \ldots, \mathbf{u}_p\}$ is an orthogonal basis for W, then $\mathbf{w}^*$ is given by

$$\mathbf{w}^* = \frac{\mathbf{v}^{\mathrm{T}}\mathbf{u}_1}{\mathbf{u}_1^{\mathrm{T}}\mathbf{u}_1}\mathbf{u}_1 + \frac{\mathbf{v}^{\mathrm{T}}\mathbf{u}_2}{\mathbf{u}_2^{\mathrm{T}}\mathbf{u}_2}\mathbf{u}_2 + \ldots + \frac{\mathbf{v}^{\mathrm{T}}\mathbf{u}_p}{\mathbf{u}_p^{\mathrm{T}}\mathbf{u}_p}\mathbf{u}_p. \tag{2.26}$$

Proof. By Theorem 2.5 we know that the vector $\mathbf{w}^*$ defined by (2.26) can also be expressed as

$$\mathbf{w}^* = \frac{\mathbf{u}_1^{\mathrm{T}}\mathbf{w}^*}{\mathbf{u}_1^{\mathrm{T}}\mathbf{u}_1}\mathbf{u}_1 + \frac{\mathbf{u}_2^{\mathrm{T}}\mathbf{w}^*}{\mathbf{u}_2^{\mathrm{T}}\mathbf{u}_2}\mathbf{u}_2 + \ldots + \frac{\mathbf{u}_p^{\mathrm{T}}\mathbf{w}^*}{\mathbf{u}_p^{\mathrm{T}}\mathbf{u}_p}\mathbf{u}_p. \tag{2.27}$$

By Exercise 12, Section 2.4, we know that the coordinate of $\mathbf{w}^*$ with respect to a basis for W are unique; so we find from (2.26) and (2.27) that

$$\boldsymbol{v}^{\mathrm{T}}\mathbf{u}_i = \mathbf{u}_i^{\mathrm{T}}\mathbf{w}^*, \ i = 1, 2, \ldots, p.$$

From this equation it follows that

$$(\boldsymbol{v} - \mathbf{w}^*)^{\mathrm{T}}\mathbf{u}_i = 0, \ i = 1, 2, \ldots, p.$$

Since every vector $\mathbf{w}$ in W is a linear combination of $\mathbf{u}_1, \mathbf{u}_2, \ldots, \mathbf{u}_p$, we know also that $\mathbf{w}^*$ as given by (2.26) satisfies the condition

$$(\boldsymbol{v} - \mathbf{w}^*)^{\mathrm{T}}\mathbf{w} = 0 \text{ for all } \mathbf{w} \text{ in } W.$$

We next show that $(\boldsymbol{v} - \mathbf{w}^*)^{\mathrm{T}}\mathbf{w} = 0$ for all $\mathbf{w}$ in W implies that $\|\boldsymbol{v} - \mathbf{w}^*\|^2 \leqslant \|\boldsymbol{v} - \mathbf{w}\|^2$ for all $\mathbf{w}$ in W (this is sufficient to prove Theorem 2.7 since minimizing $\|\boldsymbol{v} - \mathbf{w}\|^2$ is the same as minimizing $\|\boldsymbol{v} - \mathbf{w}\|$). Now as is suggested by Fig. 2.5, we write

$$\begin{aligned}\|\boldsymbol{v} - \mathbf{w}\|^2 &= \|(\boldsymbol{v} - \mathbf{w}^*) - (\mathbf{w} - \mathbf{w}^*)\|^2 \\ &= (\boldsymbol{v} - \mathbf{w}^*)^{\mathrm{T}}(\boldsymbol{v} - \mathbf{w}^*) - 2(\boldsymbol{v} - \mathbf{w}^*)^{\mathrm{T}}(\mathbf{w} - \mathbf{w}^*) + (\mathbf{w} - \mathbf{w}^*)^{\mathrm{T}}(\mathbf{w} - \mathbf{w}^*). \quad (2.28\text{a})\end{aligned}$$

We next observe that since W is a subspace, $\mathbf{w} - \mathbf{w}^*$ is in W. Furthermore we have shown above that $\boldsymbol{v} - \mathbf{w}^*$ is orthogonal to every vector in W; so $(\boldsymbol{v} - \mathbf{w}^*)^{\mathrm{T}}(\mathbf{w} - \mathbf{w}^*) = 0$. Using this fact reduces (2.28a) to

$$\|\boldsymbol{v} - \mathbf{w}\|^2 = \|\boldsymbol{v} - \mathbf{w}^*\|^2 + \|\mathbf{w} - \mathbf{w}^*\|^2. \quad (2.28\text{b})$$

Since $\|\mathbf{w} - \mathbf{w}^*\|^2 \geqslant 0$, we have that $\|\boldsymbol{v} - \mathbf{w}\|^2 \geqslant \|\boldsymbol{v} - \mathbf{w}^*\|^2$; and in fact $\|\boldsymbol{v} - \mathbf{w}\|^2 > \|\boldsymbol{v} - \mathbf{w}^*\|^2$ unless $\mathbf{w} = \mathbf{w}^*$. Therefore $\mathbf{w}^*$ as given by (2.26) is nearer to $\boldsymbol{v}$ than any other vector $\mathbf{w}$ in W is. ■

Figure 2.5 sketches schematically the proof of Theorem 2.7, which is based on (2.28b). With reference to Fig. 2.5, Eq. (2.28b) is reminiscent of the Pythagorean theorem.

Theorems 2.6 and 2.7 establish a reasonable computational procedure for finding the vector $\mathbf{w}^*$ in W that is nearest to $\boldsymbol{v}$.

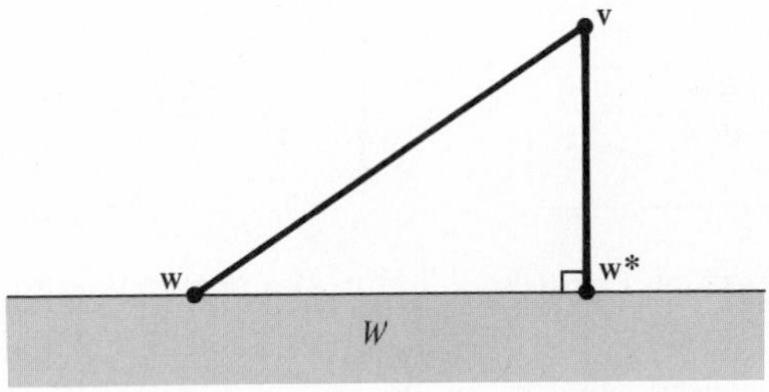

Fig. 2.5 A geometric interpretation of the vector $\mathbf{w}^*$ in W closest to $\mathbf{v}$.

1. Generate an orthogonal basis for W from (2.23).
2. Calculate $\mathbf{w}^*$ from (2.26).

EXAMPLE 2.20 In Fig. 2.4 we indicated that $\mathbf{w}^*$ was the closest vector in W to $\boldsymbol{v}$ where

$$\mathbf{w}^* = \begin{bmatrix} 2 \\ 3 \\ 0 \end{bmatrix}, \qquad \boldsymbol{v} = \begin{bmatrix} 2 \\ 3 \\ 2 \end{bmatrix}$$

and where we visualized W as the xy-plane in three-space. To verify this assertion, we calculate $\mathbf{w}^*$ according to steps (1) and (2) of the procedure outlined above. First it is clear that $\mathbf{u}_1$ and $\mathbf{u}_2$, given by

$$\mathbf{u}_1 = \begin{bmatrix} 1 \\ 0 \\ 0 \end{bmatrix}, \qquad \mathbf{u}_2 = \begin{bmatrix} 0 \\ 1 \\ 0 \end{bmatrix},$$

constitute an orthogonal basis for W; so step (1) is not necessary in this case. Following step (2), we find $\mathbf{w}^*$ in W from (2.26) and obtain

$$\mathbf{w}^* = \frac{\boldsymbol{v}^{\mathrm{T}}\mathbf{u}_1}{\mathbf{u}_1^{\mathrm{T}}\mathbf{u}_1}\mathbf{u}_1 + \frac{\boldsymbol{v}^{\mathrm{T}}\mathbf{u}_2}{\mathbf{u}_2^{\mathrm{T}}\mathbf{u}_2}\mathbf{u}_2 = \frac{2}{1}\begin{bmatrix} 1 \\ 0 \\ 0 \end{bmatrix} + \frac{3}{1}\begin{bmatrix} 0 \\ 1 \\ 0 \end{bmatrix} = \begin{bmatrix} 2 \\ 3 \\ 0 \end{bmatrix}.$$

EXAMPLE 2.21 In this example we return again to the subspace W in Example 2.18. Let $\mathbf{q}$ be the vector

$$\mathbf{q} = \begin{bmatrix} 5 \\ 3 \\ -5 \end{bmatrix};$$

and consider the problem of finding $\mathbf{w}^*$ in W such that $\|\mathbf{q} - \mathbf{w}^*\| \leqslant \|\mathbf{q} - \mathbf{w}\|$ for all $\mathbf{w}$ in W.

The first step in finding $\mathbf{w}^*$ is to construct an orthogonal basis for W. We have already carried out this step in Example 2.18 and found an orthogonal basis $\{\boldsymbol{v}_1, \boldsymbol{v}_2\}$

$$\boldsymbol{v}_1 = \begin{bmatrix} 1 \\ 1 \\ 0 \end{bmatrix}, \qquad \boldsymbol{v}_2 = \begin{bmatrix} -1 \\ 1 \\ 2 \end{bmatrix}.$$

Given this basis, we find $\mathbf{w}^*$ from (2.26),

$$\mathbf{w}^* = \frac{\mathbf{q}^{\mathrm{T}}\boldsymbol{v}_1}{\boldsymbol{v}_1^{\mathrm{T}}\boldsymbol{v}_1}\boldsymbol{v}_1 + \frac{\mathbf{q}^{\mathrm{T}}\boldsymbol{v}_2}{\boldsymbol{v}_2^{\mathrm{T}}\boldsymbol{v}_2}\boldsymbol{v}_2 = \frac{8}{2}\begin{bmatrix} 1 \\ 1 \\ 0 \end{bmatrix} + \frac{-12}{6}\begin{bmatrix} -1 \\ 1 \\ 2 \end{bmatrix} = \begin{bmatrix} 6 \\ 2 \\ -4 \end{bmatrix}.$$

An application of Theorem 2.7 is to the problem of finding best least-squares solutions to an inconsistent system $A\mathbf{x} = \mathbf{b}$. That is, if a system of linear equations $A\mathbf{x} = \mathbf{b}$ is inconsistent, then we might reasonably ask for a vector $\mathbf{x}^*$ such that the "residual vector," $\mathbf{r} = \mathbf{b} - A\mathbf{x}^*$, is as small as possible. Specifically we pose this problem as Problem A.

Problem A

Given an $(m \times n)$ system of linear equations $A\mathbf{x} = \mathbf{b}$, find $\mathbf{x}^*$ in R^n such that $\|\mathbf{b} - A\mathbf{x}^*\| \leqslant \|\mathbf{b} - A\mathbf{x}\|$ for all $\mathbf{x}$ in R^n.

In this problem the vector $\mathbf{x}^*$ is called a ***best least-squares solution*** of $A\mathbf{x} = \mathbf{b}$, and finding such a vector $\mathbf{x}^*$ represents a reasonable goal if $A\mathbf{x} = \mathbf{b}$ does not have a solution. (For a consistent system, of course, the residual vector is minimized by any solution of $A\mathbf{x} = \mathbf{b}$. Thus for a consistent system, best least-squares solutions coincide with solutions in the usual sense.) The term "best least-squares solution" is suggested by the fact that minimizing $\|\mathbf{b} - A\mathbf{x}\|$ is equivalent to minimizing $\|\mathbf{b} - A\mathbf{x}\|^2$ where

$$\|\mathbf{b} - A\mathbf{x}\|^2 = (\mathbf{b} - A\mathbf{x})^{\mathrm{T}}(\mathbf{b} - A\mathbf{x}). \tag{2.29}$$

In (2.29) the quantity $(\mathbf{b} - A\mathbf{x})^{\mathrm{T}}(\mathbf{b} - A\mathbf{x})$ represents the sum of the squares of the components of the residual vector.

We now observe that we already have the machinery to solve Problem A. To see this, we reformulate Problem A as Problem B.

Problem B

Let $A\mathbf{x} = \mathbf{b}$ be an $(m \times n)$ system of linear equations. Find $\mathbf{y}^*$ in $\mathscr{R}(A)$ such that $\|\mathbf{b} - \mathbf{y}^*\| \leqslant \|\mathbf{b} - \mathbf{y}\|$ for all $\mathbf{y}$ in $\mathscr{R}(A)$.

If we can solve Problem B, then we need only find a vector $\mathbf{x}^*$ in R^n such that $A\mathbf{x}^* = \mathbf{y}^*$ in order to solve Problem A. [Clearly if $\mathbf{y}^* = A\mathbf{x}^*$, and if $\|\mathbf{b} - \mathbf{y}^*\| \leqslant \|\mathbf{b} - \mathbf{y}\|$ for all $\mathbf{y}$ in $\mathscr{R}(A)$, then $\|\mathbf{b} - A\mathbf{x}^*\| \leqslant \|\mathbf{b} - \mathbf{y}\|$ for all $\mathbf{y}$ in $\mathscr{R}(A)$. But for any $\mathbf{x}$ in R^n, $A\mathbf{x}$ is some vector $\mathbf{y}$ in $\mathscr{R}(A)$; so $\|\mathbf{b} - A\mathbf{x}^*\| \leqslant \|\mathbf{b} - A\mathbf{x}\|$ for all $\mathbf{x}$ in R^n.] Now since $\mathscr{R}(A)$ is a subspace of R^m, we can solve Problem B by finding an orthogonal basis for $\mathscr{R}(A)$ and then calculating $\mathbf{y}^*$ from (2.26) in Theorem 2.7.

EXAMPLE 2.22 Consider the equation $A\mathbf{x} = \mathbf{b}$ where

$$A = \begin{bmatrix} 1 & 1 \\ 2 & 3 \\ 1 & 5 \end{bmatrix}, \quad \mathbf{x} = \begin{bmatrix} x_1 \\ x_2 \end{bmatrix}, \quad \mathbf{b} = \begin{bmatrix} -2 \\ 13 \\ 0 \end{bmatrix}.$$

The system $A\mathbf{x} = \mathbf{b}$ is inconsistent; so as the next best thing we seek a best least-squares solution of $A\mathbf{x} = \mathbf{b}$. Accordingly we solve Problem B by first finding an orthogonal basis for $\mathscr{R}(A)$ and then calculating the vector $\mathbf{y}^*$ in $\mathscr{R}(A)$ that is nearest $\mathbf{b}$.

It is easy to verify that the column vectors of A are linearly independent; so a basis for $\mathscr{R}(A)$ is $\{\mathbf{A}_1, \mathbf{A}_2\}$. We next use the Gram–Schmidt process to generate an orthogonal basis $\{\mathbf{u}_1, \mathbf{u}_2\}$ for $\mathscr{R}(A)$ and obtain $\mathbf{u}_1 = \mathbf{A}_1$ and

$$\mathbf{u}_2 = \begin{bmatrix} 1 \\ 3 \\ 5 \end{bmatrix} - \frac{12}{6}\begin{bmatrix} 1 \\ 2 \\ 1 \end{bmatrix} = \begin{bmatrix} -1 \\ -1 \\ 3 \end{bmatrix}.$$

Having an orthogonal basis for $\mathscr{R}(A)$, we use (2.26) of Theorem 2.7 to calculate $\mathbf{y}^*$:

$$\mathbf{y}^* = \frac{\mathbf{b}^T\mathbf{u}_1}{\mathbf{u}_1^T\mathbf{u}_1}\mathbf{u}_1 + \frac{\mathbf{b}^T\mathbf{u}_2}{\mathbf{u}_2^T\mathbf{u}_2}\mathbf{u}_2 = 4\begin{bmatrix}1\\2\\1\end{bmatrix} - \begin{bmatrix}-1\\-1\\3\end{bmatrix} = \begin{bmatrix}5\\9\\1\end{bmatrix}.$$

Thus among all possible vectors of the form $A\mathbf{x}$, the vector $\mathbf{y}^*$ is the one that is nearest to $\mathbf{b}$. To complete the solution of Problem A, we need to solve $A\mathbf{x} = \mathbf{y}^*$:

$$\begin{aligned} x_1 + x_2 &= 5 \\ 2x_1 + 3x_2 &= 9 \\ x_1 + 5x_2 &= 1; \end{aligned}$$

and we routinely solve this to obtain

$$\mathbf{x}^* = \begin{bmatrix}6\\-1\end{bmatrix}.$$

In summary as the last example illustrates, we find a best least-squares solution to $A\mathbf{x} = \mathbf{b}$ by following the steps.

1. Find an orthogonal basis for $\mathscr{R}(A)$.
2. Using (2.26), calculate $\mathbf{y}^*$ in $\mathscr{R}(A)$ that minimizes $\|\mathbf{b} - \mathbf{y}\|$.
3. Solve $A\mathbf{x} = \mathbf{y}^*$ to obtain the best least-squares solutions of $A\mathbf{x} = \mathbf{b}$.

We wish to make several comments about the procedure above. First of all, the vector $\mathbf{y}^*$ found in step (2) is always unique by Theorem 2.7. (The existence of $\mathbf{y}^*$ is also guaranteed by Theorem 2.7.) Since $\mathbf{y}^*$ is in $\mathscr{R}(A)$, there is always at least one solution to $A\mathbf{x} = \mathbf{y}^*$; so $A\mathbf{x} = \mathbf{b}$ always has at least one best least-squares solution. However as we have seen, solutions of $A\mathbf{x} = \mathbf{y}^*$ are unique if and only if the column vectors of A are linearly independent. Therefore while there is always at least one vector $\mathbf{x}^*$ that minimizes $\|\mathbf{b} - A\mathbf{x}\|$, there may be a number of minimizing vectors. [Of course if $A\mathbf{x}^* = \mathbf{y}^*$, and if $\mathbf{u}$ is any vector in $\mathscr{N}(A)$, then we also have $A(\mathbf{x}^* + \mathbf{u}) = \mathbf{y}^*$. Thus $\mathbf{x}^* + \mathbf{u}$ is another minimizing vector.]

The last result of this section is a theorem that gives a simple alternative characterization of the best least-squares solution, $\mathbf{x}^*$, of $A\mathbf{x} = \mathbf{b}$. In principle this method allows us to calculate $\mathbf{x}^*$ directly without having to use Gram–Schmidt orthogonalization.

Theorem 2.8

The vector $\mathbf{x}^*$ is a best least-squares solution of $A\mathbf{x} = \mathbf{b}$ if and only if $\mathbf{x}^*$ is a solution of

$$A^T A\mathbf{x} = A^T\mathbf{b}. \tag{2.30}$$

Proof. We first note that if $\mathbf{y}^*$ is the nearest vector in $\mathscr{R}(A)$ to $\mathbf{b}$, then $\mathbf{y}^*$ is given by (2.26); and we know that $(\mathbf{b} - \mathbf{y}^*)^T\mathbf{y} = 0$ for all $\mathbf{y}$ in $\mathscr{R}(A)$. Furthermore as is shown in the proof of Theorem 2.7, if $\hat{\mathbf{y}}$ is any vector in $\mathscr{R}(A)$ such that $(\mathbf{b}-\hat{\mathbf{y}})^{\mathrm{T}}\mathbf{y}=0$ for all $\mathbf{y}$ in $\mathscr{R}(A)$, then $\hat{\mathbf{y}}=\mathbf{y}^*$. Thus we can conclude that $\mathbf{y}^*$ is the nearest vector in $\mathscr{R}(A)$ to $\mathbf{b}$ if and only if

$$(\mathbf{b}-\mathbf{y}^*)^{\mathrm{T}}\mathbf{y}=0 \tag{2.31a}$$

for all $\mathbf{y}$ in $\mathscr{R}(A)$. Now we assume A is an $(m \times n)$ matrix and write A in column form as $A=[\mathbf{A}_1, \mathbf{A}_2, \ldots, \mathbf{A}_n]$. Since $\mathbf{y}$ is in $\mathscr{R}(A)$ if and only if $\mathbf{y}$ can be expressed as a linear combination of the columns of A, it follows from (2.31a) that $\mathbf{y}^*$ is the nearest vector in $\mathscr{R}(A)$ to $\mathbf{b}$ if and only if $\mathbf{y}^*$ satisfies the n constraints

$$\begin{aligned}(\mathbf{b}-\mathbf{y}^*)^{\mathrm{T}}\mathbf{A}_1&=0\\(\mathbf{b}-\mathbf{y}^*)^{\mathrm{T}}\mathbf{A}_2&=0\\&\vdots\\(\mathbf{b}-\mathbf{y}^*)^{\mathrm{T}}\mathbf{A}_n&=0.\end{aligned} \tag{2.31b}$$

Now (2.31b) is equivalent to $(\mathbf{b}-\mathbf{y}^*)^{\mathrm{T}}A=\boldsymbol{\theta}^{\mathrm{T}}$, or

$$A^{\mathrm{T}}(\mathbf{b}-\mathbf{y}^*)=\boldsymbol{\theta}. \tag{2.31c}$$

If $\mathbf{x}^*$ is any vector in R^n such that $A\mathbf{x}^*=\mathbf{y}^*$, then $\mathbf{x}^*$ satisfies $A^{\mathrm{T}}(\mathbf{b}-A\mathbf{x}^*)=\boldsymbol{\theta}$; and therefore $\mathbf{x}^*$ is a solution of (2.30). So any best least-squares solution of $A\mathbf{x}=\mathbf{b}$ is a solution of (2.30). Conversely if $\mathbf{x}^*$ solves (2.30), then $\mathbf{y}^*=A\mathbf{x}^*$ satisfies (2.31c), which result via (2.31b) means that $\mathbf{x}^*$ is a best least-squares solution of $A\mathbf{x}=\mathbf{b}$. ■

If A is an $(m \times n)$ matrix, then A^{T} is an $(n \times m)$ matrix; so the system $A^{\mathrm{T}}A\mathbf{x}=A^{\mathrm{T}}\mathbf{b}$ is an $(n \times n)$ system, usually called the "normal equations." Moreover as we remarked earlier, best least-squares solutions to $A\mathbf{x}=\mathbf{b}$ always exist; so the system $A^{\mathrm{T}}A\mathbf{x}=A^{\mathrm{T}}\mathbf{b}$ is always consistent. Solving (2.30) is equivalent to finding best least-squares solutions to $A\mathbf{x}=\mathbf{b}$, and (2.30) is particularly convenient for small hand calculations. However it is not usually advisable to use (2.30) in computer applications since experience has shown that the matrix $A^{\mathrm{T}}A$ may be ill-conditioned (small computational errors introduced by roundoff can lead to large errors in the final answer).

EXAMPLE 2.23 In this example we rework Example 2.22 and find the best least-squares solutions to $A\mathbf{x}=\mathbf{b}$ by solving (2.30). To use (2.30), we form $A^{\mathrm{T}}A$ and $A^{\mathrm{T}}\mathbf{b}$ and obtain

$$A^{\mathrm{T}}A=\begin{bmatrix}1&2&1\\1&3&5\end{bmatrix}\begin{bmatrix}1&1\\2&3\\1&5\end{bmatrix}=\begin{bmatrix}6&12\\12&35\end{bmatrix},$$

$$A^{\mathrm{T}}\mathbf{b}=\begin{bmatrix}1&2&1\\1&3&5\end{bmatrix}\begin{bmatrix}-2\\13\\0\end{bmatrix}=\begin{bmatrix}24\\37\end{bmatrix}.$$

According to Theorem 2.8, best least-squares solutions to $A\mathbf{x}=\mathbf{b}$ are the solutions of

$$\begin{aligned}6x_1+12x_2&=24\\12x_1+35x_2&=37;\end{aligned}$$

and we quickly find as in Example 2.22, that the only solution of $A^T A\mathbf{x} = A^T\mathbf{b}$ is $x_1 = 6$, $x_2 = -1$.

Problems requiring least-squares solutions to a linear system $A\mathbf{x} = \mathbf{b}$ occur in a number of applications, perhaps most familiarly in finding best least-squares fits to data. We consider such an example below and show how best to fit a linear expression to observations that are not necessarily distributed linearly—in statistical applications, this procedure is usually called linear regression.

EXAMPLE 2.24 Suppose we are given the following data from an experiment

x	y
1	1
4	2
8	4
11	5

where we are assuming that y is a function of x. (For example y might denote the growth of a plant furnished with x units of fertilizer, or y might denote the extension of a spring under a load of x units of force.) If we plot this data as in Fig. 2.6, we see that the data exhibit a strongly linear trend. From this empirical evidence we can plausibly postulate that y is related in a linear fashion to x—that is, we might assume y has the form

$$y = mx + b. \tag{2.32}$$

[In (2.32) we have chosen a mathematical model for the data solely on the basis of Fig. 2.6. We may not always have this luxury; for example, if the data represents results from a spring-loading experiment, and if we assume Hooke's law, then we must choose $y = mx$ instead of (2.32) as a model for the data.]

Given that we wish to model the data as $y = mx + b$, our problem becomes how to choose the appropriate constants m and b. In effect in Fig. 2.6 we would like to draw a line that best fits the data where the equation of the line is $y = mx + b$. If we

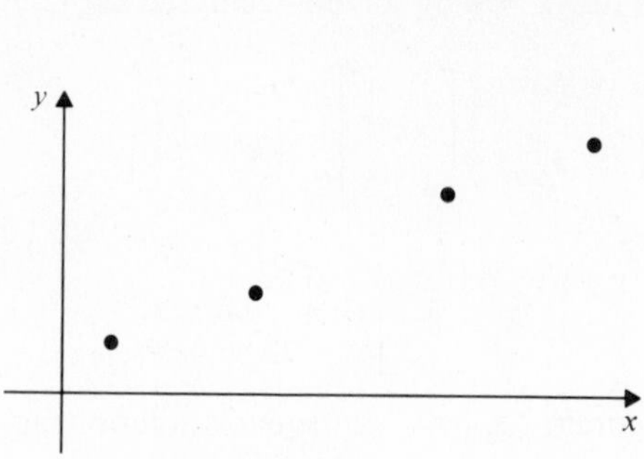

Fig. 2.6 Plot of the data points.

designate the four data points as

$$\begin{aligned}(x_1, y_1) &= (1, 1)\\ (x_2, y_2) &= (4, 2)\\ (x_3, y_3) &= (8, 4)\\ (x_4, y_4) &= (11, 5),\end{aligned}$$

then a reasonable criterion for choosing m and b is to minimize the expression

$$\sum_{i=1}^{4} [y_i - (mx_i + b)]^2. \tag{2.33}$$

A moment's reflection shows that minimizing (2.33) is a familiar problem, for we can define $\mathbf{y}$, A, and $\mathbf{u}$ by

$$\mathbf{y} = \begin{bmatrix} y_1 \\ y_2 \\ y_3 \\ y_4 \end{bmatrix}, \quad A = \begin{bmatrix} 1 & x_1 \\ 1 & x_2 \\ 1 & x_3 \\ 1 & x_4 \end{bmatrix}, \quad \mathbf{u} = \begin{bmatrix} b \\ m \end{bmatrix}. \tag{2.34}$$

With this definition $A\mathbf{u}$ is the vector

$$A\mathbf{u} = \begin{bmatrix} b + mx_1 \\ b + mx_2 \\ b + mx_3 \\ b + mx_4 \end{bmatrix},$$

and (2.33) merely represents the problem of choosing a vector $\mathbf{u}^*$ to minimize

$$(\mathbf{y} - A\mathbf{u})^{\mathrm{T}}(\mathbf{y} - A\mathbf{u}) = \|\mathbf{y} - A\mathbf{u}\|^2.$$

Thus the problem of choosing a linear expression to model the data in Fig. 2.6 can be solved by finding the best least-squares solution to the linear system $A\mathbf{u} = \mathbf{y}$.

For the data of Fig. 2.6 we have

$$\mathbf{y} = \begin{bmatrix} 1 \\ 2 \\ 4 \\ 5 \end{bmatrix}, \quad A = \begin{bmatrix} 1 & 1 \\ 1 & 4 \\ 1 & 8 \\ 1 & 11 \end{bmatrix}, \quad \mathbf{u} = \begin{bmatrix} b \\ m \end{bmatrix};$$

and we can minimize $\|\mathbf{y} - A\mathbf{u}\|^2$ by solving $A^{\mathrm{T}}A\mathbf{u} = A^{\mathrm{T}}\mathbf{b}$ where $A^{\mathrm{T}}A\mathbf{u} = A^{\mathrm{T}}\mathbf{b}$ is the consistent (2×2) system

$$\begin{aligned} 4b + 24m &= 12\\ 24b + 202m &= 96. \end{aligned}$$

The solution is $b = 15/29$, $m = 12/29$; so the best straight-line fit to the data in Fig. 2.6 is the line whose equation is

$$y = \tfrac{1}{29}(12x + 15).$$

To see how well this line fits the data, we let $L_1(x)$ denote the function

$$L_1(x) = \frac{1}{29}(12x + 15)$$

and calculate Table 2.1

TABLE 2.1

x	y	$L_1(x)$
1	1	0.931
4	2	2.172
8	4	3.828
11	5	5.069

There are several points we would like to make about the preceding example. First of all, the procedure illustrated above is not confined just to the problem of finding a best straight-line fit to data; the procedure can be used also to fit data by higher-degree polynomials or even by nonpolynomial expressions. Next we observe that in Example 2.24 we used a sum-of-squares criterion, (2.33), to define the best linear fit. Clearly there are other criteria we could have chosen. For example we could have asked that m and b be selected to minimize the quantity

$$Q_1 = \sum_{i=1}^{4} |y_i - (mx_i + b)|$$

or the quantity

$$Q_2 = \max_{1 \leqslant i \leqslant 4} |y_i - (mx_i + b)|.$$

While minimizing either Q_1 or Q_2 is a sensible condition to impose on m and b, it turns out that the condition (2.33) is usually more practical, partly because a solution to (2.33) can readily be obtained whereas minimizing Q_1 or Q_2 presents a relatively formidable computational problem.

EXAMPLE 2.25 In this example we show how to handle the general problem of finding the best nth degree polynomial fit to $m+1$ data points. Suppose we are given the data in tabular form (see Table 2.2); and suppose the data points are distributed in such a way that it is appropriate to fit the data with a polynomial $p(x)$ of degree n or less:

$$p(x) = b_n x^n + b_{n-1} x^{n-1} + \ldots + b_1 x + b_0.$$

TABLE 2.2

x	y
x_0	y_0
x_1	y_1
$\vdots$	$\vdots$
x_m	y_m

Our problem is to choose the coefficients of $p(x)$ so as to minimize

$$\sum_{i=0}^{m} [y_i - p(x_i)]^2. \tag{2.35}$$

If we write (2.35) out at length, we see that we are seeking $b_n, \ldots, b_1, b_0$ to minimize an expression of the form

$$\sum_{i=0}^{m} [y_i - (b_n(x_i)^n + \ldots + b_1 x_i + b_0)]^2;$$

and this procedure is equivalent to minimizing

$$(\mathbf{y} - A\mathbf{u})^T(\mathbf{y} - A\mathbf{u})$$

where

$$\mathbf{y} = \begin{bmatrix} y_0 \\ y_1 \\ \vdots \\ y_m \end{bmatrix}, \qquad A = \begin{bmatrix} 1 & x_0 & \ldots & x_0^n \\ 1 & x_1 & \ldots & x_1^n \\ \vdots & & & \\ 1 & x_m & \ldots & x_m^n \end{bmatrix}, \qquad \mathbf{u} = \begin{bmatrix} b_0 \\ b_1 \\ \vdots \\ b_n \end{bmatrix}. \tag{2.36}$$

Thus minimizing (2.35) is equivalent to minimizing

$$\|\mathbf{y} - A\mathbf{u}\|^2,$$

a procedure that is the same as finding the best least-squares solution to $A\mathbf{u} = \mathbf{y}$; and this is a problem we know how to solve. It is of interest to note as we see in Exercise 8, Section 2.6, that the columns of A are linearly independent when $m \geqslant n$. Thus solutions to $A^TA\mathbf{u} = A^T\mathbf{y}$ are unique. Therefore there is one and only one best least-squares polynomial fit of degree n to $m + 1$ data points when $m \geqslant n$. [Observe that $m = n$ reduces to a problem of polynomial interpolation (see Example 1.30). We already know that the problem of interpolating $(n + 1)$ values by a polynomial of degree n always has a unique solution.]

2.6 EXERCISES

1. Let W be the subspace of R^3

$$W = \{\mathbf{x} : \mathbf{x} = \begin{bmatrix} x_1 \\ x_2 \\ x_3 \end{bmatrix}, x_3 = 2x_1 - x_2\}.$$

Find an orthogonal basis for W. Find the nearest vector in W to the following.

a) $\begin{bmatrix} 2 \\ 0 \\ 2 \end{bmatrix}$, b) $\begin{bmatrix} 1 \\ 0 \\ 0 \end{bmatrix}$, c) $\begin{bmatrix} 3 \\ 1 \\ 5 \end{bmatrix}$, d) $\begin{bmatrix} 1 \\ 2 \\ 1 \end{bmatrix}$

2. Repeat Exercise 1 with the subspace W, which consists of all vectors in R^3 whose components satisfy $x_2 = x_3 - x_1$.

3. Use Theorem 2.8 to find all vectors $\mathbf{x}$ that minimize $\|A\mathbf{x} - \mathbf{b}\|$ where A and $\mathbf{b}$ are as given.

a) $A = \begin{bmatrix} 1 & 2 \\ -1 & 1 \\ 1 & 3 \end{bmatrix}$, $\mathbf{b} = \begin{bmatrix} 1 \\ 1 \\ 1 \end{bmatrix}$

b) $A = \begin{bmatrix} 1 & 2 & -1 \\ 2 & 3 & 1 \\ -1 & -1 & -2 \\ 3 & 5 & 0 \end{bmatrix}$, $\mathbf{b} = \begin{bmatrix} 1 \\ 0 \\ 1 \\ 0 \end{bmatrix}$

4. Find all vectors $\mathbf{x}$ that minimize $\|A\mathbf{x} - \mathbf{b}\|$.

a) $A = \begin{bmatrix} 1 & 2 & 4 \\ -2 & -3 & -7 \\ 1 & 3 & 5 \end{bmatrix}$, $\mathbf{b} = \begin{bmatrix} 1 \\ 1 \\ 2 \end{bmatrix}$

b) $A = \begin{bmatrix} 1 & 2 & 1 & 2 \\ 3 & 5 & 4 & 4 \\ -1 & 1 & -4 & 4 \end{bmatrix}$, $\mathbf{b} = \begin{bmatrix} 1 \\ 3 \\ 0 \end{bmatrix}$

5. As in Example 2.24, find $p(x) = mx + b$ that is a best least-squares fit to the data. In each case plot the data and the linear approximation, $p(x)$.

a)

x	y
−1	0
0	1
1	2
2	4

b)

x	y
−2	2
−1	1
1	0
2	−1

6. Prove that the columns of A are linearly independent when $n \geq 2$ and $x_1, x_2, \ldots, x_n$ are distinct.

$$A = \begin{bmatrix} 1 & x_1 \\ 1 & x_2 \\ \cdot & \\ \cdot & \\ \cdot & \\ 1 & x_n \end{bmatrix}.$$

7. Following Example 2.25, find $p(x) = ax^2 + bx + c$ that is a best least-squares fit to the data. In each case plot the data and the quadratic approximation, $p(x)$.

a)

x	y
-2	2
-1	1
1	1
2	2

b)

x	y
0	0
1	0
2	1
3	2

8. Prove that the columns of the matrix A in (2.36) are linearly independent when $x_0, x_1, \ldots, x_m$ are distinct and $m \geq n$. [Hint: If $A\mathbf{u} = \boldsymbol{\theta}$ in (2.36), what can you say about the polynomial $p(x) = b_0 + b_1 x + \ldots + b_n x^n$?]

9. Find c_0, c_1, c_2 so that the expression $q(x) = c_0 + c_1 \cos x + c_2 \sin x$ is a best least-squares fit to the data.

x	y
$-\pi/2$	1
0	0
$\pi/2$	1/2
π	1

10. Let A be an $(m \times n)$ matrix. Prove that $A^{\mathrm{T}}A$ is singular if and only if the columns of A are linearly dependent. (Hint: Consider $\mathbf{x}^{\mathrm{T}}A^{\mathrm{T}}A\mathbf{x} = \|A\mathbf{x}\|^2$.)

3

THE EIGENVALUE PROBLEM

3.1 INTRODUCTION

The ***eigenvalue problem***, the topic of this chapter, is a problem of considerable theoretical interest and wide-ranging application. The eigenvalue problem is formulated as follows (see also Definition 3.3).

For an $(n \times n)$ matrix A, find all scalars λ such that the equation

$$A\mathbf{x} = \lambda\mathbf{x} \tag{3.1}$$

has a nonzero solution, $\mathbf{x}$. Such a scalar λ is called an ***eigenvalue*** of A, and any nonzero $(n \times 1)$ vector $\mathbf{x}$ satisfying (3.1) is called an ***eigenvector*** corresponding to λ.

Equation (3.1) is obviously equivalent to the homogeneous system of equations

$$(A - \lambda I)\mathbf{x} = \boldsymbol{\theta} \tag{3.2}$$

where I is the $(n \times n)$ identity matrix. If (3.2) is to have nonzero solutions, then λ must be chosen so that the $(n \times n)$ matrix $A - \lambda I$ is singular. Therefore the eigenvalue problem consists of two parts.

1. Find all scalars λ such that $A - \lambda I$ is singular. (These are the eigenvalues of A.)
2. Given that $A - \lambda I$ is singular, find all *nonzero* vectors $\mathbf{x}$ such that $(A - \lambda I)\mathbf{x} = \boldsymbol{\theta}$. (These are the eigenvectors of A corresponding to λ.)

Clearly if we know an eigenvalue of A, then the variable-elimination techniques described in Chapter 1 provide an efficient way to find the eigenvectors. The new feature of the eigenvalue problem is in part (1), determining all scalars λ such that the matrix $A - \lambda I$ is singular.

Historically, determinant theory has been used to solve part (1) of the eigenvalue problem; and in the next section we will briefly describe some of the determinant properties that are relevant to the eigenvalue problem. Although modern computational procedures for finding eigenvalues do not usually employ determinants, we can use them to calculate eigenvalues for fairly small matrices. In Section 3.7 we will develop transformations that efficiently reduce an $(n \times n)$ matrix A to an $(n \times n)$ matrix H where H and A have the same eigenvalues and where the eigenvalues of H

are (relatively) easy to find. In addition these transformations yield important theoretical information about the eigenvalue problem. In this regard we will present the determinant approach to eigenvalues without proofs, but we will be mathematically complete in the transformation approach.

3.2 DETERMINANTS AND THE EIGENVALUE PROBLEM

Determinant theory has long intrigued mathematicians; and the reader has probably learned how to calculate determinants, at least for (2×2) and (3×3) matrices. The purpose of this section is to review briefly those aspects of determinant theory that can be used in the eigenvalue problem. A formal development of determinants, including proofs, definitions, and the important properties of determinants, can be found in Chapter 5. In this section we present two basic results: an algorithm for evaluating determinants and a characterization of singular matrices in terms of determinants. To begin, we define the determinant of a (2×2) matrix.

DEFINITION 3.1.

Let A be a (2×2) matrix,

$$A = \begin{bmatrix} a_{11} & a_{12} \\ a_{21} & a_{22} \end{bmatrix}.$$

The ***determinant*** of A, denoted by det (A), is the number

$$\det(A) = \begin{vmatrix} a_{11} & a_{12} \\ a_{21} & a_{22} \end{vmatrix} = a_{11}a_{22} - a_{21}a_{12}.$$

(As Definition 3.1 indicates, the determinant of A is designated by vertical bars when we wish to exhibit the entries of A.) As an example if

$$A = \begin{bmatrix} 2 & 4 \\ 1 & 3 \end{bmatrix},$$

then

$$\det(A) = \begin{vmatrix} 2 & 4 \\ 1 & 3 \end{vmatrix} = 2 \cdot 3 - 1 \cdot 4 = 2.$$

As another example, let

$$B = \begin{bmatrix} 2 & 4 \\ 3 & 6 \end{bmatrix}.$$

In this case det $(B) = 0$ since

$$\det(B) = \begin{vmatrix} 2 & 4 \\ 3 & 6 \end{vmatrix} = 2 \cdot 6 - 3 \cdot 4 = 0.$$

Note in the examples above that B is obviously singular while A is a nonsingular matrix. In fact the matrices A and B provide a special case of Theorem 3.1, which asserts that an $(n \times n)$ matrix Q is singular if and only if $\det(Q) = 0$.

We next give an algorithm for evaluating the determinant of an $(n \times n)$ matrix for $n > 2$. This algorithm is recursive, based on the idea of minors.

DEFINITION 3.2.

Let $A = (a_{ij})$ be an $(n \times n)$ matrix, and let A_{rs} denote the $((n-1) \times (n-1))$ matrix obtainined by deleting row r and column s from A. The number $M_{rs} = \det(A_{rs})$ is called the ***minor*** of a_{rs}.

EXAMPLE 3.1 Let A be the (3×3) matrix

$$A = \begin{bmatrix} 1 & 2 & -1 \\ 5 & 3 & 4 \\ -2 & 0 & 1 \end{bmatrix}.$$

Then M_{21} is the determinant of the (2×2) matrix obtained by deleting row two and column one from A:

$$M_{21} = \begin{vmatrix} 2 & -1 \\ 0 & 1 \end{vmatrix} = 2.$$

Similarly

$$M_{11} = \begin{vmatrix} 3 & 4 \\ 0 & 1 \end{vmatrix} = 3, \quad M_{23} = \begin{vmatrix} 1 & 2 \\ -2 & 0 \end{vmatrix} = 4, \quad M_{32} = \begin{vmatrix} 1 & -1 \\ 5 & 4 \end{vmatrix} = 9.$$

Minors can be used to state a formula for the determinant.

Evaluating det (A)

Let $A = (a_{ij})$ be an $(n \times n)$ matrix. Then for any k, $1 \leqslant k \leqslant n$,

$$\det(A) = \sum_{j=1}^{n} a_{kj}(-1)^{k+j} M_{kj} \tag{3.3a}$$

$$\det(A) = \sum_{j=1}^{n} a_{jk}(-1)^{k+j} M_{jk}. \tag{3.3b}$$

The formulas (3.3a) and (3.3b) are established in Chapter 5; for now, we wish only to discuss them. With the validity of (3.3) assumed, it follows that the determinant of an $(n \times n)$ matrix is a weighted sum of n minors. Since each minor is an $((n-1) \times (n-1))$ determinant, it follows that $\det(A)$ is a weighted sum of n determinants each of order

$n-1$. The terms of the form $(-1)^{r+s}M_{rs}$ in (3.3) are called ***cofactors*** (or signed minors), and either of (3.3a) or (3.3b) is called a cofactor expansion of det (A). Next since the index k is arbitrary, $1 \leqslant k \leqslant n$, these formulas also assert that we can use a cofactor expansion along any row of A [the kth row in (3.3a)] or along any column [the kth column in (3.3b)]; and the result will always be the same number, det (A). The next example illustrates how we can use the formulas (3.3) to reduce any $(n \times n)$ determinant to a sum of (2×2) determinants. Definition 3.1 illustrates how to evaluate (2×2) determinants, and thus in principle the cofactor expansions in (3.3) can be used recursively to evaluate a determinant of any size.

EXAMPLE 3.2 To demonstrate the cofactor expansions given in (3.3) and to show the recursive nature of these formulas, we consider the (3×3) matrix A:

$$A = \begin{bmatrix} 1 & 2 & -1 \\ 5 & 3 & 4 \\ -2 & 0 & 1 \end{bmatrix}.$$

Choosing $k = 1$ in formula (3.3a), we have

$$\det(A) = a_{11}M_{11} - a_{12}M_{12} + a_{13}M_{13}$$

or

$$\det(A) = 1 \cdot \begin{vmatrix} 3 & 4 \\ 0 & 1 \end{vmatrix} - 2 \cdot \begin{vmatrix} 5 & 4 \\ -2 & 1 \end{vmatrix} + (-1) \cdot \begin{vmatrix} 5 & 3 \\ -2 & 0 \end{vmatrix} = -29.$$

Note above that det (A) is the sum of three (2×2) determinants. To illustrate the independence of (3.3a) from the row index k, let $k = 2$ in (3.3a). Then det $(A) = -a_{21}M_{21} + a_{22}M_{22} - a_{23}M_{23}$, or

$$\det(A) = -5 \cdot \begin{vmatrix} 2 & -1 \\ 0 & 1 \end{vmatrix} + 3 \cdot \begin{vmatrix} 1 & -1 \\ -2 & 1 \end{vmatrix} - 4 \cdot \begin{vmatrix} 1 & 2 \\ -2 & 0 \end{vmatrix} = -29.$$

Finally the cofactor expansion of det (A) using the second column [$k = 2$ in (3.3b)] is

$$\det(A) = -2 \cdot \begin{vmatrix} 5 & 4 \\ -2 & 1 \end{vmatrix} + 3 \cdot \begin{vmatrix} 1 & -1 \\ -2 & 1 \end{vmatrix} - 0 \cdot \begin{vmatrix} 1 & -1 \\ 5 & 4 \end{vmatrix} = -29.$$

As a slightly more involved example, we consider how a (4×4) determinant can be broken down into a sum of four (3×3) determinants. Let B be the (4×4) matrix

$$B = \begin{bmatrix} 1 & 2 & 1 & 3 \\ 0 & 1 & 2 & 0 \\ 4 & 2 & 0 & -1 \\ -2 & 3 & 1 & 1 \end{bmatrix}.$$

Since the second row of B contains two zero entries, we naturally use a cofactor expansion of det (B) along the second row [$k = 2$ in (3.3a)] in order to reduce the computations. Thus det $(B) = M_{22} - 2M_{23}$. We again use (3.3) to evaluate the

(3 × 3) determinants M_{22} and M_{23} (expanding both M_{22} and M_{23} along the first row to avoid confusion):

$$M_{22} = \begin{vmatrix} 1 & 1 & 3 \\ 4 & 0 & -1 \\ -2 & 1 & 1 \end{vmatrix} = \begin{vmatrix} 0 & -1 \\ 1 & 1 \end{vmatrix} - \begin{vmatrix} 4 & -1 \\ -2 & 1 \end{vmatrix} + 3\begin{vmatrix} 4 & 0 \\ -2 & 1 \end{vmatrix} = 1 - 2 + 12$$
$$= 11$$

$$M_{23} = \begin{vmatrix} 1 & 2 & 3 \\ 4 & 2 & -1 \\ -2 & 3 & 1 \end{vmatrix} = \begin{vmatrix} 2 & -1 \\ 3 & 1 \end{vmatrix} - 2\begin{vmatrix} 4 & -1 \\ -2 & 1 \end{vmatrix} + 3\begin{vmatrix} 4 & 2 \\ -2 & 3 \end{vmatrix} = 5 - 4 + 48$$
$$= 49.$$

Therefore $\det(B) = M_{22} - 2M_{23} = -87$.

The example above should illustrate how the determinant of an $(n \times n)$ matrix can be computed, and how the computational details grow with n. That is, a (3 × 3) determinant is the sum of three (2 × 2) determinants while a (4 × 4) determinant is the sum of four (3 × 3) determinants, each of which is in turn the sum of three (2 × 2) determinants. Therefore evaluating a (4 × 4) determinant from (3.3) requires the evaluation of four (3 × 3) determinants and hence ultimately the evaluation of $4 \cdot 3 = 12$ determinants of size (2 × 2). From similar reasoning it is not hard to see that using (3.3) to calculate an $(n \times n)$ determinant requires the evaluation of

$$n(n-1) \ldots 4 \cdot 3 = n!/2 \tag{3.4}$$

determinants of order two. Of course if a matrix has some zero entries, some of these evaluations need not be performed. In fact one can use determinant properties based on elementary row operations to create zero entries to facilitate the evaluation of a determinant (see Chapter 5). In the next section we shall have more to say about the practical evaluation of determinants and their relation to the eigenvalue problem.

The final result of this section relates singular matrices and determinants. An indication of this result is provided by the observation that if an $(n \times n)$ matrix A has a row or column consisting entirely of zero entries, then $\det(A) = 0$. For instance if

$$A = \begin{bmatrix} 1 & 2 & 1 \\ 0 & 0 & 0 \\ 3 & 2 & 2 \end{bmatrix},$$

then using (3.3a) with $k = 2$, we have (since $a_{21} = a_{22} = a_{23} = 0$)

$$\det(A) = -a_{21}M_{21} + a_{22}M_{22} - a_{23}M_{23} = 0.$$

Furthermore a matrix (such as A above) with a zero row or column is singular. Most singular matrices do not have a zero row or column, of course; but the example above is not an isolated phenomenon; and one of the striking results of determinant theory is Theorem 3.1, which gives a precise mathematical characterization of singularity in terms of determinants.

Theorem 3.1

Let A be an $(n \times n)$ matrix. Then A is singular if and only if $\det(A) = 0$.

Theorem 3.1 is proved in Chapter 5; but it is not hard to prove a special case of Theorem 3.1; and Example 3.3 serves as a starting point.

EXAMPLE 3.3 In this example we consider how the determinant of an upper-triangular matrix might be calculated. In particular let T be the (4×4) upper-triangular matrix

$$T = \begin{bmatrix} 2 & 1 & 3 & 1 \\ 0 & 1 & 1 & 4 \\ 0 & 0 & 3 & 2 \\ 0 & 0 & 0 & 2 \end{bmatrix}.$$

If we expand $\det(T)$ along the last row, we find (since $t_{41} = t_{42} = t_{43} = 0$) that

$$\det(T) = t_{44}M_{44} = 2 \cdot \begin{vmatrix} 2 & 1 & 3 \\ 0 & 1 & 1 \\ 0 & 0 & 3 \end{vmatrix}.$$

We observe that M_{44} is again an upper-triangular determinant; so we expand M_{44} along the last row and find

$$\det(T) = 2M_{44} = 2 \cdot 3 \cdot \begin{vmatrix} 2 & 1 \\ 0 & 1 \end{vmatrix} = 2 \cdot 3 \cdot 1 \cdot 2 = 12.$$

Thus $\det(T)$ is merely the product of the main diagonal entries of T, $\det(T) = t_{11}t_{22}t_{33}t_{44}$.

Clearly this sort of procedure will work in general for any $(n \times n)$ upper- (or lower-) triangular matrix; and in Exercise 14, Section 3.2, the reader is asked to use induction to prove that the determinant of any triangular matrix is the product of the main diagonal entries. With these preliminaries we can state Theorem 3.2.

Theorem 3.2

Let $T = (t_{ij})$ be an $(n \times n)$ triangular matrix. Then T is singular if and only if $\det(T) = 0$.

Proof. From Exercises 9 and 10, Section 1.7, we know that T is singular if and only if at least one of the main diagonal entries $t_{11}, t_{22}, \ldots t_{nn}$ is zero; so Theorem 3.2 follows immediately. ∎

3.2 EXERCISES

1. Evaluate the minors M_{11}, M_{21}, M_{31}, M_{32}, M_{33} for these matrices.

a) $\begin{bmatrix} 2 & 1 & 4 \\ -3 & 2 & 1 \\ 5 & -1 & 3 \end{bmatrix}$ b) $\begin{bmatrix} -3 & 1 & 2 \\ 4 & 2 & 6 \\ 1 & -1 & 4 \end{bmatrix}$ c) $\begin{bmatrix} 4 & 2 & -1 \\ -7 & 0 & 2 \\ 1 & 3 & -2 \end{bmatrix}$

2. Evaluate the minors M_{11}, M_{21}, M_{31}, M_{32}, M_{33} for these matrices.

a) $\begin{bmatrix} 1 & 2 & 1 \\ 3 & 3 & 1 \\ 2 & 1 & 0 \end{bmatrix}$ b) $\begin{bmatrix} 2 & 1 & 2 \\ 0 & 3 & 2 \\ 0 & 0 & 5 \end{bmatrix}$ c) $\begin{bmatrix} 1 & 1 & 3 \\ 2 & 1 & -2 \\ 1 & 2 & 5 \end{bmatrix}$

3. Calculate the determinant of each matrix in Exercise 1 in two ways: expanding along the first column and along the third row.

4. Repeat Exercise 3 for each matrix in Exercise 2.

5. According to (3.1) in Section 3.1 if a matrix Q is singular, then $\lambda = 0$ is an eigenvalue of Q. For each singular matrix in Exercise 1, find the eigenvectors corresponding to $\lambda = 0$.

6. Repeat Exercise 5 for the matrices in Exercise 2.

7. Evaluate the determinant of each matrix.

$$A = \begin{bmatrix} 2 & 1 \\ -1 & 2 \end{bmatrix} \quad B = \begin{bmatrix} 1 & -1 \\ -2 & 2 \end{bmatrix} \quad C = \begin{bmatrix} 1 & 2 & 4 \\ 2 & 3 & 7 \\ 4 & 2 & 10 \end{bmatrix}$$

$$D = \begin{bmatrix} 2 & -3 & 2 \\ -1 & -2 & 1 \\ 3 & 1 & -1 \end{bmatrix}$$

8. Evaluate the determinant of each matrix.

$$A = \begin{bmatrix} 2 & 3 \\ 4 & 6 \end{bmatrix} \quad B = \begin{bmatrix} 1 & 1 \\ 2 & 1 \end{bmatrix} \quad C = \begin{bmatrix} 1 & 2 & 1 \\ 0 & 3 & 2 \\ -1 & 1 & 1 \end{bmatrix} \quad D = \begin{bmatrix} 2 & 0 & 0 \\ 1 & 3 & 2 \\ 2 & 1 & 4 \end{bmatrix}$$

9. For each singular matrix in Exercise 7 find the eigenvectors corresponding to the eigenvalue $\lambda = 0$.

10. Let $A = (a_{ij})$ be the $(n \times n)$ matrix specified thus: $a_{ij} = d$ for $i = j$, and $a_{ij} = 1$ for $i \neq j$. For $n = 2, 3, 4$, show that

$$\det(A) = (d-1)^{n-1}(d-1+n).$$

11. a) Let $A = (a_{ij})$ be a general (4×4) matrix, and let B be the (4×4) matrix obtained from A by interchanging rows one and two of A. Prove that $\det(B) = -\det(A)$. [Hint: Expand $\det(A)$ along row one and $\det(B)$ along row two.]

b) Prove that $\det(B) = -\det(A)$ when B is obtained from A by interchanging rows two and three or rows three and four.

c) Prove that $\det(B) = -\det(A)$ when B is obtained from A by interchanging any two rows of A. [Hint: By parts (a) and (b) the sign of the determinant changes when adjacent rows are switched. Next, rows one and three, for example, can be interchanged by a sequence of interchanges that involve only adjacent rows. Note: It should be clear from the proof that part (c) is valid for any $(n \times n)$ matrix.]

12. Interchange the first and second rows of the matrices in Exercise 7, and verify that the determinant changes sign.

13. a) Use part (c) of Exercise 11 to show that if any two rows of a (4×4) matrix A are equal, then $\det(A) = 0$.
    b) Let B be the (4×4) matrix obtained from A by adding a multiple of c times row one to row two. Prove that $\det(B) = \det(A)$. [Hint: Expand $\det(B)$ along the second row, and use part (a).]
    c) Use part (c) of Exercise 11 and part (b) above to show that the determinant of A is unchanged when a multiple of one row is added to any other row.

14. Prove that the determinant of an $(n \times n)$ triangular matrix is equal to the product of the diagonal entries.

15. Exercises 11 and 13 show that elementary row operations can be used to introduce zero entries, which will make the calculation of $\det(A)$ easier. For each matrix below, use elementary row operations to produce a matrix B that has zeros in positions (2, 1), (3, 1), (4, 1), (3, 2), and (4, 2). Calculate $\det(B)$ and relate it to the determinant of the original matrix A.

a) $A = \begin{bmatrix} 0 & 2 & 1 & 3 \\ 0 & 0 & 0 & 1 \\ 2 & 3 & 4 & 1 \\ 0 & 0 & 1 & 1 \end{bmatrix}$ b) $A = \begin{bmatrix} 0 & 1 & 1 & 2 \\ 1 & 2 & 1 & 1 \\ 3 & 1 & 2 & 2 \\ -1 & 2 & 0 & 1 \end{bmatrix}$

c) $A = \begin{bmatrix} 1 & 1 & -2 & 1 \\ 2 & 3 & 4 & 7 \\ -1 & 2 & 5 & 1 \\ -2 & -1 & 3 & 2 \end{bmatrix}$ d) $A = \begin{bmatrix} 2 & -1 & 3 & 1 \\ 4 & 1 & 3 & -1 \\ 6 & 2 & 4 & 1 \\ 2 & 2 & 0 & -2 \end{bmatrix}$

3.3 EIGENVALUES OF MATRICES

We now return to the central topic of this chapter, the eigenvalue problem. As was stated informally in Section 3.1, the eigenvalue problem for a matrix A consists of finding scalars λ such that the matrix $A - \lambda I$ is singular, and then finding all nonzero vectors $\mathbf{x}$ such that $(A - \lambda I)\mathbf{x} = \boldsymbol{\theta}$. Before proceeding with the theoretical development of the eigenvalue problem, we consider an example.

EXAMPLE 3.4 Let A be the (2×2) matrix

$$A = \begin{bmatrix} 5 & -2 \\ 6 & -2 \end{bmatrix}.$$

Then $A - 2I$ is a singular matrix since

$$A - 2I = \begin{bmatrix} 5 & -2 \\ 6 & -2 \end{bmatrix} - \begin{bmatrix} 2 & 0 \\ 0 & 2 \end{bmatrix} = \begin{bmatrix} 3 & -2 \\ 6 & -4 \end{bmatrix},$$

and clearly the columns of $A-2I$ are linearly dependent [equivalently see Theorem 3.1, $\det(A-2I)=0$]. Since $A-2I$ is singular, we know that the number $\lambda=2$ is an eigenvalue of A. The eigenvectors of A corresponding to $\lambda=2$ are the nonzero solutions of $(A-2I)\mathbf{x}=\boldsymbol{\theta}$, which we find by solving the linear system

$$\begin{aligned} 3x_1-2x_2&=0\\ 6x_1-4x_2&=0. \end{aligned}$$

The solutions of this system are given by $x_1=2x_2/3$ with x_2 arbitrary. Thus any vector of the form

$$\mathbf{x}=\begin{bmatrix} 2a/3\\ a \end{bmatrix}, \quad a\neq 0$$

is an eigenvector of A corresponding to $\lambda=2$. For instance with $a=3$ we obtain the eigenvector

$$\mathbf{x}=\begin{bmatrix} 2\\ 3 \end{bmatrix},$$

and it is easy to verify for this choice of $\mathbf{x}$ that $(A-2I)\mathbf{x}=\boldsymbol{\theta}$ or $A\mathbf{x}=2\mathbf{x}$.

DEFINITION 3.3

Let A be an $(n\times n)$ matrix, and let I be the $(n\times n)$ identity matrix. If λ is any scalar such that $A-\lambda I$ is singular, then λ is called an ***eigenvalue*** of A. Furthermore any nonzero vector $\mathbf{x}$ such that $(A-\lambda I)\mathbf{x}=\boldsymbol{\theta}$ is called an ***eigenvector*** of A corresponding to λ.

Again we note for emphasis that the eigenvalue problem has two parts.

1. Find all scalars λ such that $A-\lambda I$ is singular.
2. Given that $A-\lambda I$ is singular, find all nonzero vectors $\mathbf{x}$ such that $(A-\lambda I)\mathbf{x}=\boldsymbol{\theta}$.

Thus our first objective is to develop some procedure for finding scalars λ such that $A-\lambda I$ is singular. One such procedure is suggested by Theorem 3.1, and we can demonstrate this procedure by using the matrix A in Example 3.4. For the (2×2) matrix

$$A=\begin{bmatrix} 5 & -2\\ 6 & -2 \end{bmatrix},$$

$A-\lambda I$ is the matrix

$$A-\lambda I=\begin{bmatrix} 5-\lambda & -2\\ 6 & -2-\lambda \end{bmatrix}.$$

In Theorem 3.1 we have a test for singularity: $A-\lambda I$ is singular if and only if $\det(A-\lambda I)=0$. This test suggests that we might write out $\det(A-\lambda I)$ and then try to determine those values of λ that make $\det(A-\lambda I)=0$. Carrying out this procedure, we quickly find

$$\det(A-\lambda I)=-(5-\lambda)(2+\lambda)+12,$$

or $\det(A - \lambda I) = \lambda^2 - 3\lambda + 2$. We see that $\det(A - \lambda I)$ is a polynomial of degree two in the variable λ, and we see also that

$$\det(A - \lambda I) = \lambda^2 - 3\lambda + 2 = (\lambda - 2)(\lambda - 1).$$

Clearly then $\det(A - \lambda I) = 0$ only for $\lambda = 2$ and for $\lambda = 1$. Since $A - \lambda I$ is singular only when $\det(A - \lambda I) = 0$, it follows that the eigenvalues of A are precisely the numbers $\lambda = 2$ and $\lambda = 1$.

In general it can be shown that $\det(A - \lambda I)$ is a polynomial of degree n in λ when A is $(n \times n)$. Then since $A - \lambda I$ is singular if and only if $\det(A - \lambda I) = 0$, it follows that the eigenvalues of A are precisely the zeros of the polynomial $\det(A - \lambda I)$. To avoid any possible confusion between the eigenvalues λ of A and the problem of finding the zeros of this associated polynomial (called the characteristic polynomial of A), we will use the variable t instead of λ in the characteristic polynomial and write $p(t) = \det(A - tI)$. To summarize what was illustrated above, we give Theorem 3.3.

Theorem 3.3

Let A be an $(n \times n)$ matrix. Then $\det(A - tI)$ is a polynomial in t of degree n.

The proof of this theorem is not hard, but we defer the proof. Given that $\det(A - tI)$ is a polynomial, we are led to Definition 3.4.

DEFINITION 3.4.

Let A be an $(n \times n)$ matrix. The nth degree polynomial, $p(t)$, given by

$$p(t) = \det(A - tI), \tag{3.5}$$

is called the ***characteristic polynomial*** for A.

From (3.5) and Theorem 3.1 it follows that the roots of $p(t) = 0$ are the eigenvalues of A.

Theorem 3.4

Let A be an $(n \times n)$ matrix. The eigenvalues of A are precisely the zeros of the characteristic polynomial $p(t)$.

Theorem 3.4 has the effect of replacing the original problem—determining values λ for which $A - \lambda I$ is singular—by an equivalent problem—finding the roots of a polynomial equation $p(t) = 0$. Since polynomials are familiar, and since over the years an immense amount of theoretical and computational machinery has been developed for solving polynomial equations, we should feel more comfortable with the eigenvalue problem. Finally the equation $p(t) = 0$ that must be solved to find the eigenvalues of A is called the ***characteristic equation***. With regard to the characteristic equation we note that if $p(t)$ is a polynomial of degree n, then the equation $p(t) = 0$ can have no more than n distinct roots. From this observation we see that an $(n \times n)$ matrix can have no more than n distinct eigenvalues; and since every polynomial equation $p(t) = 0$ has at least one root, we see that every matrix has at least one eigenvalue.

EXAMPLE 3.5 In this example we determine the eigenvalues and corresponding eigenvectors for the (3×3) matrix

$$A = \begin{bmatrix} 1 & 1 & 1 \\ 0 & 3 & 3 \\ -2 & 1 & 1 \end{bmatrix}.$$

We first find the characteristic polynomial for A:

$$p(t) = \begin{vmatrix} 1-t & 1 & 1 \\ 0 & 3-t & 3 \\ -2 & 1 & 1-t \end{vmatrix} = (1-t)\begin{vmatrix} 3-t & 3 \\ 1 & 1-t \end{vmatrix} - \begin{vmatrix} 0 & 3 \\ -2 & 1-t \end{vmatrix} + \begin{vmatrix} 0 & 3-t \\ -2 & 1 \end{vmatrix}.$$

Thus

$$p(t) = (1-t)[(3-t)(1-t)-3] - 6 + 2(3-t)$$

or

$$p(t) = -t^3 + 5t^2 - 6t = -t(t-3)(t-2).$$

We see that $p(t)$ has three distinct zeros, namely $t = 0$, $t = 3$, and $t = 2$. Therefore A has three eigenvalues: $\lambda = 0$, $\lambda = 3$, and $\lambda = 2$.

Next for each eigenvalue λ we find the eigenvectors that correspond to λ by solving the system $(A - \lambda I)\mathbf{x} = \boldsymbol{\theta}$. For the eigenvalue $\lambda = 0$, we have $(A - 0I)\mathbf{x} = \boldsymbol{\theta}$, or $A\mathbf{x} = \boldsymbol{\theta}$, to solve:

$$\begin{aligned} x_1 + x_2 + x_3 &= 0 \\ 3x_2 + 3x_3 &= 0 \\ -2x_1 + x_2 + x_3 &= 0. \end{aligned}$$

The solution of this system is $x_1 = 0$, $x_2 = -x_3$ with x_3 arbitrary. Thus the eigenvectors of A corresponding to $\lambda = 0$ are given by

$$\mathbf{x} = \begin{bmatrix} 0 \\ -a \\ a \end{bmatrix} = a\begin{bmatrix} 0 \\ -1 \\ 1 \end{bmatrix}, \quad a \neq 0;$$

and any such vector $\mathbf{x}$ satisfies $A\mathbf{x} = 0 \cdot \mathbf{x}$. This equation illustrates that the definition of eigenvalues *does* allow the possibility that $\lambda = 0$ is an eigenvalue. We stress however that the zero vector is *never* considered an eigenvector (after all, $A\mathbf{x} = \lambda\mathbf{x}$ is always satisfied for $\mathbf{x} = \boldsymbol{\theta}$, no matter what value λ has).

The eigenvectors corresponding to the eigenvalue $\lambda = 3$ are found by solving $(A - 3I)\mathbf{x} = \boldsymbol{\theta}$:

$$\begin{aligned} -2x_1 + x_2 + x_3 &= 0 \\ 3x_3 &= 0 \\ -2x_1 + x_2 - 2x_3 &= 0. \end{aligned}$$

The solution of this system is $x_3 = 0$, $x_2 = 2x_1$ with x_1 arbitrary. Thus the *nontrivial* solutions of $(A-3I)\mathbf{x}=\boldsymbol{\theta}$ (the eigenvectors of A corresponding to $\lambda = 3$) all have the form

$$\mathbf{x}=\begin{bmatrix} a \\ 2a \\ 0 \end{bmatrix} = a\begin{bmatrix} 1 \\ 2 \\ 0 \end{bmatrix}, \quad a \neq 0.$$

Finally the eigenvectors corresponding to $\lambda = 2$ are found from $(A-2I)\mathbf{x}=\boldsymbol{\theta}$; and the solution is $x_1 = -2x_3$, $x_2 = -3x_3$ with x_3 arbitrary. So the eigenvectors corresponding to $\lambda = 2$ are of the form

$$\mathbf{x}=\begin{bmatrix} -2a \\ -3a \\ a \end{bmatrix} = a\begin{bmatrix} -2 \\ -3 \\ 1 \end{bmatrix}, \quad a \neq 0.$$

We pause here to make several comments. As Examples 3.4 and 3.5 show, there are infinitely many eigenvectors corresponding to a given eigenvalue. This comment should be obvious, for if $A-\lambda I$ is a singular matrix, there are infinitely many nontrivial solutions of $(A-\lambda I)\mathbf{x}=\boldsymbol{\theta}$. In particular if $A\mathbf{x}=\lambda\mathbf{x}$ for some nonzero vector $\mathbf{x}$, then we also have $A\mathbf{y}=\lambda\mathbf{y}$ when $\mathbf{y}=a\mathbf{x}$ with a being any scalar. Thus any nonzero multiple of an eigenvector is again an eigenvector. Next we again note that the scalar $\lambda = 0$ may be an eigenvalue of a matrix as Example 3.5 showed. We note also that if $\lambda = 0$ is an eigenvalue of A, then A is singular since $A = A - 0I$, which is singular. Conversely if A is singular, then there is a nonzero vector $\mathbf{x}$ such that $A\mathbf{x} = \boldsymbol{\theta}$; and hence $A\mathbf{x} = 0\cdot\mathbf{x}$ has a nontrivial solution. This result means that a singular matrix always has $\lambda = 0$ as an eigenvalue. Combining these two observations, we have a characterization of singular matrices in terms of eigenvalues.

Theorem 3.5

Let A be an $(n \times n)$ matrix. Then A is singular if and only if $\lambda = 0$ is an eigenvalue of A.

For some special types of matrices, the eigenvalue problem is easy to solve. The theorem below gives one such instance.

Theorem 3.6

Let $A = (a_{ij})$ be an $(n \times n)$ triangular matrix. Then the characteristic polynomial for A is $p(t) = (a_{11}-t)(a_{22}-t)\ldots(a_{nn}-t)$.

Proof. If A is a triangular matrix (either upper triangular or lower triangular), then so is $A-tI$. By Exercise 14, Section 3.2, $\det(A-tI)$ is the product of the diagonal entries of $A-tI$. Since the diagonal entries of $A-tI$ are $a_{11}-t, a_{22}-t, \ldots, a_{nn}-t$, the theorem is proved. ■

Theorem 3.6 clearly says that the eigenvalues of a *triangular* matrix are precisely the diagonal entries, and we will give some examples of this fact below.

As we note in the next section, one of the important applications of eigenvalues is in the study of solutions and stability for systems of differential equations and difference equations. A primary concern in this applications area is the number of linearly independent eigenvectors that a given matrix can have. We will address this question more fully in later sections, but we consider now an example that should illustrate some of the possibilities.

EXAMPLE 3.6 Consider the matrices

$$I = \begin{bmatrix} 1 & 0 & 0 \\ 0 & 1 & 0 \\ 0 & 0 & 1 \end{bmatrix}, \qquad A = \begin{bmatrix} 1 & 0 & 0 \\ 1 & 1 & 0 \\ 0 & 0 & 1 \end{bmatrix}, \qquad B = \begin{bmatrix} 1 & 0 & 0 \\ 1 & 1 & 0 \\ 0 & 1 & 1 \end{bmatrix}.$$

Since I is a triangular matrix, we have by Theorem 3.6 that the characteristic polynomial for I is $p(t) = (1-t)(1-t)(1-t) = (1-t)^3$. Thus $\lambda = 1$ is the only eigenvalue of I. For $\lambda = 1$ the matrix $I - \lambda I$ is the zero matrix, and hence every nonzero vector is an eigenvector of I corresponding to the eigenvalue $\lambda = 1$. In particular the unit vectors $\mathbf{e}_1$, $\mathbf{e}_2$, and $\mathbf{e}_3$ are eigenvectors; and therefore I has three linearly independent eigenvectors. As above, A is triangular; so the characteristic polynomial for A is $p(t) = (1-t)^3$, and $\lambda = 1$ is the only eigenvalue of A. The eigenvectors of A are determined by the equation $(A - I)\mathbf{x} = \boldsymbol{\theta}$:

$$\begin{bmatrix} 0 & 0 & 0 \\ 1 & 0 & 0 \\ 0 & 0 & 0 \end{bmatrix} \begin{bmatrix} x_1 \\ x_2 \\ x_3 \end{bmatrix} = \begin{bmatrix} 0 \\ 0 \\ 0 \end{bmatrix}.$$

These constraints imply that $x_1 = 0$ while x_2 and x_3 are arbitrary. Thus the eigenvectors of A are of the form

$$\mathbf{x} = \begin{bmatrix} 0 \\ a \\ b \end{bmatrix} = a \begin{bmatrix} 0 \\ 1 \\ 0 \end{bmatrix} + b \begin{bmatrix} 0 \\ 0 \\ 1 \end{bmatrix}, \quad |a| + |b| \neq 0;$$

and A is seen to have at most two linearly independent eigenvectors. For instance $\{\mathbf{e}_2, \mathbf{e}_3\}$ is a set of two linearly independent eigenvectors for A.

Finally the characteristic polynomial for B is also $p(t) = (1-t)^3$; so the only eigenvalue of B is $\lambda = 1$. The eigenvectors are found from $(B - I)\mathbf{x} = \boldsymbol{\theta}$:

$$\begin{bmatrix} 0 & 0 & 0 \\ 1 & 0 & 0 \\ 0 & 1 & 0 \end{bmatrix} \begin{bmatrix} x_1 \\ x_2 \\ x_3 \end{bmatrix} = \begin{bmatrix} 0 \\ 0 \\ 0 \end{bmatrix},$$

which insists that $x_1 = 0$, $x_2 = 0$ while x_3 is arbitrary. Thus every eigenvector of B must have the form

$$\mathbf{x} = \begin{bmatrix} 0 \\ 0 \\ a \end{bmatrix} = a \begin{bmatrix} 0 \\ 0 \\ 1 \end{bmatrix}, \quad a \neq 0.$$

---

eigenvectors → basis of the kernel

From this example we see that a (3×3) matrix may have three linearly independent eigenvectors (as with I), or a set of just two linearly independent eigenvectors (as with A), or just one linearly independent eigenvector (as with B). In general an $(n \times n)$ matrix may have as many as n independent eigenvectors or as few as one. We will show later that eigenvectors corresponding to distinct eigenvalues are linearly independent, but that the situation is more complicated with respect to "multiple" eigenvalues.

EXAMPLE 3.7 In this example we find the eigenvalues and corresponding eigenvectors of

$$A = \begin{bmatrix} -1 & 6 & 2 \\ 0 & 5 & -6 \\ 1 & 0 & -2 \end{bmatrix}.$$

We first find the characteristic polynomial by expanding $\det(A - tI)$ along the second row:

$$\det(A - tI) = (5-t)\begin{vmatrix} -1-t & 2 \\ 1 & -2-t \end{vmatrix} + 6\begin{vmatrix} -1-t & 6 \\ 1 & 0 \end{vmatrix};$$

and thus we find $p(t) = -t^3 + 2t^2 + 15t - 36$. We can easily verify for this simple example that $p(t) = -(t+4)(t-3)^2$.

Thus the (3×3) matrix A has two distinct eigenvalues, $\lambda = -4$ and $\lambda = 3$. To find the eigenvectors corresponding to $\lambda = -4$, we solve $(A + 4I)\mathbf{x} = \boldsymbol{\theta}$:

$$\begin{aligned} 3x_1 + 6x_2 + 2x_3 &= 0 \\ 9x_2 - 6x_3 &= 0 \\ x_1 \qquad + 2x_3 &= 0. \end{aligned}$$

The solution is $x_1 = -2x_3$, $x_2 = 2x_3/3$; so the eigenvectors corresponding to $\lambda = -4$ all have the form

$$\mathbf{x} = \begin{bmatrix} -2a \\ \frac{2}{3}a \\ a \end{bmatrix} = a\begin{bmatrix} -2 \\ \frac{2}{3} \\ 1 \end{bmatrix}, \quad a \neq 0.$$

The eigenvectors corresponding to $\lambda = 3$ are found by solving $(A - 3I)\mathbf{x} = \boldsymbol{\theta}$, which has solution $x_1 = 5x_3, x_2 = 3x_3$. Thus the eigenvectors of A corresponding to $\lambda = 3$ are given by

$$\mathbf{x} = \begin{bmatrix} 5a \\ 3a \\ a \end{bmatrix} = a\begin{bmatrix} 5 \\ 3 \\ 1 \end{bmatrix}, \quad a \neq 0.$$

We note that if $\mathbf{u}$ is an eigenvector of A corresponding to $\lambda = -4$, and if $\boldsymbol{v}$ is an eigenvector of A corresponding to $\lambda = 3$, then $\mathbf{u}$ and $\boldsymbol{v}$ are linearly independent. However any set of three or more eigenvectors is linearly dependent. Thus this matrix A has two linearly independent eigenvectors but no more.

Computational Considerations

In all the examples we have considered so far, it was possible to factor the characteristic polynomial and thus determine the eigenvalues by inspection. In reality we can rarely expect to be able to factor the characteristic polynomial; so we must solve the characteristic equation using numerical root-finding methods such as those discussed in Chapter 6. To be more specific about root finding, we recall that there are formulas for the roots of some polynomial equations. For instance the solution of the linear equation

$$at + b = 0,\ a \neq 0,$$

is given by

$$t = -\frac{b}{a};$$

and the roots of the quadratic equation

$$at^2 + bt + c = 0,\ a \neq 0,$$

are given by the familiar quadratic formula

$$t = \frac{-b \pm \sqrt{b^2 - 4ac}}{2a}.$$

There are similar (although more complicated) formulas for the roots of third-degree and fourth-degree polynomial equations. Unfortunately there are no such formulas for polynomials of degree five or higher [that is, formulas that express the zeros of $p(t)$ as a simple function of the coefficients of $p(t)$]. Moreover in the mid-nineteenth century, Abel *proved* that such formulas cannot exist for polynomials of degree five or higher.* This means that in general we cannot expect to find the eigenvalues of a large matrix exactly—the best we can do is to find good approximations to the eigenvalues. The eigenvalue problem differs qualitatively from the problem of solving $A\mathbf{x} = \mathbf{b}$. For a system $A\mathbf{x} = \mathbf{b}$ if we are willing to invest the effort required to solve the system by hand, we can obtain the exact solution in a finite number of steps. On the other hand we cannot in general expect to find roots of a polynomial equation in a finite number of steps.

Finding roots of the characteristic equation is not the only computational aspect of the eigenvalue problem that must be considered. In fact it is not hard to see that special techniques must be developed even to find the characteristic polynomial. To see the dimensions of this problem, consider the characteristic polynomial of an $(n \times n)$ matrix $A: p(t) = \det(A - tI)$. As we pointed out earlier in (3.4), the evaluation of det $(A - tI)$ from formula (3.3) ultimately requires the evaluation of $n!/2$ (2×2) determinants. Even for modest values of n, the number $n!/2$ is alarmingly large. For instance

$$10!/2 = 1{,}814{,}400$$

* For a historical discussion see J. E. Maxfield and M. W. Maxfield, *Abstract Algebra and Solution by Radicals*. Philadelphia: W. B. Saunders, 1971.

$\lambda^3-\lambda^2-\lambda+1 = \lambda^2(\lambda-1)-(\lambda+1)$
$= \lambda^2(\lambda-1)-(\lambda-1)$
$= (\lambda-1)(\lambda^2-1)$
$= (\lambda-1)(\lambda+1)(\lambda-1)$
$= (\lambda+1)(\lambda-1)^2$

while

$$20!/2 > 1.2 \times 10^{18}.$$

The enormous number of calculations required to compute det $(A-tI)$ means that we cannot find $p(t)$ in any practical sense by expanding det $(A-tI)$. In Chapter 5 we note that there are relatively efficient ways of finding det (A), but these techniques (which amount to using elementary row operations to triangularize A) are not useful in our problem of computing det $(A-tI)$ because of the variable t. In Section 3.7 we resolve this difficulty by using similarity transformations to transform A to a matrix H where A and H have the same characteristic polynomial, and where it is a trivial matter to calculate the characteristic polynomial for H. Moreover these transformation methods will give us some other important results as a by-product, results, such as the Cayley–Hamilton theorem, which have some practical computational significance. Finally in Chapter 6 we discuss other numerical procedures that are currently being used in software packages designed to solve the eigenvalue problem.

3.3 EXERCISES

1. Find the characteristic polynomial, the eigenvalues, and the eigenvectors of the following.

$$A=\begin{bmatrix}1 & 0\\ 2 & 3\end{bmatrix} \quad B=\begin{bmatrix}2 & -1\\ -1 & 2\end{bmatrix} \quad C=\begin{bmatrix}1 & -1\\ 1 & 3\end{bmatrix}$$

2. Repeat Exercise 1 for the following.

$$A=\begin{bmatrix}2 & 1\\ 0 & -1\end{bmatrix} \quad B=\begin{bmatrix}13 & -16\\ 9 & -11\end{bmatrix} \quad C=\begin{bmatrix}2 & 2\\ 3 & 3\end{bmatrix}$$

3. Find the characteristic polynomial, the eigenvalues, and the eigenvectors of the following.

$$A=\begin{bmatrix}-6 & -1 & 2\\ 3 & 2 & 0\\ -14 & -2 & 5\end{bmatrix} \quad B=\begin{bmatrix}-2 & -1 & 0\\ 0 & 1 & 1\\ -2 & -2 & -1\end{bmatrix} \quad C=\begin{bmatrix}3 & -1 & -1\\ -12 & 0 & 5\\ 4 & -2 & -1\end{bmatrix}$$

$$D=\begin{bmatrix}-7 & 4 & -3\\ 8 & -3 & 3\\ 32 & -16 & 13\end{bmatrix} \quad E=\begin{bmatrix}2 & 4 & 4\\ 0 & 1 & -1\\ 0 & 1 & 3\end{bmatrix}$$

4. a) Find the characteristic polynomial for the following matrices.

$$A=\begin{bmatrix}6 & 4 & 4 & 1\\ 4 & 6 & 1 & 4\\ 4 & 1 & 6 & 4\\ 1 & 4 & 4 & 6\end{bmatrix} \quad B=\begin{bmatrix}5 & 4 & 1 & 1\\ 4 & 5 & 1 & 1\\ 1 & 1 & 4 & 2\\ 1 & 1 & 2 & 4\end{bmatrix} \quad C=\begin{bmatrix}1 & -1 & -1 & -1\\ -1 & 1 & -1 & -1\\ -1 & -1 & 1 & -1\\ -1 & -1 & -1 & 1\end{bmatrix}$$

b) Verify that the roots of the characteristic equation are as follows.
For A, $t=-1$, $t=5$, $t=15$.
For B, $t=1$, $t=2$, $t=5$, $t=10$.
For C, $t=2$, $t=-2$.

5. Find the eigenvectors for the matrices in Exercise 4.

6. Let A be an $(n \times n)$ nonsingular matrix, and suppose λ is an eigenvalue of A. Prove that $1/\lambda$ is an eigenvalue of A^{-1}. (Hint: Suppose $A\mathbf{x} = \lambda\mathbf{x}$ where $\mathbf{x} \neq \boldsymbol{\theta}$. Multiply both sides by A^{-1}.)

7. Let A be an $(n \times n)$ matrix, and suppose λ is an eigenvalue of A. Prove that λ^2 is an eigenvalue of A^2. (Hint: Begin with $A\mathbf{x} = \lambda\mathbf{x}$ where $\mathbf{x} \neq \boldsymbol{\theta}$.)

8. Use induction to prove the general form of Exercise 7. That is, if λ is an eigenvalue of A, then λ^k is an eigenvalue of A^k, $k = 2, 3, 4, \ldots$.

9. Let $q(t) = t^3 - 2t^2 - t + 2$; and for any $(n \times n)$ matrix H define the "matrix polynomial" $q(H)$ by
$$q(H) = H^3 - 2H^2 - H + 2I$$
where I is the $(n \times n)$ identity matrix.
   a) Prove that if λ is an eigenvalue of H, then the number $q(\lambda)$ is an eigenvalue of the matrix $q(H)$. [Hint: Suppose $H\mathbf{x} = \lambda\mathbf{x}$ where $\mathbf{x} \neq \boldsymbol{\theta}$, and use Exercise 8 to evaluate $q(H)\mathbf{x}$.]
   b) Use part (a) to calculate the eigenvalues and eigenvectors of $q(A)$ and $q(B)$ where A and B are from Exercise 3.

10. With $q(t)$ as in Exercise 9, verify that $q(C)$ is the zero matrix where C is from Exercise 3. [Note that $q(t)$ is the characteristic polynomial for C.]

11. This problem establishes a special case of the Cayley–Hamilton theorem: if $p(t)$ is the characteristic polynomial for A, then $p(A)$ is the zero matrix.
   a) Prove that if B is a (3×3) matrix, and if $B\mathbf{x} = \boldsymbol{\theta}$ for every $\mathbf{x}$ in R^3, then B is the zero matrix. (Hint: Consider $B\mathbf{e}_1$, $B\mathbf{e}_2$, and $B\mathbf{e}_3$.)
   b) Suppose $\lambda_1, \lambda_2,$ and λ_3 are the eigenvalues of a (3×3) matrix A, and suppose $\mathbf{u}_1, \mathbf{u}_2,$ and $\mathbf{u}_3$ are corresponding eigenvectors. Prove that if $\{\mathbf{u}_1, \mathbf{u}_2, \mathbf{u}_3\}$ is a linearly independent set, and if $p(t)$ is the characteristic polynomial for A, then $p(A)$ is the zero matrix. (Hint: Any vector $\mathbf{x}$ in R^3 can be expressed as a linear combination of $\mathbf{u}_1$, $\mathbf{u}_2$, and $\mathbf{u}_3$.)

12. Verify Theorem 3.3 for $n = 2$, $n = 3$, and $n = 4$.

13. The "power method" is a numerical method that is used to estimate the "dominant" eigenvalue of a matrix A. (See Chapter 6 for the practical details. By the dominant eigenvalue we mean the one that is largest in absolute value.) The algorithm proceeds as follows.

    1. Choose any starting vector $\mathbf{x}_0$, $\mathbf{x}_0 \neq \boldsymbol{\theta}$.
    2. Let $\mathbf{x}_{k+1} = A\mathbf{x}_k$, $k = 0, 1, 2, \ldots$.
    3. Let $\beta_k = \mathbf{x}_k^T\mathbf{x}_{k+1}/\mathbf{x}_k^T\mathbf{x}_k$, $k = 0, 1, 2, \ldots$.

    Under suitable conditions it can be shown that $\{\beta_k\} \to \lambda_1$ where λ_1 is the dominant eigenvalue of A.

   a) Use the power method to estimate the dominant eigenvalue of the matrix C in Exercise 3. 3. Use the starting vector
$$\mathbf{x}_0 = \begin{bmatrix} 1 \\ 1 \\ 1 \end{bmatrix},$$
and calculate β_0, β_1, β_2, β_3, and β_4.

b) Suppose A is an $(n \times n)$ matrix and has real eigenvalues $\lambda_1, \lambda_2, \ldots, \lambda_n$ with corresponding eigenvectors $\mathbf{u}_1, \mathbf{u}_2, \ldots, \mathbf{u}_n$. Furthermore suppose $|\lambda_1| > |\lambda_2| \geqslant \ldots \geqslant |\lambda_n|$, and suppose the starting vector $\mathbf{x}_0$ satisfies $\mathbf{x}_0 = c_1\mathbf{u}_1 + c_2\mathbf{u}_2 + \ldots + c_n\mathbf{u}_n$ where $c_1 \neq 0$. Prove that

$$\lim_{k \to \infty} \beta_k = \lambda_1.$$

(Hint: Observe that $\mathbf{x}_j = A^j\mathbf{x}_0, j = 1, 2, \ldots,$ and use Exercise 8 to calculate $\mathbf{x}_{k+1}$ and $\mathbf{x}_k$. Next factor all powers of λ_1 from numerator and denominator of $\beta_k = \mathbf{x}_k^{\mathrm{T}}\mathbf{x}_{k+1}/\mathbf{x}_k^{\mathrm{T}}\mathbf{x}_k$.)

14. Prove that if λ is an eigenvalue of A, then λ is an eigenvalue of A^{T}. [Hint: If $A - \lambda I$ is singular, what about $(A - \lambda I)^{\mathrm{T}}$?]

15. Let $q(t) = t^n + a_{n-1}t^{n-1} + \ldots + a_1 t + a_0$, and define the $(n \times n)$ ***companion matrix*** by

$$A = \begin{bmatrix} -a_{n-1} & -a_{n-2} & \cdots & -a_1 & -a_0 \\ 1 & 0 & \cdots & 0 & 0 \\ 0 & 1 & \cdots & 0 & 0 \\ \vdots & & & & \vdots \\ 0 & 0 & \cdots & 1 & 0 \end{bmatrix}.$$

a) For $n = 2$ and for $n = 3$, show that $\det(A - tI) = (-1)^n q(t)$.
b) Give the companion matrix A for the polynomial $q(t) = t^4 + 3t^3 - t^2 + 2t - 2$. Verify that $q(t)$ is the characteristic polynomial for A.
c) Prove for all n that $\det(A - tI) = (-1)^n q(t)$.

3.4 APPLICATIONS OF EIGENVALUES

skip

In this section we will present several examples that show how eigenvalues can be used in solving systems of difference equations and differential equations. We begin with an example of a system of difference equations.

EXAMPLE 3.8 The problem of finding all sequences $\{s_k\}_{k=0}^{\infty}$ and $\{t_k\}_{k=0}^{\infty}$ such that

$$\begin{aligned} s_k &= 5s_{k-1} - 6t_{k-1} \\ t_k &= 3s_{k-1} - 4t_{k-1}, \quad k = 1, 2, \ldots \end{aligned} \tag{3.6}$$

is an example of a system of linear difference equations. Frequently initial values s_0 and t_0 are specified; and in this case, sequences $\{s_k\}$ and $\{t_k\}$ that satisfy (3.6) can be calculated directly. For example if $s_0 = 1$ and $t_0 = 0$, then we can use (3.6) to obtain

$$\begin{array}{llll} s_1 = 5 & s_2 = 7 & s_3 = 17 & s_4 = 31 \\ t_1 = 3 & t_2 = 3 & t_3 = 9 & t_4 = 15, \text{ etc.} \end{array} \tag{3.7}$$

† This section may be omitted without any loss of continuity.

Clearly we can write (3.6) in matrix terms as

$$\mathbf{x}_k = A\mathbf{x}_{k-1},\ k = 1, 2, \ldots \tag{3.8}$$

where

$$\mathbf{x}_k = \begin{bmatrix} s_k \\ t_k \end{bmatrix} \qquad A = \begin{bmatrix} 5 & -6 \\ 3 & -4 \end{bmatrix}. \tag{3.9}$$

Vector sequences that are generated as in (3.8) occur in a variety of applications and serve as mathematical models to describe population growth, ecological systems, radar tracking of airborne objects, digital control of chemical processes, etc. One of the objectives in such models is to describe the behavior of the sequence $\{\mathbf{x}_k\}$ in qualitative or quantitative terms.

Given a system of difference equations in the form

$$\mathbf{x}_k = A\mathbf{x}_{k-1},\ k = 1, 2, \ldots, \tag{3.10}$$

it is possible to derive a closed-form expression for $\mathbf{x}_k$ in terms of the eigenvalues and eigenvectors of A. This expression can then be used to analyze the behavior of the sequence $\{\mathbf{x}_k\}$. To keep the details simple, we will assume that A is a (2×2) matrix although clearly the discussion below can be extended to an $(n \times n)$ matrix A. To begin, let λ_1 and λ_2 be eigenvalues of A with corresponding eigenvectors $\mathbf{u}_1$ and $\mathbf{u}_2$; so

$$A\mathbf{u}_1 = \lambda_1\mathbf{u}_1 \qquad A\mathbf{u}_2 = \lambda_2\mathbf{u}_2.$$

We next make a critical assumption; we assume that $\mathbf{u}_1$ and $\mathbf{u}_2$ are linearly independent. Given this, $\{\mathbf{u}_1, \mathbf{u}_2\}$ is a basis for R^2; and we can represent any starting vector $\mathbf{x}_0$ as

$$\mathbf{x}_0 = b_1\mathbf{u}_1 + b_2\mathbf{u}_2.$$

Now $\mathbf{x}_1 = A\mathbf{x}_0$ by (3.10); so we have $\mathbf{x}_1 = A(b_1\mathbf{u}_1 + b_2\mathbf{u}_2)$, or

$$\mathbf{x}_1 = b_1A\mathbf{u}_1 + b_2A\mathbf{u}_2 = b_1\lambda_1\mathbf{u}_1 + b_2\lambda_2\mathbf{u}_2.$$

Similarly $\mathbf{x}_2 = A\mathbf{x}_1$; so $\mathbf{x}_2 = A(b_1\lambda_1\mathbf{u}_1 + b_2\lambda_2\mathbf{u}_2)$, or

$$\mathbf{x}_2 = b_1\lambda_1A\mathbf{u}_1 + b_2\lambda_2A\mathbf{u}_2 = b_1\lambda_1^2\mathbf{u}_1 + b_2\lambda_2^2\mathbf{u}_2.$$

In general since $\mathbf{x}_k = A\mathbf{x}_{k-1}$, we find

$$\mathbf{x}_k = b_1\lambda_1^k\mathbf{u}_1 + b_2\lambda_2^k\mathbf{u}_2,\ k = 0, 1, 2, 3, \ldots. \tag{3.11}$$

Thus (3.11) provides a formula in terms of the eigenvalues and eigenvectors of A for any sequence $\{\mathbf{x}_k\}_{k=0}^{\infty}$ that satisfies (3.10). Formula (3.11) also provides qualitative information about the solution of (3.10), and this information is frequently of importance in applications. For instance if $|\lambda_1| > 1$ and $|\lambda_2| > 1$, and if $\mathbf{x}_0 \neq \boldsymbol{\theta}$, then the sequence $\{\mathbf{x}_k\}$ is growing without bound. On the other hand if $\lambda_1 = 1$ and $|\lambda_2| < 1$, then

$$\lim_{k \to \infty} \mathbf{x}_k = b_1\mathbf{u}_1;$$

and this represents physically a steady state or equilibrium for the process that is modeled by (3.10). We note that the derivation of (3.11) required $\mathbf{u}_1$ and $\mathbf{u}_2$ to be linearly independent, and a more detailed analysis is necessary to solve (3.10) if A does not have two linearly independent eigenvectors. Finally it should be clear that a formula similar to (3.11) holds in the case that the matrix A in (3.10) is $(n \times n)$ and has n linearly independent eigenvectors.

To illustrate (3.11) in a specific case, let A be the (2×2) matrix in (3.9) and let $\mathbf{x}_0$ be the starting vector

$$\mathbf{x}_0 = \begin{bmatrix} 1 \\ 0 \end{bmatrix}. \tag{3.12}$$

It is easy to verify that A has eigenvalues $\lambda_1 = 2$ and $\lambda_2 = -1$ with corresponding eigenvectors

$$\mathbf{u}_1 = \begin{bmatrix} 2 \\ 1 \end{bmatrix} \quad \text{and} \quad \mathbf{u}_2 = \begin{bmatrix} 1 \\ 1 \end{bmatrix}.$$

Since $\mathbf{u}_1$ and $\mathbf{u}_2$ are linearly independent, any starting vector $\mathbf{x}_0$ can be represented as $\mathbf{x}_0 = b_1\mathbf{u}_1 + b_2\mathbf{u}_2$. By (3.11) the vectors $\mathbf{x}_k$ that satisfy $\mathbf{x}_k = A\mathbf{x}_{k-1}$, $k = 1, 2, \ldots$ are given by

$$\mathbf{x}_k = b_1 2^k \mathbf{u}_1 + b_2(-1)^k \mathbf{u}_2 \tag{3.13a}$$

where since $\mathbf{x}_0$ is given by (3.12), $b_1 = 1$ and $b_2 = -1$. Therefore the sequence $\{\mathbf{x}_k\}$ is given precisely by

$$\mathbf{x}_k = 2^k \begin{bmatrix} 2 \\ 1 \end{bmatrix} - (-1)^k \begin{bmatrix} 1 \\ 1 \end{bmatrix}, \; k = 0, 1, \ldots. \tag{3.13b}$$

[Compare (3.13b) with (3.7).]

EXAMPLE 3.9 This example shows how a system of difference equations might serve as a mathematical model for a physical process. The model is very simple so that the details do not obscure the ideas; the model should be considered illustrative rather than realistic.

Suppose that animals are being raised for market, and the grower wishes to determine how the annual rate of harvesting animals will affect the yearly size of the herd. To begin, we let $x_1(k)$ and $x_2(k)$ be the "state variables" that measure the size of the herd in the kth year of operation where

$x_1(k) =$ number of animals less than one year old at year k
$x_2(k) =$ number of animals more than one year old at year k.

We assume that animals less than one year old do not reproduce, and that animals more than one year old have a reproduction rate of b per year. Thus if the herd has $x_2(k)$ mature animals at year k, we expect to have $x_1(k+1)$ young animals at year $k+1$ where

$$x_1(k+1) = bx_2(k).$$

Next we assume that the young animals have a death rate of d_1 per year, and the mature animals have a death rate of d_2 per year. Furthermore we assume that the mature animals are harvested at a rate of h per year and that young animals are not harvested. Thus we expect to have $x_2(k+1)$ mature animals at year $k+1$ where

$$x_2(k+1) = x_1(k) + x_2(k) - d_1x_1(k) - d_2x_2(k) - hx_2(k).$$

This equation reflects the facts that an animal that is young at year k will mature by year $k+1$, an animal that is mature at year k is still mature at year $k+1$, a certain percentage of young and mature animals will die during the year, and a certain percentage of mature animals will be harvested during the year. Collecting like terms in the second equation and combining the two equations, we obtain the state equations for the herd:

$$\begin{aligned} x_1(k+1) &= bx_2(k) \\ x_2(k+1) &= (1-d_1)x_1(k) + (1-d_2-h)x_2(k). \end{aligned} \tag{3.14}$$

The state equations give the size and composition of the herd at year $k+1$ in terms of the size and composition of the herd at year k. For example if we know the initial composition of the herd at year zero, $x_1(0)$ and $x_2(0)$, we can use (3.14) to calculate the composition of the herd after one year, $x_1(1)$ and $x_2(1)$.

In matrix form, (3.14) becomes

$$\mathbf{x}(k) = A\mathbf{x}(k-1), \; k = 1, 2, 3, \ldots, \tag{3.15}$$

where

$$\mathbf{x}(k) = \begin{bmatrix} x_1(k) \\ x_2(k) \end{bmatrix} \quad \text{and} \quad A = \begin{bmatrix} 0 & b \\ (1-d_1) & (1-d_2-h) \end{bmatrix}.$$

In the context of this example the growth and composition of the herd are governed by the eigenvalues of A, and these can be controlled by varying the parameter h.

Difference equations are useful for describing the state of a physical system at discrete values of time. Mathematical models that describe the evolution of a physical system for all values of time are frequently expressed in terms of a differential equation or a system of differential equations. A simple example of a system of differential equations is

$$\begin{aligned} v'(t) &= av(t) + bw(t) \\ w'(t) &= cv(t) + dw(t). \end{aligned} \tag{3.16}$$

In (3.16) the problem is to find functions $v(t)$ and $w(t)$ that simultaneously satisfy these equations and in which initial conditions $v(0)$ and $w(0)$ may also be specified. We can express (3.16) in matrix terms if we let

$$\mathbf{x}(t) = \begin{bmatrix} v(t) \\ w(t) \end{bmatrix}.$$

Then (3.16) can be written as $\mathbf{x}'(t) = A\mathbf{x}(t)$ where

$$\mathbf{x}'(t) = \begin{bmatrix} v'(t) \\ w'(t) \end{bmatrix} \qquad A = \begin{bmatrix} a & b \\ c & d \end{bmatrix}.$$

The equation $\mathbf{x}'(t) = A\mathbf{x}(t)$ is reminiscent of the simple scalar differential equation, $y'(t) = \alpha y(t)$, which is frequently used in calculus to model problems such as radioactive decay or bacterial growth. To find a function $y(t)$ that satisfies the identity $y'(t) = \alpha y(t)$, we rewrite the equation as $y'(t)/y(t) = \alpha$. Integrating both sides with respect to t yields $\ln|y(t)| = \alpha t + \beta$, or equivalently $y(t) = y_0 e^{\alpha t}$ where $y_0 = y(0)$.

Using the scalar equation as a guide, we assume the vector equation $\mathbf{x}'(t) = A\mathbf{x}(t)$ has a solution of the form

$$\mathbf{x}(t) = e^{\lambda t}\mathbf{u} \tag{3.17}$$

where $\mathbf{u}$ is a constant vector. To see if the function $\mathbf{x}(t)$ in (3.17) can be a solution, we differentiate and get $\mathbf{x}'(t) = \lambda e^{\lambda t}\mathbf{u}$. On the other hand $A\mathbf{x}(t) = e^{\lambda t}A\mathbf{u}$; so (3.17) will be a solution of $\mathbf{x}'(t) = A\mathbf{x}(t)$ if and only if

$$e^{\lambda t}(A - \lambda I)\mathbf{u} = \boldsymbol{\theta}. \tag{3.18}$$

Now $e^{\lambda t} \neq 0$ for all values of t; so (3.18) will be satisfied only if $(A - \lambda I)\mathbf{u} = \boldsymbol{\theta}$. Therefore if λ is an eigenvalue of A, and $\mathbf{u}$ is a corresponding eigenvector, then $\mathbf{x}(t)$ given in (3.17) is a solution to $\mathbf{x}'(t) = A\mathbf{x}(t)$. (Note: The choice $\mathbf{u} = \boldsymbol{\theta}$ will also give a solution, but it is a trivial solution.)

If the (2×2) matrix A has eigenvalues λ_1 and λ_2 with corresponding eigenvectors $\mathbf{u}_1$ and $\mathbf{u}_2$, then two solutions of $\mathbf{x}'(t) = A\mathbf{x}(t)$ are $\mathbf{x}_1(t) = e^{\lambda_1 t}\mathbf{u}_1$ and $\mathbf{x}_2(t) = e^{\lambda_2 t}\mathbf{u}_2$. It is easy to verify that any linear combination of $\mathbf{x}_1(t)$ and $\mathbf{x}_2(t)$ is also a solution; so

$$\mathbf{x}(t) = a_1\mathbf{x}_1(t) + a_2\mathbf{x}_2(t) \tag{3.19}$$

will solve $\mathbf{x}'(t) = A\mathbf{x}(t)$ for any choice of scalars a_1 and a_2. Finally the ***initial-value problem*** consists of finding a solution to $\mathbf{x}'(t) = A\mathbf{x}(t)$ that satisfies an initial condition, $\mathbf{x}(0) = \mathbf{x}_0$ where $\mathbf{x}_0$ is some specified vector. Given the form of $\mathbf{x}_1(t)$ and $\mathbf{x}_2(t)$, it is clear from (3.19) that $\mathbf{x}(0) = a_1\mathbf{u}_1 + a_2\mathbf{u}_2$. If the eigenvectors $\mathbf{u}_1$ and $\mathbf{u}_2$ are linearly independent, we can always choose scalars b_1 and b_2 so that $\mathbf{x}_0 = b_1\mathbf{u}_1 + b_2\mathbf{u}_2$; and therefore $\mathbf{x}(t) = b_1\mathbf{x}_1(t) + b_2\mathbf{x}_2(t)$ is the solution of $\mathbf{x}'(t) = A\mathbf{x}(t)$, $\mathbf{x}(0) = \mathbf{x}_0$.

In general given the problem of solving

$$\mathbf{x}'(t) = A\mathbf{x}(t), \; \mathbf{x}(0) = \mathbf{x}_0, \tag{3.20}$$

where A is an $(n \times n)$ matrix, we can proceed just as above. We first find the eigenvalues $\lambda_1, \lambda_2, \ldots, \lambda_n$ of A and corresponding eigenvectors $\mathbf{u}_1, \mathbf{u}_2, \ldots, \mathbf{u}_n$. For each i, $\mathbf{x}_i(t) = e^{\lambda_i t}\mathbf{u}_i$ is a solution of $\mathbf{x}'(t) = A\mathbf{x}(t)$ as is the general expression

$$\mathbf{x}(t) = b_1\mathbf{x}_1(t) + b_2\mathbf{x}_2(t) + \ldots + b_n\mathbf{x}_n(t). \tag{3.21}$$

As before, $\mathbf{x}(0) = b_1\mathbf{u}_1 + b_2\mathbf{u}_2 + \ldots + b_n\mathbf{u}_n$; so if $\mathbf{x}_0$ can be expressed as a linear combination of $\mathbf{u}_1, \mathbf{u}_2, \ldots, \mathbf{u}_n$, then we can construct a solution to (3.20) in the form of (3.21). If the eigenvectors of A do not form a basis for R^n, we can still get a solution of the form (3.21); but a more detailed analysis is required (see Section 3.10).

EXAMPLE 3.10 As a specific example, consider the initial-value problem

$$\begin{aligned} v'(t) &= v(t) - 2w(t) \\ w'(t) &= v(t) + 4w(t) \end{aligned} \tag{3.22}$$

where $v(0) = 4$ and $w(0) = -3$. In vector form, (3.22) becomes $\mathbf{x}'(t) = A\mathbf{x}(t)$, $\mathbf{x}(0) = \mathbf{x}_0$ where

$$\mathbf{x}(t) = \begin{bmatrix} v(t) \\ w(t) \end{bmatrix}, \quad A = \begin{bmatrix} 1 & -2 \\ 1 & 4 \end{bmatrix}, \quad \mathbf{x}_0 = \begin{bmatrix} 4 \\ -3 \end{bmatrix}. \tag{3.23}$$

The eigenvalues of A are $\lambda_1 = 2$ and $\lambda_2 = 3$, and we choose corresponding eigenvectors

$$\mathbf{u}_1 = \begin{bmatrix} 2 \\ -1 \end{bmatrix} \quad \mathbf{u}_2 = \begin{bmatrix} 1 \\ -1 \end{bmatrix}.$$

As before, $\mathbf{x}_1(t) = e^{2t}\mathbf{u}_1$ and $\mathbf{x}_2(t) = e^{3t}\mathbf{u}_2$ are solutions of $\mathbf{x}'(t) = A\mathbf{x}(t)$ as is any linear combination, $\mathbf{x}(t) = b_1\mathbf{x}_1(t) + b_2\mathbf{x}_2(t)$. We now need only choose appropriate constants b_1 and b_2 so that $\mathbf{x}(0) = \mathbf{x}_0$ where we know $\mathbf{x}(0) = b_1\mathbf{u}_1 + b_2\mathbf{u}_2$. For $\mathbf{x}_0$ in (3.23) it is routine to find $\mathbf{x}_0 = \mathbf{u}_1 + 2\mathbf{u}_2$. Thus the solution of $\mathbf{x}'(t) = A\mathbf{x}(t)$, $\mathbf{x}(0) = \mathbf{x}_0$ is $\mathbf{x}(t) = \mathbf{x}_1(t) + 2\mathbf{x}_2(t)$, or

$$\mathbf{x}(t) = e^{2t}\mathbf{u}_1 + 2e^{3t}\mathbf{u}_2.$$

After $\mathbf{x}(t)$ is written out componentwise, the solution of the original problem (3.22) is seen to be $v(t) = 2e^{2t} + 2e^{3t}$, $w(t) = -e^{2t} - 2e^{3t}$.

Electrical networks, such as the one shown in Fig. 3.1, illustrate one sort of physical system that can be modeled by a system of differential equations. The network in Fig. 3.1 consists of two closed loops; and we designate arbitrary directions for the currents, I_1 and I_2, in the top and bottom loops. In this network R_1 and R_2 are known resistances, C is a known capacitance, and L is a known inductance. Given the initial state of the network, we want to determine the currents

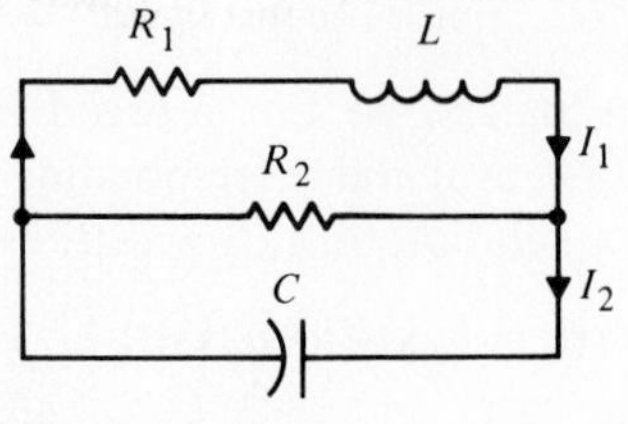

Figure 3.1

$I_1(t)$ and $I_2(t)$ for all values of time t. Two of the laws that govern the behavior of electrical networks follow.

1. In a closed loop the sum of the voltage drops is equal to the applied voltage.
2. The algebraic sum of the currents entering a node equals the sum of the currents leaving a node.

Moreover the voltage drops are as follows: RI across a resistance R, LI' across an inductance L where $I' = dI/dt$, and Q/C across a capacitance C where Q is the charge on the capacitor. Finally the charge Q is related to the current I by $Q' = I$. Therefore in a single loop we must have

$$LQ''(t) + RQ'(t) + \frac{1}{C}Q(t) = E(t) \tag{3.24}$$

where $E(t)$ is the applied voltage. A network, such as that in Fig. 3.1, consisting of several loops will give rise to a coupled system of differential equations.

:XAMPLE 3.11 For the network in Fig. 3.1 the currents [$I_1(t)$ in the upper loop and $I_2(t)$ in the lower loop] must satisfy

$$\begin{aligned} LI'_1 + R_2(I_1 - I_2) + R_1 I_1 &= 0 \\ R_2(I_2 - I_1) + \frac{1}{C}Q &= 0. \end{aligned} \tag{3.25}$$

In (3.25) the current across R_2 in the upper loop is $I_1 - I_2$ whereas with respect to the lower loop the current is $I_2 - I_1$. We could write (3.25) as a system of two second-order differential equations using (3.24). However we can simplify (3.25) by replacing the second equation by its derivative and obtain

$$\begin{aligned} LI'_1 + R_2(I_1 - I_2) + R_1 I_1 &= 0 \\ R_2(I'_2 - I'_1) + \frac{1}{C}I_2 &= 0. \end{aligned} \tag{3.26a}$$

Finally if we set $v(t) = I_1(t)$ and $w(t) = I_2(t) - I_1(t)$, then (3.26a) becomes

$$\begin{aligned} Lv' &= -R_1 v + R_2 w \\ CR_2 w' &= -v - w. \end{aligned} \tag{3.26b}$$

As a numerical example, suppose $R_1 = 4$ ohms, $R_2 = 2$ ohms, $L = 1$ henry, and $C = .5$ farads. With these values (3.26b) becomes

$$\begin{aligned} v' &= -4v + 2w \\ w' &= -v - w. \end{aligned} \tag{3.26c}$$

Expressing (3.26c) in the form $\mathbf{x}'(t) = A\mathbf{x}(t)$, we see that A has eigenvalues $\lambda_1 = -2$, $\lambda_2 = -3$ and corresponding eigenvectors

$$\mathbf{u}_1 = \begin{bmatrix} 1 \\ 1 \end{bmatrix} \qquad \mathbf{u}_2 = \begin{bmatrix} 2 \\ 1 \end{bmatrix}.$$

If $\mathbf{x}(t) = b_1\mathbf{x}_1(t) + b_2\mathbf{x}_2(t)$ where $\mathbf{x}_1(t) = e^{-2t}\mathbf{u}_1$ and $\mathbf{x}_2(t) = e^{-3t}\mathbf{u}_2$, then $\mathbf{x}(t)$ is the general solution of $\mathbf{x}'(t) = A\mathbf{x}(t)$. Writing $\mathbf{x}(t)$ componentwise, we have $v(t) = b_1e^{-2t} + 2b_2e^{-3t}$ and $w(t) = b_1e^{-2t} + b_2e^{-3t}$. In terms of the currents, $v = I_1$ and $w = I_2 - I_1$; so

$$\begin{aligned} I_1(t) &= b_1e^{-2t} + 2b_2e^{-3t} \\ I_2(t) &= 2b_1e^{-2t} + 3b_2e^{-3t}. \end{aligned} \tag{3.27}$$

When initial conditions are given for the network, we can determine b_1 and b_2. For example if the initial current in the upper loop is zero, then $I_1(0) = 0$. If the initial charge on the capacitor is $Q(0) = .01$, we can use the second equation in (3.26a) to find $I_2(0)$: $Q(0) = -CR_2(I_2(0) - I_1(0))$ or $I_2(0) = -.01$. Therefore $b_1 = -.02$, $b_2 = .01$ or

$$\begin{aligned} I_1(t) &= -.02e^{-2t} + .02e^{-3t} \\ I_2(t) &= -.04e^{-2t} + .03e^{-3t}. \end{aligned}$$

Another sort of physical system that leads to a system of differential equations is illustrated in Fig. 3.2. This figure shows a spring-mass system where $y_1 = 0$ and $y_2 = 0$ indicate the equilibrium position of the masses while $y_1(t)$ and $y_2(t)$ denote the displacements at time t. For a single spring and mass as in Fig. 3.3, we can use Hooke's law and $F = ma$ to deduce $my''(t) = -ky(t)$; that is, the restoring force of the spring is proportional to the displacement, $y(t)$, of the mass from the equilibrium position, $y = 0$. The constant of proportionality is the spring constant k, and the minus sign indicates that the force is directed towards equilibrium.

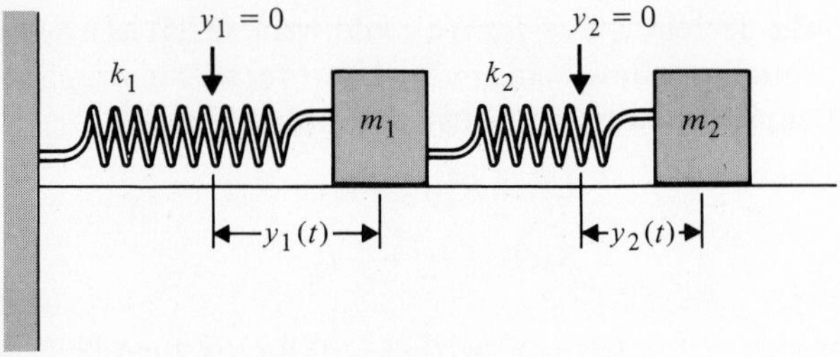

Figure 3.2

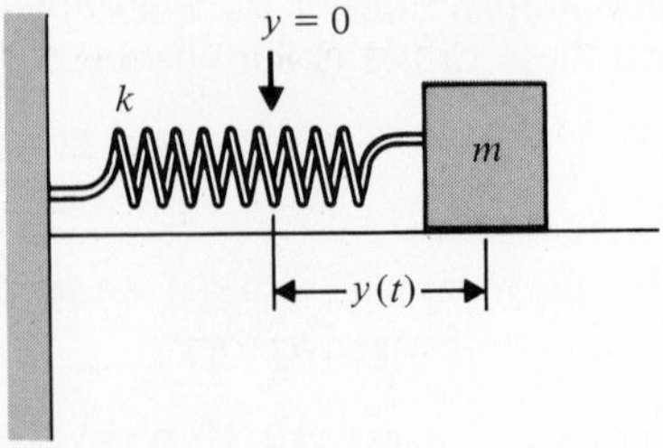

Figure 3.3

In Fig. 3.2 the spring attached to m_2 is "stretched" (or "compressed") by the amount $y_2(t)-y_1(t)$; so we can write $m_2 y_2''(t) = -k_2(y_2(t)-y_1(t))$. The mass m_1 is being "pulled" by two springs; so we have $m_1 y_1''(t) = -k_1 y_1(t)+k_2(y_2(t)-y_1(t))$. Thus the motion of the physical system is governed by

$$y_1''(t) = -\frac{k_1+k_2}{m_1} y_1(t) + \frac{k_2}{m_1} y_2(t)$$
$$y_2''(t) = \frac{k_2}{m_2} y_1(t) - \frac{k_2}{m_2} y_2(t).$$

To solve these equations, we write them in matrix form as $\mathbf{y}''(t) = A\mathbf{y}(t)$, and we use a trial solution of the form $\mathbf{y}(t)=e^{\omega t}\mathbf{u}$ where $\mathbf{u}$ is a constant vector. Since $\mathbf{y}''(t) = \omega^2 e^{\omega t}\mathbf{u}$, we will have a solution if

$$\omega^2 e^{\omega t}\mathbf{u} - e^{\omega t} A\mathbf{u} = \boldsymbol{\theta}$$

or if $(A-\omega^2 I)\mathbf{u} = \boldsymbol{\theta}$. Thus to solve $\mathbf{y}''(t) = A\mathbf{y}(t)$, we solve $(A-\lambda I)\mathbf{u} = \boldsymbol{\theta}$ and then choose ω so that $\omega^2 = \lambda$. [Note that λ will be negative and real; so ω must be a complex number. For complex ω, $e^{\omega t}$ can be expressed in terms of sines and cosines using Euler's identity. (See Section 3.5.) In fact the motion of a spring-mass system is usually oscillatory in nature; so we would expect solutions that behave like damped sine and cosine waves.]

3.4 EXERCISES

1. For $k = 0, 1, 2, 3, 4$ and $\mathbf{x}_k$ given by (3.13b), verify that (3.13b) generates the same numbers as (3.7)

2. Find values for b_1 and b_2 in (3.13a) when $\mathbf{x}_0$ is as follows.

a) $\mathbf{x}_0 = \begin{bmatrix} 1 \\ -2 \end{bmatrix}$ b) $\mathbf{x}_0 = \begin{bmatrix} 2 \\ 3 \end{bmatrix}$ c) $\mathbf{x}_0 = \begin{bmatrix} 4 \\ 2 \end{bmatrix}$

3. Write each system of difference equations in vector terms as in (3.10), and find the most general form for $\mathbf{x}_k$ as in (3.11). Next find the particular solution that satisfies the given starting values.

a) $s_k = 2s_{k-1} - t_{k-1}$
$t_k = -s_{k-1} + 2t_{k-1}$
$s_0 = 3,\ t_0 = 1$

b) $r_k = 3r_{k-1} - s_{k-1} - t_{k-1}$
$s_k = -12r_{k-1} + 5t_{k-1}$
$t_k = 4r_{k-1} - 2s_{k-1} - t_{k-1}$
$r_0 = 3,\ s_0 = -14,\ t_0 = 8$

4. Repeat Exercise 3 for these systems.

a) $s_k = s_{k-1} + 4t_{k-1}$
$t_k = s_{k-1} + t_{k-1}$
$s_0 = -1,\ t_0 = 2$

b) $r_k = -6r_{k-1} + s_{k-1} + 3t_{k-1}$
$s_k = -3r_{k-1} + 2t_{k-1}$
$t_k = -20r_{k-1} + 2s_{k-1} + 10t_{k-1}$
$r_0 = 1,\ s_0 = 1,\ t_0 = -1$

5. a) In (3.6) of Example 3.8 what starting values s_0 and t_0 will yield $s_5 = 60$ and $t_5 = 28$? [Hint: Recall (3.13a).]

b) In Exercise 3(a) what starting values s_0 and t_0 will yield $s_4 = 84$ and $t_4 = -78$?

6. In Example 3.9 let $b = .9$, $d_1 = .1$, and $d_2 = .2$.

   a) Show that a harvest rate of $h = .3$ causes the herd to grow without bound. (This is an example of "instability." Since the herd cannot actually grow without bound, the physical interpretation is that the mathematical model is not valid for large herds; and modifications to the model must be made to reflect the limitations on growth.)
   b) Show that a harvest rate of $h = .7$ causes the herd to die out.
   c) Show that a harvest rate of $h = .61$ leads to an equilibrium state for the herd.
   d) For $h = .61$, $x_1(0) = 28$, $x_2(0) = 11$, what is the eventual composition of the herd?

7. Write each system of differential equations in vector terms as $\mathbf{x}'(t) = A\mathbf{x}(t)$, and find the most general form of the solution as in (3.21). Next find the particular solution that satisfies the given initial conditions.

   a) $$\begin{aligned} u'(t) &= 3u(t) + v(t) \\ v'(t) &= 4u(t) + 3v(t) \\ u(0) &= 2,\ v(0) = 0 \end{aligned}$$

   b) $$\begin{aligned} u'(t) &= 4u(t) \qquad\quad + w(t) \\ v'(t) &= -2u(t) + v(t) \\ w'(t) &= -2u(t) \qquad + w(t) \\ u(0) &= -1,\ v(0) = 1,\ w(0) = 0 \end{aligned}$$

8. Repeat Exercise 7 for the following systems.

   a) $$\begin{aligned} u'(t) &= 3u(t) \\ v'(t) &= 8u(t) - v(t) \\ u(0) &= -3,\ v(0) = -4 \end{aligned}$$

   b) $$\begin{aligned} u'(t) &= 3u(t) + v(t) - 2w(t) \\ v'(t) &= -u(t) + 2v(t) + w(t) \\ w'(t) &= 4u(t) + v(t) - 3w(t) \\ u(0) &= -2,\ v(0) = 4,\ w(0) = -8 \end{aligned}$$

9. Consider the system

$$\begin{aligned} u'(t) &= u(t) - v(t) \\ v'(t) &= u(t) + 3v(t). \end{aligned}$$

   a) Write this system in the form $\mathbf{x}'(t) = A\mathbf{x}(t)$, and observe that there is only one solution of the form $\mathbf{x}_1(t) = e^{\lambda t}\mathbf{x}_0$. What is the solution?
   b) Having λ and $\mathbf{x}_0$, find a vector $\mathbf{y}_0$ for which $\mathbf{x}_2(t) = te^{\lambda t}\mathbf{x}_0 + e^{\lambda t}\mathbf{y}_0$ is a solution. [Hint: Substitute $\mathbf{x}_2(t)$ into $\mathbf{x}'(t) = A\mathbf{x}(t)$ to determine $\mathbf{y}_0$. The vector $\mathbf{y}_0$ is called a "generalized eigenvector." (See Section 3.10.)]
   c) Show that we can always choose constants c_1 and c_2 so that

$$\mathbf{y}(t) = c_1\mathbf{x}_1(t) + c_2\mathbf{x}_2(t)$$

   satisfies $\mathbf{y}(0) = \mathbf{x}_0$ for any $\mathbf{x}_0$ in R^2.

10. Repeat Exercise 9 for the system

$$\begin{aligned} u'(t) &= 2u(t) - v(t) \\ v'(t) &= 4u(t) + 6v(t). \end{aligned}$$

3.5 COMPLEX EIGENVALUES AND EIGENVALUES OF SYMMETRIC MATRICES

Up to now we have not considered the possibility that the characteristic equation may have complex roots. That is, a matrix A may have complex eigenvalues. We will show that the possibility of complex eigenvalues does not pose any additional problems except that the eigenvectors corresponding to complex eigenvalues will have complex components and complex arithmetic will be required to find these eigenvectors. As a simple example, let A be the (2×2) matrix

$$A = \begin{bmatrix} 3 & 1 \\ -2 & 1 \end{bmatrix}. \tag{3.28}$$

Then the characteristic polynomial for A is $p(t) = t^2 - 4t + 5$. The eigenvalues of A are the roots of $p(t) = 0$, which we can find from the quadratic formula

$$\lambda = \frac{4 \pm \sqrt{-4}}{2} = 2 \pm i$$

where $i = \sqrt{-1}$. Thus despite the fact that A is a real matrix, the eigenvalues of A are complex, $\lambda = 2 + i$ and $\lambda = 2 - i$. To find the eigenvectors of A corresponding to $\lambda = 2 + i$, we must solve $(A - (2 + i)I)\mathbf{x} = \boldsymbol{\theta}$, which leads to the (2×2) homogeneous system

$$\begin{aligned} (1 - i)x_1 \qquad &+ x_2 = 0 \\ -2x_1 - (1 + i)&x_2 = 0. \end{aligned} \tag{3.29}$$

At the end of this section we will discuss the details of how a system like (3.29) is solved. For the moment we merely say that the eigenvectors of A corresponding to $\lambda = 2 + i$ are all of the form

$$\mathbf{x} = a \begin{bmatrix} 1 + i \\ -2 \end{bmatrix}, \quad a \neq 0. \tag{3.30}$$

[This statement is easily verified by inserting $x_1 = 1 + i$ and $x_2 = -2$ into (3.29).]

Before giving the major theoretical result of this section, we briefly review several of the details of complex arithmetic. We will usually represent a complex number z in the form $z = a + ib$ where a and b are real numbers and where $i^2 = -1$. In the representation $z = a + ib$, a is called the ***real part*** of z, and b is called the ***imaginary part*** of z. If $z = a + ib$ and $w = c + id$, then $z + w = (a + c) + i(b + d)$ while $zw = (ac - bd) + i(ad + bc)$. Thus for example if $z_1 = 2 + 3i$ and $z_2 = 1 - i$, then

$$z_1 + z_2 = 3 + 2i, \qquad z_1 z_2 = 5 + i.$$

If z is the complex number $z = a + ib$, then the ***conjugate*** of z (denoted by $\bar{z}$) is defined to be $\bar{z} = a - ib$. We list several properties of the conjugate operation:

$$\begin{aligned} \overline{(z + w)} &= \bar{z} + \bar{w} \\ \overline{(zw)} &= \bar{z}\bar{w} \\ z + \bar{z} &= 2a \\ z - \bar{z} &= 2ib \\ z\bar{z} &= a^2 + b^2; \end{aligned} \tag{3.31}$$

and from the last equality we note that $z\bar{z}$ is a positive real quantity when $z \neq 0$. For example with $z_1 = 2 + 3i$ we have $\bar{z}_1 = 2 - 3i$ and $z_1\bar{z}_1 = 4 + 9 = 13$. Finally if $z = \bar{z}$, then z must be a real number (since $z = \bar{z}$ implies that $b = 0$).

In our introductory example the eigenvalue $\lambda = 2+i$ of the matrix A in (3.28) had eigenvectors with complex components. In general if A is a real matrix, and if λ is a complex eigenvalue of A, then any eigenvector corresponding to λ must have some complex components. (Obviously if λ is complex and $\mathbf{x}$ is a real vector, then $\lambda\mathbf{x}$ has complex components but $A\mathbf{x}$ has only real components. Hence we cannot have $A\mathbf{x} = \lambda\mathbf{x}$ if A and $\mathbf{x}$ are real but λ is complex.)

Also for the real matrix A in (3.28) we found the eigenvalues were $\lambda = 2+i$ and $\lambda = 2-i$. That is, the two eigenvalues are conjugates. In general we can show that when A is real, and when λ is an eigenvalue of A, then $\overline{\lambda}$ is also an eigenvalue of A (the eigenvalues of a real matrix occur in "conjugate pairs"). In a sense we now explain, we can also show that eigenvectors occur in conjugate pairs as well. To be precise if $\mathbf{x}$ is a vector with complex components, we let $\overline{\mathbf{x}}$ denote the vector obtained from $\mathbf{x}$ by taking the conjugate of each component of $\mathbf{x}$. With this notation, we can state the following.

> Let A be a *real* matrix, and let λ be an eigenvalue of A with a corresponding eigenvector $\mathbf{x}$. If λ is complex, then $\overline{\lambda}$ is also an eigenvalue of A, and $\overline{\mathbf{x}}$ is an eigenvector corresponding to $\overline{\lambda}$.

To prove this, we note that $\overline{(A\mathbf{x})} = A\overline{\mathbf{x}}$ when A is real (see Exercise 7, Section 3.5; in general this is true only when A is a real matrix). In addition it is easy to show that $\overline{(\lambda\mathbf{x})} = \overline{\lambda}\overline{\mathbf{x}}$ (see Exercise 6, Section 3.5). Given these preliminaries, let us suppose that A is a real matrix and that $A\mathbf{x} = \lambda\mathbf{x}$ where λ is complex and $\mathbf{x} \neq \boldsymbol{\theta}$. Then

$$A\overline{\mathbf{x}} = \overline{(A\mathbf{x})} = \overline{(\lambda\mathbf{x})} = \overline{\lambda}\overline{\mathbf{x}},$$

which shows that the nonzero vector $\overline{\mathbf{x}}$ satisfies

$$A\overline{\mathbf{x}} = \overline{\lambda}\overline{\mathbf{x}}.$$

For example with respect to the real matrix A in (3.28), $\lambda = 2+i$ is an eigenvalue with corresponding eigenvectors given in (3.30). Thus $\lambda = 2-i$ is also an eigenvalue of A, and one corresponding eigenvector is

$$\mathbf{x} = \begin{bmatrix} 1-i \\ -2 \end{bmatrix}.$$

Finally as the next theorem shows, there is an important class of matrices for which the possibility of complex eigenvalues is precluded.

Theorem 3.7

If A is an $(n \times n)$ real symmetric matrix, then all the eigenvalues of A are real.

Proof. We first recall that if $z = a+ib$, then $z\overline{z} = (a+ib)(a-ib) = a^2+b^2$. Furthermore if $\mathbf{x}$ is an $(n \times 1)$ vector with complex entries

$$\mathbf{x} = \begin{bmatrix} x_1 \\ x_2 \\ \vdots \\ x_n \end{bmatrix},$$

then $\bar{\mathbf{x}}$ is given by

$$\bar{\mathbf{x}} = \begin{bmatrix} \bar{x}_1 \\ \bar{x}_2 \\ \vdots \\ \bar{x}_n \end{bmatrix}.$$

Thus we see that

$$\bar{\mathbf{x}}^{\mathrm{T}}\mathbf{x} = \mathbf{x}^{\mathrm{T}}\bar{\mathbf{x}} = x_1\bar{x}_1 + x_2\bar{x}_2 + \ldots + x_n\bar{x}_n; \tag{3.32}$$

and clearly if $\mathbf{x}$ is not the zero vector, then $\bar{\mathbf{x}}^{\mathrm{T}}\mathbf{x}$ is a positive real number.

Now suppose A is a real symmetric matrix, and suppose that $A\mathbf{x} = \lambda\mathbf{x}$ where $\mathbf{x} \neq \boldsymbol{\theta}$. To isolate λ, we first note that $\bar{\mathbf{x}}^{\mathrm{T}}A\mathbf{x} = \lambda\bar{\mathbf{x}}^{\mathrm{T}}\mathbf{x}$. Regarding $A\mathbf{x}$ as a vector, we see that $\bar{\mathbf{x}}^{\mathrm{T}}A\mathbf{x} = (A\mathbf{x})^{\mathrm{T}}\bar{\mathbf{x}}$. Therefore we obtain the equalities

$$\lambda\bar{\mathbf{x}}^{\mathrm{T}}\mathbf{x} = \bar{\mathbf{x}}^{\mathrm{T}}A\mathbf{x} = (A\mathbf{x})^{\mathrm{T}}\bar{\mathbf{x}} = \mathbf{x}^{\mathrm{T}}A^{\mathrm{T}}\bar{\mathbf{x}} = \mathbf{x}^{\mathrm{T}}A\bar{\mathbf{x}} \tag{3.33}$$

with the last equality in (3.33) holding since $A = A^{\mathrm{T}}$. Since A is real, we also know that $A\bar{\mathbf{x}} = \overline{\lambda\mathbf{x}}$; and hence we deduce from (3.33) that

$$\lambda\bar{\mathbf{x}}^{\mathrm{T}}\mathbf{x} = \bar{\lambda}\mathbf{x}^{\mathrm{T}}\bar{\mathbf{x}}. \tag{3.34}$$

From (3.32) it is obvious that $\mathbf{x}^{\mathrm{T}}\bar{\mathbf{x}} = \bar{\mathbf{x}}^{\mathrm{T}}\mathbf{x}$; and since $\mathbf{x} \neq \boldsymbol{\theta}$, $\bar{\mathbf{x}}^{\mathrm{T}}\mathbf{x}$ is nonzero. Thus from (3.34) we see that $\bar{\lambda} = \lambda$, which means that λ is real. ∎

†Gauss Elimination for Systems with Complex Coefficients

The remainder of this section is concerned with the computational details of solving $(A - \lambda I)\mathbf{x} = \boldsymbol{\theta}$ when λ is complex. We will see that although the arithmetic is tiresome, we can use Gauss elimination to solve a system of linear equations that has some complex coefficients in exactly the same way we solve systems of linear equations having real coefficients. For example consider the (2×2) system

$$\begin{aligned} a_{11}x_1 + a_{12}x_2 &= b_1 \\ a_{21}x_1 + a_{22}x_2 &= b_2 \end{aligned}$$

where the coefficients a_{ij} may be complex. Just as before, we can multiply the first equation by $-a_{21}/a_{11}$, add the result to the second equation to eliminate x_1 from the second equation, and then backsolve to find x_2 and x_1. For larger systems with complex coefficients the principles of Gauss elimination are exactly the same as they are for real systems; only the computational details are different.

One computational detail that might be unfamiliar is dividing one complex number by another [the first step of Gauss elimination for the (2×2) system above is to form a_{21}/a_{11}]. To see how a complex division is carried out, let $z = a + ib$ and

† The remainder of the section may be omitted without any loss of continuity.

$w = c + id$ where $w \neq 0$. To form the quotient z/w, we multiply numerator and denominator by $\bar{w}$:

$$\frac{z}{w} = \frac{z\bar{w}}{w\bar{w}}.$$

In detail we have

$$\frac{z}{w} = \frac{z\bar{w}}{w\bar{w}} = \frac{(a+ib)(c-id)}{c^2+d^2} = \frac{(ac+bd)+i(bc-ad)}{c^2+d^2}. \tag{3.35}$$

Our objective is to express the quotient z/w in the standard form $z/w = r + is$ where r and s are real numbers; from (3.35) r and s are given by

$$r = \frac{ac+bd}{c^2+d^2} \qquad s = \frac{bc-ad}{c^2+d^2}.$$

For instance

$$\frac{2+3i}{1+2i} = \frac{(2+3i)(1-2i)}{(1+2i)(1-2i)} = \frac{8-i}{5} = \frac{8}{5} - \frac{1}{5}i.$$

EXAMPLE 3.12 As an example, we return to the complex (2×2) system in (3.29):

$$\begin{aligned} (1-i)x_1 \qquad\quad + x_2 &= 0 \\ -2x_1 - (1+i)x_2 &= 0. \end{aligned}$$

The initial step in solving this system is to multiply the first equation by $2/(1-i)$ and then add the result to the second equation. Following the discussion above, we write $2/(1-i)$ as

$$\frac{2}{1-i} = \frac{2(1+i)}{(1-i)(1+i)} = \frac{2+2i}{2} = 1+i.$$

Multiplying the first equation by $1+i$ and adding the result to the second equation produce the equivalent system

$$\begin{aligned} (1-i)x_1 + x_2 &= 0 \\ 0 &= 0, \end{aligned}$$

which leads to $x_1 = -x_2/(1-i)$. Simplifying, we obtain

$$x_1 = \frac{-x_2}{1-i} = \frac{-x_2(1+i)}{(1-i)(1+i)} = \frac{-(1+i)}{2}x_2.$$

With $x_2 = -2a$ the solutions of (3.29) are all of the form

$$\mathbf{x} = a\begin{bmatrix} 1+i \\ -2 \end{bmatrix}. \tag{3.36}$$

Note: Since we are allowing the possibility of vectors with complex components, we will also allow the parameter a in (3.36) to be complex. For example with $a = i$ we see that

$$\mathbf{x} = \begin{bmatrix} -1+i \\ -2i \end{bmatrix}$$

is also a solution of (3.29).

EXAMPLE 3.13 Let A be the (3×3) matrix

$$A = \begin{bmatrix} -2 & -2 & -9 \\ -1 & 1 & -3 \\ 1 & 1 & 4 \end{bmatrix}.$$

The characteristic polynomial of A is $p(t) = -(t-1)(t^2 - 2t + 2)$. Thus the eigenvalues of A are $\lambda = 1$, $\lambda = 1+i$, $\lambda = 1-i$.

As was noted above, the complex eigenvalues occur in conjugate pairs; and if we find an eigenvector $\mathbf{x}$ for $\lambda = 1+i$, then we immediately have that $\bar{\mathbf{x}}$ is an eigenvector for $\bar{\lambda} = 1-i$. In this example we find the eigenvectors for $\lambda = 1+i$ by reducing the augmented matrix $[A - \lambda I \,\vdots\, \boldsymbol{\theta}]$ to echelon form. Now for $\lambda = 1+i$,

$$[A - \lambda I \,\vdots\, \boldsymbol{\theta}] = \begin{bmatrix} -3-i & -2 & -9 & 0 \\ -1 & -i & -3 & 0 \\ 1 & 1 & 3-i & 0 \end{bmatrix}.$$

To introduce a zero into the (2, 1) position, we use the multiple m where

$$m = \frac{1}{-3-i} = \frac{-1}{3+i} = \frac{-(3-i)}{(3+i)(3-i)} = \frac{-3+i}{10}.$$

Multiplying row one by m and adding the result to row two, and then multiplying row one by $-m$ and adding the result to row three, we find that $[A - \lambda I \,\vdots\, \boldsymbol{\theta}]$ is row equivalent to

$$\begin{bmatrix} -3-i & -2 & -9 & 0 \\ 0 & \dfrac{6-12i}{10} & \dfrac{-3-9i}{10} & 0 \\ 0 & \dfrac{4+2i}{10} & \dfrac{3-i}{10} & 0 \end{bmatrix}.$$

Multiplying rows two and three by 10 in the matrix above, we obtain a row-equivalent matrix

$$\begin{bmatrix} -3-i & -2 & -9 & 0 \\ 0 & 6-12i & -3-9i & 0 \\ 0 & 4+2i & 3-i & 0 \end{bmatrix}.$$

Completing the reduction, we multiply row two by r and add the result to row three where r is the multiple

$$r = \frac{-(4+2i)}{6-12i} = \frac{-(4+2i)(6+12i)}{(6-12i)(6+12i)} = \frac{-60i}{180} = \frac{-i}{3}.$$

We obtain the row-equivalent matrix

$$\begin{bmatrix} -3-i & -2 & -9 & 0 \\ 0 & 6-12i & -3-9i & 0 \\ 0 & 0 & 0 & 0 \end{bmatrix};$$

and the eigenvectors of A corresponding to $\lambda = 1+i$ are found by solving

$$\begin{aligned} -(3+i)x_1 - \quad 2x_2 &= 9x_3 \\ (6-12i)x_2 &= (3+9i)x_3 \end{aligned} \tag{3.37}$$

with x_3 arbitrary, $x_3 \neq 0$. We first find x_2 from

$$x_2 = \frac{3+9i}{6-12i}x_3 = \frac{(3+9i)(6+12i)}{180}x_3 = \frac{-90+90i}{180}x_3,$$

or

$$x_2 = \frac{-1+i}{2}x_3.$$

From the first equation in (3.37) we obtain

$$-(3+i)x_1 = 2x_2 + 9x_3 = (8+i)x_3,$$

or

$$x_1 = \frac{-(8+i)}{3+i}x_3 = \frac{-(8+i)(3-i)}{10}x_3 = \frac{-25+5i}{10}x_3 = \frac{-5+i}{2}x_3.$$

Setting $x_3 = 2a$, we have $x_2 = (-1+i)a$ and $x_1 = (-5+i)a$; so the eigenvectors of A corresponding to $\lambda = 1+i$ are all of the form

$$\mathbf{x} = \begin{bmatrix} (-5+i)a \\ (-1+i)a \\ 2a \end{bmatrix} = a\begin{bmatrix} -5+i \\ -1+i \\ 2 \end{bmatrix}, \ a \neq 0.$$

Furthermore we know that eigenvectors of A corresponding to $\bar{\lambda} = 1-i$ have the form

$$\bar{\mathbf{x}} = b\begin{bmatrix} -5-i \\ -1-i \\ 2 \end{bmatrix}, \ b \neq 0.$$

As the examples above indicate, finding eigenvectors that correspond to a complex eigenvalue proceeds exactly as for a real eigenvalue except for the additional details required by complex arithmetic.

Although introducing complex eigenvalues and eigenvectors may seem an undue complication, they are in fact fairly important to applications. For instance we note (without trying to be precise) that oscillatory and periodic solutions to first-order systems of differential equations correspond to complex eigenvalues; and since many physical systems exhibit such behavior, we need some way to model them. To illustrate how complex eigenvalues lead to oscillatory solutions, let us consider the vector sequence $\{\mathbf{x}_k\}$ generated by

$$\mathbf{x}_k = A\mathbf{x}_{k-1},\ k = 1, 2, \ldots$$

where

$$A = \begin{bmatrix} 1/2 & 1/2 \\ -1/2 & 1/2 \end{bmatrix} \qquad \mathbf{x}_0 = \begin{bmatrix} 4 \\ 4 \end{bmatrix}.$$

As we saw in Section 3.4, the vectors generated this way have the form [see (3.11)]

$$\mathbf{x}_k = b_1(\lambda_1)^k\mathbf{u}_1 + b_2(\lambda_2)^k\mathbf{u}_2,\ k = 0, 1, \ldots$$

where $A\mathbf{u}_1 = \lambda_1\mathbf{u}_1$, $A\mathbf{u}_2 = \lambda_2\mathbf{u}_2$, and the scalars b_1 and b_2 are chosen to satisfy $\mathbf{x}_0 = b_1\mathbf{u}_1 + b_2\mathbf{u}_2$.

The eigenvalues of A are easily found to be

$$\lambda_1 = \frac{1+i}{2}, \qquad \lambda_2 = \frac{1-i}{2},$$

and it can be verified that corresponding eigenvectors are

$$\mathbf{u}_1 = \begin{bmatrix} 1 \\ i \end{bmatrix}, \qquad \mathbf{u}_2 = \begin{bmatrix} 1 \\ -i \end{bmatrix}.$$

Given this information, we can calculate b_1 and b_2 and find $\mathbf{x}_0 = (2-2i)\mathbf{u}_1 + (2+2i)\mathbf{u}_2$. Thus the sequence $\mathbf{x}_k$ is found to be

$$\mathbf{x}_k = (2-2i)\frac{(1+i)^k}{2^k}\mathbf{u}_1 + (2+2i)\frac{(1-i)^k}{2^k}\mathbf{u}_2,\ k = 0, 1, 2, \ldots. \tag{3.38a}$$

In order to obtain some more detailed information about the behavior of the sequence $\mathbf{x}_k$, let us write $\mathbf{x}_k$ as

$$\mathbf{x}_k = \begin{bmatrix} s_k \\ t_k \end{bmatrix}.$$

In Exercise 13, Section 3.5, the reader is asked to show that the components of $\mathbf{x}_k$ satisfy

$$\begin{aligned} s_k &= 4r^k\left(\cos\frac{k\pi}{4} + \sin\frac{k\pi}{4}\right) \\ t_k &= 4r^k\left(\cos\frac{k\pi}{4} - \sin\frac{k\pi}{4}\right) \end{aligned} \tag{3.38b}$$

where $r = 1/\sqrt{2}$. This representation shows that the components of $\mathbf{x}_k$ decay to zero as $k \to \infty$, but that the components oscillate about 0. [Also (3.38b) shows that the components of $\mathbf{x}_k$ are real, but this is obvious since A and $\mathbf{x}_0$ are both real.]

3.5 EXERCISES

1. Let $z = 3 - 2i$ and $w = 4 + i$. Calculate the following and reduce to the form $a + ib$.
a) $\bar{z}$ b) $z + \bar{w}$ c) $z + \bar{z}$ d) $z - \bar{z}$ e) $w\bar{w}$
f) $z\bar{w}$ g) z/w h) w/z^2 i) $w + iz$ j) i^3/z

2. Let $z = 1 + i$, $w = 2 - i$, $s = 1 + 2i$, and calculate the following.
a) $\bar{z}$ b) $\bar{z} + w$ c) $s\bar{w}$ d) $s^2 - w$
e) s/z f) $s/(3w + z^2)$ g) $w + is$ h) $z(z + \bar{z})$

3. Find the eigenvalues and eigenvectors of the following.

$$A = \begin{bmatrix} 6 & 8 \\ -1 & 2 \end{bmatrix} \quad B = \begin{bmatrix} 2 & 4 \\ -2 & -2 \end{bmatrix} \quad C = \begin{bmatrix} 5 & -5 & -5 \\ -1 & 4 & 2 \\ 3 & -5 & -3 \end{bmatrix}$$

4. Find the eigenvalues and eigenvectors of the following. (Hint: One eigenvalue of C is $\lambda = 1 + 5i$.)

$$A = \begin{bmatrix} -2 & -1 \\ 5 & 2 \end{bmatrix} \quad B = \begin{bmatrix} -6 & 1 & 3 \\ 10 & 1 & -4 \\ -15 & 3 & 8 \end{bmatrix} \quad C = \begin{bmatrix} 4 & -5 & 0 & 3 \\ 0 & 4 & -3 & -5 \\ 5 & -3 & 4 & 0 \\ 3 & 0 & 5 & 4 \end{bmatrix}$$

5. Solve the linear systems.

a) $(1 + i)x_1 + ix_2 = 5 + 4i$
$(1 - i)x_1 - 4x_2 = -11 + 5i$

b) $(1 - i)x_1 - (3 + i)x_2 = -5 - i$
$(2 + i)x_1 + (1 + 2i)x_2 = 1 + 6i$

6. Establish the five properties of the conjugate operation listed in (3.31).

7. Let A be an $(n \times n)$ matrix, and let B be an $(n \times p)$ matrix where the entries of A and B may be complex. Use Exercise 6 and the definition of AB (see Definition 1.3) to show that $\overline{AB} = \bar{A}\bar{B}$. (By $\bar{A}$, we mean the matrix whose ijth entry is the conjugate of the ijth entry of A.) If A is a real matrix, and $\mathbf{x}$ is an $(n \times 1)$ vector, show that $\overline{A\mathbf{x}} = A\bar{\mathbf{x}}$.

8. Let A be an $(m \times n)$ matrix where the entries of A may be complex. It is customary to use the symbol A^* to denote the matrix

$$A^* = (\bar{A})^{\mathrm{T}}.$$

Suppose A is an $(m \times n)$ matrix and B is an $(n \times p)$ matrix. Use Exercise 7 and the properties of the transpose operation to give a quick proof that $(AB)^* = B^*A^*$.

9. An $(n \times n)$ matrix A is called ***Hermitian*** if $A^* = A$.
a) Prove that a Hermitian matrix A has only real eigenvalues.
(Hint: Observing that $\bar{\mathbf{x}}^{\mathrm{T}}\mathbf{x} = \mathbf{x}^*\mathbf{x}$, modify the proof of Theorem 3.7.)
b) Let $A = (a_{ij})$ be an $(n \times n)$ Hermitian matrix. Show that a_{ii} is real for $1 \leqslant i \leqslant n$.

10. Let $p(t) = a_0 + a_1 t + \ldots + a_n t^n$ where the coefficients $a_0, a_1, \ldots, a_n$ are all real.
a) Prove that if r is a complex root of $p(t) = 0$, then $\bar{r}$ is also a root of $p(t) = 0$.
b) If $p(t)$ has degree three, argue that $p(t)$ must have at least one real root.

c) If A is a (3×3) real matrix, argue that A must have at least one real eigenvalue.

11. An $(n \times n)$ real matrix A is called ***orthogonal*** if $A^{\mathrm{T}}A = I$. Let λ be an eigenvalue of an orthogonal matrix A where $\lambda = r + is$. Prove that $\lambda\bar{\lambda} = r^2 + s^2 = 1$. (Hint: How are the eigenvalues of A and of A^{T} related?)

12. A real symmetric $(n \times n)$ matrix A is called ***positive definite*** if $\mathbf{x}^{\mathrm{T}}A\mathbf{x} > 0$ for all $\mathbf{x}$ in R^n, $\mathbf{x} \neq \boldsymbol{\theta}$. Prove that the eigenvalues of a real symmetric positive-definite matrix A are all positive.

13. With polar coordinates $x = r\cos\theta$ and $y = r\sin\theta$, the complex number $z = x + iy$ can be written as $z = r(\cos\theta + i\sin\theta)$.

a) Using induction and some trigonometric identities, prove that

$$(\cos\theta + i\sin\theta)^k = \cos k\theta + i\sin k\theta$$

for $k = 2, 3, \ldots$.

b) In (3.38a) $\mathbf{x}_k = b_1\lambda^k\mathbf{u}_1 + \bar{b}_1\bar{\lambda}^k\bar{\mathbf{u}}_1$ where $b_1 = 2 - 2i$ and $\lambda = (1+i)/2$. Find r and θ to express λ in "polar form" as

$$\lambda = r(\cos\theta + i\sin\theta).$$

c) By part (a), $\lambda^k = r^k(\cos k\theta + i\sin k\theta)$. Use this equation to verify (3.38b).

†**14.** We know that if λ is real, and if $A\mathbf{s} = \lambda\mathbf{s}$, $\mathbf{s} \neq \boldsymbol{\theta}$, then $\mathbf{w}(t) = e^{\lambda t}\mathbf{s}$ is a solution of $\mathbf{x}'(t) = A\mathbf{x}(t)$. If λ is complex, $\lambda = a + ib$, then $e^{\lambda t}$ is defined by $e^{\lambda t} = e^{at}(\cos bt + i\sin bt)$ where t is a real variable. Suppose A is a real matrix, and suppose $A\mathbf{z} = \lambda\mathbf{z}$ where $\mathbf{z} \neq \boldsymbol{\theta}$, $\mathbf{z} = \mathbf{r} + i\mathbf{s}$, $\lambda = a + ib$, $b \neq 0$. Set $\mathbf{w}(t) = e^{\lambda t}\mathbf{z}$, and express $\mathbf{w}(t)$ in the form $\mathbf{w}(t) = \mathbf{u}(t) + i\mathbf{v}(t)$ where $\mathbf{u}(t)$ and $\mathbf{v}(t)$ are real functions.

a) Verify directly that $\mathbf{u}(t)$ and $\mathbf{v}(t)$ are solutions to $\mathbf{x}'(t) = A\mathbf{x}(t)$.

b) Use the fact that $b \neq 0$ to verify that $\mathbf{u}(0)$ and $\mathbf{v}(0)$ are linearly independent vectors.

†**15.** Use Exercise 14 to construct n real solutions to $\mathbf{x}'(t) = A\mathbf{x}(t)$ where A is the following $(n \times n)$ matrix.

a) $A = \begin{bmatrix} 1 & -3 \\ 3 & 1 \end{bmatrix}$ b) $A = \begin{bmatrix} 3 & 13 \\ -2 & 1 \end{bmatrix}$ c) $A = \begin{bmatrix} -8 & -1 & 3 \\ 4 & 1 & -1 \\ -21 & -3 & 8 \end{bmatrix}$

3.6 SIMILARITY TRANSFORMATIONS AND DIAGONALIZATION

If A is an $(n \times n)$ matrix, then the eigenvalues of A are the zeros of the characteristic polynomial $p(t) = \det(A - tI)$. As we have mentioned, calculating $\det(A - tI)$ is a formidable task unless A has some special structure; for example if A is a triangular matrix, then it is easy to calculate $\det(A - tI)$. Therefore given the problem of determining the eigenvalues of A, one logically asks if there is another matrix B that has the same eigenvalues as A, but that has some special structure that simplifies the eigenvalue problem for B.

Now λ is an eigenvalue of an $(n \times n)$ matrix A if and only if $A - \lambda I$ is a singular matrix. Furthermore if S is *any* nonsingular $(n \times n)$ matrix, then we can write $S^{-1}S = I$. Therefore λ is an eigenvalue of A if and only if

$$A - \lambda S^{-1}S$$

is a singular matrix. Next we can rewrite $A - \lambda S^{-1}S$ as $S^{-1}(SAS^{-1} - \lambda I)S$ and obtain the identity

$$A - \lambda I = S^{-1}(SAS^{-1} - \lambda I)S. \tag{3.39}$$

From Theorem 1.13 we know that the product of several matrices is singular if and only if at least one of the factors is singular. In (3.39), S^{-1} and S are nonsingular; so $S^{-1}(SAS^{-1} - \lambda I)S$ is singular if and only if $SAS^{-1} - \lambda I$ is singular. Thus from (3.39) we conclude that λ is an eigenvalue of A if and only if λ is an eigenvalue of SAS^{-1}.

The matrices A and SAS^{-1} are called similar; and since similar matrices have the same eigenvalues, it seems reasonable that we might be able to choose a matrix S so that SAS^{-1} has a form that simplifies the calculation of the characteristic polynomial for SAS^{-1}. We will develop this idea in the next section. Formally we have Definition 3.5.

DEFINITION 3.5.

Let A and B be $(n \times n)$ matrices. A and B are said to be ***similar*** if there is a nonsingular $(n \times n)$ matrix S such that $B = SAS^{-1}$.

With this definition we can rephrase the discussion following (3.39) as Theorem 3.8.

Theorem 3.8

If A and B are similar $(n \times n)$ matrices, then A and B have the same eigenvalues.

While similar matrices always have the same set of eigenvalues (in fact in Chapter 5 we show that similar matrices have the same characteristic polynomial), it is not true that two matrices with the same set of eigenvalues are necessarily similar. As a simple example, consider the two matrices

$$A = \begin{bmatrix} 1 & 0 \\ 1 & 1 \end{bmatrix}, \quad I = \begin{bmatrix} 1 & 0 \\ 0 & 1 \end{bmatrix}.$$

Now $p(t) = (1 - t)^2$ is the characteristic polynomial for both A and I; so A and I have the same set of eigenvalues. However if A and I were similar, there would be a (2×2) matrix S such that

$$I = SAS^{-1}.$$

But the equation $I = SAS^{-1}$ is equivalent to $S = SA$, which is in turn equivalent to $S^{-1}S = A$ or $I = A$. Thus I and A cannot be similar. (A repetition of this argument shows that the only matrix similar to the identity matrix is I itself.) In this respect similarity is a more fundamental concept for the eigenvalue problem than is the characteristic polynomial; two matrices can have exactly the same characteristic polynomial without being similar; so similarity leads to a more finely detailed way of distinguishing matrices.

The process of forming $B = SAS^{-1}$ is called a ***similarity transformation***; and two easily established properties of the similarity transformation are that like powers of similar matrices are similar, and inverses of nonsingular similar matrices are similar. For example if $B = SAS^{-1}$, then we can show that B^2 is similar to A^2:

$$B^2 = (SAS^{-1})(SAS^{-1}) = (SA)(S^{-1}S)(AS^{-1}) = SA^2S^{-1}.$$

If A is similar to B, and A is nonsingular, then B is nonsingular (if 0 is not an eigenvalue of A, then 0 is not an eigenvalue of B). So if $B = SAS^{-1}$, then B^{-1} is similar to A^{-1} since

$$B^{-1} = (SAS^{-1})^{-1} = SA^{-1}S^{-1}$$

by Theorem 1.14. Finally we note that although they have the same eigenvalues, similar matrices do not generally have the same eigenvectors. For example if $B = SAS^{-1}$, and if $B\mathbf{x} = \lambda\mathbf{x}$, then

$$SAS^{-1}\mathbf{x} = \lambda\mathbf{x} \quad \text{or} \quad A(S^{-1}\mathbf{x}) = \lambda(S^{-1}\mathbf{x}).$$

Thus if $\mathbf{x}$ is an eigenvector for B corresponding to λ, then $S^{-1}\mathbf{x}$ is an eigenvector for A corresponding to λ.

Similarity transformations are important for a number of reasons, but for now we are concerned only with using them to solve the eigenvalue problem. In this context if we are given a matrix A we would like to produce a matrix B that is similar to A, but in which the eigenvalue problem for B is relatively simple. A particularly nice sort of matrix with respect to the eigenvalue problem is a diagonal matrix. For example if D is the $(n \times n)$ diagonal matrix

$$D = \begin{bmatrix} d_1 & 0 & 0 & \dots & 0 \\ 0 & d_2 & 0 & \dots & 0 \\ 0 & 0 & d_3 & \dots & 0 \\ \vdots & & & & \vdots \\ 0 & 0 & 0 & \dots & d_n \end{bmatrix},$$

then the eigenvalues of D are $d_1, d_2, d_3, \dots, d_n$; and the corresponding eigenvectors are the unit vectors $\mathbf{e}_1, \mathbf{e}_2, \mathbf{e}_3, \dots, \mathbf{e}_n$. Whenever an $(n \times n)$ matrix A is similar to a diagonal matrix D, we say that A is ***diagonalizable***; and we give a characterization of diagonalizable matrices below. We will see later that the property of being similar to a diagonal matrix is useful also for problems involving matrix representations for linear transformations.

Theorem 3.9

An $(n \times n)$ matrix A is diagonalizable if and only if A possesses a set of n linearly independent eigenvectors.

Proof. Suppose that $\{\mathbf{u}_1, \mathbf{u}_2, \dots, \mathbf{u}_n\}$ is a set of n linearly independent eigenvectors for A; so

$$A\mathbf{u}_k = \lambda_k\mathbf{u}_k, \; k = 1, 2, \dots, n.$$

Let S be the $(n \times n)$ matrix whose column vectors are the eigenvectors of A:

$$S = [\mathbf{u}_1, \mathbf{u}_2, \dots, \mathbf{u}_n].$$

Now S is a nonsingular matrix; so S^{-1} exists where

$$S^{-1}S = [S^{-1}\mathbf{u}_1, S^{-1}\mathbf{u}_2, \ldots, S^{-1}\mathbf{u}_n] = [\mathbf{e}_1, \mathbf{e}_2, \ldots, \mathbf{e}_n] = I. \tag{3.40}$$

Furthermore since $A\mathbf{u}_k = \lambda_k \mathbf{u}_k$, we have

$$AS = [A\mathbf{u}_1, A\mathbf{u}_2, \ldots, A\mathbf{u}_n] = [\lambda_1 \mathbf{u}_1, \lambda_2 \mathbf{u}_2, \ldots, \lambda_n \mathbf{u}_n];$$

and so from (3.40)

$$S^{-1}AS = [\lambda_1 S^{-1}\mathbf{u}_1, \lambda_2 S^{-1}\mathbf{u}_2, \ldots, \lambda_n S^{-1}\mathbf{u}_n] = [\lambda_1 \mathbf{e}_1, \lambda_2 \mathbf{e}_2, \ldots, \lambda_n \mathbf{e}_n].$$

Therefore $S^{-1}AS$ has the form

$$S^{-1}AS = \begin{bmatrix} \lambda_1 & 0 & 0 & \ldots & 0 \\ 0 & \lambda_2 & 0 & \ldots & 0 \\ 0 & 0 & \lambda_3 & \ldots & 0 \\ \vdots & & & & \vdots \\ 0 & 0 & 0 & \ldots & \lambda_n \end{bmatrix} = D;$$

and we have shown that if A has n linearly independent eigenvectors, then A is similar to a diagonal matrix.

Now suppose that $C^{-1}AC = D$ where C is nonsingular and D is a diagonal matrix. Let us write C and D in column form as

$$C = [\mathbf{C}_1, \mathbf{C}_2, \ldots, \mathbf{C}_n], \qquad D = [d_1 \mathbf{e}_1, d_2 \mathbf{e}_2, \ldots, d_n \mathbf{e}_n].$$

From $C^{-1}AC = D$ we obtain $AC = CD$, and we write both of these in column form as

$$AC = [A\mathbf{C}_1, A\mathbf{C}_2, \ldots, A\mathbf{C}_n].$$
$$CD = [d_1 C\mathbf{e}_1, d_2 C\mathbf{e}_2, \ldots, d_n C\mathbf{e}_n].$$

But since $C\mathbf{e}_k = \mathbf{C}_k$ for $k = 1, 2, \ldots, n$, we see that $AC = CD$ implies

$$A\mathbf{C}_k = d_k \mathbf{C}_k, \; k = 1, 2, \ldots, n.$$

Since C is nonsingular, the vectors $\mathbf{C}_1, \mathbf{C}_2, \ldots, \mathbf{C}_n$ are linearly independent (and in particular no $\mathbf{C}_k$ is the zero vector). Thus the diagonal entries of D are the eigenvalues of A, and the column vectors of C are a set of n linearly independent eigenvectors. ∎

The theorem above appears to be of only theoretical importance since as the proof indicates, we cannot actually diagonalize a matrix A unless we have the eigenvectors of A, and the presupposition is that we know the eigenvalues of A. However diagonalization is an important concept in many applications and has computational as well as theoretical significance. Furthermore there are some types of matrices that are known to be diagonalizable; for example every real symmetric matrix is diagonalizable (see Exercises, Section 3.9). Another important instance is given in the corollary to Theorem 3.10. As a final point about Theorem 3.9, we note that the proof of the theorem shows how to construct the matrix that diagonalizes A.

That is, to produce a matrix S such that $S^{-1}AS = D$, we choose the columns of S to be linearly independent eigenvectors of A (see Example 3.14).

Theorem 3.10

Let A be an $(n \times n)$ matrix, and let $\lambda_1, \lambda_2, \ldots, \lambda_k$ be distinct eigenvalues of A with corresponding eigenvectors $\mathbf{x}_1, \mathbf{x}_2, \ldots, \mathbf{x}_k$. Then $\{\mathbf{x}_1, \mathbf{x}_2, \ldots, \mathbf{x}_k\}$ is a linearly independent set of vectors.

Proof. If $\{\mathbf{x}_1, \mathbf{x}_2, \ldots, \mathbf{x}_k\}$ could be a linearly dependent set, then there exists an integer r, $1 \leqslant r \leqslant k-1$, such that $S_1 = \{\mathbf{x}_1, \ldots, \mathbf{x}_r\}$ is linearly independent, and $S_2 = \{\mathbf{x}_1, \ldots, \mathbf{x}_r, \mathbf{x}_{r+1}\}$ is linearly dependent. Thus there exist constants $a_1, \ldots, a_r, a_{r+1}$, not all of which are zero, such that

$$a_1\mathbf{x}_1 + \ldots + a_r\mathbf{x}_r + a_{r+1}\mathbf{x}_{r+1} = \boldsymbol{\theta}. \tag{3.41a}$$

Since S_1 is linearly independent, we know $a_{r+1} \neq 0$. Therefore (3.41a) can be written as

$$\mathbf{x}_{r+1} = b_1\mathbf{x}_1 + \ldots + b_r\mathbf{x}_r \tag{3.41b}$$

where $b_i = (-a_i/a_{r+1})$, $1 \leqslant i \leqslant r$. Multiplying this equation by A leads to

$$\lambda_{r+1}\mathbf{x}_{r+1} = b_1\lambda_1\mathbf{x}_1 + \ldots + b_r\lambda_r\mathbf{x}_r.$$

Multiplying (3.41b) by λ_{r+1} leads to

$$\lambda_{r+1}\mathbf{x}_{r+1} = b_1\lambda_{r+1}\mathbf{x}_1 + \ldots + b_r\lambda_{r+1}\mathbf{x}_r;$$

and subtracting these, we obtain

$$b_1(\lambda_1 - \lambda_{r+1})\mathbf{x}_1 + \ldots + b_r(\lambda_r - \lambda_{r+1})\mathbf{x}_r = \boldsymbol{\theta}.$$

By the linear independence of S_1, $b_i(\lambda_i - \lambda_{r+1}) = 0$ for $1 \leqslant i \leqslant r$; and therefore each b_i in (3.41b) is zero. In turn this means that $\mathbf{x}_{r+1} = \boldsymbol{\theta}$, which is a contradiction. Therefore $\{\mathbf{x}_1, \mathbf{x}_2, \ldots, \mathbf{x}_k\}$ cannot be a linearly dependent set. ■

Corollary

If A is an $(n \times n)$ matrix with n distinct eigenvalues, then A is diagonalizable.

This corollary is immediate from Theorem 3.10 and Theorem 3.9 (if A has n distinct eigenvalues, then A has n linearly independent eigenvectors). The converse of the corollary is not correct; there are many diagonalizable matrices that do not have distinct eigenvalues.

EXAMPLE 3.14 The eigenvalues of

$$A = \begin{bmatrix} 5 & -2 \\ 6 & -2 \end{bmatrix}$$

are $\lambda_1 = 2$ and $\lambda_2 = 1$ with corresponding eigenvectors

$$\mathbf{x}_1 = \begin{bmatrix} 2 \\ 3 \end{bmatrix} \qquad \mathbf{x}_2 = \begin{bmatrix} 1 \\ 2 \end{bmatrix}.$$

By the corollary to Theorem 3.10, A is diagonalizable; and the proof of Theorem 3.9 shows how to diagonalize A. Thus $S^{-1}AS = D$ where

$$S = \begin{bmatrix} 2 & 1 \\ 3 & 2 \end{bmatrix} \qquad D = \begin{bmatrix} 2 & 0 \\ 0 & 1 \end{bmatrix}.$$

As another example, the matrix

$$B = \begin{bmatrix} 1 & 1 \\ -1 & 1 \end{bmatrix}$$

has eigenvalues $\lambda_1 = 1 + i$ and $\lambda_2 = 1 - i$ with corresponding eigenvectors

$$\mathbf{x}_1 = \begin{bmatrix} 1 \\ i \end{bmatrix} \qquad \mathbf{x}_2 = \begin{bmatrix} 1 \\ -i \end{bmatrix}.$$

Thus B is diagonalizable, and $S^{-1}BS = D$ where

$$S = \begin{bmatrix} 1 & 1 \\ i & -i \end{bmatrix} \qquad D = \begin{bmatrix} 1+i & 0 \\ 0 & 1-i \end{bmatrix}.$$

The matrix B in Example 3.14 indicates that complex matrices may be required to diagonalize some real matrices. The theory for complex matrices is exactly the same as the theory for real matrices as outlined in Chapter 1. Although we do not intend to treat complex matrices in any depth, we do wish to note that they are necessary to the rich theory of matrices. The situation is analogous to the relation between the real number system and the complex number system. For example if we restrict ourselves to the reals, some polynomial equations have no solutions—the equation $x^2 + 1 = 0$ is one such. On the other hand if we will only admit the possibility of complex numbers, then every polynomial equation has a solution.

3.6 EXERCISES

1. Find a matrix S such that $S^{-1}AS = D$ where D is diagonal

$$A = \begin{bmatrix} 1 & 0 & 0 \\ 1 & 2 & 0 \\ -1 & -1 & -2 \end{bmatrix}.$$

2. Find a matrix R such that $R^{-1}BR = E$ where E is diagonal

$$B = \begin{bmatrix} -7 & 1 & 3 \\ 9 & 1 & -3 \\ -16 & 2 & 7 \end{bmatrix}.$$

3. Show that A and B in Exercises 1 and 2 are similar. (Hint: If necessary, rearrange the columns of R so that $E = D$.)
4. Prove that if A is similar to B, and if B is similar to C, then A is similar to C.
5. Which of the following are diagonalizable? For each diagonalizable matrix construct a diagonalizing matrix S.

$$A = \begin{bmatrix} 3 & -2 & -4 \\ 8 & -7 & -16 \\ -3 & 3 & 7 \end{bmatrix} \quad B = \begin{bmatrix} -1 & -1 & -4 \\ -8 & -3 & -16 \\ 1 & 2 & 7 \end{bmatrix} \quad C = \begin{bmatrix} 3 & -1 & -1 \\ -12 & 0 & 5 \\ 4 & -2 & -1 \end{bmatrix}$$

6. Use Theorem 3.9 to show that A is diagonalizable. (That is, show A has a set of three linearly independent eigenvectors.)

$$A = \begin{bmatrix} 1 & 1 & -1 \\ 0 & 2 & -1 \\ 0 & 0 & 1 \end{bmatrix}$$

7. Calculate B^{10} where B is the (3×3) matrix in Exercise 2. (Hint: Powers of a diagonal matrix are easy to calculate.)
8. Prove that if A is similar to B, and if A is diagonalizable, then B is diagonalizable.
9. Prove that if A is similar to B, then A^T is similar to B^T.
10. Consider the system of linear differential equations $\mathbf{x}'(t) = A\mathbf{x}(t)$ as in Section 3.4. Suppose that $S^{-1}AS = D$, and let $\mathbf{y}(t) = S^{-1}\mathbf{x}(t)$. It is easy to see that this "change of variables" transforms $\mathbf{x}'(t) = A\mathbf{x}(t)$ into $\mathbf{y}'(t) = D\mathbf{y}(t)$. Thus the components of $\mathbf{y}(t)$ are easily found, and $\mathbf{x}(t)$ can be recovered from $\mathbf{x}(t) = S\mathbf{y}(t)$. Use this technique to solve the systems in Exercise 7, Section 3.4.
11. An $(n \times n)$ ***permutation*** matrix is a matrix obtained by any rearrangement of the columns of the $(n \times n)$ identity matrix I. For example one (3×3) permutation matrix is $P = [\mathbf{e}_3, \mathbf{e}_2, \mathbf{e}_1]$.
    a) In column form list all the possible (3×3) permutation matrices; there are six, counting the identity I.
    b) How many different $(n \times n)$ permutation matrices are there? (Hint: How many positions can $\mathbf{e}_1$ occupy? Once $\mathbf{e}_1$ is fixed, how many rearrangements of the remaining $n-1$ columns are there?)
12. Let P be any of the six (3×3) permutation matrices. Convince yourself that P is orthogonal; that is, $P^TP = I$.
13. Prove that if P is an $(n \times n)$ permutation matrix, then $P^TP = I$. (Hint: Let $P = [\mathbf{P}_1, \mathbf{P}_2, \ldots, \mathbf{P}_n]$, and write P^T in row form as

$$P^T = \begin{bmatrix} \mathbf{P}_1^T \\ \mathbf{P}_2^T \\ \vdots \\ \mathbf{P}_n^T \end{bmatrix}$$

What is the ijth entry of P^TP?)

14. a) Prove that if A is $(n \times n)$ and P is an $(n \times n)$ permutation matrix, then forming AP amounts to rearranging the columns of A using the same pattern of rearrangement that produced P from I.

b) Convince yourself that forming $P^{T}A$ rearranges the rows of A and uses the same pattern that produced P.

15. If P is a permutation matrix, then $P^{T}AP$ is a similarity transformation. Consider the matrix

$$A = \begin{bmatrix} a_{11} & a_{12} & a_{13} \\ 0 & a_{22} & a_{23} \\ a_{31} & a_{32} & a_{33} \end{bmatrix}.$$

a) Using Exercise 14, find a permutation matrix P so that forming $P^{T}A$ interchanges the second and third rows of A.

b) From $P^{T}AP$, and verify that A is similar to a matrix B where the (2, 1) entry of B is a_{31}.

c) Let A be an $(n \times n)$ matrix. For any k, $2 \leqslant k \leqslant n$, convince yourself that A is similar to a matrix B where the (2, 1) entry of B is equal to a_{k1}. (We will use such transformations in the next section to reduce a matrix A to Hessenberg form.)

3.7 TRANSFORMATION TO HESSENBERG FORM AND ELEMENTARY SIMILARITY TRANSFORMATIONS

In order to find the eigenvalues of an $(n \times n)$ matrix A, we would like to find a matrix H that has the same eigenvalues as A but in which the eigenvalues of H are relatively easy to determine. We already know from Theorem 3.8 that similar matrices have the same eigenvalues; so we shall look for a matrix H such that

$$H = SAS^{-1}$$

and such that H has some special sort of form that facilitates finding the characteristic polynomial for H.

We might hope that we could choose H to be a diagonal or triangular matrix since this choice would make the eigenvalue problem for H trivial. Unfortunately we cannot expect easily to reduce an arbitrary matrix A to a similar matrix H where H is triangular or diagonal. To see why, we recall that if $p(t)$ is any polynomial, then we can construct a matrix B for which $p(t)$ is the characteristic polynomial of B (see Exercise 15, Section 3.3). If it were easy to reduce B to a similar matrix H that was triangular or diagonal, then we would have an easy means of finding the roots of $p(t) = 0$. But as we have commented, Abel showed that finding the roots of a polynomial equation cannot be an "easy" problem. Since we cannot expect to find an efficient procedure to transform an $(n \times n)$ matrix A into a similar matrix H that is triangular, we ask for the next best thing—a way to transform A into an almost triangular or Hessenberg matrix. In this section we will establish the details of reduction to Hessenberg form, and then in the next section we will state an algorithm that can be used to find the characteristic polynomial of a Hessenberg matrix.

We will also prove that this algorithm is mathematically sound, and in the process we will develop more of the theoretical foundation for the eigenvalue problem.

To begin, we say that an $(n \times n)$ matrix $H = (h_{ij})$ is a ***Hessenberg*** matrix if $h_{ij} = 0$ whenever $i > j+1$. Thus H is a Hessenberg matrix if all the entries below the subdiagonal of H are zero where the ***subdiagonal*** of H means the entries $h_{21}, h_{32}, h_{43}, \ldots, h_{n,n-1}$. For example a (6×6) Hessenberg matrix has the form

$$H = \begin{bmatrix} \times & \times & \times & \times & \times & \times \\ \times & \times & \times & \times & \times & \times \\ 0 & \times & \times & \times & \times & \times \\ 0 & 0 & \times & \times & \times & \times \\ 0 & 0 & 0 & \times & \times & \times \\ 0 & 0 & 0 & 0 & \times & \times \end{bmatrix}.$$

Note that the definition of a Hessenberg matrix insists only that the entries below the subdiagonal are zero; it is irrelevant whether the other entries are zero or not. Thus for example diagonal and upper-triangular matrices are in Hessenberg form; and as an extreme example, the $(n \times n)$ zero matrix is a Hessenberg matrix. Every (2×2) matrix is (trivially) a Hessenberg matrix since there are no entries below the subdiagonal. We will see shortly that Hessenberg form plays the same role for the eigenvalue problem as echelon form does for the problem of solving $A\mathbf{x} = \mathbf{b}$.

EXAMPLE 3.15 The following matrices are in Hessenberg form:

$$H_1 = \begin{bmatrix} 1 & 2 \\ 3 & 1 \end{bmatrix} \quad H_2 = \begin{bmatrix} 1 & 2 & 1 \\ 2 & 3 & 1 \\ 0 & 4 & 2 \end{bmatrix} \quad H_3 = \begin{bmatrix} 1 & 2 & 0 & 3 \\ 2 & 0 & 1 & 4 \\ 0 & 1 & 3 & 2 \\ 0 & 0 & 0 & 5 \end{bmatrix}.$$

Our approach to finding the eigenvalues of A has two parts.

Find a Hessenberg matrix H that is similar to A.

Calculate the characteristic polynomial for H.

As we show below, both of these steps are (relatively) easy. Transforming A to Hessenberg form is accomplished by simple row and column operations that resemble the operations used previously to reduce a matrix to echelon form. Next the characteristic polynomial for a Hessenberg matrix can be found simply by solving a triangular system of equations. The main theoretical result of this section is Theorem 3.11, which asserts that every $(n \times n)$ matrix is similar to a Hessenberg matrix. The proof is constructive and shows how the similarity transformation is made.

In order to make the $(n \times n)$ case easier to understand, we begin by showing how a (4×4) matrix can be reduced to Hessenberg form. Let A be a (4×4) matrix:

$$A = \begin{bmatrix} a_{11} & a_{12} & a_{13} & a_{14} \\ a_{21} & a_{22} & a_{23} & a_{24} \\ a_{31} & a_{32} & a_{33} & a_{34} \\ a_{41} & a_{42} & a_{43} & a_{44} \end{bmatrix}; \tag{3.42}$$

and suppose for the moment that $a_{21} \neq 0$. Define a matrix Q_1 by

$$Q_1 = \begin{bmatrix} 1 & 0 & 0 & 0 \\ 0 & 1 & 0 & 0 \\ 0 & \dfrac{-a_{31}}{a_{21}} & 1 & 0 \\ 0 & \dfrac{-a_{41}}{a_{21}} & 0 & 1 \end{bmatrix}, \tag{3.43a}$$

and observe that Q_1^{-1} is given by

$$Q_1^{-1} = \begin{bmatrix} 1 & 0 & 0 & 0 \\ 0 & 1 & 0 & 0 \\ 0 & \dfrac{a_{31}}{a_{21}} & 1 & 0 \\ 0 & \dfrac{a_{41}}{a_{21}} & 0 & 1 \end{bmatrix}. \tag{3.43b}$$

(That is, Q_1^{-1} is obtained from Q_1 by changing the sign of the off-diagonal entries of Q_1; equivalently $Q_1 + Q_1^{-1} = 2I$.)

It is easy to see that forming the product $Q_1 A$ has the effect of adding a multiple of $-a_{31}/a_{21}$ times row two of A to row three and adding a multiple of $-a_{41}/a_{21}$ times row two of A to row four of A. Thus $Q_1 A$ has zeros in the (3, 1) and (4, 1) positions. The matrix $Q_1 A Q_1^{-1}$ is similar to A, and we note that the zeros in the (3, 1) and (4, 1) positions are not disturbed when the product $(Q_1 A)Q_1^{-1}$ is formed. (This fact is easy to see since Q_1^{-1} has the form $Q_1^{-1} = [\mathbf{e}_1, \mathbf{q}, \mathbf{e}_3, \mathbf{e}_4]$; so the first, third, and fourth columns of $Q_1 A$ are not disturbed when $(Q_1 A)Q_1^{-1}$ is formed.) In summary when Q_1 and Q_1^{-1} are defined by (3.43), then $A_1 = Q_1 A Q_1^{-1}$ has the form

$$A_1 = \begin{bmatrix} b_{11} & b_{12} & b_{13} & b_{14} \\ b_{21} & b_{22} & b_{23} & b_{24} \\ 0 & b_{32} & b_{33} & b_{34} \\ 0 & b_{42} & b_{43} & b_{44} \end{bmatrix}. \tag{3.44}$$

The matrix A_1 is similar to A and represents the first step in Hessenberg reduction. As a point of interest that will be elaborated later, we note that there is an easy way to see how to construct Q_1. That is, if we wished to zero the (3, 1) and (4, 1) entries of A using elementary row operations, we could multiply row two by $-a_{31}/a_{21}$ and add the result to row three, and next multiply row two by $-a_{41}/a_{21}$ and add the result to row four. The matrix Q_1 is formed from the (4×4) identity I by performing these same row operations on I. [It is not usually possible to use row one to zero the (2, 1), (3, 1), and (4, 1) positions and still produce a similar matrix.]

The next step in Hessenberg reduction is analogous to the first. We can introduce a zero into the (4, 2) position of A_1 if we multiply row three of A_1 by $-b_{42}/b_{32}$ and add the result to row four. Following the discussion above, we define Q_2 to be the matrix

$$Q_2 = \begin{bmatrix} 1 & 0 & 0 & 0 \\ 0 & 1 & 0 & 0 \\ 0 & 0 & 1 & 0 \\ 0 & 0 & \dfrac{-b_{42}}{b_{32}} & 1 \end{bmatrix}, \tag{3.45a}$$

and we note as before that Q_2^{-1} is obtained from Q_2 by changing the sign of the off-diagonal entries:

$$Q_2^{-1} = \begin{bmatrix} 1 & 0 & 0 & 0 \\ 0 & 1 & 0 & 0 \\ 0 & 0 & 1 & 0 \\ 0 & 0 & \dfrac{b_{42}}{b_{32}} & 1 \end{bmatrix}. \tag{3.45b}$$

By a direct multiplication, it is easy to see that $H = Q_2 A_1 Q_2^{-1}$ is a Hessenberg matrix. Since H is similar to A_1, and A_1 is similar to A, we see that H is similar to A. In fact $H = Q_2 A_1 Q_2^{-1} = Q_2(Q_1 A Q_1^{-1})Q_2^{-1} = (Q_2 Q_1)A(Q_2 Q_1)^{-1}$.

Except for the possibility that $a_{21} = 0$ and/or $b_{32} = 0$, this discussion shows how to reduce an arbitrary (4×4) matrix to Hessenberg form. The general case is exactly parallel to this special case, and we will describe it and show how to handle zero pivot elements after an example.

EXAMPLE 3.16 Consider the (4×4) matrix A given by

$$A = \begin{bmatrix} 1 & -2 & 4 & 1 \\ 2 & 0 & 5 & 2 \\ 2 & -2 & 9 & 3 \\ -6 & -1 & -16 & -6 \end{bmatrix}.$$

Following (3.43a) and (3.43b), we define Q_1 and Q_1^{-1} to be

$$Q_1 = \begin{bmatrix} 1 & 0 & 0 & 0 \\ 0 & 1 & 0 & 0 \\ 0 & -1 & 1 & 0 \\ 0 & 3 & 0 & 1 \end{bmatrix}, \quad Q_1^{-1} = \begin{bmatrix} 1 & 0 & 0 & 0 \\ 0 & 1 & 0 & 0 \\ 0 & 1 & 1 & 0 \\ 0 & -3 & 0 & 1 \end{bmatrix}.$$

Given this definition, Q_1A is

$$Q_1A = \begin{bmatrix} 1 & -2 & 4 & 1 \\ 2 & 0 & 5 & 2 \\ 0 & -2 & 4 & 1 \\ 0 & -1 & -1 & 0 \end{bmatrix},$$

and $A_1 = (Q_1A)Q_1^{-1}$ is

$$A_1 = Q_1AQ_1^{-1} = \begin{bmatrix} 1 & -1 & 4 & 1 \\ 2 & -1 & 5 & 2 \\ 0 & -1 & 4 & 1 \\ 0 & -2 & -1 & 0 \end{bmatrix}.$$

The final step of Hessenberg reduction is to use (3.45a) and (3.45b) to define Q_2 and Q_2^{-1}:

$$Q_2 = \begin{bmatrix} 1 & 0 & 0 & 0 \\ 0 & 1 & 0 & 0 \\ 0 & 0 & 1 & 0 \\ 0 & 0 & -2 & 1 \end{bmatrix}, \qquad Q_2^{-1} = \begin{bmatrix} 1 & 0 & 0 & 0 \\ 0 & 1 & 0 & 0 \\ 0 & 0 & 1 & 0 \\ 0 & 0 & 2 & 1 \end{bmatrix}.$$

We obtain $H = Q_2A_1Q_2^{-1}$

$$H = \begin{bmatrix} 1 & -1 & 6 & 1 \\ 2 & -1 & 9 & 2 \\ 0 & -1 & 6 & 1 \\ 0 & 0 & -13 & -1 \end{bmatrix};$$

and H is a Hessenberg matrix that is similar to A.

To complete our discussion of how to reduce a (4×4) matrix to Hessenberg form, we must show how to proceed when $a_{21} = 0$ in (3.42) or when $b_{32} = 0$ in (3.44). This situation is easily handled by using one of the permutation matrices (see Exercises 11–15, Section 3.6):

$$P_1 = \begin{bmatrix} 1 & 0 & 0 & 0 \\ 0 & 0 & 1 & 0 \\ 0 & 1 & 0 & 0 \\ 0 & 0 & 0 & 1 \end{bmatrix} \quad P_2 = \begin{bmatrix} 1 & 0 & 0 & 0 \\ 0 & 0 & 0 & 1 \\ 0 & 0 & 1 & 0 \\ 0 & 1 & 0 & 0 \end{bmatrix} \quad P_3 = \begin{bmatrix} 1 & 0 & 0 & 0 \\ 0 & 1 & 0 & 0 \\ 0 & 0 & 0 & 1 \\ 0 & 0 & 1 & 0 \end{bmatrix}.$$

Each of these matrices is its own inverse: $P_1P_1 = I$, $P_2P_2 = I$, $P_3P_3 = I$. Thus P_1AP_1 is similar to A as are P_2AP_2 and P_3AP_3. The action of these similarity transformations is easy to visualize; for example forming P_1A has the effect of interchanging rows two and three of A while forming $(P_1A)P_1$ switches columns two and three of P_1A. In detail P_1AP_1 is given by

$$P_1AP_1 = \begin{bmatrix} a_{11} & a_{13} & a_{12} & a_{14} \\ a_{31} & a_{33} & a_{32} & a_{34} \\ a_{21} & a_{23} & a_{22} & a_{24} \\ a_{41} & a_{43} & a_{42} & a_{44} \end{bmatrix}.$$

If $a_{21} = 0$ in (3.42), but $a_{31} \neq 0$, then P_1AP_1 is a matrix similar to A with a nonzero entry in the (2, 1) position. We can clearly carry out the first stage of Hessenberg reduction on P_1AP_1. If $a_{21} = a_{31} = 0$ in (3.42), but $a_{41} \neq 0$, then P_2AP_2 has a nonzero entry in the (2, 1) position; and we can now carry out the first stage of Hessenberg reduction. Finally if $a_{21} = a_{31} = a_{41} = 0$, the first stage is not necessary. In A_1 in (3.44) if $b_{32} = 0$, but $b_{42} \neq 0$, then forming $P_3A_1P_3$ will produce a similar matrix with a nonzero entry in the (3, 2) position. Moreover the first column of A_1 will be left unchanged, and so the second step of Hessenberg reduction can be executed. (In general, interchanging two rows of a matrix A and then interchanging the same two columns produce a matrix similar to A. Also note that the permutation matrices P_1, P_2, and P_3 are derived from the identity matrix I by performing the desired row-interchange operations on I.)

The discussion above proves that every (4×4) matrix is similar to a Hessenberg matrix H and also shows how to construct H. The situation with respect to $(n \times n)$ matrices is exactly analogous, and we can now state the main result of this section.

Theorem 3.11

Let A be an $(n \times n)$ matrix. Then there is a nonsingular $(n \times n)$ matrix Q such that $QAQ^{-1} = H$ where H is a Hessenberg matrix.

Before sketching the proof of this theorem, we wish to interpret reduction to Hessenberg form in terms of elementary similarity transformations, which are analogous to the elementary row operations used to reduce a matrix to echelon form, and as such, are useful for theoretical analysis and in computation.

†Elementary Similarity Transformations

In what follows, we wish to demonstrate how Hessenberg reduction can be accomplished by a sequence of elementary row and column operations. We begin by defining two elementary similarity transformations, and we will prove that these operations do indeed produce similar matrices. Let A be an $(n \times n)$ matrix. The following are called ***elementary similarity transformations***.

---

† The reader may wish to accept Theorem 3.11 and proceed directly to Section 3.8. The remainder of this section deals with the computational details of Hessenberg reduction.

1. Multiply row i of A by a constant m, and add the result to row k; then multiply column k by $-m$, and add this result to column i (where $i \neq k$).

2. Interchange row i and row k of A; then interchange column i and column k (where $i \neq k$).

To show that these operations produce a matrix that is similar to A, we construct a matrix realization for each of them. For the operation (1), we define an $(n \times n)$ matrix S by

$$S = [\mathbf{e}_1, \mathbf{e}_2, \ldots, \mathbf{e}_{i-1}, \boldsymbol{v}, \mathbf{e}_{i+1}, \ldots, \mathbf{e}_n] \tag{3.46a}$$

where $(n \times 1)$ column vector $\boldsymbol{v}$ has 1 as the ith component, m as the kth component, and zeros elsewhere. That is, S is the $(n \times n)$ identity except that the (k, i)th entry of S is m. It is easy to show (Exercise 8, Section 3.7) that $S^{-1} = 2I - S$; so S^{-1} is obtained from S by replacing m by $-m$. Also it is easy to show (Exercise 8, Section 3.7) that SAS^{-1} produces the same matrix as the operation (1). (As a point of interest, we note that S is an example of what is usually called an elementary matrix. These elementary matrices result from performing an elementary row operation on the identity matrix.)

Operation (2) can be produced by applying a permutation matrix to A. Let P_{ik} be the $(n \times n)$ matrix formed from the $(n \times n)$ identity by interchanging columns i and k:

$$P_{ik} = [\mathbf{e}_1, \mathbf{e}_2, \ldots, \mathbf{e}_{i-1}, \mathbf{e}_k, \mathbf{e}_{i+1}, \ldots, \mathbf{e}_{k-1}, \mathbf{e}_i, \mathbf{e}_{k+1}, \ldots, \mathbf{e}_n]. \tag{3.46b}$$

It is easy to show that $P_{ik}^{-1} = P_{ik}$ and that $P_{ik}AP_{ik}$ produces the same matrix as operation (2). [Any matrix P derived from I by rearranging the columns of I is called a ***permutation*** matrix. Permutation matrices have a number of interesting properties. For example if P is a permutation matrix, then $P^{-1} = P^{\mathrm{T}}$. (See Exercise 13, Section 3.6.) The matrix P_{ik} above is symmetric; so $P_{ik}^{-1} = P_{ik}^{\mathrm{T}} = P_{ik}$.] We will use the permutation matrix P_{ik} to get around the problem of a zero pivot element in Hessenberg reduction.

The example below illustrates how the elementary similarity transformations can be applied to reduce a matrix to Hessenberg form. Observe that we perform only row and column operations, we do not actually form any matrix products.

EXAMPLE 3.17 Let A be the (4×4) matrix

$$A = \begin{bmatrix} 1 & 1 & 8 & -2 \\ 0 & 3 & 5 & -1 \\ 1 & -1 & -3 & 2 \\ 3 & -1 & -4 & 9 \end{bmatrix}.$$

We would like to use multiples of the second row to zero the (3, 1) and (4, 1) positions of A. However the pivot element [the (2, 1) entry of A] is zero. To overcome this problem, we interchange rows two and three and then interchange columns two and three—this is an example of the elementary similarity transformation (2) listed above. Carrying this transformation out, we interchange rows two and three and obtain

$$\begin{bmatrix} 1 & 1 & 8 & -2 \\ 1 & -1 & -3 & 2 \\ 0 & 3 & 5 & -1 \\ 3 & -1 & -4 & 9 \end{bmatrix}.$$

We complete the elementary similarity transformation by interchanging columns two and three and produce the matrix A_1:

$$A_1 = \begin{bmatrix} 1 & 8 & 1 & -2 \\ 1 & -3 & -1 & 2 \\ 0 & 5 & 3 & -1 \\ 3 & -4 & -1 & 9 \end{bmatrix}.$$

Now A_1 is similar to A, and in matrix terms $A_1 = P_{23}AP_{23}^{-1}$ where

$$P_{23} = P_{23}^{-1} = \begin{bmatrix} 1 & 0 & 0 & 0 \\ 0 & 0 & 1 & 0 \\ 0 & 1 & 0 & 0 \\ 0 & 0 & 0 & 1 \end{bmatrix}.$$

We zero the (4, 1) entry of A_1 by adding a multiple of -3 times row two to row four and produce

$$\begin{bmatrix} 1 & 8 & 1 & -2 \\ 1 & -3 & -1 & 2 \\ 0 & 5 & 3 & -1 \\ 0 & 5 & 2 & 3 \end{bmatrix}.$$

We complete the elementary similarity transformation (1) by adding a multiple of 3 times column four to column two. This transformation gives us the matrix A_2:

$$A_2 = \begin{bmatrix} 1 & 2 & 1 & -2 \\ 1 & 3 & -1 & 2 \\ 0 & 2 & 3 & -1 \\ 0 & 14 & 2 & 3 \end{bmatrix}.$$

To finish the Hessenberg reduction, multiply row three by -7 and add the result to row four; then multiply column four by 7 and add the result to column three. The result is a matrix H in Hessenberg form:

$$H = \begin{bmatrix} 1 & 2 & -13 & -2 \\ 1 & 3 & 13 & 2 \\ 0 & 2 & -4 & -1 \\ 0 & 0 & 51 & 10 \end{bmatrix};$$

and H is similar to A.

We are now in a position to prove Theorem 3.11, showing that any $(n \times n)$ matrix A is similar to a Hessenberg matrix.

Proof of Theorem 3.11. Let A be the $(n \times n)$ matrix

$$A = \begin{bmatrix} a_{11} & a_{12} & \cdots & a_{1n} \\ a_{21} & a_{22} & \cdots & a_{2n} \\ a_{31} & a_{32} & \cdots & a_{3n} \\ \vdots & & & \\ a_{n1} & a_{n2} & \cdots & a_{nn} \end{bmatrix}. \tag{3.47}$$

If $a_{21} \neq 0$, let $m_k = -a_{k1}/a_{21}$ where $3 \leqslant k \leqslant n$. The elementary transformation of adding a multiple of m_k times row two to row k and then adding a multiple of $-m_k$ times column k to column two produces a similar matrix with a zero in the $(k, 1)$ position. Moreover this transformation does not affect any entry in the first column except the $(k, 1)$ entry. Thus if a_{21} is nonzero, we can use a sequence of $n-3$ elementary similarity transformations to produce a matrix A_1, which is similar to A and which has the form

$$A_1 = \begin{bmatrix} b_{11} & b_{12} & b_{13} & \cdots & b_{1n} \\ b_{21} & b_{22} & b_{23} & \cdots & b_{2n} \\ 0 & b_{32} & b_{33} & \cdots & b_{3n} \\ 0 & b_{42} & b_{43} & \cdots & b_{4n} \\ \vdots & & & & \\ 0 & b_{n2} & b_{n3} & \cdots & b_{nn} \end{bmatrix}. \tag{3.48}$$

If $a_{21} = 0$, but $a_{k1} \neq 0$ for some k, $3 \leqslant k \leqslant n$, then we interchange row two and row k, and then interchange column two and column k to produce a similar matrix with a nonzero entry in the $(2, 1)$ position. Proceeding as above, we can produce a matrix of the form of A_1 in (3.48) after $n-3$ elementary similarity transformations. Finally if $a_{21} = a_{31} = \ldots = a_{n1} = 0$, then A is already in the form of A_1.

Continuing, if $b_{32} \neq 0$, we can add a multiple of row three to row k to zero the $(k, 2)$ entry, $4 \leqslant k \leqslant n$. To complete the transformation, we add the appropriate multiple of column k to column three and observe that this procedure does not affect the zero entries in column one or the zeros we have introduced in column two. Finally suppose $b_{32} = 0$, but $b_{k2} \neq 0$ for some k, $4 \leqslant k \leqslant n$. In this event we interchange row three and row k, then column three and column k. Again this procedure does not disturb the zeros in column one. The process can obviously be continued until we have arrived at a Hessenberg matrix that is similar to A. ■

Computational Considerations

A variety of similarity transformations, besides the elementary ones we have described, have been developed to reduce a matrix to Hessenberg form. Householder transformations are particularly effective—these are a sequence of explicitly defined transformations involving orthogonal matrices (see Chapter 6 and Exercise 9, Section 3.7).

While a reduction process like transformation to Hessenberg form may seem quite tedious, we will show in the next section that it is easy to calculate the characteristic polynomial of a Hessenberg matrix. Also, however tedious Hessenberg reduction may seem, the alternative of calculating the characteristic polynomial from $p(t) = \det(A - tI)$ is worse. To illustrate this point, we note that in order to gauge the efficiency of an algorithm (particularly an algorithm that will be implemented on a computer), operations counts are frequently used as a first approximation. By an operations count, we mean a count of the number of multiplications and additions that must be performed in order to execute the algorithm. Given an $(n \times n)$ matrix A, it is not hard to show (see Chapter 6) that a total of approximately n^3 multiplies and n^3 adds are needed to reduce A to Hessenberg form and then to find the characteristic polynomial. By contrast if A is $(n \times n)$, calculating $p(t)$ from

$$p(t) = \det(A - tI)$$

requires on the order of $n!$ multiplies and $n!$ adds. In the language of computer science, reduction to Hessenberg form is a "polynomial-time" algorithm while computing $\det(A - tI)$ is an "exponential-time" algorithm. In a polynomial-time algorithm, execution time grows at a rate proportional to n^k as n grows (where k is a constant) while in an exponential-time algorithm, execution time grows at least as fast as b^n (where b is a constant larger than 1). The distinction is more than academic since exponential-time algorithms can be used on only the smallest problems, and the basic question is whether or not we can produce acceptable answers to practical problems in a reasonable amount of time. In fact in some areas of applications the only known algorithms are exponential-time algorithms, and hence realistic problems cannot be solved except by an inspired guess. (An example of such a problem is the "traveling salesman's" problem, which arises in operations research.)

TABLE 3.1

n	n^3	$n!$
3	27	6
4	64	24
5	125	120
6	216	720
7	343	5,040
8	512	40,320
9	729	362,880
10	1,000	3,628,800
11	1,331	39,916,800
12	1,728	479,001,600

Table 3.1 should illustrate the difference between polynomial time and exponential time for the problem of calculating the characteristic polynomial. We

can draw some rough conclusions from this table. For instance if an algorithm requiring n^3 operations is used on a (12×12) matrix, and if the algorithm executes in 1 second, then we would expect any algorithm requiring $n!$ operations to take on the order of 77 hours to execute when applied to the same (12×12) matrix. For larger values of n, the comparison between polynomial-time and exponential-time algorithms borders on the absurd. For example if an algorithm requiring n^3 operations executes in 1 second for $n = 20$, we would suspect that an algorithm requiring 20! operations would take something like 8×10^{10} hours, or approximately 9,000,000 years.

3.7 EXERCISES

1. We have seen explicit matrices that can be used to transform a (4×4) matrix A to Hessenberg form. This problem will treat the case in which A is a (3×3) matrix

$$A = \begin{bmatrix} a_{11} & a_{12} & a_{13} \\ a_{21} & a_{22} & a_{23} \\ a_{31} & a_{32} & a_{33} \end{bmatrix}.$$

Suppose $a_{21} \neq 0$, and let Q be the (3×3) matrix obtained from Q_1 in (3.43a) by removing row four and column four from Q_1.

a) Reverse the sign of the off-diagonal entry of Q, and prove by direct multiplication that this matrix is Q^{-1}.

b) Verify by direct multiplication that QAQ^{-1} is a Hessenberg matrix.
Suppose $a_{21} = 0$, and $a_{31} \neq 0$. Let P be the matrix obtained from the (3×3) identity matrix I by interchanging rows two and three of I.

c) Verify that $PP = I$.

d) Verify that PAP is in Hessenberg form.

2. Let $A = (a_{ij})$ be a (5×5) matrix where $a_{21} \neq 0$. Construct a (5×5) matrix Q such that QAQ^{-1} has zeros in the (3, 1), (4, 1), (5, 1) positions. [Hint: Add an appropriate row and column to make Q_1 in (3.43a) a (5×5) matrix.]

3. Reduce each of the matrices to Hessenberg form, and display the matrix Q used in the similarity transformation.

$$\text{a)} \begin{bmatrix} -7 & 4 & -3 \\ 8 & -3 & 3 \\ 32 & -15 & 13 \end{bmatrix} \qquad \text{b)} \begin{bmatrix} -6 & 3 & -14 \\ -1 & 2 & -2 \\ 2 & 0 & 5 \end{bmatrix} \qquad \text{c)} \begin{bmatrix} 1 & 3 & 1 \\ 0 & 2 & 4 \\ 1 & 1 & 3 \end{bmatrix}$$

4. Repeat Exercise 3 for the following matrices.

$$\text{a)} \begin{bmatrix} 1 & 2 & -1 \\ 3 & 2 & 1 \\ -6 & 1 & 3 \end{bmatrix} \qquad \text{b)} \begin{bmatrix} 4 & 0 & 3 \\ 0 & 1 & 2 \\ 3 & 2 & 1 \end{bmatrix} \qquad \text{c)} \begin{bmatrix} 3 & -1 & -1 \\ 4 & -1 & -2 \\ -12 & 5 & 0 \end{bmatrix}$$

5. Reduce each of the following matrices to Hessenberg form.

$$\text{a)} \begin{bmatrix} 1 & -1 & -1 & -1 \\ -1 & 1 & -1 & -1 \\ -1 & -1 & 1 & -1 \\ -1 & -1 & -1 & 1 \end{bmatrix} \qquad \text{b)} \begin{bmatrix} 6 & 1 & 4 & 4 \\ 1 & 6 & 4 & 4 \\ 4 & 4 & 6 & 1 \\ 4 & 4 & 1 & 6 \end{bmatrix} \qquad \text{c)} \begin{bmatrix} 4 & 1 & 2 & 3 \\ 2 & 3 & -1 & 1 \\ 4 & 2 & -1 & 0 \\ -2 & 3 & 2 & 1 \end{bmatrix}$$

6. Reduce each of the following matrices to Hessenberg form.

a) $\begin{bmatrix} 5 & 1 & 4 & 1 \\ 1 & 4 & 1 & 2 \\ 4 & 1 & 5 & 1 \\ 1 & 2 & 1 & 4 \end{bmatrix}$ b) $\begin{bmatrix} 1 & 2 & -1 & 3 & 4 \\ -3 & 1 & -3 & 2 & 1 \\ 3 & 2 & 1 & -1 & 2 \\ 0 & 2 & -1 & 1 & 4 \\ 6 & 8 & 7 & -5 & 2 \end{bmatrix}$

7. Use elementary similarity transformations to reduce the following matrices to Hessenberg form.

a) $\begin{bmatrix} 1 & 5 & 1 & 1 \\ 2 & 4 & 1 & 1 \\ -2 & 6 & 1 & 3 \\ -4 & -14 & -4 & -4 \end{bmatrix}$ b) $\begin{bmatrix} -2 & -9 & -3 & 1 \\ -1 & -8 & -3 & 1 \\ 2 & 10 & 4 & -1 \\ -3 & -37 & -12 & 5 \end{bmatrix}$

c) $\begin{bmatrix} 2 & -2 & 0 & -1 \\ -1 & -1 & -2 & 1 \\ 2 & 2 & 1 & 4 \\ 1 & 1 & -3 & 9 \end{bmatrix}$

8. For S as in (3.46a), show that S^{-1} is obtained by changing the sign of the off-diagonal entry of S. Show also that SAS^{-1} produces the result of the elementary similarity transformation (1).

9. (***Householder transformations***) Let $\boldsymbol{v}$ be a vector in R^n such that $\boldsymbol{v}^T\boldsymbol{v} = 1$, and define the $(n \times n)$ matrix Q by $Q = I - 2\boldsymbol{v}\boldsymbol{v}^T$. (The matrix Q is called a Householder transformation.)

a) Prove that Q is symmetric.
b) Prove that $QQ = I$.
c) Let $\mathbf{x}$ be any vector in R^n. Prove that $\|Q\mathbf{x}\| = \|\mathbf{x}\|$.
d) If A is symmetric, prove that QAQ is also symmetric.

10. Let $A = [\mathbf{A}_1, \mathbf{A}_2, \ldots, \mathbf{A}_n]$ be an $(n \times n)$ matrix, and let Q be as in Exercise 9. Verify that the ith column of QA is equal to $\mathbf{A}_i - c\boldsymbol{v}$ where c is a scalar. How many multiplies are required to form QA? How many multiplies are needed to form BA when B is an arbitrary $(n \times n)$ matrix?

11. Let $\mathbf{x}$ and $\mathbf{y}$ be any vectors in R^n such that $\mathbf{x}^T\mathbf{x} = \mathbf{y}^T\mathbf{y}$. Let $\boldsymbol{v} = (\mathbf{x} - \mathbf{y})/a$ where $a = \|\mathbf{x} - \mathbf{y}\|$.

a) Use $\boldsymbol{v}$ to define a Householder transformation Q, and show that $Q\mathbf{x} = \mathbf{y}$.
b) Let $\mathbf{x}$ and $\mathbf{y}$ be given by

$$\mathbf{x} = \begin{bmatrix} x_1 \\ \vdots \\ x_{p-1} \\ x_p \\ x_{p+1} \\ \vdots \\ x_n \end{bmatrix} \qquad \mathbf{y} = \begin{bmatrix} x_1 \\ \vdots \\ x_{p-1} \\ s \\ 0 \\ \vdots \\ 0 \end{bmatrix}$$

where $s^2 = x_p^2 + x_{p+1}^2 + \dots + x_n^2$. Verify that $\mathbf{x}^T\mathbf{x} = \mathbf{y}^T\mathbf{y}$. Construct Q as in part (a), and verify that $a^2 = 2s(s - x_p)$. Therefore $Q = I + b\mathbf{w}\mathbf{w}^T$ is a Householder transformation that zeros the last $n-p$ components of $\mathbf{x}$; $Q\mathbf{x} = \mathbf{y}$ where $b = 1/s(x_p - s)$ and $\mathbf{w} = \mathbf{x} - \mathbf{y}$.

12. Using the ideas in Exercise 11, construct a Householder transformation Q_1 that zeros the (3, 1) and (4, 1) positions of A when Q_1AQ_1 is formed and where A is the matrix in Exercise 6. Construct a Householder transformation Q_2 such that $Q_2Q_1AQ_1Q_2$ is in Hessenberg form.

3.8 EIGENVALUES OF HESSENBERG MATRICES

From Theorem 3.11 we know that every $(n \times n)$ matrix A is similar to a Hessenberg matrix H. Moreover given A, we have a relatively efficient way to find H by using elementary similarity transformations (although for computer applications several modifications must be made in order to obtain a transformation procedure that is not too sensitive to roundoff errors). Since A and H have the same eigenvalues, and since H has a particularly simple form, it is logical to study the eigenvalue problem for Hessenberg matrices. In particular with a given $(n \times n)$ Hessenberg matrix H, our first task is to find the characteristic polynomial, $p(t)$, for H. Since H has quite a few zero entries, we could (for small n) calculate $p(t)$ from

$$p(t) = \det(H - tI);$$

but we would like to take a somewhat different approach in this section. In particular we will state an algorithm below that can be used to find the characteristic polynomial; this algorithm requires us to set up and solve an $(n \times n)$ *triangular* system of linear equations where the solution of this system will be the coefficients of $p(t)$. There are two reasons for this approach. First of all, we will learn some more of the important theoretical aspects of the eigenvalue problem. Second, the algorithm we present requires about n^2 arithmetic operations while calculating $p(t)$ from $p(t) = \det(H - tI)$ requires on the order of 2^n arithmetic operations when H is a Hessenberg matrix. For large n of course n^2 is much smaller than 2^n; for example if $n = 20$, then $n^2 = 400$ while $2^n = 1{,}048{,}576$.

Before stating the algorithm, we introduce one new term. We say an $(n \times n)$ Hessenberg matrix $H = (h_{ij})$ is a ***standard Hessenberg matrix*** if $h_{k,k-1} \neq 0$, $k = 2, 3, \dots, n$. In other words a Hessenberg matrix is standard if none of the subdiagonal entries is zero. We will show how to find the characteristic polynomial for a standard Hessenberg matrix; and we will show later that if H is not a standard Hessenberg matrix, then the eigenvalue problem for H uncouples into two or more simpler problems involving standard Hessenberg matrices. Thus if we know how to find eigenvalues for a standard Hessenberg matrix, then we can find the eigenvalues of any Hessenberg matrix and so via Theorem 3.11 can solve the eigenvalue problem for any matrix A.

Algorithm for determining the characteristic polynomial of a standard Hessenberg matrix

Let H be an $(n \times n)$ standard Hessenberg matrix, and let $\mathbf{w}_0$ denote the $(n \times 1)$ unit vector $\mathbf{e}_1$. Define the vectors $\mathbf{w}_1, \mathbf{w}_2, \ldots, \mathbf{w}_n$ by

$$\mathbf{w}_{i+1} = H\mathbf{w}_i,\ i = 0, 1, \ldots, n-1; \tag{3.49}$$

and let $a_0, a_1, \ldots, a_{n-1}$ be the unique solution of

$$a_0\mathbf{w}_0 + a_1\mathbf{w}_1 + \ldots + a_{n-1}\mathbf{w}_{n-1} + \mathbf{w}_n = \boldsymbol{\theta}. \tag{3.50}$$

Given the scalars $a_0, a_1, \ldots, a_{n-1}$ defined by (3.50), let $p(t)$ be the polynomial

$$p(t) = t^n + a_{n-1}t^{n-1} + \ldots + a_1 t + a_0. \tag{3.51}$$

Then λ is an eigenvalue of H if and only if λ is a root of $p(t) = 0$.

After giving several examples to illustrate the mechanics of this algorithm, we will prove that it is an effective algorithm. These examples show why, as the algorithm states, there is always a unique solution $a_0, a_1, \ldots, a_{n-1}$ to Eq. (3.50); and the examples also illustrate that Eq. (3.50) is always a triangular system of equations. In the next section we will prove that if $p(t)$ is given by (3.51), then the zeros of $p(t)$ are precisely the eigenvalues of H. [We will call the polynomial $p(t)$ defined by (3.51) the "characteristic polynomial" for H even though this usage is not strictly in accordance with Definition 3.4. Specifically we will show in Chapter 5 that $p(t)$ as defined by (3.51) is equal to $\det(tI - H)$ whereas in Definition 3.4 the characteristic polynomial is $\det(H - tI)$. However it is not hard to show that $\det(tI - A) = (-1)^n \det(A - tI)$ for any $(n \times n)$ matrix A; so $p(t)$ as given by (3.51) is a multiple of $\det(H - tI)$. This variation in terminology is not important since the central feature is that there is a polynomial $p(t)$ whose zeros are precisely the eigenvalues of H; and clearly the zeros of $p(t)$ and $(-1)^n p(t)$ are the same.]

EXAMPLE 3.18 Consider the (2×2) standard Hessenberg matrix

$$H = \begin{bmatrix} 5 & -2 \\ 6 & -2 \end{bmatrix}.$$

To obtain the polynomial $p(t)$ defined by the algorithm above, we form the vectors $\mathbf{w}_0, \mathbf{w}_1, \mathbf{w}_2$ from

$$\mathbf{w}_0 = \begin{bmatrix} 1 \\ 0 \end{bmatrix}, \quad \mathbf{w}_1 = H\mathbf{w}_0 = \begin{bmatrix} 5 \\ 6 \end{bmatrix}, \quad \mathbf{w}_2 = H\mathbf{w}_1 = \begin{bmatrix} 13 \\ 18 \end{bmatrix}.$$

Writing the system $a_0\mathbf{w}_0 + a_1\mathbf{w}_1 + \mathbf{w}_2 = \boldsymbol{\theta}$ in the form $a_0\mathbf{w}_0 + a_1\mathbf{w}_1 = -\mathbf{w}_2$, we have

$$a_0\begin{bmatrix} 1 \\ 0 \end{bmatrix} + a_1\begin{bmatrix} 5 \\ 6 \end{bmatrix} = -\begin{bmatrix} 13 \\ 18 \end{bmatrix},$$

which is the same as

$$\begin{aligned} a_0 + 5a_1 &= -13 \\ 6a_1 &= -18. \end{aligned}$$

Solving this triangular system, we obtain $a_1 = -3$ and $a_0 = 2$. Thus the polynomial $p(t)$ given in (3.51) is

$$p(t) = t^2 - 3t + 2.$$

Since $p(t) = (t-2)(t-1)$, the algorithm asserts that the eigenvalues of H are $\lambda = 1$ and $\lambda = 2$. [Note that

$$\det(H - tI) = \begin{vmatrix} 5-t & -2 \\ 6 & -2-t \end{vmatrix} = t^2 - 3t + 2;$$

so $p(t)$ as given by (3.51) agrees with Definition 3.4.] We can see also in this example why there is a unique solution to $a_0 \mathbf{w}_0 + a_1 \mathbf{w}_1 + \mathbf{w}_2 = \boldsymbol{\theta}$; the reason is that $\mathbf{w}_0$ and $\mathbf{w}_1$ are linearly independent and hence constitute a basis for R^2.

EXAMPLE 3.19 Consider the (3×3) standard Hessenberg matrix

$$H = \begin{bmatrix} 2 & 2 & -1 \\ -1 & -1 & 1 \\ 0 & 2 & 1 \end{bmatrix}.$$

To find the characteristic polynomial $p(t)$, we form

$$\mathbf{w}_0 = \begin{bmatrix} 1 \\ 0 \\ 0 \end{bmatrix}, \qquad \mathbf{w}_1 = H\mathbf{w}_0 = \begin{bmatrix} 2 \\ -1 \\ 0 \end{bmatrix}, \qquad \mathbf{w}_2 = H\mathbf{w}_1 = \begin{bmatrix} 2 \\ -1 \\ -2 \end{bmatrix},$$

$$\mathbf{w}_3 = H\mathbf{w}_2 = \begin{bmatrix} 4 \\ -3 \\ -4 \end{bmatrix};$$

and we solve the triangular system $a_0 \mathbf{w}_0 + a_1 \mathbf{w}_1 + a_2 \mathbf{w}_2 = -\mathbf{w}_3$. In detail this system is

$$\begin{aligned} a_0 + 2a_1 + 2a_2 &= -4 \\ -a_1 - a_2 &= 3 \\ -2a_2 &= 4. \end{aligned}$$

Solving this system, we find $a_2 = -2$, $a_1 = -1$, $a_0 = 2$; so $p(t)$ is the polynomial

$$p(t) = t^3 - 2t^2 - t + 2.$$

[The reader may verify that $p(t) = -\det(H - tI)$ although as we have said, this is not central to the discussion in this section.] Again we observe that $\{\mathbf{w}_0, \mathbf{w}_1, \mathbf{w}_2\}$ is a linearly independent set of vectors since $W = [\mathbf{w}_0, \mathbf{w}_1, \mathbf{w}_2]$ is an upper-triangular matrix with nonzero diagonal entries. Consequently since $\{\mathbf{w}_0, \mathbf{w}_1, \mathbf{w}_2\}$ is a basis for R^3, there is a unique solution to $a_0 \mathbf{w}_0 + a_1 \mathbf{w}_1 + a_2 \mathbf{w}_2 + \mathbf{w}_3 = \boldsymbol{\theta}$. Finally since $p(t) = t^3 - 2t^2 - t + 2 = (t+1)(t-1)(t-2)$, the algorithm asserts that the eigenvalues of H are $\lambda = -1$, $\lambda = 1$, $\lambda = 2$.

EXAMPLE 3.20 Consider the (4×4) standard Hessenberg matrix

$$H = \begin{bmatrix} 1 & 1 & 1 & 1 \\ 2 & 0 & 1 & 1 \\ 0 & -1 & -2 & -2 \\ 0 & 0 & 2 & 2 \end{bmatrix}.$$

As above, we form $\mathbf{w}_0, \mathbf{w}_1, \mathbf{w}_2, \mathbf{w}_3$, and find

$$\mathbf{w}_0 = \begin{bmatrix} 1 \\ 0 \\ 0 \\ 0 \end{bmatrix}, \quad \mathbf{w}_1 = \begin{bmatrix} 1 \\ 2 \\ 0 \\ 0 \end{bmatrix}, \quad \mathbf{w}_2 = \begin{bmatrix} 3 \\ 2 \\ -2 \\ 0 \end{bmatrix}, \quad \mathbf{w}_3 = \begin{bmatrix} 3 \\ 4 \\ 2 \\ -4 \end{bmatrix},$$

$$\mathbf{w}_4 = \begin{bmatrix} 5 \\ 4 \\ 0 \\ -4 \end{bmatrix}.$$

The system $a_0\mathbf{w}_0 + a_1\mathbf{w}_1 + a_2\mathbf{w}_2 + a_3\mathbf{w}_3 = -\mathbf{w}_4$ is

$$\begin{aligned} a_0 + a_1 + 3a_2 + 3a_3 &= -5 \\ 2a_1 + 2a_2 + 4a_3 &= -4 \\ -2a_2 + 2a_3 &= 0 \\ -4a_3 &= 4; \end{aligned}$$

and the solution is $a_3 = -1$, $a_2 = -1$, $a_1 = 1$, $a_0 = 0$. Thus $p(t)$ is the polynomial

$$p(t) = t^4 - t^3 - t^2 + t = t(t-1)^2(t+1).$$

The algorithm asserts that the eigenvalues of H are $\lambda = 0$, $\lambda = 1$, $\lambda = -1$. In particular if $\lambda = 0$ is an eigenvalue of H, then H must be singular; and we observe that H is indeed singular since the last two columns of H are identical. This example should emphasize that even though the columns of H may be linearly dependent, the vectors $\mathbf{w}_0, \mathbf{w}_1, \ldots, \mathbf{w}_{n-1}$ are linearly independent when H is an $(n \times n)$ standard Hessenberg matrix.

EXAMPLE 3.21 As a final example that illustrates why we insist on standard Hessenberg matrices, consider the (4×4) Hessenberg matrix

$$H = \begin{bmatrix} 1 & 2 & 1 & 3 \\ 2 & 1 & 1 & 1 \\ 0 & 0 & 2 & 1 \\ 0 & 0 & 1 & 1 \end{bmatrix}.$$

Note that H is not a standard Hessenberg matrix since the subdiagonal entry in the (3, 2) position is zero. If we form $\mathbf{w}_0, \mathbf{w}_1, \mathbf{w}_2, \mathbf{w}_3$, and $\mathbf{w}_4$ as above, we find

$$\mathbf{w}_0 = \begin{bmatrix} 1 \\ 0 \\ 0 \\ 0 \end{bmatrix}, \quad \mathbf{w}_1 = \begin{bmatrix} 1 \\ 2 \\ 0 \\ 0 \end{bmatrix}, \quad \mathbf{w}_2 = \begin{bmatrix} 5 \\ 4 \\ 0 \\ 0 \end{bmatrix}, \quad \mathbf{w}_3 = \begin{bmatrix} 13 \\ 14 \\ 0 \\ 0 \end{bmatrix},$$

$$\mathbf{w}_4 = \begin{bmatrix} 41 \\ 40 \\ 0 \\ 0 \end{bmatrix}.$$

Thus while $\mathbf{w}_0$ and $\mathbf{w}_1$ are linearly independent, all the succeeding vectors $\mathbf{w}_2, \mathbf{w}_3, \mathbf{w}_4$ are linear combinations of $\mathbf{w}_0$ and $\mathbf{w}_1$. Consequently the equation $a_0\mathbf{w}_0 + a_1\mathbf{w}_1 + a_2\mathbf{w}_2 + a_3\mathbf{w}_3 + \mathbf{w}_4 = \boldsymbol{\theta}$ does not have a unique solution, and we cannot use (3.51) to specify the characteristic polynomial.

The matrix H does have an interesting property however. In particular if we let V denote the subspace of R^4 spanned by $\mathbf{w}_0$ and $\mathbf{w}_1$, then we can show that $H\boldsymbol{v}$ is in V whenever $\boldsymbol{v}$ is in V. To see this, note that $H\mathbf{w}_1 = 3\mathbf{w}_0 + 2\mathbf{w}_1$. Thus if $\boldsymbol{v}$ is a linear combination of $\mathbf{w}_0$ and $\mathbf{w}_1$, say

$$\boldsymbol{v} = b_0\mathbf{w}_0 + b_1\mathbf{w}_1,$$

then $H\boldsymbol{v} = b_0 H\mathbf{w}_0 + b_1 H\mathbf{w}_1 = b_0\mathbf{w}_1 + b_1(3\mathbf{w}_0 + 2\mathbf{w}_1)$, which shows that $H\boldsymbol{v}$ is also a linear combination of $\mathbf{w}_0$ and $\mathbf{w}_1$. A subspace V such that $H\boldsymbol{v}$ is in V whenever $\boldsymbol{v}$ is in V is called an ***invariant subspace*** for H.

Before proving that the algorithm given above is an effective procedure for finding the eigenvalues of a standard Hessenberg matrix, we prove a theorem that shows how the eigenvalue problem for H uncouples into smaller problems when H is not a standard Hessenberg matrix. The theorem is based on the observation that a Hessenberg matrix that has a zero subdiagonal entry can be partitioned in a natural and useful way. To illustrate, we consider a (5×5) nonstandard Hessenberg matrix:

$$H = \begin{bmatrix} 2 & 1 & 3 & 5 & 7 \\ 6 & 2 & 1 & 3 & 8 \\ 0 & 1 & 2 & 1 & 3 \\ 0 & 0 & 0 & 4 & 1 \\ 0 & 0 & 0 & 1 & 6 \end{bmatrix}. \tag{3.52}$$

We can partition H into four submatrices H_{11}, H_{12}, H_{22}, and $\mathcal{O}$ as indicated below:

$$H = \left[\begin{array}{ccc|cc} 2 & 1 & 3 & 5 & 7 \\ 6 & 2 & 1 & 3 & 8 \\ 0 & 1 & 2 & 1 & 3 \\ \hline 0 & 0 & 0 & 4 & 1 \\ 0 & 0 & 0 & 1 & 6 \end{array}\right] = \begin{bmatrix} H_{11} & H_{12} \\ \mathcal{O} & H_{22} \end{bmatrix}.$$

A matrix written in partitioned form, such as

$$H = \begin{bmatrix} H_{11} & H_{12} \\ \mathcal{O} & H_{22} \end{bmatrix},$$

is usually called a "block" matrix—the entries in H are blocks, or submatrices, of H. In fact H is called "block upper triangular" since the only block below the diagonal blocks is a zero block.

When some care is exercised to see that all the products are defined, the blocks in a block matrix can be treated as though they were scalars when forming the product of two block matrices. For example suppose Q is a (5×5) matrix partitioned in the same fashion as H:

$$Q = \begin{bmatrix} Q_{11} & Q_{12} \\ Q_{21} & Q_{22} \end{bmatrix}$$

so that Q_{11} is (3×3), Q_{12} is (3×2), Q_{21} is (2×3), and Q_{22} is (2×2). Then it is not hard to show that the product HQ is also given in block form as

$$HQ = \begin{bmatrix} H_{11}Q_{11} + H_{12}Q_{21} & H_{11}Q_{12} + H_{12}Q_{22} \\ H_{22}Q_{21} & H_{22}Q_{22} \end{bmatrix}.$$

(Note that all the products make sense in the block representation of HQ.)

With these preliminaries we now state an important theorem.

Theorem 3.12

Let B be an $(n \times n)$ matrix of the form

$$B = \begin{bmatrix} B_{11} & B_{12} \\ \mathcal{O} & B_{22} \end{bmatrix}$$

where B_{11} is $(k \times k)$, B_{12} is $(k \times (n-k))$, $\mathcal{O}$ is the $((n-k) \times k)$ zero matrix, and B_{22} is $((n-k) \times (n-k))$. Then λ is an eigenvalue of B if and only if λ is an eigenvalue either of B_{11} or of B_{22}.

Proof. Let $\mathbf{x}$ be any $(n \times 1)$ vector, and write $\mathbf{x}$ in partitioned form as

$$\mathbf{x} = \begin{bmatrix} \mathbf{u} \\ \boldsymbol{v} \end{bmatrix} \tag{3.53}$$

where $\mathbf{u}$ is $(k \times 1)$ and $\boldsymbol{v}$ is $((n-k) \times 1)$. It is easy to see that the equation $B\mathbf{x} = \lambda\mathbf{x}$ is equivalent to

$$\begin{aligned} B_{11}\mathbf{u} + B_{12}\boldsymbol{v} &= \lambda\mathbf{u} \\ B_{22}\boldsymbol{v} &= \lambda\boldsymbol{v}. \end{aligned} \tag{3.54}$$

Suppose first that λ is an eigenvalue of B. Then there is a vector $\mathbf{x}$, $\mathbf{x} \neq \boldsymbol{\theta}$, such that $B\mathbf{x} = \lambda\mathbf{x}$. If $\boldsymbol{v} \neq \boldsymbol{\theta}$ in (3.53), then we see from (3.54) that λ is an eigenvalue of B_{22}. On the other hand if $\boldsymbol{v} = \boldsymbol{\theta}$ in (3.53), then we must have $\mathbf{u} \neq \boldsymbol{\theta}$; and (3.54) guarantees that λ is an eigenvalue of B_{11}.

Conversely if λ is an eigenvalue of B_{11}, then there is a nonzero vector $\mathbf{u}_1$ such that $B_{11}\mathbf{u}_1 = \lambda\mathbf{u}_1$. In (3.53) we set $\mathbf{u} = \mathbf{u}_1$ and $\boldsymbol{v} = \boldsymbol{\theta}$ to produce a solution of (3.54), and this result shows that any eigenvalue of B_{11} is also an eigenvalue of B. Finally suppose λ is not an eigenvalue of B_{11}, but λ is an eigenvalue of B_{22}. Then there is a nonzero vector $\boldsymbol{v}_1$ such that $B_{22}\boldsymbol{v}_1 = \lambda\boldsymbol{v}_1$; and so $\boldsymbol{v}_1$ satisfies the last equation in (3.54). To satisfy the first equation in (3.54), we must solve

$$(B_{11} - \lambda I)\mathbf{u} = -B_{12}\boldsymbol{v}_1. \tag{3.55}$$

But since λ is not an eigenvalue of B_{11}, we know that $B_{11} - \lambda I$ is nonsingular; and so we can solve (3.55). Thus every eigenvalue of B_{22} is also an eigenvalue of B. ∎

As an application of this theorem, we recall the (5 × 5) Hessenberg matrix in (3.52), which is in the form required by Theorem 3.12; that is, H is a block upper-triangular matrix, and the diagonal blocks are square. Thus by Theorem 3.12 the eigenvalues of H are precisely the eigenvalues of H_{11} and H_{22} where

$$H_{11} = \begin{bmatrix} 2 & 1 & 3 \\ 6 & 2 & 1 \\ 0 & 1 & 2 \end{bmatrix}, \qquad H_{22} = \begin{bmatrix} 4 & 1 \\ 1 & 6 \end{bmatrix}.$$

The matrix H illustrates the general principle that the eigenvalue problem for a nonstandard Hessenberg matrix can be recast as several eigenvalue problems for smaller Hessenberg matrices, each of which is in standard form.

EXAMPLE 3.22 Consider the (7 × 7) Hessenberg matrix

$$H = \begin{bmatrix} 2 & 3 & 1 & 6 & -1 & 3 & 8 \\ 5 & 7 & 2 & 8 & 2 & 2 & 1 \\ 0 & 0 & 4 & 1 & 3 & -5 & 2 \\ 0 & 0 & 6 & 1 & 2 & 4 & 3 \\ 0 & 0 & 0 & 4 & 1 & 2 & 1 \\ 0 & 0 & 0 & 0 & 0 & 6 & 5 \\ 0 & 0 & 0 & 0 & 0 & 7 & 3 \end{bmatrix}.$$

We first partition H as

$$H = \begin{bmatrix} H_{11} & H_{12} \\ \mathcal{O} & H_{22} \end{bmatrix}$$

where H_{11} is the upper (2 × 2) standard block

$$H_{11} = \begin{bmatrix} 2 & 3 \\ 5 & 7 \end{bmatrix}, \qquad H_{22} = \begin{bmatrix} 4 & 1 & 3 & -5 & 2 \\ 6 & 1 & 2 & 4 & 3 \\ 0 & 4 & 1 & 2 & 1 \\ 0 & 0 & 0 & 6 & 5 \\ 0 & 0 & 0 & 7 & 3 \end{bmatrix}.$$

Now the eigenvalues of H are precisely the eigenvalues of H_{11} and H_{22}. The block H_{11} is in standard form; so we can apply the algorithm to find the characteristic polynomial for H_{11}. However H_{22} is not in standard form; so we partition H_{22} as

$$H_{22} = \begin{bmatrix} C_{11} & C_{12} \\ \mathcal{O} & C_{22} \end{bmatrix}$$

where C_{11} and C_{22} are standard blocks:

$$C_{11} = \begin{bmatrix} 4 & 1 & 3 \\ 6 & 1 & 2 \\ 0 & 4 & 1 \end{bmatrix}, \qquad C_{22} = \begin{bmatrix} 6 & 5 \\ 7 & 3 \end{bmatrix}.$$

The eigenvalues of H_{22} are precisely the eigenvalues of C_{11} and C_{22}, and we can apply the algorithm to find the characteristic polynomial for C_{11} and C_{22}. In summary the eigenvalue problem for H has uncoupled into three eigenvalue problems for the standard Hessenberg matrices H_{11}, C_{11}, and C_{22}.

We conclude this section with a theorem that shows that the algorithm given above always produces a triangular system of equations $a_0\mathbf{w}_0 + a_1\mathbf{w}_1 + \ldots + a_{n-1}\mathbf{w}_{n-1} + \mathbf{w}_n = \boldsymbol{\theta}$, which has a unique solution. In the next section we show that the zeros of $p(t)$ are the eigenvalues of H. We show this by proving a sequence of simple theorems, which are interesting in their own right and which provide a deeper insight into the eigenvalue problem.

Theorem 3.13

Let $H = (h_{ij})$ be an $(n \times n)$ standard Hessenberg matrix, and let $\mathbf{w}_0$ denote the $(n \times 1)$ unit vector $\mathbf{e}_1$. Then the vectors $\mathbf{w}_0, \mathbf{w}_1, \ldots, \mathbf{w}_{n-1}$, defined by

$$\mathbf{w}_i = H\mathbf{w}_{i-1}, \; i = 1, 2, \ldots, n-1,$$

form a basis for R^n.

Proof. Since any set of n linearly independent vectors in R^n is a basis for R^n, we can prove this theorem by showing that $\{\mathbf{w}_0, \mathbf{w}_1, \ldots, \mathbf{w}_{n-1}\}$ is a linearly independent set of vectors. To prove this, we observe first that $\mathbf{w}_0$ and $\mathbf{w}_1$ are given by

$$\mathbf{w}_0 = \begin{bmatrix} 1 \\ 0 \\ 0 \\ \vdots \\ 0 \end{bmatrix}, \qquad \mathbf{w}_1 = \begin{bmatrix} h_{11} \\ h_{21} \\ 0 \\ \vdots \\ 0 \end{bmatrix}.$$

Forming $\mathbf{w}_2 = H\mathbf{w}_1$, we find that

$$\mathbf{w}_2 = \begin{bmatrix} h_{11}h_{11} + h_{12}h_{21} \\ h_{21}h_{11} + h_{22}h_{21} \\ h_{32}h_{21} \\ 0 \\ \vdots \\ 0 \end{bmatrix}.$$

Since H was given as a *standard* Hessenberg matrix, the second component of $\mathbf{w}_1$ and the third component of $\mathbf{w}_2$ are nonzero.

In general (see Exercise 6, Section 3.8) it can be shown that the ith component of $\mathbf{w}_{i-1}$ is the product $h_{i,i-1}h_{i-1,i-2} \ldots h_{32}h_{21}$ while the kth component of $\mathbf{w}_{i-1}$ is zero for $k = i+1, i+2, \ldots, n$. Thus the $(n \times n)$ matrix

$$W = [\mathbf{w}_0, \mathbf{w}_1, \ldots, \mathbf{w}_{n-1}]$$

is upper triangular, and the diagonal entries of W are all nonzero. In light of this we conclude that $\{\mathbf{w}_0, \mathbf{w}_1, \ldots, \mathbf{w}_{n-1}\}$ is a set of n linearly independent vectors in R^n and is hence a basis for R^n. ■

3.8 EXERCISES

1. Using the algorithm described in (3.49)–(3.51), find the characteristic polynomial and the eigenvalues of these matrices.

$$\text{a)}\ \begin{bmatrix} 3 & -8 & 4 \\ -1 & -2 & -2 \\ 0 & 6 & 1 \end{bmatrix} \qquad \text{b)}\ \begin{bmatrix} -6 & 31 & -14 \\ -1 & 6 & -2 \\ 0 & 2 & 1 \end{bmatrix} \qquad \text{c)}\ \begin{bmatrix} 1 & 1 & 1 \\ 1 & 1 & 1 \\ 0 & 1 & 1 \end{bmatrix}$$

2. Repeat Exercise 1 for these matrices.

$$\text{a)}\ \begin{bmatrix} -2 & 0 & -2 \\ -1 & 1 & -2 \\ 0 & 1 & -1 \end{bmatrix} \qquad \text{b)}\ \begin{bmatrix} 1 & 2 & 1 \\ -1 & 1 & 2 \\ 0 & 2 & 2 \end{bmatrix} \qquad \text{c)}\ \begin{bmatrix} -2 & 0 & -1 \\ -2 & -1 & -2 \\ 0 & 1 & 1 \end{bmatrix}$$

3. Partition the matrix B into blocks as in Theorem 3.12. Find the eigenvalues of the diagonal blocks, and for each distinct eigenvalue of B find an eigenvector as in (3.54).

$$\text{a)}\ B = \begin{bmatrix} 1 & -1 & 1 & 4 \\ 1 & 3 & -2 & 1 \\ 0 & 0 & 2 & -1 \\ 0 & 0 & -1 & 2 \end{bmatrix} \qquad \text{b)}\ B = \begin{bmatrix} 1 & 1 & 2 & 1 \\ 1 & 1 & 1 & 3 \\ 0 & 0 & 3 & 0 \\ 0 & 0 & 1 & 4 \end{bmatrix}$$

$$\text{c)}\ B = \begin{bmatrix} -2 & 0 & -2 & 1 \\ -1 & 1 & -2 & 3 \\ 0 & 1 & -1 & -2 \\ 0 & 0 & 0 & 2 \end{bmatrix}$$

4. Find the characteristic polynomial of H according to (3.49)–(3.51) and by evaluating $\det(H - tI)$. Verify that these results are the same.

$$H = \begin{bmatrix} 2 & 3 & -21 & -3 \\ 2 & 7 & -41 & -5 \\ 0 & 1 & -5 & -1 \\ 0 & 0 & 4 & 4 \end{bmatrix}$$

5. Let H be an $(n \times n)$ standard Hessenberg matrix, let λ be an eigenvalue of H, and suppose $\mathbf{u}$ is a corresponding eigenvector.
   a) Prove that the nth component of $\mathbf{u}$ is nonzero. [Hint: Consider $(H - \lambda I)\mathbf{u} = \boldsymbol{\theta}$, and suppose the nth component is zero.]
   b) Prove that $\dim(\mathcal{N}(H - \lambda I)) = 1$. [Hint: Suppose $\mathbf{v} \neq \boldsymbol{\theta}$ and $(H - \lambda I)\mathbf{v} = \boldsymbol{\theta}$. Choose a scalar b so that the last components of $\mathbf{u}$ and $b\mathbf{v}$ agree.]
   c) Prove that an $(n \times n)$ standard Hessenberg matrix H is diagonalizable if and only if H has n distinct eigenvalues. (Hint: Recall Theorem 3.9.)
6. Complete the proof of Theorem 3.13 by showing that the ith component of $\mathbf{w}_{i-1}$ is nonzero.

3.9 MATRIX POLYNOMIALS AND THE CAYLEY–HAMILTON THEOREM

To complete our discussion of the algorithm presented in the Section 3.8 it is convenient to introduce the idea of a matrix polynomial. By way of example, consider the polynomial

$$q(t) = t^2 + 3t - 2.$$

If A is an $(n \times n)$ matrix, then we can define a matrix expression corresponding to $q(t)$:

$$q(A) = A^2 + 3A - 2I$$

where I is the $(n \times n)$ identity. In effect we have inserted A for t in $q(t) = t^2 + 3t - 2$ and defined A^0 by $A^0 = I$. In general if $q(t)$ is the kth degree polynomial

$$q(t) = b_k t^k + \ldots + b_2 t^2 + b_1 t + b_0,$$

and if A is an $(n \times n)$ matrix, we define $q(A)$ by

$$q(A) = b_k A^k + \ldots + b_2 A^2 + b_1 A + b_0 I$$

where I is the $(n \times n)$ identity matrix. Since $q(A)$ is obviously an $(n \times n)$ matrix, we might ask for the eigenvalues and eigenvectors of $q(A)$. It is easy to show that if λ is an eigenvalue of A, then $q(\lambda)$ is an eigenvalue of $q(A)$. [Note that $q(\lambda)$ is the scalar obtained by substituting the value $t = \lambda$ into $q(t)$.]

Theorem 3.14

Suppose $q(t)$ is a kth degree polynomial, and suppose A is an $(n \times n)$ matrix such that $A\mathbf{x} = \lambda\mathbf{x}$ where $\mathbf{x} \neq \boldsymbol{\theta}$. Then $q(A)\mathbf{x} = q(\lambda)\mathbf{x}$.

Proof. Suppose that $A\mathbf{x} = \lambda\mathbf{x}$ where $\mathbf{x} \neq \boldsymbol{\theta}$. As we know, the implication is that $A^2\mathbf{x} = \lambda^2\mathbf{x}$ and in general

$$A^i\mathbf{x} = \lambda^i\mathbf{x}, \qquad i = 2, 3, \ldots .$$

Therefore if $q(t) = b_k t^k + \ldots + b_2 t^2 + b_1 t + b_0$, then

$$\begin{aligned} q(A)\mathbf{x} &= (b_k A^k + \ldots + b_2 A^2 + b_1 A + b_0 I)\mathbf{x} \\ &= b_k A^k\mathbf{x} + \ldots + b_2 A^2\mathbf{x} + b_1 A\mathbf{x} + b_0\mathbf{x} \\ &= b_k \lambda^k\mathbf{x} + \ldots + b_2\lambda^2\mathbf{x} + b_1\lambda\mathbf{x} + b_0\mathbf{x} \\ &= q(\lambda)\mathbf{x}. \end{aligned}$$

Thus if λ is an eigenvalue of A, then $q(\lambda)$ is an eigenvalue of $q(A)$. ■

An interesting special case of this theorem is provided when $q(t)$ is the characteristic polynomial for A. In particular suppose λ is an eigenvalue of A, and suppose $p(t)$ is the characteristic polynomial for A so that $p(\lambda) = 0$. Since $p(\lambda)$ is an eigenvalue of $p(A)$, and since $p(\lambda) = 0$, we conclude that zero is an eigenvalue for $p(A)$; that is, $p(A)$ is a singular matrix. In fact we will be able to prove more than this; we will show that $p(A)$ is the zero matrix [$p(A) = \mathcal{O}$ is the conclusion of the Cayley–Hamilton theorem].

EXAMPLE 3.23 Consider the (2×2) matrix

$$A = \begin{bmatrix} 1 & -2 \\ 2 & 3 \end{bmatrix}.$$

Since A is a standard Hessenberg matrix, we can employ the algorithm of Section 3.8 to determine the characteristic polynomial. In particular

$$\mathbf{w}_0 = \begin{bmatrix} 1 \\ 0 \end{bmatrix}, \qquad \mathbf{w}_1 = A\mathbf{w}_0 = \begin{bmatrix} 1 \\ 2 \end{bmatrix}, \qquad \mathbf{w}_2 = A\mathbf{w}_1 = \begin{bmatrix} -3 \\ 8 \end{bmatrix}.$$

Thus we find $p(t) = t^2 + a_1 t + a_0$ from

$$\begin{aligned} a_0 + a_1 &= 3 \\ 2a_1 &= -8, \end{aligned}$$

which leads to $p(t) = t^2 - 4t + 7$. Now $p(A) = A^2 - 4A + 7I$ is given by

$$p(A) = \begin{bmatrix} -3 & -8 \\ 8 & 5 \end{bmatrix} - \begin{bmatrix} 4 & -8 \\ 8 & 12 \end{bmatrix} + \begin{bmatrix} 7 & 0 \\ 0 & 7 \end{bmatrix} = \begin{bmatrix} 0 & 0 \\ 0 & 0 \end{bmatrix},$$

and we have a particular instance of the Cayley–Hamilton theorem; $p(A)$ is the zero matrix.

The theorems that follow show that the algorithm given in Section 3.8 leads to a polynomial $p(t)$ whose zeros are the eigenvalues of H. In the process of verifying this, we will prove an interesting version of the Cayley–Hamilton theorem that is applicable to a standard Hessenberg matrix. Before beginning, we make an observation about the sequence of vectors $\mathbf{w}_0, \mathbf{w}_1, \mathbf{w}_2, \ldots$ defined in (3.49) from

$$\mathbf{w}_i = H\mathbf{w}_{i-1}, \; i = 1, 2, \ldots.$$

Since $\mathbf{w}_0 = \mathbf{e}_1$, we have that $\mathbf{w}_1 = H\mathbf{w}_0 = H\mathbf{e}_1$. Given that $\mathbf{w}_1 = H\mathbf{e}_1$, we see that

$$\mathbf{w}_2 = H\mathbf{w}_1 = H(H\mathbf{e}_1) = H^2\mathbf{e}_1;$$

and in general $\mathbf{w}_k = H^k\mathbf{e}_1$. Thus we can interpret the sequence $\mathbf{w}_0, \mathbf{w}_1, \mathbf{w}_2, \ldots, \mathbf{w}_n$ as being given by $\mathbf{e}_1, H\mathbf{e}_1, H^2\mathbf{e}_1, \ldots, H^n\mathbf{e}_1$.

With this interpretation we rewrite the equation $a_0\mathbf{w}_0 + a_1\mathbf{w}_1 + \ldots + a_{n-1}\mathbf{w}_{n-1} + \mathbf{w}_n = \boldsymbol{\theta}$ given in (3.50) as

$$a_0\mathbf{e}_1 + a_1 H\mathbf{e}_1 + a_2 H^2\mathbf{e}_1 + \ldots + a_{n-1}H^{n-1}\mathbf{e}_1 + H^n\mathbf{e}_1 = \boldsymbol{\theta}; \tag{3.56}$$

or by regrouping, (3.56) is the same as

$$(a_0 I + a_1 H + a_2 H^2 + \ldots + a_{n-1}H^{n-1} + H^n)\mathbf{e}_1 = \boldsymbol{\theta}. \tag{3.57}$$

Now Theorem 3.13 asserts that if H is a standard $(n \times n)$ Hessenberg matrix, then the vectors $\mathbf{e}_1, H\mathbf{e}_1, H^2\mathbf{e}_1, \ldots, H^{n-1}\mathbf{e}_1$ are linearly independent and that there is a unique set of scalars $a_0, a_1, \ldots, a_{n-1}$ that satisfy (3.56). Defining $p(t)$ from (3.56) as

$$p(t) = a_0 + a_1 t + a_2 t^2 + \ldots a_{n-1}t^{n-1} + t^n,$$

we see that (3.57) says $p(H)\mathbf{e}_1 = \boldsymbol{\theta}$. With these preliminaries we prove the following result.

Theorem 3.15

Let H be an $(n \times n)$ standard Hessenberg matrix; let $a_0, a_1, \ldots, a_{n-1}$ be the unique scalars satisfying

$$a_0\mathbf{e}_1 + a_1 H\mathbf{e}_1 + a_2 H^2\mathbf{e}_1 + \ldots + a_{n-1}H^{n-1}\mathbf{e}_1 + H^n\mathbf{e}_1 = \boldsymbol{\theta};$$

and let $p(t) = a_0 + a_1 t + a_2 t^2 + \ldots + a_{n-1}t^{n-1} + t^n$.

1. $p(H)$ is the zero matrix.
2. If $q(t) = b_0 + b_1 t + b_2 t^2 + \ldots + b_{k-1}t^{k-1} + t^k$ is any monic kth degree polynomial, and if $q(H)$ is the zero matrix, then $k \geqslant n$. Moreover if $k = n$, then $q(t) \equiv p(t)$.

Proof. For (1) since $\{\mathbf{e}_1, H\mathbf{e}_1, H^2\mathbf{e}_1, \ldots, H^{n-1}\mathbf{e}_1\}$ is a basis for R^n, we can express any vector $\mathbf{y}$ in R^n as a linear combination

$$\mathbf{y} = c_0\mathbf{e}_1 + c_1 H\mathbf{e}_1 + c_2 H^2\mathbf{e}_1 + \ldots + c_{n-1}H^{n-1}\mathbf{e}_1.$$

Therefore $p(H)\mathbf{y}$ is the vector

$$p(H)\mathbf{y} = c_0 p(H)\mathbf{e}_1 + c_1 p(H)H\mathbf{e}_1 + \ldots. + c_{n-1}p(H)H^{n-1}\mathbf{e}_1. \tag{3.58}$$

Now while matrix products do not normally commute, it is easy to see that $p(H)H^i = H^i p(H)$. Therefore from (3.58) we can represent $p(H)\mathbf{y}$ as

$$p(H)\mathbf{y} = c_0 p(H)\mathbf{e}_1 + c_1 Hp(H)\mathbf{e}_1 + \ldots + c_{n-1}H^{n-1}p(H)\mathbf{e}_1;$$

and since $p(H)\mathbf{e}_1 = \boldsymbol{\theta}$ [see (3.56) and (3.57)], we have that $p(H)\mathbf{y} = \boldsymbol{\theta}$ for *any* $\mathbf{y}$ in R^n. In particular, $p(H)\mathbf{e}_j = \boldsymbol{\theta}$ for $j = 1, 2, \ldots, n$; and since $p(H)\mathbf{e}_j$ is the jth column of $p(H)$, it follows that $p(H) = \mathcal{O}$.

For the proof of part (2) suppose $q(H)$ is the zero matrix where

$$q(H) = b_0 + b_1 H + b_2 H^2 + \ldots + b_{k-1}H^{k-1} + H^k.$$

Then $q(H)\mathbf{y} = \boldsymbol{\theta}$ for every $\mathbf{y}$ in R^n, and in particular for $\mathbf{y} = \mathbf{e}_1$ we have $q(H)\mathbf{e}_1 = \boldsymbol{\theta}$ or

$$b_0\mathbf{e}_1 + b_1 H\mathbf{e}_1 + b_2 H^2\mathbf{e}_1 + \ldots + b_{k-1}H^{k-1}\mathbf{e}_1 + H^k\mathbf{e}_1 = \boldsymbol{\theta}. \tag{3.59}$$

However the vectors $\mathbf{e}_1, H\mathbf{e}_1, \ldots, H^k\mathbf{e}_1$ are linearly independent when $k \leqslant n-1$; so (3.59) can hold only if $k \geqslant n$ [recall that the leading coefficient of $q(t)$ is 1; so we are excluding the possibility that $q(t)$ is the zero polynomial]. Moreover if $k = n$, we can satisfy (3.59) only with the choice $b_0 = a_0, b_1 = a_1, \ldots, b_{n-1} = a_{n-1}$ by Theorem 3.13; so if $k = n$, then $q(t) \equiv p(t)$. ∎

Since it was shown above that $p(H)$ is the zero matrix whenever $p(t)$ is the polynomial defined by (3.51) in the algorithm of Section 3.8, it is now an easy matter to show that the zeros of $p(t)$ are precisely the eigenvalues of H.

Theorem 3.16

Let H be an $(n \times n)$ standard Hessenberg matrix, and let $p(t)$ be the polynomial defined by (3.51). Then λ is a root of $p(t) = 0$ if and only if λ is an eigenvalue of H.

Proof. We show first that every eigenvalue of H is a zero of $p(t)$. Thus we suppose that $H\mathbf{x} = \lambda\mathbf{x}$ where $\mathbf{x} \neq \boldsymbol{\theta}$. By Theorem 3.14 we know that

$$p(H)\mathbf{x} = p(\lambda)\mathbf{x};$$

and since $p(H)$ is the zero matrix by Theorem 3.15, we must also conclude that

$$\boldsymbol{\theta} = p(\lambda)\mathbf{x}.$$

But since $\mathbf{x} \neq \boldsymbol{\theta}$, the equality $\boldsymbol{\theta} = p(\lambda)\mathbf{x}$ implies that $p(\lambda) = 0$. Thus every eigenvalue of H is a zero of $p(t)$.

Conversely suppose λ is a zero of $p(t)$. Then we can write $p(t)$ in the form

$$p(t) = (t - \lambda)q(t) \tag{3.60}$$

where $q(t)$ is a monic polynomial of degree $n-1$. Now equating coefficients of like powers shows that if $u(t) = r(t)s(t)$ where u, r, and s are polynomials, then we also have a corresponding matrix identity

$$u(A) = r(A)s(A)$$

for any square matrix A. Thus from (3.60) we may assert that

$$p(H) = (H - \lambda I)q(H). \tag{3.61}$$

If $H - \lambda I$ were nonsingular, we could rewrite (3.61) as

$$(H - \lambda I)^{-1}p(H) = q(H);$$

and since $p(H)$ is the zero matrix by part (1) of Theorem 3.15, we would have that $q(H)$ is the zero matrix also. However $q(t)$ is a monic polynomial of degree $n-1$; so Theorem 3.15, part (2), assures us that $q(H)$ is *not* the zero matrix. Thus if λ is a root of $p(t) = 0$, we know that $H - \lambda I$ must be singular; and hence λ is an eigenvalue of H. ∎

We conclude this section by outlining a proof of the Cayley–Hamilton theorem for an arbitrary matrix. If H is a Hessenberg matrix of the form

$$H = \begin{bmatrix} H_{11} & H_{12} & \cdots & H_{1r} \\ \mathcal{O} & H_{22} & \cdots & H_{2r} \\ \vdots & & & \\ \mathcal{O} & \mathcal{O} & \cdots & H_{rr} \end{bmatrix} \tag{3.62}$$

where $H_{11}, H_{22}, \ldots, H_{rr}$ are standard Hessenberg blocks, then we define $p(t)$ to be the characteristic polynomial for H where

$$p(t) = p_1(t)p_2(t)\ldots p_r(t)$$

and where $p_i(t)$ is the characteristic polynomial for H_{ii}, $1 \leqslant i \leqslant r$.

Theorem 3.17

If $p(t)$ is the characteristic polynomial for a Hessenberg matrix H, then $p(H)$ is the zero matrix.

Proof. We sketch the proof for the case $r = 2$. If H has the form

$$H = \begin{bmatrix} H_{11} & H_{12} \\ \mathcal{O} & H_{22} \end{bmatrix}$$

where H_{11} and H_{22} are square blocks, then it can be shown that H^k is a block matrix of the form

$$H^k = \begin{bmatrix} H_{11}^k & V_k \\ \mathcal{O} & H_{22}^k \end{bmatrix}.$$

Given this, it follows that if $q(t)$ is any polynomial, then $q(H)$ is a block matrix of the form

$$q(H) = \begin{bmatrix} q(H_{11}) & W \\ \mathcal{O} & q(H_{22}) \end{bmatrix}.$$

From these preliminaries if H_{11} and H_{22} are standard blocks, then

$$p(H) = p_1(H)p_2(H) = \begin{bmatrix} p_1(H_{11}) & R \\ \mathcal{O} & p_1(H_{22}) \end{bmatrix} \begin{bmatrix} p_2(H_{11}) & S \\ \mathcal{O} & p_2(H_{22}) \end{bmatrix};$$

and since $p_1(H_{11})$ and $p_2(H_{22})$ are zero blocks, it is easy to see that $p(H)$ is the zero matrix. This argument can be repeated inductively to show that $p(H)$ is the zero matrix when H has the form (3.62) for $r > 2$. ∎

Finally we note that the essential features of polynomial expressions are preserved by similarity transformations. For example if $H = SAS^{-1}$, and if $q(t)$ is any polynomial, then (Exercise 5, Section 3.9)

$$q(H) = Sq(A)S^{-1}.$$

Thus if A is similar to H, and if $p(H)$ is the zero matrix, then $p(A)$ is the zero matrix as well.

3.9 EXERCISES

1. Let $q(t) = t^2 - 4t + 3$. Calculate the matrices $q(A)$, $q(B)$, and $q(C)$.

$$A = \begin{bmatrix} 1 & -1 \\ 1 & 3 \end{bmatrix}, \quad B = \begin{bmatrix} 2 & -1 \\ -1 & 2 \end{bmatrix}, \quad C = \begin{bmatrix} -2 & 0 & -2 \\ -1 & 1 & -2 \\ 0 & 1 & -1 \end{bmatrix}$$

2. The polynomial $p(t) = (t-1)^3 = t^3 - 3t^2 + 3t - 1$ is the characteristic polynomial for A, B, C, and I.

$$A = \begin{bmatrix} 1 & 0 & 0 \\ 1 & 1 & 0 \\ 0 & 0 & 1 \end{bmatrix}, \quad B = \begin{bmatrix} 1 & 0 & 0 \\ 0 & 1 & 0 \\ 0 & 1 & 1 \end{bmatrix}, \quad C = \begin{bmatrix} 1 & 0 & 0 \\ 1 & 1 & 0 \\ 0 & 1 & 1 \end{bmatrix}, \quad I = \begin{bmatrix} 1 & 0 & 0 \\ 0 & 1 & 0 \\ 0 & 0 & 1 \end{bmatrix}$$

a) Verify that $p(A)$, $p(B)$, $p(C)$, and $p(I)$ are each the zero matrix.
b) For A and B find a quadratic polynomial $q(t)$ such that $q(A) = q(B) = \mathcal{O}$.

3. Suppose $q(t)$ is any polynomial and $p(t)$ is the characteristic polynomial for a matrix A. If we divide $p(t)$ into $q(t)$, we obtain an identity

$$q(t) = s(t)p(t) + r(t)$$

where the degree of $r(t)$ is less than the degree of $p(t)$. From this result, it can be shown that $q(A) = s(A)p(A) + r(A)$; and since $p(A) = \mathcal{O}$, $q(A) = r(A)$.

a) Let $p(t) = t^2 - 4t + 3$ and $q(t) = t^5 - 4t^4 + 4t^3 - 5t^2 + 8t - 1$. Find $s(t)$ and $r(t)$ so that $q(t) = s(t)p(t) + r(t)$.
b) Observe that $p(t)$ is the characteristic polynomial for the matrix B in Exercise 1. Calculate the matrix $q(B)$ without forming the powers B^5, B^4, etc.

4. Consider the (7×7) Hessenberg matrix H of Example 3.22, Section 3.8, where H is partitioned with three standard diagonal blocks, H_{11}, H_{22}, and H_{33}. Verify that $\det(H - tI) = -p_1(t)p_2(t)p_3(t)$ where $p_1(t)$, $p_2(t)$, and $p_3(t)$ are the characteristic polynomials for H_{11}, H_{22}, and H_{33} as given by (3.49)–(3.51).

5. Suppose $H = SAS^{-1}$, and suppose $q(t)$ is any polynomial. Show that $q(H) = Sq(A)S^{-1}$. (Hint: Show that $H^k = SA^kS^{-1}$ by direct multiplication.)

† *Exercises 6–8 show that a symmetric matrix is diagonalizable.*

6. Let A be an $(n \times n)$ symmetric matrix. Let λ_1 and λ_2 be distinct eigenvalues of A with corresponding eigenvectors $\mathbf{u}_1$ and $\mathbf{u}_2$. Prove that $\mathbf{u}_1^T\mathbf{u}_2 = 0$. (Hint: Given that $A\mathbf{u}_1 = \lambda_1\mathbf{u}_1$ and $A\mathbf{u}_2 = \lambda_2\mathbf{u}_2$, show that $\mathbf{u}_1^TA\mathbf{u}_2 = \mathbf{u}_2^TA\mathbf{u}_1$.)

7. Let W be a subspace of R^n where $\dim(W) = d$, $d \geq 1$. Let A be an $(n \times n)$ matrix, and suppose that $A\mathbf{x}$ is in W whenever $\mathbf{x}$ is in W.

a) Let $\mathbf{x}_0$ be any fixed vector in W. Prove that $A^j\mathbf{x}_0$ is in W for $j = 1, 2, \ldots$. There is a smallest value k for which the set of vectors $\{\mathbf{x}_0, A\mathbf{x}_0, A^2\mathbf{x}_0, \ldots, A^k\mathbf{x}_0\}$ is linearly dependent; and thus there are unique scalars $a_0, a_1, \ldots, a_{k-1}$ such that $a_0\mathbf{x}_0 + a_1A\mathbf{x}_0 + \ldots + a_{k-1}A^{k-1}\mathbf{x}_0 + A^k\mathbf{x}_0 = \boldsymbol{\theta}$. Use these scalars to define the polynomial $m(t)$ where $m(t) = t^k + a_{k-1}t^{k-1} + \ldots + a_1t + a_0$. Observe that $m(A)\mathbf{x}_0 = \boldsymbol{\theta}$; $m(t)$ is called the "minimal annihilating polynomial" for $\mathbf{x}_0$—by construction there is no monic polynomial $q(t)$ where $q(t)$ has degree less than k and $q(A)\mathbf{x}_0 = \boldsymbol{\theta}$.
b) Let r be a root of $m(t) = 0$ so that $m(t) = (t - r)s(t)$. Prove that r is an eigenvalue of A. [Hint: Is the vector $s(A)\mathbf{x}_0$ nonzero?] Note: Part (b) shows that every root of $m(t) = 0$ is an eigenvalue of A. If A is symmetric, then $m(t) = 0$ has only real roots; so $s(A)\mathbf{x}_0$ is in W.

8. Exercise 6 shows that eigenvectors of a symmetric matrix belonging to distinct eigenvalues are orthogonal. We now show that if A is a symmetric $(n \times n)$ matrix, then A has a set of n orthogonal eigenvectors. Let $\{\mathbf{u}_1, \mathbf{u}_2, \ldots, \mathbf{u}_k\}$ be a set of k eigenvectors for A, $1 \leq k < n$ where $\mathbf{u}_i^T\mathbf{u}_j = 0$, $i \neq j$. Let W be the subset of R^n defined by

$$W = \{\mathbf{x}: \mathbf{x}^T\mathbf{u}_i = 0, i = 1, 2, \ldots, k\}.$$

From the Gram–Schmidt theorem the subset W contains nonzero vectors.

a) Prove that W is a subspace of R^n.
b) Suppose A is $(n \times n)$ and symmetric. Prove that $A\mathbf{x}$ is in W whenever $\mathbf{x}$ is in W.

From Exercise 7, A has an eigenvector, $\mathbf{u}$, in W. If we label $\mathbf{u}$ as $\mathbf{u}_{k+1}$, then by construction $\{\mathbf{u}_1, \mathbf{u}_2, \ldots, \mathbf{u}_k, \mathbf{u}_{k+1}\}$ is a set of orthogonal eigenvectors for A. It follows that A has a set of n orthogonal eigenvectors, $\mathbf{u}_1, \mathbf{u}_2, \ldots, \mathbf{u}_n$. Using these, we can form Q so that $Q^TAQ = D$ where D is diagonal and $Q^TQ = I$.

†3.10 GENERALIZED EIGENVECTORS AND SOLUTIONS OF SYSTEMS OF DIFFERENTIAL EQUATIONS

In this section we will develop the idea of a generalized eigenvector in order to give the complete solution to the system of differential equations $\mathbf{x}' = A\mathbf{x}$. When an $(n \times n)$ matrix A has real eigenvalues, the eigenvectors and generalized eigenvectors of A form a basis for R^n; and we will show how to construct the complete solution of $\mathbf{x}' = A\mathbf{x}$ from this special basis. (When some of the eigenvalues of A are complex, a few modifications are necessary to obtain the complete solution of $\mathbf{x}' = A\mathbf{x}$ in a real form. In any event the eigenvectors and generalized eigenvectors of A form a basis for C^n where C^n denotes the set of all n-dimensional vectors with real or complex components.)

To begin, let A be an $(n \times n)$ matrix. The problem we wish to solve is called an "initial-value problem" and is formulated as follows: given a vector $\mathbf{x}_0$ in R^n, find a function $\mathbf{x}(t)$ such that

$$\begin{aligned} \mathbf{x}(0) &= \mathbf{x}_0 \\ \mathbf{x}'(t) &= A\mathbf{x}(t) \text{ for all } t. \end{aligned} \tag{3.63}$$

If we can find n functions $\mathbf{x}_1(t), \mathbf{x}_2(t), \ldots, \mathbf{x}_n(t)$ that satisfy

$$\mathbf{x}_1'(t) = A\mathbf{x}_1(t),\ \mathbf{x}_2'(t) = A\mathbf{x}_2(t), \ldots, \mathbf{x}_n'(t) = A\mathbf{x}_n(t)$$

and such that $\{\mathbf{x}_1(0), \mathbf{x}_2(0), \ldots, \mathbf{x}_n(0)\}$ is linearly independent, then we can always solve (3.63). To show why, we merely note that there must be constants $c_1, c_2, \ldots, c_n$ such that

$$\mathbf{x}_0 = c_1\mathbf{x}_1(0) + c_2\mathbf{x}_2(0) + \ldots + c_n\mathbf{x}_n(0),$$

and then note that the function

$$\mathbf{y}(t) = c_1\mathbf{x}_1(t) + c_2\mathbf{x}_2(t) + \ldots + c_n\mathbf{x}_n(t)$$

satisfies the requirements of (3.63). Thus to solve $\mathbf{x}' = A\mathbf{x}$, $\mathbf{x}(0) = \mathbf{x}_0$, we are led to search for n solutions $\mathbf{x}_1(t), \mathbf{x}_2(t), \ldots, \mathbf{x}_n(t)$ of $\mathbf{x}' = A\mathbf{x}$ for which $\{\mathbf{x}_1(0), \mathbf{x}_2(0), \ldots, \mathbf{x}_n(0)\}$ is linearly independent.

If A has a set of k linearly independent eigenvectors $\{\mathbf{u}_1, \mathbf{u}_2, \ldots, \mathbf{u}_k\}$ where

$$A\mathbf{u}_i = \lambda_i\mathbf{u}_i, \quad i = 1, 2, \ldots, k,$$

then as in Section 3.4, we can immediately construct k solutions to $\mathbf{x}' = A\mathbf{x}$, namely

$$\mathbf{x}_1(t) = e^{\lambda_1 t}\mathbf{u}_1,\ \mathbf{x}_2(t) = e^{\lambda_2 t}\mathbf{u}_2, \ldots, \mathbf{x}_k(t) = e^{\lambda_k t}\mathbf{u}_k.$$

Also since $\mathbf{x}_i(0) = \mathbf{u}_i$, it follows that $\{\mathbf{x}_1(0), \mathbf{x}_2(0), \ldots, \mathbf{x}_k(0)\}$ is a linearly independent set. The difficulty arises when $k < n$, for then we must produce an additional set of $n-k$ solutions of $\mathbf{x}' = A\mathbf{x}$. In this connection an $(n \times n)$ matrix A is called

† This section may be omitted without any loss of continuity.

defective if A has fewer than n linearly independent eigenvectors. [Note: Distinct eigenvalues give rise to linearly independent eigenvectors; so A can be defective only if the characteristic equation $p(t) = 0$ has fewer than n distinct roots.]

A complete analysis of the initial-value problem is simplified considerably if we assume A is a Hessenberg matrix. If A is not a Hessenberg matrix, then a simple change of variables can be used to convert $\mathbf{x}' = A\mathbf{x}$ to an equivalent problem $\mathbf{y}' = H\mathbf{y}$ where H is a Hessenberg matrix. In particular suppose $QAQ^{-1} = H$, and let $\mathbf{y}(t) = Q\mathbf{x}(t)$. Therefore we see that $\mathbf{x}(t) = Q^{-1}\mathbf{y}(t)$ and $\mathbf{x}'(t) = Q^{-1}\mathbf{y}'(t)$. Therefore $\mathbf{x}'(t) = A\mathbf{x}(t)$ is the same as

$$Q^{-1}\mathbf{y}'(t) = AQ^{-1}\mathbf{y}(t).$$

Multiplying both sides by Q, we obtain the related equation $\mathbf{y}' = H\mathbf{y}$ where $H = QAQ^{-1}$ is a Hessenberg matrix. Given that we can always make this change of variables, we will focus for the remainder of this section on the problem of solving

$$\mathbf{x}'(t) = H\mathbf{x}(t), \qquad \mathbf{x}(0) = \mathbf{x}_0. \tag{3.64}$$

As we know, if H is $(n \times n)$ and has n linearly independent eigenvectors, we can always solve (3.64). To see how to solve (3.64) when H is defective, let us suppose that $p(t)$ is the characteristic polynomial for H. If we write $p(t)$ in factored form as

$$p(t) = (t-\lambda_1)^{m_1}(t-\lambda_2)^{m_2}\ldots(t-\lambda_k)^{m_k}$$

where $m_1 + m_2 + \ldots + m_k = n$, then we say that the eigenvalue λ_i has ***algebraic multiplicity*** m_i. Given λ_i, we want to construct m_i solutions of $\mathbf{x}' = H\mathbf{x}$ that are associated with λ_i. For example suppose λ is an eigenvalue of H of algebraic multiplicity 2. We have one solution of $\mathbf{x}' = H\mathbf{x}$, namely $\mathbf{x}(t) = e^{\lambda t}\mathbf{u}$ where $H\mathbf{u} = \lambda\mathbf{u}$; and we would like another solution. To find this additional solution, we note that the theory from elementary differential equations suggests that we should look for another solution to $\mathbf{x}' = H\mathbf{x}$ that is of the form $\mathbf{x}(t) = te^{\lambda t}\mathbf{a} + e^{\lambda t}\mathbf{b}$ where $\mathbf{a} \neq \boldsymbol{\theta}$, $\mathbf{b} \neq \boldsymbol{\theta}$. To see what conditions $\mathbf{a}$ and $\mathbf{b}$ must satisfy, we calculate

$$\begin{aligned}\mathbf{x}'(t) &= t\lambda e^{\lambda t}\mathbf{a} + e^{\lambda t}\mathbf{a} + \lambda e^{\lambda t}\mathbf{b}\\ H\mathbf{x}(t) &= te^{\lambda t}H\mathbf{a} + e^{\lambda t}H\mathbf{b}.\end{aligned}$$

After we equate $\mathbf{x}'(t)$ with $H\mathbf{x}(t)$, and group like powers of t, our guess leads to the conditions.

$$\begin{aligned}t\lambda e^{\lambda t}\mathbf{a} &= te^{\lambda t}H\mathbf{a}\\ e^{\lambda t}(\mathbf{a} + \lambda\mathbf{b}) &= e^{\lambda t}H\mathbf{b}.\end{aligned} \tag{3.65}$$

If (3.65) is to hold for all t, we will need

$$\begin{aligned}\lambda\mathbf{a} &= H\mathbf{a}\\ \mathbf{a} + \lambda\mathbf{b} &= H\mathbf{b},\end{aligned}$$

or equivalently

$$\begin{aligned}(H - \lambda I)\mathbf{a} &= \boldsymbol{\theta}\\ (H - \lambda I)\mathbf{b} &= \mathbf{a}\end{aligned} \tag{3.66}$$

where $\mathbf{a}$ and $\mathbf{b}$ are nonzero vectors. From (3.66) we see that $\mathbf{a}$ is an eigenvector and that $(H - \lambda I)^2\,\mathbf{b} = \boldsymbol{\theta}$, but $(H - \lambda I)\mathbf{b} \neq \boldsymbol{\theta}$. We will call $\mathbf{b}$ a "generalized eigenvector of order 2." If we can find vectors **a** and **b** that satisfy (3.66), then we have two solutions of $\mathbf{x}' = H\mathbf{x}$ associated with λ, namely

$$\mathbf{x}_1(t) = e^{\lambda t}\mathbf{a}$$
$$\mathbf{x}_2(t) = te^{\lambda t}\mathbf{a} + e^{\lambda t}\mathbf{b}.$$

Moreover $\mathbf{x}_1(0) = \mathbf{a}$, $x_2(0) = \mathbf{b}$, and it is easy to see that $\mathbf{x}_1(0)$ and $\mathbf{x}_2(0)$ are linearly independent. [If $c_1\mathbf{a} + c_2\mathbf{b} = \boldsymbol{\theta}$, then $(H - \lambda I)(c_1\mathbf{a} + c_2\mathbf{b}) = \boldsymbol{\theta}$. Since $(H - \lambda I)\mathbf{a} = \boldsymbol{\theta}$, it follows that $c_2(H - \lambda I)\mathbf{b} = c_2\mathbf{a} = \boldsymbol{\theta}$, which shows that $c_2 = 0$. Finally if $c_2 = 0$, then $c_1\mathbf{a} = \boldsymbol{\theta}$, which means that $c_1 = 0$.] The discussion leads to a definition.

DEFINITION 3.6.

Let A be an $(n \times n)$ matrix. A nonzero vector $\mathbf{v}$ such that

$$(A - \lambda I)^j\mathbf{v} = \boldsymbol{\theta}$$
$$(A - \lambda I)^{j-1}\mathbf{v} \neq \boldsymbol{\theta}$$

is called a ***generalized eigenvector of order*** j corresponding to λ.

Note that an eigenvector can be regarded as a generalized eigenvector of order 1.

If a matrix H has a generalized eigenvector $\mathbf{v}_m$ of order m corresponding to λ, then the following sequence of vectors can be defined:

$$\begin{aligned}(H - \lambda I)\mathbf{v}_m &= \mathbf{v}_{m-1}\\ (H - \lambda I)\mathbf{v}_{m-1} &= \mathbf{v}_{m-2}\\ &\vdots\\ (H - \lambda I)\mathbf{v}_2 &= \mathbf{v}_1.\end{aligned} \tag{3.67}$$

It is easy to show that each vector $\mathbf{v}_r$ in (3.67) is a generalized eigenvector of order r and that $\{\mathbf{v}_1, \mathbf{v}_2, \ldots, \mathbf{v}_m\}$ is a linearly independent set (see Exercise 6, Section 3.10). In addition each generalized eigenvector $\mathbf{v}_r$ leads to a solution $\mathbf{x}_r(t)$ of $\mathbf{x}' = H\mathbf{x}$ where

$$\mathbf{x}_r(t) = e^{\lambda t}\left(\mathbf{v}_r + t\mathbf{v}_{r-1} + \ldots + \frac{t^{r-1}}{(r-1)!}\mathbf{v}_1\right) \tag{3.68}$$

(see Exercise 7, Section 3.10).

We begin the analysis by proving two lemmas that show that an $(n \times n)$ standard Hessenberg matrix H has a set of n linearly independent eigenvectors and generalized eigenvectors. Then following several examples, we treat the general case.

Lemma 1

Let H be an $(n \times n)$ standard Hessenberg matrix, and let λ be an eigenvalue of H where λ has algebraic multiplicity m. Then H has a generalized eigenvector of order m corresponding to λ.

Proof. Let $p(t) = (t - \lambda)^m q(t)$ be the characteristic polynomial for H where $q(\lambda) \neq 0$. Let $\mathbf{v}_m$ be the vector $\mathbf{v}_m = q(H)\mathbf{e}_1$. By Theorem 3.15

$(H-\lambda I)^{m-1}q(H)\mathbf{e}_1 \neq \boldsymbol{\theta}$; so $(H-\lambda I)^{m-1}\boldsymbol{v}_m \neq \boldsymbol{\theta}$. Also by Theorem 3.15 $(H-\lambda I)^m \boldsymbol{v}_m = (H-\lambda I)^m q(H)\mathbf{e}_1 = p(H)\mathbf{e}_1 = \boldsymbol{\theta}$; so we see that $\boldsymbol{v}_m$ is a generalized eigenvector of order m. ■

Lemma 2

Let H be an $(n \times n)$ standard Hessenberg matrix. There is a set $\{\mathbf{u}_1, \mathbf{u}_2, \ldots, \mathbf{u}_n\}$ of linearly independent vectors where each $\mathbf{u}_i$ is an eigenvector or a generalized eigenvector of H.

Proof. Suppose H has eigenvalues $\lambda_1, \lambda_2, \ldots, \lambda_k$ where λ_i has multiplicity m_i. Thus the characteristic polynomial has the form $p(t) = (t-\lambda_1)^{m_1}(t-\lambda_2)^{m_2} \ldots (t-\lambda_k)^{m_k}$ where $m_1 + m_2 + \ldots + m_k = n$. By Lemma 1 each eigenvalue λ_i has an associated generalized eigenvector of order m_i. For each eigenvalue λ_i we can use (3.67) to generate a set of m_i generalized eigenvectors having order $1, 2, \ldots, m_i$. Let us denote this collection of n generalized eigenvectors as

$$\boldsymbol{v}_1, \boldsymbol{v}_2, \ldots, \boldsymbol{v}_{m_1}, \mathbf{w}_1, \mathbf{w}_2, \ldots, \mathbf{w}_r \tag{3.69}$$

where $m_1 + r = n$. In (3.69), $\boldsymbol{v}_j$ is a generalized eigenvector of order j corresponding to the eigenvalue λ_1 while each of the vectors $\mathbf{w}_j$ is a generalized eigenvector for one of $\lambda_2, \lambda_3, \ldots, \lambda_k$.

To show that the vectors in (3.69) are linearly independent, consider

$$a_1\boldsymbol{v}_1 + a_2\boldsymbol{v}_2 + \ldots + a_{m_1}\boldsymbol{v}_{m_1} + b_1\mathbf{w}_1 + b_2\mathbf{w}_2 + \ldots + b_r\mathbf{w}_r = \boldsymbol{\theta}. \tag{3.70}$$

Now for $q(t) = (t-\lambda_2)^{m_2} \ldots (t-\lambda_k)^{m_k}$ and for $1 \leqslant j \leqslant r$, we note that

$$q(H)\mathbf{w}_j = \boldsymbol{\theta}$$

since $\mathbf{w}_j$ is a generalized eigenvector of order m_i or less corresponding to some λ_i. [That is, $(H-\lambda_i I)^{m_i}\mathbf{w}_j = \boldsymbol{\theta}$ for some λ_i and $(H-\lambda_i I)^{m_i}$ is one of the factors of $q(H)$. Thus $q(H)\mathbf{w}_j = \boldsymbol{\theta}$ for any j, $1 \leqslant j \leqslant r$.]

Now multiplying both sides of (3.70) by $q(H)$, we obtain

$$a_1 q(H)\boldsymbol{v}_1 + a_2 q(H)\boldsymbol{v}_2 + \ldots + a_{m_1} q(H)\boldsymbol{v}_{m_1} = \boldsymbol{\theta}. \tag{3.71}$$

Finally we can use (3.67) to show that $a_1, a_2, \ldots, a_{m_1}$ are all zero in (3.71) (see Exercise 8, Section 3.10). Since we could have made this argument for any of the eigenvalues $\lambda_2, \lambda_3, \ldots, \lambda_k$, it follows that all the coefficients b_j in (3.70) are also zero. ■

Before giving some examples, we wish to make a point about calculating generalized eigenvectors for a standard Hessenberg matrix H. Let λ be an eigenvalue of multiplicity m for H. Since $\dim(\mathcal{N}(H-\lambda I)) = 1$ (see Exercise 5, Section 3.8), there is, up to a scalar multiple, only one eigenvector corresponding to λ. Thus if $H\boldsymbol{v}_1 = \lambda\boldsymbol{v}_1$ where $\boldsymbol{v}_1 \neq \boldsymbol{\theta}$, it is not hard to show that we can work backwards through (3.67) to find the generalized eigenvectors of H corresponding to λ. That is, we can solve $(H-\lambda I)\mathbf{x} = \boldsymbol{v}_1$ to obtain $\boldsymbol{v}_2$, then solve $(H-\lambda I)\mathbf{x} = \boldsymbol{v}_2$ to obtain $\boldsymbol{v}_3$, etc. [Lemma 1 guarantees the existence of the chain of vectors in (3.67); and since $\boldsymbol{v}_1$ is essentially unique, we can work backwards.]

EXAMPLE 3.24 Consider the problem of solving $\mathbf{x}' = H\mathbf{x}$ where

$$H = \begin{bmatrix} 1 & 0 & 0 \\ 1 & 3 & 0 \\ 0 & 1 & 1 \end{bmatrix}.$$

The characteristic polynomial is $p(t) = (t-1)^2(t-3)$; so $\lambda = 1$ is an eigenvalue of multiplicity 2 while $\lambda = 3$ is an eigenvalue of multiplicity 1. Eigenvectors corresponding to $\lambda = 1$ and $\lambda = 3$ are (respectively)

$$\mathbf{v}_1 = \begin{bmatrix} 0 \\ 0 \\ 1 \end{bmatrix}, \qquad \mathbf{w}_1 = \begin{bmatrix} 0 \\ 2 \\ 1 \end{bmatrix}.$$

Thus we have two solutions of $\mathbf{x}' = H\mathbf{x}$, namely $\mathbf{x}(t) = e^t\mathbf{v}_1$ and $\mathbf{x}(t) = e^{3t}\mathbf{w}_1$. We need one more solution to be able to solve the initial-value problem for any $\mathbf{x}_0$ in R^3. To find a third solution, we need a vector $\mathbf{v}_2$ that is a generalized eigenvector of order 2 corresponding to $\lambda = 1$. According to the remarks above, we solve $(H-I)\mathbf{x} = \mathbf{v}_1$ and obtain

$$\mathbf{v}_2 = \begin{bmatrix} -2 \\ 1 \\ 0 \end{bmatrix}.$$

By (3.68) a third solution to $\mathbf{x}' = H\mathbf{x}$ is given by $\mathbf{x}(t) = e^t(\mathbf{v}_2 + t\mathbf{v}_1)$. Clearly $\{\mathbf{v}_1, \mathbf{v}_2, \mathbf{w}_1\}$ is a basis for R^3; so if $\mathbf{x}_0 = c_1\mathbf{v}_1 + c_2\mathbf{v}_2 + c_3\mathbf{w}_1$, then

$$\mathbf{x}(t) = c_1e^t\mathbf{v}_1 + c_2e^t(\mathbf{v}_2 + t\mathbf{v}_1) + c_3e^{3t}\mathbf{w}_1$$

will satisfy $\mathbf{x}' = H\mathbf{x}$, $\mathbf{x}(0) = \mathbf{x}_0$.

EXAMPLE 3.25 Consider the problem of solving $\mathbf{x}' = A\mathbf{x}$, $\mathbf{x}(0) = \mathbf{x}_0$ where

$$A = \begin{bmatrix} -1 & -8 & 1 \\ -1 & -3 & 2 \\ -4 & -16 & 7 \end{bmatrix}.$$

Reducing A to Hessenberg form, we have $H = QAQ^{-1}$ where H, Q, and Q^{-1} are

$$H = \begin{bmatrix} -1 & -4 & 1 \\ -1 & 5 & 2 \\ 0 & -8 & -1 \end{bmatrix}, \qquad Q = \begin{bmatrix} 1 & 0 & 0 \\ 0 & 1 & 0 \\ 0 & -4 & 1 \end{bmatrix}, \qquad Q^{-1} = \begin{bmatrix} 1 & 0 & 0 \\ 0 & 1 & 0 \\ 0 & 4 & 1 \end{bmatrix}.$$

The change of variables $\mathbf{y}(t) = Q\mathbf{x}(t)$ converts $\mathbf{x}' = A\mathbf{x}$, $\mathbf{x}(0) = \mathbf{x}_0$ to the problem $\mathbf{y}' = H\mathbf{y}$, $\mathbf{y}(0) = Q\mathbf{x}_0$.

The characteristic polynomial for H is $p(t) = (t-1)^3$; so $\lambda = 1$ is an eigenvalue of multiplicity 3 of H. Up to a scalar multiple, the only eigenvector of H is

$$\mathbf{v}_1 = \begin{bmatrix} 4 \\ -1 \\ 4 \end{bmatrix}.$$

We obtain two generalized eigenvectors for H by solving $(H-I)\mathbf{x}=\mathbf{v}_1$ to get $\mathbf{v}_2$, and $(H-I)\mathbf{x}=\mathbf{v}_2$ to get $\mathbf{v}_3$. These generalized eigenvectors are

$$\mathbf{v}_2=\begin{bmatrix}1\\-1\\2\end{bmatrix},\qquad \mathbf{v}_3=\begin{bmatrix}3\\-1\\3\end{bmatrix}.$$

Thus the general solution of $\mathbf{y}'=H\mathbf{y}$ is

$$\mathbf{y}(t)=e^t\left[c_1\mathbf{v}_1+c_2(\mathbf{v}_2+t\mathbf{v}_1)+c_3\left(\mathbf{v}_3+t\mathbf{v}_2+\frac{t^2}{2}\mathbf{v}_1\right)\right],$$

and we can recover $\mathbf{x}(t)$ from $\mathbf{x}(t)=Q^{-1}\mathbf{y}(t)$.

---

†Generalized Eigenvectors for Nonstandard Hessenberg Matrices

From Lemmas 1 and 2 we know how to solve $\mathbf{x}'=H\mathbf{x}$ when H is a standard Hessenberg matrix. When H is not in standard form, the analysis is a bit more complicated. We will treat the general case by partitioning H as

$$H=\begin{bmatrix}H_{11} & H_{12}\\ \mathcal{O} & H_{22}\end{bmatrix}$$

where H_{11} and H_{22} are Hessenberg blocks. We will show how any generalized eigenvector for H_{22} can be "extended" to be a generalized eigenvector for H.

The lemma below is basic to our construction and says essentially this: if H is an $(n\times n)$ standard Hessenberg matrix with λ an eigenvalue of multiplicity m, and if $\mathbf{v}_m$ is a generalized eigenvector of order m corresponding to λ, then $\mathbf{v}_m$ together with the first $(n-1)$ columns of $H-\lambda I$ is a set of n linearly independent vectors. Put another way, given any n-dimensional vector $\mathbf{b}$, either $(H-\lambda I)\mathbf{x}=\mathbf{b}$ is consistent or $(H-\lambda I)\mathbf{x}=\mathbf{b}+\beta\mathbf{v}_m$ is consistent for some scalar β.

Lemma 3

Let H be an $(n\times n)$ standard Hessenberg matrix, and let λ be an eigenvalue of H of algebraic multiplicity m. Let $\mathbf{v}_m$ be a generalized eigenvector of order m corresponding to λ, and let $\mathbf{b}$ be any n-dimensional vector. Then there is a scalar α and a vector $\mathbf{w}$ such that $\mathbf{b}=\mathbf{w}+\alpha\mathbf{v}_m$ where $\mathbf{w}$ is in $\mathcal{R}(H-\lambda I)$.

Proof. (We note that $\mathbf{b}$ can have complex components although this is not central to an understanding of the lemma.) Since H is a standard Hessenberg matrix, so is $H-\lambda I$. If $H-\lambda I=[\mathbf{P}_1,\mathbf{P}_2,\ldots,\mathbf{P}_n]$, then it is easy to show that $\{\mathbf{P}_1,\mathbf{P}_2,\ldots,\mathbf{P}_{n-1}\}$ is linearly independent (Exercise 9, Section 3.10). So (Exercise

---

† The remainder of this section deals with solving $\mathbf{x}'=H\mathbf{x}$ for arbitrary H and may be omitted.

9, Section 3.10) if $\{\mathbf{P}_1, \mathbf{P}_2, \ldots, \mathbf{P}_{n-1}, \boldsymbol{v}_m\}$ were linearly dependent, there would be a nonzero vector $\mathbf{a}$ such that

$$(H - \lambda I)\mathbf{a} = \boldsymbol{v}_m. \tag{3.72}$$

However the characteristic polynomial for H is of the form $p(t) = (t-\lambda)^m q(t)$ where $q(\lambda) \neq 0$; and if (3.72) were to hold, then $(H-\lambda I)^m \mathbf{a} = (H-\lambda I)^{m-1}\boldsymbol{v}_m = \boldsymbol{v}_1$ where $\boldsymbol{v}_1$ is an eigenvector of H corresponding to λ. Multiplying $(H-\lambda I)^m\mathbf{a} = \boldsymbol{v}_1$ by $q(H)$, we get

$$p(H)\mathbf{a} = q(H)(H-\lambda I)^m \mathbf{a} = q(H)\boldsymbol{v}_1 = q(\lambda)\boldsymbol{v}_1.$$

However by Theorem 3.15, $p(H)$ is the zero matrix; so $q(\lambda)\boldsymbol{v}_1 = \boldsymbol{\theta}$. But $q(\lambda) \neq 0$, $\boldsymbol{v}_1 \neq \boldsymbol{\theta}$; so (3.72) cannot possibly hold, and therefore $\{\mathbf{P}_1, \mathbf{P}_2, \ldots, \mathbf{P}_{n-1}, \boldsymbol{v}_m\}$ cannot be linearly dependent.

Thus we can represent any n-dimensional vector $\mathbf{b}$ as

$$\mathbf{b} = c_1\mathbf{P}_1 + c_2\mathbf{P}_2 + \ldots + c_{n-1}\mathbf{P}_{n-1} + \alpha\boldsymbol{v}_m;$$

and since $\mathbf{w} = c_1\mathbf{P}_1 + c_2\mathbf{P}_2 + \ldots + c_{n-1}\mathbf{P}_{n-1}$ is in $\mathscr{R}(H-\lambda I)$, the lemma is proved. ∎

Now to treat the general case of $\mathbf{x}' = H\mathbf{x}$, suppose H is an $(n \times n)$ Hessenberg matrix. Write H as

$$H = \begin{bmatrix} H_{11} & H_{12} \\ \mathcal{O} & H_{22} \end{bmatrix}$$

where H_{11} is an $(s \times s)$ standard Hessenberg matrix while H_{22} is $(r \times r)$ but not necessarily in standard form. We will show that any eigenvector or generalized eigenvector of H_{22} can be used to produce an eigenvector or generalized eigenvector of H. To see this, suppose $(H_{22} - \lambda I)\boldsymbol{v} = \boldsymbol{\theta}$, $\boldsymbol{v} \neq \boldsymbol{\theta}$. If λ is not an eigenvalue of H_{11}, we can choose an s-dimensional vector $\mathbf{x}_1$ such that

$$\begin{aligned} (H_{11} - \lambda I)\mathbf{x}_1 + H_{12}\boldsymbol{v} &= \boldsymbol{\theta} \\ (H_{22} - \lambda I)\boldsymbol{v} &= \boldsymbol{\theta}. \end{aligned} \tag{3.73}$$

Thus the nonzero vector

$$\mathbf{z} = \begin{bmatrix} \mathbf{x}_1 \\ \boldsymbol{v} \end{bmatrix} \tag{3.74}$$

satisfies $(H - \lambda I)\mathbf{z} = \boldsymbol{\theta}$; so $\mathbf{z}$ is an eigenvector of H. On the other hand if λ is an eigenvalue of multiplicity m for H_{11}, we may not be able to solve (3.73) since $H_{11} - \lambda I$ is singular. However by Lemma 3 we can write $H_{12}\boldsymbol{v} = \mathbf{w} + \alpha\mathbf{u}_m$ where $\mathbf{u}_m$ is a generalized eigenvector of order m for H_{11}, and where $\mathbf{w}$ is in $\mathscr{R}(H_{11} - \lambda I)$. Thus while we may not be able to solve (3.73), we can certainly find a vector $\mathbf{x}_1$ such that

$$\begin{aligned} (H_{11} - \lambda I)\mathbf{x}_1 + H_{12}\boldsymbol{v} &= \alpha\mathbf{u}_m \\ (H_{22} - \lambda I)\boldsymbol{v} &= \boldsymbol{\theta}. \end{aligned} \tag{3.75}$$

[$H_{12}\boldsymbol{v} = \mathbf{w} + \alpha\mathbf{u}_m$ and $(H_{11} - \lambda I)\mathbf{x} = -\mathbf{w}$ are solvable.] It is easy to show (Exercise 10, Section 3.10) that a vector $\mathbf{z}$ of the form (3.74) that satisfies (3.75) is a generalized eigenvector of order $m+1$ for H corresponding to λ when $\alpha \neq 0$ and is an eigenvector of H when $\alpha = 0$.

We see above that every eigenvector of H_{22} will lead to either an eigenvector or a generalized eigenvector of H. We now continue inductively and suppose that $\boldsymbol{v}$ is a generalized eigenvector of order j for H_{22} where $j \geq 2$. Then $\mathbf{y} = (H_{22} - \lambda I)\boldsymbol{v}$ is a generalized eigenvector of order $j-1$ for H_{22}; and let us suppose that there is a vector $\mathbf{z}$ of the form

$$\mathbf{z} = \begin{bmatrix} \mathbf{x}_1 \\ \mathbf{y} \end{bmatrix}$$

where $\mathbf{z}$ is an eigenvector or a generalized eigenvector of H (we know such a vector $\mathbf{z}$ exists when $j = 2$ by the arguments above). If λ is not an eigenvalue of H_{11}, we can find a vector $\mathbf{x}_2$ such that

$$\begin{aligned} (H_{11} - \lambda I)\mathbf{x}_2 + \qquad H_{12}\boldsymbol{v} &= \mathbf{x}_1 \\ (H_{22} - \lambda I)\boldsymbol{v} &= \mathbf{y}. \end{aligned}$$

Thus the vector $\mathbf{t}$ given by

$$\mathbf{t} = \begin{bmatrix} \mathbf{x}_2 \\ \boldsymbol{v} \end{bmatrix}$$

satisfies $(H - \lambda I)\mathbf{t} = \mathbf{z}$; so $\mathbf{t}$ is also a generalized eigenvector for H. On the other hand if λ is an eigenvalue of multiplicity m for H_{11}, then we can find a vector $\mathbf{x}_2$ such that

$$\begin{aligned} (H_{11} - \lambda I)\mathbf{x}_2 + H_{12}\boldsymbol{v} &= \mathbf{x}_1 + \alpha\mathbf{u}_m \\ (H_{22} - \lambda I)\boldsymbol{v} &= \mathbf{y}. \end{aligned} \tag{3.76}$$

[By Lemma 3 we can solve $(H_{11} - \lambda I)\mathbf{x} = -H_{12}\boldsymbol{v} + \mathbf{x}_1 + \alpha\mathbf{u}_m$ for some α.] By (3.76) we see that

$$(H - \lambda I)\begin{bmatrix} \mathbf{x}_2 \\ \boldsymbol{v} \end{bmatrix} = \begin{bmatrix} \mathbf{x}_1 \\ \mathbf{y} \end{bmatrix} + \alpha \begin{bmatrix} \mathbf{u}_m \\ \boldsymbol{\theta} \end{bmatrix}.$$

Now if $\mathbf{z}$ is a generalized eigenvector of order k, then $\mathbf{t}$ is a generalized eigenvector of order $k+1$ if $\alpha = 0$, and $\mathbf{t}$ is a generalized eigenvector of order $= \max\{k+1, m+1\}$ if $\alpha \neq 0$.

All of the preceding arguments show that if $\boldsymbol{v}$ is any eigenvector or generalized eigenvector for H_{22}, then we can find an eigenvector or generalized eigenvector for H of the form (3.74). To complete the argument, we know by Lemma 2 that H_{11} has a set of s vectors $\mathbf{u}_1, \mathbf{u}_2, \ldots, \mathbf{u}_s$ such that each $\mathbf{u}_i$ is an eigenvector or a generalized eigenvector of H_{11}. Thus the s vectors

$$\begin{bmatrix} \mathbf{u}_1 \\ \boldsymbol{\theta} \end{bmatrix}, \quad \begin{bmatrix} \mathbf{u}_2 \\ \boldsymbol{\theta} \end{bmatrix}, \ldots, \begin{bmatrix} \mathbf{u}_s \\ \boldsymbol{\theta} \end{bmatrix} \tag{3.77}$$

form a collection of s linearly independent eigenvectors and generalized eigenvectors of H. If H_{22} is a standard Hessenberg matrix, then H_{22} has a set of r eigenvectors and generalized eigenvectors $\{\boldsymbol{v}_1, \boldsymbol{v}_2, \ldots, \boldsymbol{v}_r\}$, which are linearly independent. From the discussion above, we can use each of these to produce another collection of r eigenvectors or generalized eigenvectors for H:

$$\begin{bmatrix} \mathbf{x}_1 \\ \boldsymbol{v}_1 \end{bmatrix}, \quad \begin{bmatrix} \mathbf{x}_2 \\ \boldsymbol{v}_2 \end{bmatrix}, \ldots, \begin{bmatrix} \mathbf{x}_r \\ \boldsymbol{v}_r \end{bmatrix}. \tag{3.78}$$

Clearly the $n = s+r$ vectors from (3.77) and (3.78) are linearly independent. If H_{22} is not a standard Hessenberg matrix, we can split H_{22} into a block (2×2) matrix and continue inductively. Thus we have proved Theorem 3.18.

Theorem 3.18

Every $(n \times n)$ Hessenberg matrix H has a set of n linearly independent eigenvectors and generalized eigenvectors.

EXAMPLE 3.26 In this example we find all the eigenvectors and generalized eigenvectors of

$$H = \begin{bmatrix} 1 & 0 & 1 & 2 \\ 1 & 1 & 2 & 1 \\ 0 & 0 & 2 & 0 \\ 0 & 0 & 1 & 1 \end{bmatrix}.$$

Partition H so that H_{11} is the upper-left (2×2) block and H_{22} is the lower-right (2×2) block. The unit vectors in R^2 are generalized eigenvectors for H_{11}; note that $(H_{11}-I)\mathbf{e}_2 = \boldsymbol{\theta}$ and $(H_{11}-I)\mathbf{e}_1 = \mathbf{e}_2$. From these we get [as in (3.77)] two generalized eigenvectors for H, $\mathbf{w}_1$ and $\mathbf{w}_2$ where $(H-I)\mathbf{w}_1 = \boldsymbol{\theta}$, $(H-I)\mathbf{w}_2 = \mathbf{w}_1$:

$$\mathbf{w}_1 = \begin{bmatrix} 0 \\ 1 \\ 0 \\ 0 \end{bmatrix}, \quad \mathbf{w}_2 = \begin{bmatrix} 1 \\ 0 \\ 0 \\ 0 \end{bmatrix}.$$

We use the eigenvectors of H_{22} to generate two more generalized eigenvectors for H. With respect to H_{22}, $(H_{22}-2I)\mathbf{v}_1 = \boldsymbol{\theta}$ and $(H_{22}-I)\mathbf{v}_2 = \boldsymbol{\theta}$ where

$$\mathbf{v}_1 = \begin{bmatrix} 1 \\ 1 \end{bmatrix}, \quad \mathbf{v}_2 = \begin{bmatrix} 0 \\ 1 \end{bmatrix}.$$

As in (3.73), we solve $(H_{11}-2I)\mathbf{x} = -H_{12}\mathbf{v}_1$ to obtain an eigenvector of H corresponding to $\lambda = 2$. As in (3.78), $\mathbf{w}_3$ satisfies $(H-2I)\mathbf{w}_3 = \boldsymbol{\theta}$ where

$$\mathbf{w}_3 = \begin{bmatrix} 3 \\ 6 \\ 1 \\ 1 \end{bmatrix}.$$

Finally the eigenvector $\mathbf{v}_2$ of H_{22} will not lead to an eigenvector of H since we cannot solve $(H_{11}-I)\mathbf{x} = -H_{12}\mathbf{v}_2$ in (3.73). However we can solve $(H_{11}-I)\mathbf{x} = -H_{12}\mathbf{v}_2 + 2\mathbf{e}_1$, and the solution is

$$\mathbf{x} = \begin{bmatrix} -1 \\ 0 \end{bmatrix}.$$

Therefore the vector $\mathbf{w}_4$ satisfies $(H - I)\mathbf{w}_4 = \mathbf{w}_2$ where

$$\mathbf{w}_4 = \begin{bmatrix} -1/2 \\ 0 \\ 0 \\ 1/2 \end{bmatrix}.$$

Clearly the vectors $\{\mathbf{w}_1, \mathbf{w}_2, \mathbf{w}_3, \mathbf{w}_4\}$ are linearly independent, and the example illustrates how the proof of Theorem 3.18 can be used to construct a set of linearly independent generalized eigenvectors for a Hessenberg matrix.

3.10 EXERCISES

1. Find a full set of eigenvectors and generalized eigenvectors for the following.

$$\text{a)} \begin{bmatrix} 1 & -1 \\ 1 & 3 \end{bmatrix} \quad \text{b)} \begin{bmatrix} -2 & 0 & -2 \\ -1 & 1 & -2 \\ 0 & 1 & -1 \end{bmatrix} \quad \text{c)} \begin{bmatrix} -6 & 31 & -14 \\ -1 & 6 & -2 \\ 0 & 2 & 1 \end{bmatrix}$$

2. Find a full set of eigenvectors and generalized eigenvectors for the following. (Note: $\lambda = 2$ is the only eigenvalue of B.)

$$A = \begin{bmatrix} 1 & 0 & 0 & 0 \\ 1 & 1 & 0 & 0 \\ 0 & 1 & 1 & 0 \\ 0 & 0 & 1 & 1 \end{bmatrix} \quad B = \begin{bmatrix} 2 & 3 & -21 & -3 \\ 2 & 7 & -41 & -5 \\ 0 & 1 & -5 & -1 \\ 0 & 0 & 4 & 4 \end{bmatrix}$$

3. Solve $\mathbf{x}' = A\mathbf{x}$, $\mathbf{x}(0) = \mathbf{x}_0$ by transforming A to Hessenberg form where

$$\mathbf{x}_0 = \begin{bmatrix} -1 \\ -1 \\ 1 \end{bmatrix}, \text{ and (a) } A = \begin{bmatrix} 8 & -6 & 21 \\ 1 & -1 & 3 \\ -3 & 2 & -8 \end{bmatrix}, \quad \text{(b) } A = \begin{bmatrix} 2 & 1 & -1 \\ -3 & -1 & 1 \\ 9 & 3 & -4 \end{bmatrix},$$

$$\text{(c) } A = \begin{bmatrix} 1 & 1 & -1 \\ -3 & -2 & 1 \\ 9 & 3 & -5 \end{bmatrix}.$$

4. Give the general solution of $\mathbf{x}' = A\mathbf{x}$ where A is from Exercise 2.

5. Find a full set of eigenvectors and generalized eigenvectors for the following.

$$\text{a)} \begin{bmatrix} 2 & 0 & 1 & 0 & 1 \\ 1 & 1 & 1 & 0 & 0 \\ 0 & 0 & 2 & 0 & 0 \\ 0 & 0 & 1 & 2 & 0 \\ 0 & 0 & 0 & 1 & 1 \end{bmatrix} \quad \text{b)} \begin{bmatrix} 1 & 0 & 0 & 1 & 1 \\ 1 & 1 & 0 & 0 & 0 \\ 0 & 1 & 2 & 0 & 0 \\ 0 & 0 & 0 & 1 & 0 \\ 0 & 0 & 0 & 1 & 2 \end{bmatrix}$$

6. Prove that each vector $\mathbf{v}_r$ in (3.67) is a generalized eigenvector of order r and that $\{\mathbf{v}_1, \mathbf{v}_2, \ldots, \mathbf{v}_m\}$ is linearly independent.
7. Prove that the functions $\mathbf{x}_r(t)$ defined in (3.68) are solutions of $\mathbf{x}' = H\mathbf{x}$.
8. Prove that the coefficients $a_1, a_2, \ldots$ in (3.71) are all zero. [Hint: Multiply (3.71) by $(H - \lambda_1 I)^{m_1 - 1}$.]
9. a) Prove that $\{\mathbf{P}_1, \mathbf{P}_2, \ldots, \mathbf{P}_{n-1}\}$ in the proof of Lemma 3 is a linearly independent set. (Hint: Multiply $a_1\mathbf{P}_1 + a_2\mathbf{P}_2 + \ldots + a_{n-1}\mathbf{P}_{n-1} = \mathbf{0}$ by $\mathbf{e}_n^{\mathrm{T}}$.)
   b) Verify (3.72).
10. Prove that if $\mathbf{z}$ given in (3.74) satisfies (3.75), then $\mathbf{z}$ is a generalized eigenvector of order $m+1$ when $\alpha \neq 0$, and an eigenvector when $\alpha = 0$.
11. Find the eigenvectors and generalized eigenvectors for the following.

$$H = \begin{bmatrix} 1 & 0 & 1 & 1 & 0 & 1 \\ 1 & 2 & 0 & 0 & 0 & 0 \\ 0 & 0 & 2 & 0 & 1 & 1 \\ 0 & 0 & 1 & 1 & 0 & 0 \\ 0 & 0 & 0 & 0 & 1 & 0 \\ 0 & 0 & 0 & 0 & 1 & 1 \end{bmatrix}$$

[Hint: There are three diagonal blocks. First find four generalized eigenvectors for the lower right (4×4) portion of H.]

4

VECTOR SPACES AND LINEAR TRANSFORMATIONS

4.1 INTRODUCTION

This chapter will be devoted to extending the theory for matrices and R^n to the more general setting of linear transformations and vector spaces. In this setting, R^n will serve as a model for a general vector space while linear transformations play a role analogous to matrices. Most of the elementary concepts, such as subspace, basis, dimension, etc., that are important to understanding vector spaces are immediate generalizations of those same concepts from R^n.

Although the theory of vector spaces is relatively abstract, the vector-space structure provides a unifying framework of great flexibility; and we will give several examples of important practical problems that fit naturally into the vector-space framework. A basic feature of vector spaces is that they possess both an algebraic character and a geometric character. In this regard the geometric character frequently gives a pictorial insight into how a particular problem can be solved while the algebraic character is used actually to calculate a solution.

4.2 VECTOR SPACES

We begin our study of vector spaces by reviewing a familiar example, namely R^3, the set of all three-dimensional vectors with real components:

$$R^3 = \{\boldsymbol{v}: \boldsymbol{v} = \begin{bmatrix} v_1 \\ v_2 \\ v_3 \end{bmatrix}, v_1, v_2, v_3 \text{ are real numbers}\}.$$

This example will serve as a model for a general vector space; so we start by recalling some of the properties of R^3. First of all, there are two algebraic operations associated with this set of vectors; that is, vectors in R^3 can be added, and any vector in R^3 can be multiplied by a real number. Precisely if

$$\mathbf{u} = \begin{bmatrix} u_1 \\ u_2 \\ u_3 \end{bmatrix} \quad \text{and} \quad \boldsymbol{v} = \begin{bmatrix} v_1 \\ v_2 \\ v_3 \end{bmatrix},$$

and if a is any scalar, then

$$\mathbf{u} + \boldsymbol{v} = \begin{bmatrix} u_1 + v_1 \\ u_2 + v_2 \\ u_3 + v_3 \end{bmatrix} \quad \text{and} \quad a\mathbf{u} = \begin{bmatrix} au_1 \\ au_2 \\ au_3 \end{bmatrix}.$$

These operations provide R^3 with an algebraic structure and allow us to interpret R^3 as more than just the collection of all points in three-dimensional space. Similarly a general vector space is essentially a set of objects (called vectors), which has an algebraic structure similar to that of R^3. The algebraic structure of R^3 has a number of essential features. For instance given vectors $\mathbf{u}$ and $\boldsymbol{v}$ in R^3, the sum $\mathbf{u} + \boldsymbol{v}$ is uniquely defined and is also a vector in R^3. Similarly given any real number a and any vector $\boldsymbol{v}$ in R^3, the "scalar multiple" $a\boldsymbol{v}$ is uniquely defined and is also a vector in R^3. Besides producing well-defined results, these operations of addition and scalar multiplication also obey certain rules. For example $(\mathbf{u} + \boldsymbol{v}) + \mathbf{w} = \mathbf{u} + (\boldsymbol{v} + \mathbf{w})$, $a(b\mathbf{u}) = (ab)\mathbf{u}$, $(a + b)\mathbf{u} = a\mathbf{u} + b\mathbf{u}$, etc. The algebraic operations in R^3 obey other rules as well, but this brief list gives the flavor of what we should want in any generalization of R^3.

Drawing on the discussion above, we see that a general vector space should consist of a set of elements (or vectors), V, and a set of scalars, S, together with two algebraic operations:

1. an "addition," which is defined between any two elements of V and which produces a sum that is in V; and
2. a "scalar multiplication," which defines how to multiply any element of V by a scalar from S.

In practice the set V can consist of any collection of objects for which meaningful operations of addition and scalar multiplication can be defined. For example V might be the set of all (2×3) matrices, the set R^4 of all four-dimensional vectors, a set of functions, a set of polynomials, or the set of all solutions to a linear homogeneous differential equation. We will take the set S of scalars to be the set of real numbers although for added flexibility other sets of scalars may be used (for example, S could be the set of complex numbers). Throughout this chapter the term "scalar" will always denote a real number. Using R^3 as a model, we now define a general vector space. Note that the definition says nothing about the set V but rather specifies rules that the algebraic operations must satisfy.

DEFINITION 4.1.

A set of elements V is said to be a ***vector space*** over a scalar field S if there is an addition operation defined between any two elements of V and a scalar multiplication operation defined between any element of S and any vector in V where these operations satisfy the following conditions. If $\mathbf{u}$, $\mathbf{v}$, and $\mathbf{w}$ are any vectors in V, and if a and b are any two scalars, then

c1) $\mathbf{u}+\mathbf{v}$ is a vector in V;
c2) $a\mathbf{v}$ is a vector in V;
a1) $\mathbf{u}+\mathbf{v}=\mathbf{v}+\mathbf{u}$;
a2) $\mathbf{u}+(\mathbf{v}+\mathbf{w})=(\mathbf{u}+\mathbf{v})+\mathbf{w}$;
a3) there is a vector $\boldsymbol{\theta}$ in V such that $\mathbf{v}+\boldsymbol{\theta}=\mathbf{v}$ for all $\mathbf{v}$ in V;
a4) given a $\mathbf{v}$ in V, there is a vector $-\mathbf{v}$ in V such that $\mathbf{v}+(-\mathbf{v})=\boldsymbol{\theta}$;
m1) $a(b\mathbf{v})=(ab)\mathbf{v}$;
m2) $a(\mathbf{u}+\mathbf{v})=a\mathbf{u}+a\mathbf{v}$;
m3) $(a+b)\mathbf{v}=a\mathbf{v}+b\mathbf{v}$; and
m4) $1\mathbf{v}=\mathbf{v}$ for all $\mathbf{v}$ in V.

The first two conditions, (c1) and (c2), in Definition 4.1 are called "closure properties." They ensure that the sum of any two vectors in V remains in V and that any scalar multiple of a vector in V remains in V. In condition (a3), $\boldsymbol{\theta}$ is naturally called the ***zero vector*** (or the additive identity). In (a4) the vector $-\mathbf{v}$ is called the ***additive inverse*** of $\mathbf{v}$, and (a4) asserts that the equation $\mathbf{v}+\mathbf{x}=\boldsymbol{\theta}$ has a solution in V. When the set of scalars S is the set of real numbers, V is called a ***real vector space***; and as we have said, we will consider only real vector spaces. We already have two familiar examples of vector spaces, namely R^n and the set of all $(m \times n)$ matrices. It is easy to verify that these are vector spaces, and the verification is sketched in the first two examples below. Example 4.1 may strike the reader as being a little unusual since we are considering matrices as elements in a vector space. However, the example illustrates the flexibility of the vector-space concept: any set of entities that has addition and scalar multiplication operations can be a vector space, provided the addition and scalar multiplication satisfy the requirements of Definition 4.1.

EXAMPLE 4.1 In this example we verify that the set of all (2×3) matrices with real entries is a real vector space. To verify this, let A and B be any (2×3) matrices, and let addition and scalar multiplication be defined as in Definitions 1.4 and 1.5. Therefore $A+B$ and aA are defined by

$$A+B=\begin{bmatrix} a_{11} & a_{12} & a_{13} \\ a_{21} & a_{22} & a_{23} \end{bmatrix}+\begin{bmatrix} b_{11} & b_{12} & b_{13} \\ b_{21} & b_{22} & b_{23} \end{bmatrix}$$

$$=\begin{bmatrix} a_{11}+b_{11} & a_{12}+b_{12} & a_{13}+b_{13} \\ a_{21}+b_{21} & a_{22}+b_{22} & a_{23}+b_{23} \end{bmatrix}$$

$$aA=a\begin{bmatrix} a_{11} & a_{12} & a_{13} \\ a_{21} & a_{22} & a_{23} \end{bmatrix}=\begin{bmatrix} aa_{11} & aa_{12} & aa_{13} \\ aa_{21} & aa_{22} & aa_{23} \end{bmatrix}.$$

matrices not determinants

From these definitions it is obvious that both the sum $A+B$ and the scalar multiple aA are again (2×3) matrices; so (c1) and (c2) of Definition 4.1 hold. Properties (a1), (a2), (a3), and (a4) follow from Theorem 1.5; and (m1), (m2), and (m3) are proved in Theorems 1.6 and 1.7. Property (m4) is immediate from the definition of scalar multiplication [clearly $1\cdot A = A$ for any (2×3) matrix A]. For emphasis we recall that the zero element in this vector space is the matrix

$$\mathcal{O} = \begin{bmatrix} 0 & 0 & 0 \\ 0 & 0 & 0 \end{bmatrix},$$

and clearly $A+\mathcal{O} = A$ for any (2×3) matrix A. We further observe that $(-1)\cdot A$ is the additive inverse for A since

$$A+(-1)\cdot A = \mathcal{O}.$$

[That is, $(-1)\cdot A$ is a matrix we can add to A to produce the zero element $\mathcal{O}$.] A duplication of these arguments shows that for any m and n the set of all $(m\times n)$ matrices with real entries is a real vector space.

EXAMPLE 4.2 For any positive integer n, R^n is a real vector space. This can be verified easily by using the same reasoning that was employed in Example 4.1.

EXAMPLE 4.3 This example shows that certain sets of functions have a natural vector-space structure. We let $\mathcal{P}_2$ denote the set of all real polynomials of degree two or less, and we note that there is a natural addition associated with polynomials. For example let $p(x)$ and $q(x)$ be the polynomials

$$p(x) = 2x^2 - x + 3 \quad \text{and} \quad q(x) = x^2 + 2x - 1.$$

Then the sum $r(x) = p(x) + q(x)$ is the polynomial $r(x) = 3x^2 + x + 2$. Scalar multiplication is defined similarly; so if $s(x) = 2q(x)$, then

$$s(x) = 2x^2 + 4x - 2.$$

Given this natural addition and scalar multiplication associated with the set $\mathcal{P}_2$, it seems reasonable to expect that $\mathcal{P}_2$ is a real vector space.

To establish this rigorously, we must be a bit more careful. To begin, we define $\mathcal{P}_2$ to be the set of all expressions (or functions) of the form

$$p(x) = a_2x^2 + a_1x + a_0 \tag{4.1}$$

where a_2, a_1, and a_0 are any real constants. Thus the following polynomials are *vectors* in $\mathcal{P}_2$:

$$\begin{array}{lll} p_1(x) = x^2 - x + 3, & p_2(x) = x^2 + 1, & p_3(x) = x - 2 \\ p_4(x) = 2x, & p_5(x) = 7, & p_6(x) = 0. \end{array}$$

For instance we see that $p_2(x)$ has the form of (4.1) with $a_2 = 1$, $a_1 = 0$, $a_0 = 1$. Similarly $p_4(x)$ is in $\mathcal{P}_2$ since $p_4(x)$ is a function of the form (4.1), and $p_4(x)$ has

$a_2 = 0$, $a_1 = 2$, $a_0 = 0$. Finally $p_6(x)$ has the form (4.1) with $a_2 = 0$, $a_1 = 0$ and $a_0 = 0$. To define addition precisely, let

$$p(x) = a_2x^2 + a_1x + a_0 \quad \text{and} \quad q(x) = b_2x^2 + b_1x + b_0$$

be two vectors in $\mathscr{P}_2$. We define the sum $r(x) = p(x) + q(x)$ to be the polynomial

$$r(x) = (a_2 + b_2)x^2 + (a_1 + b_1)x + (a_0 + b_0);$$

and we define the scalar multiple $s(x) = cp(x)$ to be the polynomial

$$s(x) = (ca_2)x^2 + (ca_1)x + (ca_0).$$

We leave it to the reader to verify that these algebraic operations meet the requirements of Definition 4.1; we note only that we choose the zero vector to be the polynomial that is identically zero. That is, the zero element in $\mathscr{P}_2$ is the polynomial $\theta(x)$ where $\theta(x) \equiv 0$; or in terms of (4.1), $\theta(x)$ is defined by

$$\theta(x) = 0x^2 + 0x + 0.$$

EXAMPLE 4.4 In this example we take $\mathscr{P}_n$ to be the set of all real polynomials of degree n or less. That is, $\mathscr{P}_n$ consists of all functions $p(x)$ of the form

$$p(x) = a_nx^n + a_{n-1}x^{n-1} + \ldots + a_2x^2 + a_1x + a_0$$

where $a_n, a_{n-1}, \ldots, a_2, a_1, a_0$ are any real constants. With addition and scalar multiplication defined as in Example 4.3, it is easy to show that $\mathscr{P}_n$ is a real vector space.

EXAMPLE 4.5 One of the important vector spaces in applications is a vector space of functions called $C[a, b]$, which consists of all the functions that are defined, real valued, and continuous on the interval $[a, b]$:

$$C[a, b] = \{f(x): f(x) \text{ is a real-valued continuous function, } a \leqslant x \leqslant b\}.$$

$C[a, b]$ has a natural addition just as $\mathscr{P}_n$. If f and g are vectors in $C[a, b]$, then we define the sum $h = f + g$ to be function $h(x)$ given by

$$h(x) = f(x) + g(x), \qquad a \leqslant x \leqslant b.$$

Similarly if c is a scalar, then the scalar multiple $q = cf$ is the function

$$q(x) = cf(x), \qquad a \leqslant x \leqslant b.$$

As a concrete example if $f(x) = e^x$ and $g(x) = \sin x$, then $3f + g$ is the function r where the action of r is defined by $r(x) = 3e^x + \sin x$. Note that the closure properties, (c1) and (c2), follow from elementary results of calculus—sums and scalar multiples of continuous functions are again continuous functions. The remaining eight properties of Definition 4.1 are easily seen to hold in $C[a, b]$; the verification proceeds exactly as in $\mathscr{P}_n$.

Note that any polynomial can be regarded as a continuous function on any interval $[a, b]$. Thus for any given positive integer n, $\mathscr{P}_n$ is not only a subset of $C[a, b]$ but also a vector space contained in the vector space $C[a, b]$. This concept of a vector space that contains a smaller vector space (or a "vector subspace") is quite important and is one topic of the next section.

The algebraic operations in a vector space have additional properties that can be derived from the ten fundamental properties listed in Definition 4.1; and some of these additional properties are summarized in Theorem 4.1.

Theorem 4.1

If V is a vector space, then

1. the zero vector $\boldsymbol{\theta}$ is unique;
2. for each $\boldsymbol{v}$ the additive inverse $-\boldsymbol{v}$ is unique;
3. $0\boldsymbol{v} = \boldsymbol{\theta}$ for every $\boldsymbol{v}$ in V where 0 is the zero scalar;
4. $a\boldsymbol{\theta} = \boldsymbol{\theta}$ for every scalar a;
5. if $a\boldsymbol{v} = \boldsymbol{\theta}$, then $a = 0$ or $\boldsymbol{v} = \boldsymbol{\theta}$; and
6. $(-1)\boldsymbol{v} = -\boldsymbol{v}$.

Proof. [We prove parts (1), (4), and (6), and leave the remaining parts as exercises.] We first prove part (1). Suppose $\boldsymbol{\zeta}$ is a vector in V such that $\boldsymbol{v} + \boldsymbol{\zeta} = \boldsymbol{v}$ for all $\boldsymbol{v}$ in V. Then setting $\boldsymbol{v} = \boldsymbol{\theta}$, we would have

$$\boldsymbol{\theta} + \boldsymbol{\zeta} = \boldsymbol{\theta}.$$

By (a3) of Definition 4.1 we know also that

$$\boldsymbol{\zeta} + \boldsymbol{\theta} = \boldsymbol{\zeta} \tag{4.3}$$

But from (a1) we know that $\boldsymbol{\zeta} + \boldsymbol{\theta} = \boldsymbol{\theta} + \boldsymbol{\zeta}$; so using (4.2), (a1), and (4.3), we conclude that

$$\boldsymbol{\theta} = \boldsymbol{\theta} + \boldsymbol{\zeta} = \boldsymbol{\zeta} + \boldsymbol{\theta} = \boldsymbol{\zeta}$$

or $\boldsymbol{\theta} = \boldsymbol{\zeta}$.

We next prove part (4). We prove this by observing that $\boldsymbol{\theta} + \boldsymbol{\theta} = \boldsymbol{\theta}$ from (a3). Therefore if a is any scalar, we have from (m2)

$$a\boldsymbol{\theta} = a(\boldsymbol{\theta} + \boldsymbol{\theta}) = a\boldsymbol{\theta} + a\boldsymbol{\theta}. \tag{4.4}$$

Now let $(-a\boldsymbol{\theta})$ be the additive inverse of $a\boldsymbol{\theta}$; so we have the following sequence of identities [starting with (4.4)]:

$$\begin{aligned} a\boldsymbol{\theta} &= a\boldsymbol{\theta} + a\boldsymbol{\theta} \\ a\boldsymbol{\theta} + (-a\boldsymbol{\theta}) &= (a\boldsymbol{\theta} + a\boldsymbol{\theta}) + (-a\boldsymbol{\theta}) \\ \boldsymbol{\theta} &= a\boldsymbol{\theta} + (a\boldsymbol{\theta} + (-a\boldsymbol{\theta})) \\ \boldsymbol{\theta} &= a\boldsymbol{\theta} + \boldsymbol{\theta} \\ \boldsymbol{\theta} &= a\boldsymbol{\theta}. \end{aligned}$$

This proof that $a\boldsymbol{\theta} = \boldsymbol{\theta}$ uses a number of properties from Definition 4.1: (a3) and (m2) to obtain (4.4); and then (a2), (a3), and (a4) to obtain $\boldsymbol{\theta} = a\boldsymbol{\theta}$ from (4.4).

Finally we outline a proof for part (6) by displaying a sequence of equalities [the last equality is based on part (3), which is an exercise]:

$$\boldsymbol{v} + (-1)\boldsymbol{v} = (1)\boldsymbol{v} + (-1)\boldsymbol{v} = (1 + (-1))\boldsymbol{v} = 0\boldsymbol{v} = \boldsymbol{\theta}.$$

Thus $(-1)\boldsymbol{v}$ is a solution to the equation $\boldsymbol{v} + \mathbf{x} = \boldsymbol{\theta}$. But from part (2) the additive inverse $-\boldsymbol{v}$ is the only solution of $\boldsymbol{v} + \mathbf{x} = \boldsymbol{\theta}$; so we must have $(-1)\boldsymbol{v} = -\boldsymbol{v}$. Thus part (6) constitutes a formula for the additive inverse. This formula is not totally unexpected, but neither is it so obvious as it might seem since a number of vector-space properties were required to prove it. ■

EXAMPLE 4.6 We conclude this section by introducing the ***zero vector space***. The zero vector space contains only one vector, $\boldsymbol{\theta}$; and the arithmetic operations are defined by

$$\boldsymbol{\theta} + \boldsymbol{\theta} = \boldsymbol{\theta}$$
$$k\boldsymbol{\theta} = \boldsymbol{\theta}.$$

It is easy to verify that the set $\{\boldsymbol{\theta}\}$ with the operations defined above is a vector space.

4.2 EXERCISES

1. For $\mathbf{u}$, $\boldsymbol{v}$, and $\mathbf{w}$ given below, calculate $\mathbf{u} - 2\boldsymbol{v}$, $\mathbf{u} - (2\boldsymbol{v} - 3\mathbf{w})$, $-2\mathbf{u} - \boldsymbol{v} + 3\mathbf{w}$.
   a) In the vector space of (2×3) matrices

$$\mathbf{u} = \begin{bmatrix} 2 & 1 & 3 \\ -1 & 1 & 2 \end{bmatrix} \quad \boldsymbol{v} = \begin{bmatrix} 1 & 4 & -1 \\ 5 & 2 & 7 \end{bmatrix} \quad \mathbf{w} = \begin{bmatrix} 4 & -5 & 11 \\ -13 & -1 & -1 \end{bmatrix}.$$

   b) In the vector space $\mathscr{P}_2$

$$\mathbf{u} = x^2 - 2 \qquad \boldsymbol{v} = x^2 + 2x - 1 \qquad \mathbf{w} = 2x + 1.$$

   c) In the vector space $C[0, 1]$

$$\mathbf{u} = e^x \qquad \boldsymbol{v} = \sin x \qquad \mathbf{w} = \sqrt{x^2 + 1}.$$

2. For $\mathbf{u}$, $\boldsymbol{v}$, $\mathbf{w}$ in part (b) of Exercise 1, find nonzero scalars c_1, c_2, c_3 such that $c_1\mathbf{u} + c_2\boldsymbol{v} + c_3\mathbf{w} = \boldsymbol{\theta}$. Are there nonzero scalars c_1, c_2, c_3 such that $c_1\mathbf{u} + c_2\boldsymbol{v} + c_3\mathbf{w} = \boldsymbol{\theta}$ for $\mathbf{u}$, $\boldsymbol{v}$, and $\mathbf{w}$ in part (a) of Exercise 1?
3. The following sets are subsets of a vector space. Which of these subsets are also vector spaces in their own right? To answer this question, determine whether the subset satisfies the ten conditions of Definition 4.1. [Note: Because these sets are subsets of a vector space, conditions (a1), (a2), (m1), (m2), (m3), and (m4) are automatically satisfied.]
   a) $S = \{\boldsymbol{v} \text{ in } R^4 : v_1 + v_4 = 0\}$
   b) $S = \{\boldsymbol{v} \text{ in } R^4 : v_1 + v_4 = 1\}$
   c) $P = \{p(x) \text{ in } \mathscr{P}_2 : p(0) = 0\}$
   d) $P = \{p(x) \text{ in } \mathscr{P}_2 : p''(0) \neq 0\}$
   e) $P = \{p(x) \text{ in } \mathscr{P}_2 : p(x) = p(-x) \text{ for all } x\}$

4. Let V be the vector space of all real (3×4) matrices. Which of the following subsets of V are also vector spaces?

   a) $S = \{A \text{ in } V: a_{11} = 0\}$
   b) $S = \{A \text{ in } V: a_{11} + a_{23} = 0\}$
   c) $S = \{A \text{ in } V: |a_{11}| + |a_{21}| = 1\}$
   d) $S = \{A \text{ in } V: a_{32} \neq 0\}$

5. Let Q denote the set of all (2×2) nonsingular matrices with the usual matrix addition and scalar multiplication. Show that Q is not a vector space by exhibiting specific matrices in Q that violate (c1) of Definition 4.1. Also show that conditions (c2) and (a3) are not met.
6. Let Q denote the set of all (2×2) symmetric matrices with the usual matrix addition and scalar multiplication. Verify that Q is a vector space.
7. Prove parts (2), (3) and (5) of Theorem 4.1.
8. Prove that the zero vector space, defined in Example 4.6, is indeed a vector space.
9. The following are subsets of $C[-1, 1]$. Which of these are also vector spaces?

   a) $F = \{f(x) \text{ in } C[-1, 1]: f(-1) = f(1)\}$
   b) $F = \{f(x) \text{ in } C[-1, 1]: f(x) = 0 \text{ for } -\frac{1}{2} \leqslant x \leqslant \frac{1}{2}\}$
   c) $F = \{f(x) \text{ in } C[-1, 1]: f(1) = 1\}$
   d) $F = \{f(x) \text{ in } C[-1, 1]: f(1) = 0\}$
   e) $F = \{f(x) \text{ in } C[-1, 1]: \int_{-1}^{1} f(x)\,dx = 0\}$

10. The set $C^2[a, b]$ is defined to be the set of all real-valued functions $f(x)$ defined on $[a, b]$ where $f(x), f'(x)$, and $f''(x)$ are continuous on $[a, b]$. Verify that $C^2[a, b]$ is a vector space by citing the appropriate theorems on continuity and differentiability from calculus.
11. The following are subsets of the vector space $C^2[-1, 1]$. Which of these are vector spaces?

    a) $F = \{f(x) \text{ in } C^2[-1, 1]: f''(x) + f(x) = 0, \ -1 \leqslant x \leqslant 1\}$
    b) $F = \{f(x) \text{ in } C^2[-1, 1]: f''(x) + f(x) = x^2, \ -1 \leqslant x \leqslant 1\}$

4.3 SUBSPACES

Chapter 2 demonstrated that whenever W is a p-dimensional subspace of R^n, then W behaves essentially like R^p (for instance any set of $p+1$ vectors in W is linearly dependent). The situation is much the same in a general vector space V. In this setting, certain subsets of V inherit the vector-space structure of V and are vector spaces in their own right.

DEFINITION 4.2.

If V and W are real vector spaces, and if W is a nonempty subset of V, then W is called a ***subspace*** of V.

While Definition 4.2 does not quite coincide with the subspace definition given in Section 2.2, we will show that they are essentially the same.

Subspaces have considerable practical importance and are useful in problems involving approximation, optimization, differential equations, etc. The vector-space/subspace framework allows us to pose and rigorously answer questions such as "How can we find good polynomial approximations to complicated functions?" and "How can we generate good approximate solutions to differential equations?" Questions such as these are at the heart of many technical problems; and vector-space techniques, together with the computational power of the digital computer, are useful in helping to answer them.

It is fairly easy to recognize when a subset of a vector space is a subspace. The theorem below provides a characterization of subspaces and is suggested by what we know of the structure of R^n. Note that Theorem 4.2 guarantees that subspaces of R^n (as defined in Definition 2.1) correspond exactly with subspaces as defined in Definition 4.2.

Theorem 4.2

Let W be a nonempty subset of a vector space V. Then W is a subspace of V if and only if the following two conditions are met:

s1) $\mathbf{u}+\boldsymbol{v}$ is in W whenever $\mathbf{u}$ and $\boldsymbol{v}$ are in W, and

s2) $a\mathbf{u}$ is in W whenever $\mathbf{u}$ is in W and a is any scalar.

Proof. Suppose W satisfies conditions (s1) and (s2). To show that W is a subspace of V, we must show that W meets the ten requirements of Definition 4.1. Requirements (a1), (a2), (m1), (m2), (m3), and (m4) are clearly satisfied in any subset of V and so hold also in W. Requirements (c1) and (c2) are satisfied in W by the hypotheses (s1) and (s2); so the only conditions that must be verified are (a3) and (a4). Condition (a3) is met if we can show that the zero vector $\boldsymbol{\theta}$ is in W. This follows immediately from (s2) and Theorem 4.1, for if $\boldsymbol{v}$ is some vector in W, then the scalar multiple $0\boldsymbol{v}$ is in W by (s2). But by Theorem 4.1, $0\boldsymbol{v} = \boldsymbol{\theta}$; so the zero vector is in W. Similarly if $\boldsymbol{v}$ is in W, then the additive inverse $-\boldsymbol{v}$ is also in W since $-\boldsymbol{v} = (-1)\boldsymbol{v}$ by Theorem 4.1; and the scalar multiple $(-1)\boldsymbol{v}$ is in W by (s2).

On the other hand if W is a subspace of V, then (s1) and (s2) must hold since W must satisfy (c1) and (c2) of Definition 4.1. ∎

If we are given that W is a subset of a known vector space V, Theorem 4.2 simplifies the task of determining whether or not W is itself a vector space. Instead of testing all ten properties of Definition 4.1, Theorem 4.2 states that we need test only the two closure properties, (s1) and (s2). Furthermore just as in Chapter 2, a subset W of V will be specified by certain defining relationships that tell whether a vector $\mathbf{u}$ is in W or is not in W. To check (s1) and (s2), we select two arbitrary vectors, say $\mathbf{u}$ and $\boldsymbol{v}$, which satisfy the defining relationships of W (that is, $\mathbf{u}$ and $\boldsymbol{v}$ are in W). We then test $\mathbf{u}+\boldsymbol{v}$ and $a\mathbf{u}$ to see if they also satisfy the defining relationships of W. (That is, do $\mathbf{u}+\boldsymbol{v}$ and $a\mathbf{u}$ belong to W?)

EXAMPLE 4.7 Let V be the vector space of all real (2×2) matrices, and let W be the subset of V specified by

$$W = \{A : A = \begin{bmatrix} 0 & a_{12} \\ a_{21} & 0 \end{bmatrix}, a_{12} \text{ and } a_{21} \text{ any real scalars}\}.$$

If A and B are any two vectors in W, then A and B have the form

$$A = \begin{bmatrix} 0 & a_{12} \\ a_{21} & 0 \end{bmatrix}, \qquad B = \begin{bmatrix} 0 & b_{12} \\ b_{21} & 0 \end{bmatrix}.$$

Clearly $A + B$ and aA have this same form since

$$A + B = \begin{bmatrix} 0 & a_{12} + b_{12} \\ a_{21} + b_{21} & 0 \end{bmatrix}, \qquad aA = \begin{bmatrix} 0 & aa_{12} \\ aa_{21} & 0 \end{bmatrix}.$$

Therefore $A + B$ and aA are in W, and we conclude that W is a subspace of the set of all real (2×2) matrices.

EXAMPLE 4.8 If $n < m$, then $\mathscr{P}_n$ is a subspace of $\mathscr{P}_m$. We can verify this assertion directly from Definition 4.2 since we have already shown that $\mathscr{P}_n$ and $\mathscr{P}_m$ are each real vector spaces, and $\mathscr{P}_n$ is a subset of $\mathscr{P}_m$.

Similarly for any n, $\mathscr{P}_n$ is a subspace of $C[a, b]$. Again this assertion follows directly from Definition 4.2 since any polynomial is continuous on any interval $[a, b]$. Therefore $\mathscr{P}_n$ can be considered a subspace of $C[a, b]$ as well as a vector space in its own right.

In our study of subspaces in R^n we saw that a useful concept was a spanning set for a subspace. The vector-space structure as given in Definition 4.1 guarantees that the notion of a linear combination makes sense in a general vector space, and we can define spanning sets exactly as before.

DEFINITION 4.3.

Let V be a vector space, and let $Q = \{\mathbf{v}_1, \mathbf{v}_2, \ldots, \mathbf{v}_m\}$ be a set of vectors in V. If every vector $\mathbf{v}$ in V is a linear combination of vectors in Q,

$$\mathbf{v} = a_1\mathbf{v}_1 + a_2\mathbf{v}_2 + \ldots + a_m\mathbf{v}_m,$$

then we say that Q is a ***spanning set*** for V.

For many vector spaces V, it is relatively easy to find a "natural" spanning set.

EXAMPLE 4.9 Let V be the vector space of all real (2×2) matrices, and let W be the subspace (see Example 4.7)

$$W = \{A : A = \begin{bmatrix} 0 & a_{12} \\ a_{21} & 0 \end{bmatrix}, a_{12} \text{ and } a_{21} \text{ any real scalars}\}.$$

One obvious spanning set for W is seen to be the set of vectors $Q = \{A_1, A_2\}$ where

$$A_1 = \begin{bmatrix} 0 & 1 \\ 0 & 0 \end{bmatrix}, \; A_2 = \begin{bmatrix} 0 & 0 \\ 1 & 0 \end{bmatrix}.$$

To verify this assertion, suppose A is in W where

$$A = \begin{bmatrix} 0 & a_{12} \\ a_{21} & 0 \end{bmatrix}.$$

Then clearly $A = a_{12}A_1 + a_{21}A_2$, and therefore Q is a spanning set for W. Similarly it is easy to see that a natural spanning set for all of V is the set $\{B_1, B_2, B_3, B_4\}$ where

$$B_1 = \begin{bmatrix} 1 & 0 \\ 0 & 0 \end{bmatrix}, \quad B_2 = \begin{bmatrix} 0 & 1 \\ 0 & 0 \end{bmatrix}, \quad B_3 = \begin{bmatrix} 0 & 0 \\ 1 & 0 \end{bmatrix}, \quad B_4 = \begin{bmatrix} 0 & 0 \\ 0 & 1 \end{bmatrix}.$$

EXAMPLE 4.10 A natural spanning set for $\mathscr{P}_2$,

$$\mathscr{P}_2 = \{p(x) : p(x) = a_2x^2 + a_1x + a_0\},$$

is the set of polynomials (vectors in $\mathscr{P}_2$) $Q = \{1, x, x^2\}$. As we have seen before, spanning sets are not unique; and the set Q_1 is also a spanning set for $\mathscr{P}_2$:

$$Q_1 = \{1, x+1, (x+1)^2\}.$$

In Exercise 6, Section 4.3, the reader is asked to show that every vector in $\mathscr{P}_2$ can be written as a linear combination of 1, $x+1$, and $(x+1)^2$. For example if $p(x) = x^2 + 4x - 7$, then (in terms of Q_1)

$$p(x) = x^2 + 4x - 7 = (x+1)^2 + 2(x+1) - 10.$$

In practice the choice of a particular spanning set for a vector space is usually a matter of convenience, drawn from the particular problem under consideration.

In the general vector-space setting it is convenient to have a notation to indicate the set consisting of all linear combinations of a particular set of vectors. In particular let V be a vector space, and let $Q = \{\mathbf{v}_1, \mathbf{v}_2, \ldots, \mathbf{v}_k\}$ be a set of vectors in V. The ***span*** of Q, denoted $\mathrm{Sp}(Q)$, is the set of all linear combinations of $\mathbf{v}_1, \mathbf{v}_2, \ldots, \mathbf{v}_k$;

$$\mathrm{Sp}(Q) = \{\mathbf{v} : \mathbf{v} = a_1\mathbf{v}_1 + a_2\mathbf{v}_2 + \ldots + a_k\mathbf{v}_k\}.$$

From the closure properties (c1) and (c2) of Definition 4.1 it is obvious that $\mathrm{Sp}(Q)$ is a subset of V whenever Q is a subset of V. In fact as the next theorem shows, $\mathrm{Sp}(Q)$ is a subspace of V.

Theorem 4.3

If V is a vector space, and $Q = \{\mathbf{v}_1, \mathbf{v}_2, \ldots, \mathbf{v}_k\}$ is a set of vectors in V, then $\mathrm{Sp}(Q)$ is a subspace of V.

Proof. As usual we establish this theorem by showing that Sp(Q) satisfies conditions (s1) and (s2) of Theorem 4.2. Let $\mathbf{u}$ and $\boldsymbol{v}$ be any vectors in Sp(Q). From the definition of Sp(Q) we know that $\mathbf{u}$ and $\boldsymbol{v}$ have the form

$$\mathbf{u} = a_1\boldsymbol{v}_1 + a_2\boldsymbol{v}_2 + \ldots + a_k\boldsymbol{v}_k \quad \text{and} \quad \boldsymbol{v} = b_1\boldsymbol{v}_1 + b_2\boldsymbol{v}_2 + \ldots + b_k\boldsymbol{v}_k.$$

Adding $\mathbf{u}$ and $\boldsymbol{v}$ and grouping common terms, we have

$$\mathbf{u} + \boldsymbol{v} = (a_1 + b_1)\boldsymbol{v}_1 + (a_2 + b_2)\boldsymbol{v}_2 + \ldots + (a_k + b_k)\boldsymbol{v}_k;$$

so clearly the sum $\mathbf{u} + \boldsymbol{v}$ is a linear combination of vectors from Q; and hence $\mathbf{u} + \boldsymbol{v}$ is in Sp(Q). Similarly $c\mathbf{u}$ is in Sp(Q) since

$$c\mathbf{u} = (ca_1)\boldsymbol{v}_1 + (ca_2)\boldsymbol{v}_2 + \ldots + (ca_k)\boldsymbol{v}_k. \qquad \blacksquare$$

The connection between spanning sets and the span of a set is fairly obvious. In particular if W is a subspace of V and $Q \subseteq W$, then Q is a spanning set for W if and only if $W = \text{Sp}(Q)$.

EXAMPLE 4.11 A natural spanning set for $\mathscr{P}_n$ is the set of vectors

$$Q = \{1, x, x^2, \ldots x^n\}.$$

A natural spanning set for R^n is the set of n-dimensional unit vectors $Q = \{\mathbf{e}_1, \mathbf{e}_2, \ldots, \mathbf{e}_n\}$, and $R^n = \text{Sp}(Q)$.

EXAMPLE 4.12 Let W be the subset of $\mathscr{P}_2$ defined by

$$W = \{p(x) : p(x) \text{ is in } \mathscr{P}_2,\ p(0) = 0\}.$$

It is easy to show that W is a subspace of $\mathscr{P}_2$. By considering the conditions that a vector $p(x)$ must meet in order to be in W, we can find a spanning set for W. That is, if

$$p(x) = a_2x^2 + a_1x + a_0,$$

then $p(x)$ is in W if and only if $p(0) = 0$. But $p(0) = a_0$; so $p(x)$ is in W if and only if $a_0 = 0$. Therefore we see that $W = \text{Sp}(Q)$ where $Q = \{x, x^2\}$.

EXAMPLE 4.13 A square matrix A is called ***skew-symmetric*** if $A^T = -A$. It is easy to show that the set W of all (3×3) skew-symmetric matrices is a subspace of the set of all (3×3) matrices. To find a spanning set for W, we can equate A^T and $-A$ to determine the conditions that $A = (a_{ij})$ must meet in order for A to be in W. We find that the entries in A must satisfy $a_{ji} = -a_{ij}$, $1 \leqslant i, j \leqslant 3$ (note that the diagonal entries of A must all be zero). Thus a natural spanning set for W consists of

$$A_1 = \begin{bmatrix} 0 & 1 & 0 \\ -1 & 0 & 0 \\ 0 & 0 & 0 \end{bmatrix}, \quad A_2 = \begin{bmatrix} 0 & 0 & 1 \\ 0 & 0 & 0 \\ -1 & 0 & 0 \end{bmatrix}, \quad A_3 = \begin{bmatrix} 0 & 0 & 0 \\ 0 & 0 & 1 \\ 0 & -1 & 0 \end{bmatrix}.$$

Finally we note that in Definition 4.3 we have implicitly assumed that spanning sets are finite. This is not a required assumption, and frequently Sp(Q) is defined as the set of all *finite* linear combinations of vectors from Q where Q may be either an infinite set or a finite set. We do not need this full generality, and we will explore this idea no further than to note later that one contrast between the vector space R^n and a general vector space V is that V might not possess a finite spanning set. One example of a vector space where the most natural spanning set is infinite is the vector space $\mathscr{P}$, consisting of *all* polynomials (we place no upper limit on the degree). Then for instance $\mathscr{P}_n$ is a subspace of $\mathscr{P}$ for each n, $n = 1, 2, 3, \ldots$. A natural spanning set for $\mathscr{P}$ (in the generalized sense described above) is the infinite set

$$Q = \{1, x, x^2, \ldots, x^k, \ldots\}.$$

4.3 EXERCISES

1. Let V be the vector space of all (2×3) matrices. Which of the following subsets are subspaces of V?

a) $W = \{A \text{ in } V: a_{11} + a_{13} = 1\}$

b) $W = \{A \text{ in } V: a_{11} - a_{12} + 2a_{13} = 0\}$

c) $W = \{A \text{ in } V: a_{11} - a_{12} = 0,\ a_{12} + a_{13} = 0, \text{ and } a_{23} = 0\}$

d) $W = \{A \text{ in } V: a_{11}a_{12}a_{13} = 0\}$

2. Which of the following subsets of $\mathscr{P}_2$ are subspaces of $\mathscr{P}_2$?

a) $W = \{p(x) \text{ in } \mathscr{P}_2: p(0) + p(2) = 0\}$

b) $W = \{p(x) \text{ in } \mathscr{P}_2: p(1) = p(3)\}$

c) $W = \{p(x) \text{ in } \mathscr{P}_2: p(1)p(3) = 0\}$

d) $W = \{p(x) \text{ in } \mathscr{P}^2: p(1) = -p(-1)\}$

3. Let V be the vector space of all (2×2) matrices. The subset W of V defined by

$$W = \{A \text{ in } V: a_{11} - a_{12} = 0,\ a_{12} + a_{22} = 0\}$$

is a subspace of V. Find a spanning set for W. (Hint: Observe that A is in W if and only if A has the form

$$A = \begin{bmatrix} a_{11} & a_{11} \\ a_{21} & -a_{11} \end{bmatrix}$$

where a_{11} and a_{21} are arbitrary.)

4. Let W be the subset of $\mathscr{P}_3$ defined by

$$W = \{p(x) \text{ in } \mathscr{P}_3: p(1) = p(-1) \text{ and } p(2) = p(-2)\}.$$

Show that W is a subspace of $\mathscr{P}_3$, and find a spanning set for W.

5. Find a spanning set for each of the subsets that is a subspace in Exercises 1 and 2.

6. Let $p(x) = a_0 + a_1x + a_2x^2$ be a vector in $\mathscr{P}_3$. Find b_0, b_1, and b_2 in terms of a_0, a_1, and a_2 so that $p(x) = b_0 + b_1(x+1) + b_2(x+1)^2$. (Hint: Equate the coefficients of like powers of x.) Represent $q(x) = 1 - x + 2x^2$ and $r(x) = 2 - 3x + x^2$ in terms of the spanning set $\{1, x+1, (x+1)^2\}$.

7. Show that the set W of all symmetric (3×3) matrices is a subspace of the vector space of all (3×3) matrices. Find a spanning set for W.
8. Let A be an $(n \times n)$ matrix. Show that $B = (A + A^{\mathrm{T}})/2$ is symmetric and that $C = (A - A^{\mathrm{T}})/2$ is skew symmetric.
9. Use Exercise 8 to show that every $(n \times n)$ matrix can be expressed as the sum of a symmetric matrix and a skew-symmetric matrix.
10. Use Exercises 7 and 9, and Example 4.13 to construct a spanning set for the vector space of all (3×3) matrices where the spanning set consists entirely of symmetric and skew-symmetric matrices. Specify how a (3×3) matrix $A = (a_{ij})$ can be expressed by using this spanning set.
11. Let V be the set of all (3×3) upper-triangular matrices, and note that V is a vector space. Each of the subsets W is a subspace of V; find a spanning set for W.

a) $W = \{A \text{ in } V: a_{11} = 0,\ a_{22} = 0,\ a_{33} = 0\}$

b) $W = \{A \text{ in } V: a_{11} + a_{22} + a_{33} = 0,\ a_{12} + a_{23} = 0\}$

c) $W = \{A \text{ in } V: a_{11} = a_{12},\ a_{13} = a_{23},\ a_{22} = a_{33}\}$

d) $W = \{A \text{ in } V: a_{11} = a_{22},\ a_{22} - a_{33} = 0,\ a_{12} + a_{23} = 0\}$

4.4 LINEAR INDEPENDENCE, BASES, AND COORDINATES

One of the central ideas of Chapters 1 and 2 was linear independence. This concept generalizes directly to vector spaces.

DEFINITION 4.4.

Let V be a vector space, and let $\{\mathbf{v}_1, \mathbf{v}_2, \ldots, \mathbf{v}_p\}$ be a set of vectors in V. This set is ***linearly dependent*** if there are scalars $a_1, a_2, \ldots, a_p$, not all of which are zero, such that

$$a_1\mathbf{v}_1 + a_2\mathbf{v}_2 + \ldots + a_p\mathbf{v}_p = \boldsymbol{\theta}.$$

The set $\{\mathbf{v}_1, \mathbf{v}_2, \ldots, \mathbf{v}_p\}$ is ***linearly independent*** if it is not linearly dependent.

As before, it is easy to prove that a set $\{\mathbf{v}_1, \mathbf{v}_2, \ldots, \mathbf{v}_p\}$ is linearly dependent if and only if some $\mathbf{v}_i$ is a linear combination of the other $p - 1$ vectors in the set. The only real distinction between linear independence/dependence in R^n and in a general vector space is that we do not always have a simple test for dependence such as (1.25). That is, in a general vector space we may have to go directly to the defining equation

$$a_1\mathbf{v}_1 + a_2\mathbf{v}_2 + \ldots + a_p\mathbf{v}_p = \boldsymbol{\theta}$$

and attempt to determine whether there are nontrivial solutions.

EXAMPLE 4.14 Let V be the vector space of (2×2) matrices, and W be the subspace

$$W = \left\{A : A = \begin{bmatrix} 0 & a_{12} \\ a_{21} & 0 \end{bmatrix},\ a_{12} \text{ and } a_{21} \text{ any real scalars}\right\}.$$

As an example of a set of linearly dependent vectors in W, consider the set $\{B_1, B_2, B_3\}$ where

$$B_1 = \begin{bmatrix} 0 & 2 \\ 1 & 0 \end{bmatrix}, \quad B_2 = \begin{bmatrix} 0 & 1 \\ 0 & 0 \end{bmatrix}, \quad B_3 = \begin{bmatrix} 0 & 2 \\ 3 & 0 \end{bmatrix}.$$

According to Definition 4.4, this set is linearly dependent provided there exist nontrivial solutions to the equation

$$a_1 B_1 + a_2 B_2 + a_3 B_3 = \mathcal{O} \tag{4.5}$$

where $\mathcal{O}$ is the zero element in V [that is, $\mathcal{O}$ is the (2×2) zero matrix]. Writing equation (4.5) in detail, we see that a_1, a_2, a_3 are solutions of (4.5) if

$$\begin{bmatrix} 0 & 2a_1 \\ a_1 & 0 \end{bmatrix} + \begin{bmatrix} 0 & a_2 \\ 0 & 0 \end{bmatrix} + \begin{bmatrix} 0 & 2a_3 \\ 3a_3 & 0 \end{bmatrix} = \begin{bmatrix} 0 & 0 \\ 0 & 0 \end{bmatrix}.$$

With corresponding entries equated, a_1, a_2, a_3 must satisfy

$$2a_1 + a_2 + 2a_3 = 0 \quad \text{and} \quad a_1 + 3a_3 = 0.$$

This (2×3) homogeneous system has nontrivial solutions by Theorem 1.2, and one such solution is $a_1 = -3, a_2 = 4, a_3 = 1$. In particular $-3B_1 + 4B_2 + B_3 = \mathcal{O}$; so the set $\{B_1, B_2, B_3\}$ is a linearly dependent set of vectors in W.

On the other hand the set $\{B_1, B_2\}$ is linearly independent, for if a_1 and a_2 are scalars such that $a_1 B_1 + a_2 B_2 = \mathcal{O}$, then we must have

$$2a_1 + a_2 = 0 \quad \text{and} \quad a_1 = 0.$$

Hence $a_1 = 0$ and $a_2 = 0$; so if $a_1 B_1 + a_2 B_2 = \mathcal{O}$, then $a_1 = a_2 = 0$. Thus $\{B_1, B_2\}$ is a linearly independent set of vectors in W.

EXAMPLE 4.15 Establishing linear independence/dependence in a vector space of functions such as $\mathscr{P}_n$ or $C[a, b]$ may sometimes require techniques from calculus. We illustrate one such technique here and show that $\{1, x, x^2\}$ is a linearly independent set in $\mathscr{P}_2$ (see Exercise 3, Section 4.4, for another procedure).

Suppose a_0, a_1, a_2 are any scalars that satisfy the defining equation

$$a_0 + a_1 x + a_2 x^2 = \theta(x) \tag{4.6a}$$

where $\theta(x)$ is the zero polynomial. If (4.6a) is to be an identity holding for all values of x, then we can differentiate both sides of (4.6a) to obtain [since $\theta'(x) = \theta(x)$]

$$a_1 + 2a_2 x = \theta(x). \tag{4.6b}$$

Similarly differentiating both sides of (4.6b), we get

$$2a_2 = \theta(x). \tag{4.6c}$$

From (4.6c) we must have $a_2 = 0$. If $a_2 = 0$, then (4.6b) requires $a_1 = 0$; and hence in

(4.6a), $a_0 = 0$ as well. Therefore the only scalars that satisfy (4.6a) are $a_0 = a_1 = a_2 = 0$, and thus $\{1, x, x^2\}$ is linearly independent in $\mathscr{P}_2$. (Also see the material on Wronskians, Section 5.5.)

With the concepts of linear independence and spanning sets, it is easy to see that the idea of a basis will generalize directly to our vector-space setting. Moreover a basis can be used to produce a "coordinate system" for certain vector spaces. One consequence is that every real vector space with a basis of n vectors behaves "essentially" like R^n.

DEFINITION 4.5.

Let V be a vector space, and let $V = \mathrm{Sp}(B)$ where $B = \{\boldsymbol{v}_1, \boldsymbol{v}_2, \ldots, \boldsymbol{v}_p\}$. If B is linearly independent, than B is called a ***basis*** for V.

Thus as before, a basis for V is a linearly independent spanning set for V. (Again we note the implicit assumption that a basis contains only a finite number of vectors.) Finally with the definition of a basis for V, it would be natural to show that if V has a basis, then all bases for V have the same number of vectors. Given this result, we would define the dimension of V to be the number of vectors in a basis for V. We will in fact carry out this program shortly. First however, we wish to introduce another useful concept, that of coordinates.

As we have noted in Chapter 2, a basis is a minimal spanning set; and as such, a basis contains no redundant information. This lack of redundancy is an important feature of a basis in the general vector-space setting, and allows every vector to be represented unambiguously in terms of the basis (see Theorem 4.4 below). We cannot make such an assertion of unique representation about a spanning set that is linearly dependent.

Theorem 4.4

Let V be a vector space, and let $B = \{\boldsymbol{v}_1, \boldsymbol{v}_2, \ldots, \boldsymbol{v}_p\}$ be a basis for V. For each vector $\mathbf{w}$ in V, there exists a unique set of scalars $w_1, w_2, \ldots, w_p$ such that

$$\mathbf{w} = w_1\boldsymbol{v}_1 + w_2\boldsymbol{v}_2 + \ldots + w_p\boldsymbol{v}_p.$$

This theorem is already familiar from Chapter 2; so we leave the proof as an exercise. Now let V be a vector space with a basis $B = \{\boldsymbol{v}_1, \boldsymbol{v}_2, \ldots, \boldsymbol{v}_p\}$. Given that each vector $\mathbf{w}$ in V has a unique representation in terms of B as

$$\mathbf{w} = w_1\boldsymbol{v}_1 + w_2\boldsymbol{v}_2 + \ldots + w_p\boldsymbol{v}_p, \tag{4.7}$$

it follows that the scalars $w_1, w_2, \ldots, w_p$ serve to characterize $\mathbf{w}$ completely in terms of the basis B. In particular we can identify $\mathbf{w}$ unambiguously with the vector $[\mathbf{w}]_B$ in R^p where

$$[\mathbf{w}]_B = \begin{bmatrix} w_1 \\ w_2 \\ \vdots \\ w_p \end{bmatrix}.$$

We will call the unique scalars $w_1, w_2, \ldots, w_p$ in (4.7) the ***coordinates*** of $\mathbf{w}$ with respect to the basis B, and we will call the vector $[\mathbf{w}]_B$ in R^p the ***coordinate vector*** of $\mathbf{w}$ with respect to B. This idea is a useful one; and for example we will show that a set of vectors $\{\mathbf{u}_1, \mathbf{u}_2, \ldots, \mathbf{u}_r\}$ in V is linearly independent if and only if the coordinate vectors $[\mathbf{u}_1]_B, [\mathbf{u}_2]_B, \ldots, [\mathbf{u}_r]_B$ are linearly independent in R^n. Since we know how to determine whether vectors in R^p are linearly independent or not, we can use the idea of coordinates to reduce a problem of linear independence/dependence in a general vector space to an equivalent problem in R^p, which we can work. Finally we note that the subscript B is necessary when we write $[\mathbf{w}]_B$ since the coordinate vector for $\mathbf{w}$ changes when we change the basis.

EXAMPLE 4.16 Let V be the vector space of all real (2×3) matrices. An obvious basis for V is the set $B = \{A_1, A_2, A_3, A_4, A_5, A_6\}$ where

$$A_1 = \begin{bmatrix} 1 & 0 & 0 \\ 0 & 0 & 0 \end{bmatrix} \quad A_2 = \begin{bmatrix} 0 & 1 & 0 \\ 0 & 0 & 0 \end{bmatrix} \quad A_3 = \begin{bmatrix} 0 & 0 & 1 \\ 0 & 0 & 0 \end{bmatrix}$$

$$A_4 = \begin{bmatrix} 0 & 0 & 0 \\ 1 & 0 & 0 \end{bmatrix} \quad A_5 = \begin{bmatrix} 0 & 0 & 0 \\ 0 & 1 & 0 \end{bmatrix} \quad A_6 = \begin{bmatrix} 0 & 0 & 0 \\ 0 & 0 & 1 \end{bmatrix}.$$

To illustrate the idea of coordinates, we suppose A is the (2×3) matrix

$$A = \begin{bmatrix} 2 & 7 & 6 \\ 4 & 8 & 5 \end{bmatrix}.$$

Then $A = 2A_1 + 7A_2 + 6A_3 + 4A_4 + 8A_5 + 5A_6$, and we can associate A with the vector $[A]_B$ in R^6 where

$$[A]_B = \begin{bmatrix} 2 \\ 7 \\ 6 \\ 4 \\ 8 \\ 5 \end{bmatrix}.$$

That is, we have made an identification between a (2×3) matrix and an element of R^6 by using the basis B as a point of reference. Readers may choose another (2×3) matrix C and convince themselves that $[A + C]_B = [A]_B + [C]_B$. (This hints at the idea of an isomorphism that will be developed in detail later.)

EXAMPLE 4.17 It was shown in Example 4.15 that the natural spanning set $B = \{1, x, x^2\}$ for $\mathscr{P}_2$ is linearly independent. Thus B is a basis for $\mathscr{P}_2$ and can be used to define coordinate vectors in R^3 for each p in $\mathscr{P}_2$. For example if $p(x) = x^2 + 4x - 7$, then

$$[p]_B = \begin{bmatrix} -7 \\ 4 \\ 1 \end{bmatrix}.$$

In Example 4.10 we saw that $Q = \{1, x+1, (x+1)^2\}$ was also a spanning set for $\mathscr{P}_2$, and (see Example 4.18) it is easy to show that Q is a basis for $\mathscr{P}_2$. However with respect to the basis Q, the polynomial $p(x) = x^2 + 4x - 7$ has a different coordinate vector. In particular $p(x) = (x+1)^2 + 2(x+1) - 10$; so

$$[p]_Q = \begin{bmatrix} -10 \\ 2 \\ 1 \end{bmatrix}.$$

Our next theorem will use coordinates to obtain a useful tool for determining linear independence in a vector space with a basis.

Theorem 4.5

Suppose V is a vector space, and suppose $B = \{\mathbf{v}_1, \mathbf{v}_2, \ldots, \mathbf{v}_p\}$ is a basis for V. Then $\{\mathbf{u}_1, \mathbf{u}_2, \ldots, \mathbf{u}_m\}$ is a linearly independent set in V if and only if $\{[\mathbf{u}_1]_B, [\mathbf{u}_2]_B, \ldots, [\mathbf{u}_m]_B\}$ is linearly independent in R^p.

Proof. Since B is a basis for V, we can express each $\mathbf{u}_i$ as a linear combination of vectors from B:

$$\begin{aligned} \mathbf{u}_1 &= a_{11}\mathbf{v}_1 + a_{21}\mathbf{v}_2 + \ldots + a_{p1}\mathbf{v}_p \\ \mathbf{u}_2 &= a_{12}\mathbf{v}_1 + a_{22}\mathbf{v}_2 + \ldots + a_{p2}\mathbf{v}_p \\ &\vdots \\ \mathbf{u}_m &= a_{1m}\mathbf{v}_1 + a_{2m}\mathbf{v}_2 + \ldots + a_{pm}\mathbf{v}_p. \end{aligned} \tag{4.8}$$

Now $\{\mathbf{u}_1, \mathbf{u}_2, \ldots, \mathbf{u}_m\}$ is linearly dependent if and only if there is a nontrivial solution of

$$c_1\mathbf{u}_1 + c_2\mathbf{u}_2 + \ldots + c_m\mathbf{u}_m = \boldsymbol{\theta}. \tag{4.9}$$

Using (4.8), we rewrite (4.9) in terms of the basis B as

$$\begin{aligned} &c_1(a_{11}\mathbf{v}_1 + a_{21}\mathbf{v}_2 + \ldots + a_{p1}\mathbf{v}_p) + \\ &\quad c_2(a_{12}\mathbf{v}_1 + a_{22}\mathbf{v}_2 + \ldots + a_{p2}\mathbf{v}_p) + \\ &\quad \ldots + c_m(a_{1m}\mathbf{v}_1 + a_{2m}\mathbf{v}_2 + \ldots + a_{pm}\mathbf{v}_p) = \boldsymbol{\theta}. \end{aligned} \tag{4.10a}$$

Collecting common terms in (4.10a), we obtain

$$\begin{aligned} &(c_1a_{11} + c_2a_{12} + \ldots + c_ma_{1m})\mathbf{v}_1 + \\ &\quad (c_1a_{21} + c_2a_{22} + \ldots + c_ma_{2m})\mathbf{v}_2 + \\ &\quad \ldots + (c_1a_{p1} + c_2a_{p2} + \ldots + c_ma_{pm})\mathbf{v}_p = \boldsymbol{\theta}. \end{aligned} \tag{4.10b}$$

Now finding $c_1, c_2, \ldots, c_m$ that satisfy (4.10b) is the same as finding $c_1, c_2, \ldots, c_m$ that satisfy (4.9). Furthermore since $\{\mathbf{v}_1, \mathbf{v}_2, \ldots, \mathbf{v}_p\}$ is a linearly independent set, (4.10b) can be satisfied if and only if

$$\begin{aligned} c_1a_{11} + c_2a_{12} + \ldots + c_ma_{1m} &= 0 \\ c_1a_{21} + c_2a_{22} + \ldots + c_ma_{2m} &= 0 \\ &\vdots \\ c_1a_{p1} + c_2a_{p2} + \ldots + c_ma_{pm} &= 0. \end{aligned} \tag{4.11}$$

When we observe from (4.8) that

$$[\mathbf{u}_1]_B = \begin{bmatrix} a_{11} \\ a_{21} \\ \vdots \\ a_{p1} \end{bmatrix}, \quad [\mathbf{u}_2]_B = \begin{bmatrix} a_{12} \\ a_{22} \\ \vdots \\ a_{p2} \end{bmatrix}, \ldots, [\mathbf{u}_m]_B = \begin{bmatrix} a_{1m} \\ a_{2m} \\ \vdots \\ a_{pm} \end{bmatrix},$$

it follows that (4.11) holds if and only if

$$c_1[\mathbf{u}_1]_B + c_2[\mathbf{u}_2]_B + \ldots + c_m[\mathbf{u}_m]_B = \boldsymbol{\theta}_p \qquad (4.12)$$

where $\boldsymbol{\theta}_p$ denotes the zero vector in R^p.

In summary, (4.12) is equivalent to (4.11). Furthermore (4.11) is equivalent to (4.10b) since B is a basis. Finally (4.10b) is the same as (4.9) by virtue of (4.8); so we have shown that (4.12) and (4.9) are equivalent problems; and this equivalence proves the theorem. ■

EXAMPLE 4.18 In this example we use Theorem 4.5 to show that $\{1, x+1, (x+1)^2\}$ is a linearly independent set in $\mathscr{P}_2$ by using the fact (Example 4.15) that $B = \{1, x, x^2\}$ is a basis for $\mathscr{P}_2$. The application of Theorem 4.5 to this special case is straightforward; we merely calculate the coordinate vectors of 1, $x+1$, and $(x+1)^2 = x^2 + 2x + 1$ with respect to B. Doing this calculation, we have

$$[1]_B = \begin{bmatrix} 1 \\ 0 \\ 0 \end{bmatrix}, \quad [x+1]_B = \begin{bmatrix} 1 \\ 1 \\ 0 \end{bmatrix}, \quad [(x+1)^2]_B = \begin{bmatrix} 1 \\ 2 \\ 1 \end{bmatrix};$$

and clearly the coordinate vectors are linearly independent in R^3; so $\{1, x+1, (x+1)^2\}$ is a linearly independent set in $\mathscr{P}_2$.

We should note two things that are apparent from the example above. First of all in Example 4.18 we wrote the vectors in the basis B in the order $1, x, x^2$; and the ordering of the basis vectors determined the ordering of the components of the coordinate vectors. A basis (a set of vectors) with such an implicitly fixed ordering is usually called an ***ordered basis***. While we do not intend to dwell on this point, we do have to be careful to work with a fixed ordering in a basis. Another point illustrated in Example 4.18 is that while Theorem 4.5 shows that questions of linear independence/dependence in V can be translated to an equivalent problem in R^p, we do need one basis for V as a point of reference. In $\mathscr{P}_2$ once we know that $B = \{1, x, x^2\}$ is a basis, we can answer any question of dependence/independence in $\mathscr{P}_2$ via Theorem 4.5. But in order to obtain the first basis B we cannot use Theorem 4.5.

4.4 EXERCISES

1. Let V be the vector space of all (2×3) matrices, and let B be the basis for V given in Example 4.16. Find the coordinate vector $[A]_B$ for the following.

a) $A = \begin{bmatrix} 2 & 1 & 1 \\ 1 & 2 & 1 \end{bmatrix}$ b) $A = \begin{bmatrix} 1 & 0 & 0 \\ 0 & 1 & 0 \end{bmatrix}$ c) $A = \begin{bmatrix} 3. & 4 & 2 \\ 1 & 0 & 8 \end{bmatrix}$

2. With respect to the basis $B = \{1, x, x^2\}$ for $\mathscr{P}_2$, find the coordinate vector for the following.

a) $p(x) = x^2 - x + 1$ b) $p(x) = x^2 + 4x - 1$ c) $p(x) = 2x + 5$

3. Prove that $\{1, x, x^2\}$ is a linearly independent set in $\mathscr{P}_2$. [Hint: Let $p(x) = a_0 + a_1x + a_2x^2$, and suppose $p(x) = \theta(x)$; that is, $p(x) \equiv 0$. In this event, $p(-1) = 0$, $p(0) = 0$, and $p(1) = 0$; and these three equations can be used to show that $a_0 = a_1 = a_2 = 0$.]

4. Prove that $\{1, x, x^2, \ldots, x^n\}$ is a linearly independent set in $\mathscr{P}_n$ by supposing that $p(x) = \theta(x)$ where $p(x) = a_0 + a_1x + \ldots + a_nx^n$. Next take successive derivatives as in Example 4.15.

5. Let V be the set of all (2×2) matrices. Show that $B = \{B_1, B_2, B_3, B_4\}$ is a basis for V where

$$B_1 = \begin{bmatrix} 1 & 0 \\ 0 & 0 \end{bmatrix} \quad B_2 = \begin{bmatrix} 0 & 1 \\ 0 & 0 \end{bmatrix} \quad B_3 = \begin{bmatrix} 0 & 0 \\ 1 & 0 \end{bmatrix} \quad B_4 = \begin{bmatrix} 0 & 0 \\ 0 & 1 \end{bmatrix}.$$

Calculate the coordinate vectors for C_1, C_2, C_3, and use Theorem 4.5 to show that $\{C_1, C_2, C_3\}$ is linearly independent in V where

$$C_1 = \begin{bmatrix} 1 & 0 \\ 1 & 1 \end{bmatrix} \quad C_2 = \begin{bmatrix} 1 & 1 \\ 0 & 1 \end{bmatrix} \quad C_3 = \begin{bmatrix} 1 & 1 \\ 0 & 0 \end{bmatrix}.$$

6. Use Exercise 5 and Theorem 4.5 to test for linear independence in the vector space of (2×2) matrices.

a) $A_1 = \begin{bmatrix} 2 & 1 \\ 2 & 1 \end{bmatrix}$, $A_2 = \begin{bmatrix} 3 & 0 \\ 0 & 2 \end{bmatrix}$, $A_3 = \begin{bmatrix} 1 & 1 \\ 2 & 1 \end{bmatrix}$

b) $A_1 = \begin{bmatrix} 1 & 3 \\ 2 & 1 \end{bmatrix}$, $A_2 = \begin{bmatrix} 4 & -2 \\ 0 & 6 \end{bmatrix}$, $A_3 = \begin{bmatrix} 6 & 4 \\ 4 & 8 \end{bmatrix}$

c) $A_1 = \begin{bmatrix} 2 & 2 \\ 1 & 3 \end{bmatrix}$, $A_2 = \begin{bmatrix} 1 & 4 \\ 0 & 5 \end{bmatrix}$, $A_3 = \begin{bmatrix} 4 & 10 \\ 1 & 13 \end{bmatrix}$

7. Use Exercise 4 and Theorem 4.5 to test for linear independence in $\mathscr{P}_3$.

a) $\{x^3 - x, x^2 - 1, x + 4\}$

b) $\{x^2 + 2x - 1, x^2 - 5x + 2, 3x^2 - x\}$

c) $\{x^3 - x^2, x^2 - x, x - 1, x^3 - 1\}$

d) $\{x^3 + 1, x^2 + 1, x + 1, 1\}$

8. Prove Theorem 4.4.

9. Use Theorem 4.5 and Exercise 4 to show that $\{1, x + 1, x^2 - 2, (x + 1)^3\}$ is a linearly independent set in $\mathscr{P}_3$.

10. Find a basis for the subspace V of $\mathscr{P}_4$ where $V = \{p(x) \text{ in } \mathscr{P}_4\text{: } p(0) = 0,\ p'(1) = 0,\ p''(-1) = 0\}$.

11. Prove that the set of all real (2×2) symmetric matrices is a subspace of the vector space of all real (2×2) matrices. Find a basis for this subspace.

12. Prove that if $Q = \{\mathbf{v}_1, \mathbf{v}_2, \ldots, \mathbf{v}_m\}$ is a linearly independent subset of a vector space V, and if $\mathbf{w}$ is a vector in V such that $\mathbf{w}$ is not in Sp(Q), then $\{\mathbf{v}_1, \mathbf{v}_2, \ldots, \mathbf{v}_m, \mathbf{w}\}$ is also a linearly independent set in V. [Note: $\boldsymbol{\theta}$ is always in Sp(Q).]

4.5 DIMENSION

We now use Theorem 4.5 to generalize the idea of dimension to the general vector-space setting. We begin with two theorems that will be needed to show that dimension is a well-defined concept. These theorems are direct applications of Theorem 4.5, and the proofs are left to the exercises since these proofs are essentially the same as those of the analogous theorems from Chapter 2, Section 2.4.

Theorem 4.6

If V is a vector space, and if $B = \{\mathbf{v}_1, \mathbf{v}_2, \ldots, \mathbf{v}_p\}$ is a basis of V, then any set of $p+1$ vectors in V is linearly dependent.

Theorem 4.7

Let V be a vector space, and let $B = \{\mathbf{v}_1, \mathbf{v}_2, \ldots, \mathbf{v}_p\}$ be a basis for V. If $Q = \{\mathbf{u}_1, \mathbf{u}_2, \ldots, \mathbf{u}_m\}$ is also a basis for V, then $m = p$.

If V is a vector space that has a basis of p vectors, then no ambiguity can arise if we define the dimension of V to be p (since the number of vectors in a basis for V is an invariant property of V by Theorem 4.7). There is however one extreme case, which is also included below in Definition 4.6. That is, there may not be a *finite* set of vectors that spans V; and in this case we call V an infinite-dimensional vector space.

DEFINITION 4.6

Let V be a vector space.

1. If V has a basis $B = \{\mathbf{v}_1, \mathbf{v}_2, \ldots, \mathbf{v}_n\}$ of n vectors, then V has ***dimension*** n, and we write $\dim(V) = n$. (If $V = \{\boldsymbol{\theta}\}$, then $\dim(V) = 0$.)
2. If V does not have a basis containing a finite number of vectors, then V is an ***infinite-dimensional*** vector space.

EXAMPLE 4.19 The vector space $\mathscr{P}_2$ has a basis with three vectors, namely $B = \{1, x, x^2\}$. Therefore $\dim(\mathscr{P}_2) = 3$, and similarly $\dim(\mathscr{P}_n) = n+1$.

If W is a subspace of a vector space V, and if $\dim(W) = k$, then it is almost obvious that $\dim(V) \geq \dim(W) = k$ (we leave the proof of this as an exercise). This observation can be used to show that $C[a, b]$ is an infinite-dimensional vector space.

EXAMPLE 4.20 To show that $C[a, b]$ is not a finite-dimensional vector space, we merely note that $\mathscr{P}_n$ is a subspace of $C[a, b]$ for every n. By Example 4.19, $\dim(\mathscr{P}_n) = n+1$; and so $C[a, b]$ contains subspaces of arbitrarily large dimension. Thus, $C[a, b]$ must be an infinite-dimensional vector space.

The next two theorems summarize some of the properties of a p-dimensional vector space V and show how properties of R^p carry over into V.

Theorem 4.8

Let V be a finite-dimensional vector space with $\dim(V) = p$.

1. Any set of $p+1$ or more vectors in V is linearly dependent.
2. Any set of p linearly independent vectors in V is a basis for V.

This theorem is a direct generalization from R^p (Exercise 9, Section 4.5). To complete our discussion of finite-dimensional vector spaces, we state the following lemma.

Lemma

Let V be a vector space, and let $Q = \{\mathbf{u}_1, \mathbf{u}_2, \ldots, \mathbf{u}_p\}$ be a spanning set for V. Then there is a subset Q' of Q that is a basis for V.

Proof. (We only sketch the proof of this lemma since the proof follows familiar lines.) If Q is linearly independent, then Q itself is a basis for V. If Q is linearly dependent, we can express some vector from Q in terms of the other $p-1$ vectors in Q. Without loss of generality, let us suppose we can express $\mathbf{u}_1$ in terms of $\mathbf{u}_2, \mathbf{u}_3, \ldots, \mathbf{u}_p$. In that event we have

$$\mathrm{Sp}\{\mathbf{u}_2, \mathbf{u}_3, \ldots, \mathbf{u}_p\} = \mathrm{Sp}\{\mathbf{u}_1, \mathbf{u}_2, \mathbf{u}_3, \ldots, \mathbf{u}_p\} = V;$$

and if $\{\mathbf{u}_2, \mathbf{u}_3, \ldots, \mathbf{u}_p\}$ is linearly independent, it is a basis for V. If $\{\mathbf{u}_2, \mathbf{u}_3, \ldots, \mathbf{u}_p\}$ is linearly dependent, we continue discarding redundant vectors until we obtain a linearly independent spanning set, Q'. ■

The following theorem is a companion of Theorem 4.8.

Theorem 4.9

Let V be a finite-dimensional vector space with $\dim(V) = p$.

1. Any spanning set for V must contain at least p vectors.
2. Any set of p vectors that spans V is a basis for V.

Proof. Part (1) follows immediately from the lemma above, for if there were a spanning set Q for V that contained fewer than p vectors, then we could find a subset

Q' of Q that is a basis for V containing fewer than p vectors. This finding would contradict Theorem 4.7; so part (1) must be valid.

Part (2) also follows from the lemma, for we know there is a subset Q' of Q such that Q' is a basis for V. Since $\dim(V) = p$, Q' must have p vectors; and since $Q' \subseteq Q$ where Q has p vectors, we must have that $Q' = Q$. ■

EXAMPLE 4.21 Let V be the vector space of all real (2×2) matrices. A natural basis for V is the set $\{A_1, A_2, A_3, A_4\}$ where

natural basis

$$A_1 = \begin{bmatrix} 1 & 0 \\ 0 & 0 \end{bmatrix}, \quad A_2 = \begin{bmatrix} 0 & 1 \\ 0 & 0 \end{bmatrix}, \quad A_3 = \begin{bmatrix} 0 & 0 \\ 1 & 0 \end{bmatrix}, \quad A_4 = \begin{bmatrix} 0 & 0 \\ 0 & 1 \end{bmatrix}.$$

Thus $\dim(V) = 4$; so any set of five or more (2×2) matrices is linearly dependent while any set of four linearly independent (2×2) matrices is a basis for V. In addition any spanning set for V must contain at least four (2×2) matrices, and any spanning set of four (2×2) matrices is a basis for V.

4.5 EXERCISES

1. Let V be the set of all real (3×3) matrices, and let V_1 and V_2 be subsets of V where V_1 consists of all the (3×3) lower-triangular matrices and V_2 consists of all the (3×3) upper-triangular matrices.
   a) Show that V_1 and V_2 are subspaces of V.
   b) Find bases for V_1 and V_2.
   c) Calculate $\dim(V)$, $\dim(V_1)$, and $\dim(V_2)$.
2. Suppose V_1 and V_2 are subspaces of a vector space V. Show that $V_1 \cap V_2$ is also a subspace of V. It is not necessarily true that $V_1 \cup V_2$ is a subspace of V; let $V = R^2$, and find two subspaces of R^2 whose union is not a subspace of R^2.
3. Let V, V_1, and V_2 be as in Exercise 1. By Exercise 2, $V_1 \cap V_2$ is a subspace of V. Describe $V_1 \cap V_2$ and calculate its dimension.
4. Let V be as in Exercise 1, and let W be the subset of all the (3×3) symmetric matrices in V. Clearly W is a subspace of V; what is $\dim(W)$?
5. Let W be the subspace of $\mathscr{P}_4$ defined thus: $p(x)$ is in W if and only if $p(1) + p(-1) = 0$ and $p(2) + p(-2) = 0$. What is $\dim(W)$?
6. Let V be a vector space, and let W be a subspace of V where $\dim(W) = k$. Prove that if V is finite dimensional, then $\dim(V) \geqslant k$. (Hint: W must contain a set of k linearly independent vectors.)
7. A square matrix A is called skew symmetric if $A^T = -A$. Let V be as in Exercise 1, and let W be the subset of all the (3×3) skew-symmetric matrices in V. Calculate $\dim(W)$.
8. Prove Theorems 4.6 and 4.7.
9. Prove Theorem 4.8.
10. Let W be a subspace of a finite-dimensional vector space V where W contains at least one nonzero vector. Prove that W has a basis and that $\dim(W) \leqslant \dim(V)$. (Hint: Use Exercise 12, Section 4.4, to show that W has a basis.)

11. (Change of basis; see also Section 4.10.) Let V be a vector space where $\dim(V) = n$, and let $B = \{\mathbf{v}_1, \mathbf{v}_2, \ldots, \mathbf{v}_n\}$ and $C = \{\mathbf{u}_1, \mathbf{u}_2, \ldots, \mathbf{u}_n\}$ be two bases for V. Let $\mathbf{w}$ be any vector in V, and suppose that $\mathbf{w}$ has these representations in terms of the bases B and C:

$$\mathbf{w} = d_1\mathbf{v}_1 + d_2\mathbf{v}_2 + \ldots + d_n\mathbf{v}_n$$
$$\mathbf{w} = c_1\mathbf{u}_1 + c_2\mathbf{u}_2 + \ldots + c_n\mathbf{u}_n.$$

By considering the proof of Theorem 4.5, convince yourself that the coordinate vectors for $\mathbf{w}$ satisfy

$$[\mathbf{w}]_B = A[\mathbf{w}]_C$$

where A is the $(n \times n)$ matrix whose ith column is equal to $[\mathbf{u}_i]_B$, $1 \leqslant i \leqslant n$. [To see this point, replace $\boldsymbol{\theta}$ in (4.9) by $\mathbf{w} = d_1\mathbf{v}_1 + d_2\mathbf{v}_2 + \ldots + d_n\mathbf{v}_n$.] As an application, consider the two bases for $\mathscr{P}_2$: $C = \{1, x, x^2\}$ and $B = \{1, x+1, (x+1)^2\}$.

a) Calculate the (3×3) matrix A described above.
b) Using the identity $[p]_B = A[p]_C$, calculate the coordinate vector of $p(x) = x^2 + 4x + 8$ with respect to B.
c) Use the result of part (b) to express $p(x)$ in the form $p(x) = a_0 + a_1(x+1) + a_2(x+1)^2$.

12. The matrix A in Exercise 11 is called a "transition matrix" and shows how to transform a representation with respect to one basis into a representation with respect to another. Use the matrix in part(a) of Exercise 11 to convert $p(x) = c_0 + c_1x + c_2x^2$ to the form $p(x) = a_0 + a_1(x+1) + a_2(x+1)^2$ where

a) $p(x) = x^2 + 3x - 2$, b) $p(x) = 2x^2 - 5x + 8$
c) $p(x) = -x^2 - 2x + 3$, d) $p(x) = x - 9$.

13. By Theorem 4.5 an $(n \times n)$ transition matrix is always nonsingular. Thus if $[\mathbf{w}]_B = A[\mathbf{w}]_C$, then $[\mathbf{w}]_C = A^{-1}[\mathbf{w}]_B$. Calculate A^{-1} for the matrix in part (a) of Exercise 11, and use the result to transform the polynomials below to the form $a_0 + a_1x + a_2x^2$:

a) $p(x) = 2 - 3(x+1) + 7(x+1)^2$ b) $p(x) = 1 + 4(x+1) - (x+1)^2$
c) $p(x) = 4 + (x+1)$ d) $p(x) = 9 - (x+1)^2$

14. Find a matrix A such that $[p]_B = A[p]_C$ for all $p(x)$ in $\mathscr{P}_3$ where $C = \{1, x, x^2, x^3\}$ and $B = \{1, x, x(x-1), x(x-1)(x-2)\}$. Use A to convert the following to the form $p(x) = a_0 + a_1x + a_2x(x-1) + a_3x(x-1)(x-2)$.

a) $p(x) = x^3 - 2x^2 + 5x - 9$ b) $p(x) = x^2 + 7x - 2$
c) $p(x) = x^3 + 1$ d) $p(x) = x^3 + 2x^2 + 2x + 3$

†4.6 INNER-PRODUCT SPACES, ORTHOGONAL BASES, AND PROJECTIONS

skip

Up to now we have considered a vector space solely as an entity with an algebraic structure. However we know that R^n possesses more than just an algebraic structure; and in particular we know that we can measure the "size" or "length" of a vector $\mathbf{x}$ in R^n by the quantity $\|\mathbf{x}\| = \sqrt{\mathbf{x}^T\mathbf{x}}$. Similarly we can define the distance from $\mathbf{x}$ to $\mathbf{y}$ to be $\|\mathbf{x} - \mathbf{y}\|$; and the ability to measure distances means that R^n has a "topological"

† This section may be omitted with no loss of continuity.

structure, which supplements the algebraic structure. The topological structure can be employed of course to study problems of convergence, continuity, etc. In this section we will briefly describe how a suitable measure of distance might be imposed on a general vector space. Our development will be sketchy, and we will leave most of the details to the reader; but the ideas parallel those in Sections 2.5 and 2.6.

To begin, we observe that the topological structure for R^n is based on the scalar product $\mathbf{x}^T\mathbf{y}$. Essentially the scalar product is a real-valued function of two vector variables—given $\mathbf{x}$ and $\mathbf{y}$ in R^n, the scalar product produces a number $\mathbf{x}^T\mathbf{y}$. Thus to derive a topological structure for a vector space V, we should look for a generalization of the scalar-product function. A consideration of the properties of the scalar-product function leads to the definition of an "inner-product" function for a vector space. (With reference to Definition 4.7 below, we note that the expression $\mathbf{u}^T\mathbf{v}$ does not make sense in a general vector space V. Thus not only does the nomenclature change—scalar product becomes inner product—but also the notation changes as well, with $\langle \mathbf{u}, \mathbf{v} \rangle$ denoting the inner product of $\mathbf{u}$ and $\mathbf{v}$.)

DEFINITION 4.7.

An ***inner product*** on a real vector space V is a function that assigns a real number, $\langle \mathbf{u}, \mathbf{v} \rangle$, to each pair of vectors $\mathbf{u}$ and $\mathbf{v}$ in V where this function satisfies these conditions:

1) $\langle \mathbf{u}, \mathbf{u} \rangle \geqslant 0$ and $\langle \mathbf{u}, \mathbf{u} \rangle = 0$ if and only if $\mathbf{u} = \boldsymbol{\theta}$,
2) $\langle \mathbf{u}, \mathbf{v} \rangle = \langle \mathbf{v}, \mathbf{u} \rangle$,
3) $\langle a\mathbf{u}, \mathbf{v} \rangle = a\langle \mathbf{u}, \mathbf{v} \rangle$, and
4) $\langle \mathbf{u}, \mathbf{v}+\mathbf{w} \rangle = \langle \mathbf{u}, \mathbf{v} \rangle + \langle \mathbf{u}, \mathbf{w} \rangle$.

EXAMPLE 4.22 The usual scalar product in R^n is an inner product in the sense of Definition 4.7 where $\langle \mathbf{x}, \mathbf{y} \rangle = \mathbf{x}^T\mathbf{y}$. To illustrate the flexibility of Definition 4.7, we also note that there are other sorts of inner products for R^n. As an example, let V be the vector space R^2, and let A be the (2×2) matrix

$$A = \begin{bmatrix} 3 & 2 \\ 2 & 4 \end{bmatrix}.$$

In Exercise 2, Section 4.6, the reader is asked to verify that the function $\langle \mathbf{u}, \mathbf{v} \rangle = \mathbf{u}^T A\mathbf{v}$ is an inner product for R^2.

EXAMPLE 4.23 This example illustrates two different inner products for $\mathscr{P}_2$. First for $p(x)$ and $q(x)$ in $\mathscr{P}_2$, define

$$\langle p, q \rangle = p(0)q(0) + p(1)q(1) + p(2)q(2).$$

Yet another inner product for $\mathscr{P}_2$ is

$$\langle p, q \rangle = \int_0^1 p(t)q(t)dt,$$

and we leave the verification to the exercises.

After the key step of defining a vector-space analog of the scalar product, the rest is routine. For purposes of reference we call a vector space with an inner product an ***inner-product space***. As in R^n, we can use the inner product as a measure of size: if V is an inner-product space, then for each $\mathbf{v}$ in V we define $\|\mathbf{v}\|$ (the ***norm*** of v) as

$$\|\mathbf{v}\| = \sqrt{\langle \mathbf{v}, \mathbf{v} \rangle}.$$

Note that $\langle \mathbf{v}, \mathbf{v} \rangle \geqslant 0$ for all $\mathbf{v}$ in V; so the norm function is always defined. If $\mathbf{u}$ and $\mathbf{v}$ are vectors in an inner-product space V, we say $\mathbf{u}$ and $\mathbf{v}$ are ***orthogonal*** if $\langle \mathbf{u}, \mathbf{v} \rangle = 0$. Similarly $B = \{\mathbf{v}_1, \mathbf{v}_2, \ldots, \mathbf{v}_p\}$ is an ***orthogonal set*** in V if $\langle \mathbf{v}_i, \mathbf{v}_j \rangle = 0$ when $i \neq j$. In addition if an orthogonal set of vectors B is a basis for V, we call B an ***orthogonal basis***. The next two theorems correspond to their analogs in R^n, and we leave the proofs to the exercises. (See Theorems 2.5 and 2.6.)

Theorem 4.10

Let $B = \{\mathbf{v}_1, \mathbf{v}_2, \ldots, \mathbf{v}_n\}$ be an orthogonal basis for an inner-product space V. If $\mathbf{u}$ is any vector in V, then

$$\mathbf{u} = \frac{\langle \mathbf{v}_1, \mathbf{u} \rangle}{\langle \mathbf{v}_1, \mathbf{v}_1 \rangle}\mathbf{v}_1 + \frac{\langle \mathbf{v}_2, \mathbf{u} \rangle}{\langle \mathbf{v}_2, \mathbf{v}_2 \rangle}\mathbf{v}_2 + \ldots + \frac{\langle \mathbf{v}_n, \mathbf{u} \rangle}{\langle \mathbf{v}_n, \mathbf{v}_n \rangle}\mathbf{v}_n.$$

Theorem 4.11 (Gram–Schmidt orthogonalization)

Let V be an inner-product space, and let $\{\mathbf{u}_1, \mathbf{u}_2, \ldots, \mathbf{u}_n\}$ be a basis for V. Let $\mathbf{v}_1 = \mathbf{u}_1$, and for $2 \leqslant k \leqslant n$ define $\mathbf{v}_k$ by

$$\mathbf{v}_k = \mathbf{u}_k - \sum_{j=1}^{k-1} \frac{\langle \mathbf{u}_k, \mathbf{v}_j \rangle}{\langle \mathbf{v}_j, \mathbf{v}_j \rangle}\mathbf{v}_j.$$

Then $\{\mathbf{v}_1, \mathbf{v}_2, \ldots, \mathbf{v}_n\}$ is an orthogonal basis for V.

Before continuing, we pause to give a simple example that illustrates one way in which the inner-product space framework is used in practice.

EXAMPLE 4.24 One of the many inner products for the vector space $C[0, 1]$ is

$$\langle f, g \rangle = \int_0^1 f(x)g(x)dx$$

If f is a relatively complicated function in $C[0, 1]$, we might wish to approximate f by a simpler function, say a polynomial. For definiteness suppose we want to find a polynomial p in $\mathscr{P}_2$ that is a good approximation to f. The phrase "good approximation" is too vague to be used in any calculation, but the inner-product-space framework allows us to measure size and thus allows us to pose some meaningful problems. In particular we can ask for a polynomial p^* *in* $\mathscr{P}_2$ such that

$$\|f-p^*\| \leqslant \|f-p\|$$

for all p in $\mathscr{P}_2$. Finding such a polynomial p^* in this setting is equivalent to minimizing

$$\int_0^1 [f(x)-p(x)]^2\,dx$$

among all p in $\mathscr{P}_2$. We will present a procedure for doing this shortly.

Note that the problem posed in the example above amounts to determining a vector p^* in a subspace of an inner-product space where p^* is "closer" to f than any other vector in the subspace. The essential aspects of this problem can be stated formally as the following general problem.

Let V be an inner-product space and let W be a subspace of V.

Given a vector $\mathbf{v}$ in V, find a vector $\mathbf{w}^*$ in W such that

$$\|\mathbf{v}-\mathbf{w}^*\| \leqslant \|\mathbf{v}-\mathbf{w}\| \text{ for all } \mathbf{w} \text{ in } W. \tag{4.13}$$

A vector $\mathbf{w}^*$ in W satisfying (4.13) is called the ***projection*** of $\mathbf{v}$ onto W, or (frequently) the ***best least-squares approximation*** to $\mathbf{v}$. Intuitively $\mathbf{w}^*$ is the nearest vector in W to $\mathbf{v}$.

The solution process for the problem above is almost exactly the same as that for the least-squares problem in R^n. One distinction in our general setting is that the subspace W might not be finite dimensional. If W is an infinite-dimensional subspace of V, then there may or may not be a projection of $\mathbf{v}$ onto W. If W is finite dimensional, then a projection always exists, is unique, and can be found explicitly. The next two theorems outline this concept, and again we leave the proofs to the reader since they parallel the proof of Theorem 2.7.

Theorem 4.12

Let V be an inner-product space and let W be a subspace of V. Let $\mathbf{v}$ be a vector in V, and suppose $\mathbf{w}^*$ is a vector in W such that

$$\langle \mathbf{v}-\mathbf{w}^*, \mathbf{w} \rangle = 0 \text{ for all } \mathbf{w} \text{ in } W.$$

Then $\|\mathbf{v}-\mathbf{w}^*\| \leqslant \|\mathbf{v}-\mathbf{w}\|$ for all $\mathbf{w}$ in W with equality holding only for $\mathbf{w}=\mathbf{w}^*$.

Theorem 4.13

Let V be an inner-product space and let $\mathbf{v}$ be a vector in V. Let W be an n-dimensional subspace of V, and let $\{\mathbf{u}_1, \mathbf{u}_2, \ldots, \mathbf{u}_n\}$ be an orthogonal basis for W. Then

$$\|\mathbf{v}-\mathbf{w}^*\| \leqslant \|\mathbf{v}-\mathbf{w}\| \text{ for all } \mathbf{w} \text{ in } W$$

if and only if

$$\mathbf{w}^* = \frac{\langle \mathbf{v}, \mathbf{u}_1 \rangle}{\langle \mathbf{u}_1, \mathbf{u}_1 \rangle}\mathbf{u}_1 + \frac{\langle \mathbf{v}, \mathbf{u}_2 \rangle}{\langle \mathbf{u}_2, \mathbf{u}_2 \rangle}\mathbf{u}_2 + \ldots + \frac{\langle \mathbf{v}, \mathbf{u}_n \rangle}{\langle \mathbf{u}_n, \mathbf{u}_n \rangle}\mathbf{u}_n. \tag{4.14}$$

In view of Theorem 4.13, it follows that when W is a finite-dimensional subspace of an inner-product space V, we can always find projections by first finding an orthogonal basis for W (by using Theorem 4.11) and then calculating the projection $\mathbf{w}^*$ from (4.14).

To illustrate the process of finding a projection, we return to the inner-product space of Example 4.24 with the subspace $\mathscr{P}_2$. As a specific, but rather unrealistic function, $f(x)$, we choose $f(x) = \cos x$, x in radians. The inner product is as in Example 4.23; and we use the natural basis for $\mathscr{P}_2$, $\{1, x, x^2\}$, as a starting point for Gram–Schmidt orthogonalization. If we let $\{p_0, p_1, p_2\}$ denote the orthogonal basis, we have $p_0(x) = 1$ and find $p_1(x)$ from

$$p_1(x) = x - \frac{\langle p_0, x \rangle}{\langle p_0, p_0 \rangle} p_0(x).$$

We calculate

$$\langle p_0, x \rangle = \int_0^1 x dx = \tfrac{1}{2} \quad \text{and} \quad \langle p_0, p_0 \rangle = \int_0^1 dx = 1;$$

so $p_1(x) = x - \frac{1}{2}$. The next step of the Gram–Schmidt orthogonalization process is to form

$$p_2(x) = x^2 - \frac{\langle p_1, x^2 \rangle}{\langle p_1, p_1 \rangle} p_1(x) - \frac{\langle p_0, x^2 \rangle}{\langle p_0, p_0 \rangle} p_0(x);$$

and the required constants are

$$\langle p_1, x^2 \rangle = \int_0^1 \left(x^3 - \frac{x^2}{2} \right) dx = \tfrac{1}{12}, \quad \langle p_1, p_1 \rangle = \int_0^1 (x^2 - x + \tfrac{1}{4}) dx = \tfrac{1}{12}$$

$$\langle p_0, x^2 \rangle = \int_0^1 x^2 dx = \tfrac{1}{3}; \quad \langle p_0, p_0 \rangle = \int_0^1 dx = 1.$$

Therefore $p_2(x) = x^2 - p_1(x) - p_0(x)/3 = x^2 - x + 1/6$, and we have that $\{p_0, p_1, p_2\}$ is an orthogonal basis for $\mathscr{P}_2$ with respect to the inner product.

Continuing, to find the nearest point p^* in $\mathscr{P}_2$ to f, we form

$$p^*(x) = \frac{\langle f, p_0 \rangle}{\langle p_0, p_0 \rangle} p_0(x) + \frac{\langle f, p_1 \rangle}{\langle p_1, p_1 \rangle} p_1(x) + \frac{\langle f, p_2 \rangle}{\langle p_2, p_2 \rangle} p_2(x)$$

where

$$\langle f, p_0 \rangle = \int_0^1 \cos(x) dx \simeq .841471$$

$$\langle f, p_1 \rangle = \int_0^1 (x - \tfrac{1}{2}) \cos(x) dx \simeq -.038962$$

$$\langle f, p_2 \rangle = \int_0^1 (x^2 - x + \tfrac{1}{6})\cos(x)\,dx \simeq -.002394$$

$$\langle p_2, p_2 \rangle = \int_0^1 (x^2 - x + \tfrac{1}{6})^2\,dx = \tfrac{1}{180}.$$

Therefore $p^*(x)$ is given by

$$\begin{aligned} p^*(x) &= \langle f, p_0 \rangle p_0(x) + 12\langle f, p_1 \rangle p_1(x) + 180\,\langle f, p_2 \rangle p_2(x) \\ &\simeq .841471\ p_0(x) - .467544\ p_1(x) - .430920\ p_2(x). \end{aligned}$$

In order to assess how well $p^*(x)$ approximates $\cos x$ in $[0, 1]$, we can tabulate $p^*(x)$ and $\cos x$ at various values of x (see Table 4.1).

TABLE 4.1

x	$p^*(x)$	$\cos x$	$p^*(x) - \cos x$
0.0	1.0034	1.0000	.0034
0.2	.9789	.9801	−.0012
0.4	.9198	.9211	−.0013
0.6	.8263	.8253	.0010
0.8	.6983	.6967	.0016
1.0	.5359	.5403	−.0044

4.6 EXERCISES

1. A real $(n \times n)$ symmetric matrix A is called "positive definite" if $\mathbf{x}^T A\mathbf{x} > 0$ for all $\mathbf{x}$ in R^n, $\mathbf{x} \neq \boldsymbol{\theta}$. Let A be a symmetric positive-definite matrix, and verify that

$$\langle \mathbf{x}, \mathbf{y} \rangle = \mathbf{x}^T A\mathbf{y}$$

defines an inner product on R^n; that is, verify that the four conditions of Definition 4.7 are satisfied.

2. Prove that the symmetric matrix A in Example 4.22 is positive definite. Prove this by choosing an arbitrary vector $\mathbf{x}$ in R^2, $\mathbf{x} \neq \boldsymbol{\theta}$, and calculating $\mathbf{x}^T A\mathbf{x}$.
3. Prove that $\langle p, q \rangle = p(0)q(0) + p(1)q(1) + p(2)q(2)$ is an inner product for $\mathscr{P}_2$.
4. Show that the function defined in Exercise 3 is not an inner product for $\mathscr{P}_3$. [Hint: Find $p(x)$ in $\mathscr{P}_3$ such that $\langle p, p \rangle = 0$ but $p \neq \theta$.]
5. Prove that if $\{\mathbf{v}_1, \mathbf{v}_2, \ldots, \mathbf{v}_k\}$ is an orthogonal set of nonzero vectors in an inner-product space, then this set is linearly independent.
6. Prove Theorem 4.10.
7. Use Theorem 4.11 to calculate an orthogonal basis for $\mathscr{P}_2$ with respect to the inner product in Exercise 3. Start with the natural basis $\{1, x, x^2\}$ for $\mathscr{P}_2$.

8. Starting with the natural basis $\{1, x, x^2, x^3, x^4\}$, generate an orthogonal basis for $\mathscr{P}_4$ with respect to the inner product

$$\langle p, q \rangle = \sum_{i=-2}^{2} p(i)q(i).$$

9. After Theorem 4.13 we found $p_0(x)$, $p_1(x)$, and $p_2(x)$, which are orthogonal with respect to

$$\langle f, g \rangle = \int_0^1 f(x)g(x)\,dx.$$

Continue the process and find $p_3(x)$ and $p_4(x)$ so that $\{p_0, p_1, p_2, p_3, p_4\}$ is an orthogonal basis for $\mathscr{P}_4$. (Clearly there is an infinite sequence of these polynomials p_0, $p_1, \ldots, p_n, \ldots$ that satisfy

$$\int_0^1 p_i(x)p_j(x)\,dx = 0, \qquad i \neq j.$$

These are called the ***Legendre polynomials***.) Given the orthogonal basis for $\mathscr{P}_4$, use Theorem 4.13 to find the projection of $f(x) = \cos x$ in $\mathscr{P}_4$. Construct a table similar to Table 4.1 and note the improvement.

10. An inner product for $C[-1, 1]$ is

$$\langle f, g \rangle = \frac{2}{\pi}\int_{-1}^{1} \frac{f(x)g(x)}{\sqrt{1-x^2}}\,dx.$$

Starting with the set $\{1, x, x^2, x^3, \ldots,\}$, use the Gram–Schmidt process to find polynomials $T_0(x)$, $T_1(x)$, $T_2(x)$, $T_3(x)$ such that $\langle T_i, T_j \rangle = 0$ when $i \neq j$. These polynomials are called the ***Chebyshev polynomials of the first kind***. (Hint: Make a change of variables $x = \cos\theta$.)

11. A sequence of orthogonal polynomials usually satisfies a "three-term recurrence relation." For example the Chebyshev polynomials are related by

$$T_{n+1}(x) = 2xT_n(x) - T_{n-1}(x), \; n = 1, 2, \ldots \qquad (R)$$

where $T_0(x) = 1$ and $T_1(x) = x$. Verify that the polynomials *defined* by the relation (R) above are indeed orthogonal in $C[-1, 1]$ with respect to the inner product in Exercise 10. Verify this as follows.

a) Make the change of variables $x = \cos\theta$, and use induction to show that $T_k(\cos\theta) = \cos k\theta$, $k = 0, 1, \ldots$ where $T_k(x)$ is defined by (R).
b) Using part (a), show that $\langle T_i, T_j \rangle = 0$ when $i \neq j$.
c) Use induction to show that $T_k(x)$ is a polynomial of degree k, $k = 0, 1, \ldots$.
d) Use (R) to calculate T_2, T_3, T_4, and T_5.

12. Let $C[-1, 1]$ have the inner product of Exercise 10, and let f be in $C[-1, 1]$. Use Theorem 4.13 to prove that $\|f - p^*\| \leq \|f - p\|$ for all p in $\mathscr{P}_n$ if

$$p^*(x) = \frac{a_0}{2} + \sum_{j=1}^{n} a_j T_j(x)$$

where $a_j = \langle f, T_j \rangle$, $j = 0, 1, \ldots, n$.

13. The iterated trapezoid rule provides a good estimate of $\int_a^b f(x)dx$ when $f(x)$ is periodic in $[a, b]$. In particular let N be a positive integer, and let $h = (b-a)/N$. Next define x_i by $x_i = a + ih$, $i = 0, 1, \ldots, N$; and suppose $f(x)$ is in $C[a, b]$. If we define $A(f)$ by

$$A(f) = \frac{h}{2} f(x_0) + h \sum_{j=1}^{N-1} f(x_j) + \frac{h}{2} f(x_N),$$

then $A(f)$ is the iterated trapezoid rule applied to $f(x)$. Using the result in Exercise 12, write a computer program that generates a good approximation to $f(x)$ in $C[-1, 1]$. That is, for an input function $f(x)$ and a specified value of n, calculate estimates of $a_0, a_1, \ldots, a_n$ where

$$a_k = \langle f, T_k \rangle \simeq A(fT_k).$$

To do this calculation, make the usual change of variables $x = \cos\theta$ so that

$$a_k = \frac{2}{\pi} \int_0^{\pi} f(\cos\theta) \cos(k\theta)\, d\theta,\ k = 0, 1, \ldots, n;$$

and use the iterated trapezoid rule to estimate each a_k. Test your program on $f(x) = e^{2x}$ and note that (R) can be used to evaluate $p^*(x)$ at any point x in $[-1, 1]$.

4.7 LINEAR TRANSFORMATIONS

As we have remarked before, we can regard an $(m \times n)$ matrix A as defining a function from R^n to R^m. That is, we can think of the relation $\mathbf{y} = A\mathbf{x}$ as assigning each vector $\mathbf{x}$ in R^n to a unique vector $\mathbf{y}$ in R^m. This aspect of matrices can be easily generalized to a vector-space setting. The essential feature of matrices in this context is that their action is governed by the two linearity properties.

$$\begin{aligned} A(\mathbf{x} + \mathbf{w}) &= A\mathbf{x} + A\mathbf{w}. \\ A(c\mathbf{x}) &= cA\mathbf{x}. \end{aligned} \tag{4.15}$$

To derive an analog appropriate to vector spaces, we are led to ask for a function that satisfies these two linearity properties. In Definition 4.8, we use the standard notation $T: U \to V$ to denote a function whose domain is U and whose range is contained in V—thus for each vector $\mathbf{u}$ in U, $T(\mathbf{u})$ is a unique vector in V.

DEFINITION 4.8.

Let U and V be vector spaces, and let T be a function from U to V, $T: U \to V$. We say that T is a ***linear transformation*** if for all $\mathbf{u}$ and $\mathbf{w}$ in U and all scalars a

$$\begin{aligned} T(\mathbf{u} + \mathbf{w}) &= T(\mathbf{u}) + T(\mathbf{w}), \text{ and} \\ T(a\mathbf{u}) &= aT(\mathbf{u}). \end{aligned}$$

EXAMPLE 4.25 Let A be an $(m \times n)$ matrix, and let $T: R^n \to R^m$ be defined by $T(\mathbf{x}) = A\mathbf{x}$. Then T is a linear transformation; so Definition 4.8 is consistent with our earlier treatment of matrices. Some other examples of linear transformations are given below, with the verification left to the exercises.

EXAMPLE 4.26 If $T: \mathscr{P}_2 \to R^1$ is defined by $T(p) = p(2)$, then T is a linear transformation. [For example if $p(x) = x^2 - 3x + 1$, then $T(p) = -1$.] In general if W is any subspace of $C[a, b]$, and if α is any number in $[a, b]$, then the function $T: W \to R^1$ defined by $T(f) = f(\alpha)$ is a linear transformation. (In this context we are regarding the set of real numbers as a vector space.)

EXAMPLE 4.27 Let V be a vector space, and let $B = \{\boldsymbol{v}_1, \boldsymbol{v}_2, \ldots, \boldsymbol{v}_p\}$ be a basis for V. If we define $T: V \to R^p$ by $T(\boldsymbol{v}) = [\boldsymbol{v}]_B$ where $[\boldsymbol{v}]_B$ is the coordinate vector of $\boldsymbol{v}$ with respect to B, then T is a linear transformation.

EXAMPLE 4.28 Let V be a vector space, and let $T: V \to V$ be given by $T(\boldsymbol{v}) = \boldsymbol{v}$. This linear transformation is called the ***identity*** transformation.

EXAMPLE 4.29 Let U and V be vector spaces, and let $\boldsymbol{\theta}_V$ denote the zero vector in V. The linear transformation defined by $T(\mathbf{u}) = \boldsymbol{\theta}_V$ for all $\mathbf{u}$ in U is called the ***zero*** transformation.

EXAMPLE 4.30 Let $T: C[0, 1] \to R^1$ be defined by $T(f) = \int_0^1 f(t)\,dt$; then T is a linear transformation.

EXAMPLE 4.31 Let $\phi(x)$ be a fixed function in $C[a, b]$ where $\phi(x) > 0$, $a \leqslant x \leqslant b$, and let $T: C[a, b] \to C[a, b]$ be given by $T(f) = g$ where $g(x) = \phi(x)f(x)$ for $a \leqslant x \leqslant b$. Again T is a linear transformation.

EXAMPLE 4.32 Let $C^1[0, 1]$ denote the set of all functions that have a continuous first derivative in $[0, 1]$. (Note that $C^1[0, 1]$ is a subspace of $C[0, 1]$.) Let $a(x)$ be a fixed function in $C[0, 1]$, and let $T: C^1[0, 1] \to C[0, 1]$ be defined by $T(f) = f' + af$. [For example if $a(x) = x^2$ and $f(x) = \sin x$, then $T(f)$ is the function $T(f) = \cos x + x^2 \sin x$.] This linear transformation is an example of a differential operator. We will return to differential operators later and only mention here that the name "operator" is traditional in the study of differential equations. Operator is another term for function or transformation, and we could equally well speak of T as a "differential transformation."

One of the important features of the two linearity conditions in Definition 4.8 is that if $T: U \to V$ is a linear transformation and if U is a finite-dimensional vector space, then the action of T on U is completely determined by the action of T on a basis for U. To see why this statement is true, suppose U has a basis $B = \{\mathbf{u}_1, \mathbf{u}_2, \ldots, \mathbf{u}_p\}$. Then given any $\mathbf{u}$ in U, we know that $\mathbf{u}$ can be expressed uniquely as

$$\mathbf{u} = a_1\mathbf{u}_1 + a_2\mathbf{u}_2 + \ldots + a_p\mathbf{u}_p.$$

From this expression it follows that $T(\mathbf{u})$ is given by

$$T(\mathbf{u}) = T(a_1\mathbf{u}_1 + a_2\mathbf{u}_2 + \ldots + a_p\mathbf{u}_p) = a_1T(\mathbf{u}_1) + a_2T(\mathbf{u}_2) + \ldots + a_pT(\mathbf{u}_p). \tag{4.16}$$

Clearly (4.16) shows that if we know the vectors $T(\mathbf{u}_1)$, $T(\mathbf{u}_2), \ldots, T(\mathbf{u}_p)$, then we know $T(\mathbf{u})$ for any $\mathbf{u}$ in U; T is completely determined once T is defined on the basis. This situation is reminiscent of that with respect to matrices. In particular suppose A is an $(m \times n)$ matrix with $A = [\mathbf{A}_1, \mathbf{A}_2, \ldots, \mathbf{A}_n]$. Given any $\mathbf{x}$ in R^n, we can express $\mathbf{x}$ in terms of the natural basis $B = \{\mathbf{e}_1, \mathbf{e}_2, \ldots, \mathbf{e}_n\}$ as $\mathbf{x} = x_1\mathbf{e}_1 + x_2\mathbf{e}_2 + \ldots + x_n\mathbf{e}_n$. Thus the product $A\mathbf{x}$ is given by $A\mathbf{x} = x_1A\mathbf{e}_1 + x_2A\mathbf{e}_2 + \ldots + x_nA\mathbf{e}_n$; and as we know, $A\mathbf{e}_i = \mathbf{A}_i$. Therefore we are led to the familiar relationship $A\mathbf{x} = x_1\mathbf{A}_1 + x_2\mathbf{A}_2 + \ldots + x_n\mathbf{A}_n$.

Linear transformations have many of the same properties as those of the function defined by $A\mathbf{x}$. In order to summarize some of these, we make several definitions. Let $T: U \to V$ be a linear transformation, and for clarity let us denote the zero vectors in U and V as $\boldsymbol{\theta}_U$ and $\boldsymbol{\theta}_V$, respectively. The ***null space*** (or ***kernel***) of T, denoted by $\mathcal{N}(T)$, is the subset of U defined by

$$\mathcal{N}(T) = \{\mathbf{u}: T(\mathbf{u}) = \boldsymbol{\theta}_V\}.$$

The ***range*** of T, denoted by $\mathcal{R}(T)$, is the subset of V defined by

$$\mathcal{R}(T) = \{\boldsymbol{v}: \boldsymbol{v} = T(\mathbf{u}) \text{ for some } \mathbf{u} \text{ in } U\}.$$

Finally we say a linear transformation is ***one to one*** if $T(\mathbf{u}) = T(\mathbf{w})$ implies $\mathbf{u} = \mathbf{w}$ for all $\mathbf{u}$ and $\mathbf{w}$ in U. (The idea of one-to-one transformations corresponds to nonsingularity with respect to matrices. We will be able to show that if T is one to one, then T has an inverse, the inverse is also a linear transformation, solutions to an equation $T(\mathbf{x}) = \boldsymbol{v}$ are unique, etc.) Some of the elementary properties of linear transformations are given in the theorem below.

Theorem 4.14

Let $T: U \to V$ be a linear transformation. Then

1. $T(\boldsymbol{\theta}_U) = \boldsymbol{\theta}_V$,
2. $\mathcal{N}(T)$ is a subspace of U,
3. $\mathcal{R}(T)$ is a subspace of V, and
4. T is one to one if and only if $\mathcal{N}(T) = \{\boldsymbol{\theta}_U\}$.

Proof. We prove only (1) and leave the other parts as an exercise. To prove (1), we need only note that $0\boldsymbol{\theta}_U = \boldsymbol{\theta}_U$; so

$$T(\boldsymbol{\theta}_U) = T(0\boldsymbol{\theta}_U) = 0T(\boldsymbol{\theta}_U) = \boldsymbol{\theta}_V. \qquad \blacksquare$$

[When $T: R^n \to R^m$ is given by $T(\mathbf{x}) = A\mathbf{x}$ with A an $(m \times n)$ matrix, then this theorem is already familiar. That is, $\mathcal{N}(T)$ is the null space of A, $\mathcal{R}(T)$ is the range space of A, and $A\mathbf{x} = \mathbf{b}$ has a unique solution if and only if $A\mathbf{x} = \boldsymbol{\theta}$ has a unique solution.] If $T: U \to V$ is a linear transformation and if U is finite dimensional, then we can deduce some additional properties of T.

Theorem 4.15

Let $T: U \to V$ be a linear transformation, and let U be p-dimensional where $B = \{\mathbf{u}_1, \mathbf{u}_2, \ldots, \mathbf{u}_p\}$ is a basis for U.

1. $\mathscr{R}(T) = \mathrm{Sp}\{T(\mathbf{u}_1), T(\mathbf{u}_2), \ldots, T(\mathbf{u}_p)\}$.
2. T is one to one if and only if $\{T(\mathbf{u}_1), T(\mathbf{u}_2), \ldots, T(\mathbf{u}_p)\}$ is linearly independent in V.
3. $\dim(\mathscr{N}(T)) + \dim(\mathscr{R}(T)) = p$.

Proof. Part (1) is immediate from (4.16). That is, if $\mathbf{v}$ is in $\mathscr{R}(T)$, then $\mathbf{v} = T(\mathbf{u})$ for some $\mathbf{u}$ in U. But B is a basis for U; so $\mathbf{u}$ is of the form $\mathbf{u} = a_1\mathbf{u}_1 + a_2\mathbf{u}_2 + \ldots + a_p\mathbf{u}_p$; and hence $T(\mathbf{u}) = \mathbf{v} = a_1T(\mathbf{u}_1) + a_2T(\mathbf{u}_2) + \ldots + a_pT(\mathbf{u}_p)$. Therefore $\mathbf{v}$ is in $\mathrm{Sp}\{T(\mathbf{u}_1), T(\mathbf{u}_2), \ldots, T(\mathbf{u}_p)\}$.

To prove part (2),we can use part (4) of Theorem 4.14: T is one to one if and only if $\boldsymbol{\theta}_U$ is the only vector in $\mathscr{N}(T)$. In particular let us suppose that $\mathbf{u}$ is some vector in $\mathscr{N}(T)$ where $\mathbf{u} = b_1\mathbf{u}_1 + b_2\mathbf{u}_2 + \ldots + b_p\mathbf{u}_p$. Then $T(\mathbf{u}) = \boldsymbol{\theta}_V$, or

$$b_1T(\mathbf{u}_1) + b_2T(\mathbf{u}_2) + \ldots + b_pT(\mathbf{u}_p) = \boldsymbol{\theta}_V. \tag{4.17}$$

If $\{T(\mathbf{u}_1), T(\mathbf{u}_2), \ldots, T(\mathbf{u}_p)\}$ is a linearly independent set in V, then the only scalars satisfying (4.17) are $b_1 = b_2 = \ldots = b_p = 0$. Therefore $\mathbf{u}$ must be $\boldsymbol{\theta}_U$; so T is one to one. On the other hand if T is one to one, then there cannot be a nontrivial solution to (4.17); for if there were, $\mathscr{N}(T)$ would contain some nonzero vector $\mathbf{u}$.

To prove part (3), we first note that $0 \leqslant \dim(\mathscr{R}(T)) \leqslant p$ by part (1). We leave the two extreme cases, $\dim(\mathscr{R}(T)) = p$ and $\dim(\mathscr{R}(T)) = 0$, to the exercises and consider only $0 < \dim(\mathscr{R}(T)) < p$. [Note that $\dim(\mathscr{R}(T)) < p$ implies that $\dim(\mathscr{N}(T)) \geqslant 1$ since T is not one to one. We mention this point because we will need to choose a basis for $\mathscr{N}(T)$ below.]

It is conventional to let $r = \dim(\mathscr{R}(T))$; so let us suppose $\mathscr{R}(T)$ has a basis of r vectors, $\{\mathbf{v}_1, \mathbf{v}_2, \ldots, \mathbf{v}_r\}$. From the definition of of $\mathscr{R}(T)$ we know there are vectors $\mathbf{w}_1, \mathbf{w}_2, \ldots, \mathbf{w}_r$ in U such that

$$T(\mathbf{w}_i) = \mathbf{v}_i, \ 1 \leqslant i \leqslant r. \tag{4.18}$$

Now it is easy to show that $\{\mathbf{w}_1, \mathbf{w}_2, \ldots, \mathbf{w}_r\}$ is a linearly independent set in U. Next let us suppose that $\dim(\mathscr{N}(T)) = k$ and that $\{\mathbf{x}_1, \mathbf{x}_2, \ldots, \mathbf{x}_k\}$ is a basis for $\mathscr{N}(T)$. We now show that the set

$$Q = \{\mathbf{x}_1, \mathbf{x}_2, \ldots, \mathbf{x}_k, \mathbf{w}_1, \mathbf{w}_2, \ldots, \mathbf{w}_r\}$$

is a basis for U [therefore $k + r = p$, which proves part (3).]

We first establish that Q is a linearly independent set in U by considering

$$c_1\mathbf{x}_1 + c_2\mathbf{x}_2 + \ldots + c_k\mathbf{x}_k + a_1\mathbf{w}_1 + a_2\mathbf{w}_2 + \ldots + a_r\mathbf{w}_r = \boldsymbol{\theta}_U. \tag{4.19}$$

Applying T to both sides of (4.19), and recalling (4.18) and that the vectors $\mathbf{x}_i$ are in $\mathscr{N}(T)$, we obtain

$$\boldsymbol{\theta} + \boldsymbol{\theta} + \ldots + \boldsymbol{\theta} + a_1\mathbf{v}_1 + a_2\mathbf{v}_2 + \ldots + a_r\mathbf{v}_r = \boldsymbol{\theta}_V.$$

But since $\{\boldsymbol{v}_1, \boldsymbol{v}_2, \ldots, \boldsymbol{v}_r\}$ is linearly independent in V, we know that we must have $a_1 = a_2 = \ldots = a_r = 0$ in (4.19). Given this equality, and given that $\{\mathbf{x}_1, \mathbf{x}_2, \ldots, \mathbf{x}_k\}$ is a linearly independent set in U, we know also that we must have $c_1 = c_2 = \ldots = c_k = 0$ if (4.19) holds. Therefore Q is a linearly independent set.

To complete the argument, we need to show that Q is a spanning set for U. So let $\mathbf{u}$ be any vector in U. Then $\boldsymbol{v} = T(\mathbf{u})$ is a vector in $\mathscr{R}(T)$; so

$$T(\mathbf{u}) = b_1\boldsymbol{v}_1 + b_2\boldsymbol{v}_2 + \ldots + b_r\boldsymbol{v}_r.$$

Consider an associated vector $\mathbf{x}$ in U where $\mathbf{x}$ is defined by

$$\mathbf{x} = b_1\mathbf{w}_1 + b_2\mathbf{w}_2 + \ldots + b_r\mathbf{w}_r. \tag{4.20}$$

We observe that $T(\mathbf{u} - \mathbf{x}) = \boldsymbol{\theta}_V$; so obviously $\mathbf{u} - \mathbf{x}$ is in $\mathscr{N}(T)$ and can be written as

$$\mathbf{u} - \mathbf{x} = d_1\mathbf{x}_1 + d_2\mathbf{x}_2 + \ldots + d_k\mathbf{x}_k. \tag{4.21}$$

Placing $\mathbf{x}$ on the right-hand side of (4.21) and using (4.20), we have shown that $\mathbf{u}$ is a linear combination of vectors in Q. Thus Q is a basis for U, and part (3) is proved since $k + r$ must equal p. ■

The usual terminology is that the ***nullity*** of T is $\dim(\mathscr{N}(T))$ and the ***rank*** of T is $\dim(\mathscr{R}(T))$. Thus Theorem 4.15 asserts that when $T: U \to V$, and U is a finite-dimensional vector space, then nullity plus rank equals $\dim(U)$. The same terminology is used extensively in matrix theory; but we can say a bit more about the transformation $R^n \to R^m$, $T(\mathbf{x}) = A\mathbf{x}$ where A is an $(m \times n)$ matrix. That is, if $A = [\mathbf{A}_1, \mathbf{A}_2, \ldots, \mathbf{A}_n]$, then a spanning set for the range is the set of column vectors of A; so the rank of A is equal to the number of linearly independent columns of A while the nullity of A is (by Theorem 4.15) equal to n minus the rank of A. We see also that it is now easy to prove a theorem mentioned in Example 1.23: $A\mathbf{x} = \mathbf{b}$ is consistent if and only if the rank of A is equal to the rank of $[A \,\vdots\, \mathbf{b}]$ (see Exercise 12, Section 4.7). In addition the special structure inherent in matrix theory allows us also to state the following theorem.

Theorem

If $A = (a_{ij})$ is an $(m \times n)$ matrix, then the rank of A is equal to the rank of A^{T}.

Proof. (Before sketching a proof, we note that the columns of A^{T} correspond to rows of A; so the theorem asserts that the number of linearly independent columns in A is equal to the number of linearly independent rows in A, or "row rank" = "column rank.")

Let $\{\mathbf{R}_1, \mathbf{R}_2, \ldots, \mathbf{R}_m\}$ be the row vectors of A; so $\mathbf{R}_i = [a_{i1}, a_{i2}, \ldots, a_{in}]$, $1 \leqslant i \leqslant m$. Let $W = \mathrm{Sp}\{\mathbf{R}_1, \mathbf{R}_2, \ldots, \mathbf{R}_m\}$; then W has a basis $\{\mathbf{w}_1, \mathbf{w}_2, \ldots, \mathbf{w}_k\}$ where $k \leqslant m$, and where each $\mathbf{w}_j$ is a $(1 \times n)$ vector. Writing each $\mathbf{R}_i$ in terms of this basis yields

$$\begin{aligned}
[a_{11}, a_{12}, \ldots, a_{1n}] &= \mathbf{R}_1 = c_{11}\mathbf{w}_1 + c_{12}\mathbf{w}_2 + \ldots + c_{1k}\mathbf{w}_k \\
[a_{21}, a_{22}, \ldots, a_{2n}] &= \mathbf{R}_2 = c_{21}\mathbf{w}_1 + c_{22}\mathbf{w}_2 + \ldots + c_{2k}\mathbf{w}_k \\
&\ \ \vdots \\
[a_{m1}, a_{m2}, \ldots, a_{mn}] &= \mathbf{R}_m = c_{m1}\mathbf{w}_1 + c_{m2}\mathbf{w}_2 + \ldots + c_{mk}\mathbf{w}_k.
\end{aligned}$$

For a fixed j, $1 \leqslant j \leqslant n$, let w_{ij} denote the jth component of $\mathbf{w}_i$. Equating the jth component of the left- and right-hand sides yields

$$\begin{aligned} a_{1j} &= c_{11}w_{1j} + c_{12}w_{2j} + \ldots + c_{1k}w_{kj} \\ a_{2j} &= c_{21}w_{1j} + c_{22}w_{2j} + \ldots + c_{2k}w_{kj} \\ &\vdots \\ a_{mj} &= c_{m1}w_{1j} + c_{m2}w_{2j} + \ldots + c_{mk}w_{kj}, \end{aligned}$$

or

$$\mathbf{A}_j = \begin{bmatrix} a_{1j} \\ a_{2j} \\ \vdots \\ a_{mj} \end{bmatrix} = w_{1j}\begin{bmatrix} c_{11} \\ c_{21} \\ \vdots \\ c_{m1} \end{bmatrix} + w_{2j}\begin{bmatrix} c_{12} \\ c_{22} \\ \vdots \\ c_{m2} \end{bmatrix} + \ldots + w_{kj}\begin{bmatrix} c_{1k} \\ c_{2k} \\ \vdots \\ c_{mk} \end{bmatrix}, \quad 1 \leqslant j \leqslant n.$$

Since each column vector $\mathbf{A}_j$, $1 \leqslant j \leqslant n$, can be written as a linear combination of k vectors as above, and since $\mathscr{R}(A) = \mathrm{Sp}\{\mathbf{A}_1, \mathbf{A}_2, \ldots, \mathbf{A}_n\}$, then $\dim(\mathscr{R}(A)) \leqslant k = \dim(\mathscr{R}(A^{\mathrm{T}}))$. We can repeat this argument interchanging the roles of A and A^{T} and get $\dim(\mathscr{R}(A^{\mathrm{T}})) \leqslant \dim(\mathscr{R}(A))$. ■

To illustrate Theorem 4.15 in a setting other than R^n, we consider the following example.

EXAMPLE 4.33 Let T be the linear transformation defined on $\mathscr{P}_4$ by $T(p) = p''(x)$. Thus if p is given by $p(x) = a_0 + a_1x + a_2x^2 + a_3x^3 + a_4x^4$, then

$$T(p) = 2a_2 + 6a_3x + 12a_4x^2;$$

so $T: \mathscr{P}_4 \to \mathscr{P}_2$. In particular $\{T(x^2), T(x^3), T(x^4)\}$ is a basis for $\mathscr{R}(T)$; so $\dim(\mathscr{R}(T)) = 3$; and in fact $\mathscr{R}(T) = \mathscr{P}_2$. Since $\dim(\mathscr{N}(T)) + 3 = 5$ by part (3) of Theorem 4.15, $\mathscr{N}(T)$ has dimension 2. A basis for $\mathscr{N}(T)$ is clearly given by $\{1, x\}$.

Note that T is not one to one [since $\mathscr{N}(T)$ contains nonzero vectors]. As an illustration, $T(x^2) = 2$, $T(x^2 + 1) = 2$, $T(x^2 + 4x) = 2$, and in general $T(x^2 + ax + b) = 2$.

4.7 EXERCISES

1. Prove that T in Example 4.26 is a linear transformation.
2. Prove that T in each of Examples 4.28 and 4.29 is a linear transformation.
3. Prove that T in Example 4.27 is a linear transformation.
4. Prove that T in Example 4.30 is a linear transformation.
5. Prove that T in Example 4.31 is a linear transformation.
6. Prove that T in Example 4.32 is a linear transformation.
7. Suppose that $T: \mathscr{P}_2 \to \mathscr{P}_4$ is a linear transformation where $T(1) = x^4$, $T(x+1) = x^3 - 2x$, $T((x+1)^2) = x$. Find $T(p)$ and $T(q)$ where $p(x) = x^2 + 5x - 1$ and $q(x) = x^2 + 9x + 5$.

[T(p)] = [P] [?]

8. Suppose that $T: R^3 \leftrightarrow R^4$ is a linear transformation. Find $T(\mathbf{u})$ and $T(\mathbf{v})$ if you know the following.

$$T(\mathbf{e}_1) = \begin{bmatrix} 2 \\ 1 \\ 3 \\ 2 \end{bmatrix}, \quad T(\mathbf{e}_2) = \begin{bmatrix} 1 \\ 1 \\ 0 \\ 0 \end{bmatrix}, \quad T(\mathbf{e}_3) = \begin{bmatrix} 0 \\ 1 \\ 2 \\ 1 \end{bmatrix}, \quad \mathbf{u} = \begin{bmatrix} 2 \\ 1 \\ 5 \end{bmatrix}, \quad \mathbf{v} = \begin{bmatrix} 1 \\ 1 \\ 1 \end{bmatrix}.$$

9. Identify $\mathcal{N}(T)$ for each of the transformations in Exercises 1, 2, and 3. Is T one to one?
10. Which of the transformations in Exercises 4, 5, and 6 are one to one?
11. Prove parts (2), (3), and (4) of Theorem 4.14.
12. Let A be an $(m \times n)$ matrix, and let $\mathbf{b}$ be a vector in R^m. Prove that $A\mathbf{x} = \mathbf{b}$ is consistent if and only if the rank of A is equal to the rank of $[A \,\vdots\, \mathbf{b}]$.
13. Suppose that $T: \mathcal{P}_4 \to \mathcal{P}_2$ is a linear transformation. Enumerate the various possibilities for $\dim(\mathcal{R}(T))$ and $\dim(\mathcal{N}(T))$. Can T possibly be one to one?
14. Let $T: U \to V$ be a linear transformation, and let U be finite dimensional. Prove that if $\dim(U) > \dim(V)$, then T cannot be one to one.
15. Identify $\mathcal{R}(T)$ for each of the transformations in Exercises 1, 2, 3, 4, and 5.
16. Consider $T: C^1[0, 1] \to C[0, 1]$ given by $T(f) = f' + 2xf$. Find $\mathcal{N}(T)$ and verify that the nullity of T is 1. Let g be the function defined by $g(x) = 3x$. Find all solutions to $T(f) = g$.

4.8 OPERATIONS WITH LINEAR TRANSFORMATIONS

We know that there is a useful arithmetic structure associated with matrices—matrices can be added and multiplied, nonsingular matrices have inverses, etc. Much of this structure is available also for linear transformations; for our explanation we will need some definitions. Let U and V be vector spaces, and let T_1 and T_2 be linear transformations where $T_1: U \to V$ and $T_2: U \to V$. By the ***sum*** $T_3 = T_1 + T_2$ we mean the function $T_3: U \to V$ where $T_3(\mathbf{u}) = T_1(\mathbf{u}) + T_2(\mathbf{u})$ for all $\mathbf{u}$ in U. By the multiple aT_1 where a is a scalar, we mean the function $aT_1: U \to V$ where $aT_1(\mathbf{u}) = a(T_1(\mathbf{u}))$ for all $\mathbf{u}$ in U. It is easy to see that $T_1 + T_2$ and aT_1 are also linear transformations. If U, V, and W are vector spaces, and if S and T are linear transformations where $S: U \to V$ and $T: V \to W$, then we define the ***composition*** $L = T \circ S$ to be the function $L: U \to W$ where $L(\mathbf{u}) = T(S(\mathbf{u}))$ for all $\mathbf{u}$ in U. Again it is easy to verify that $T \circ S$ is a linear transformation. [In this context, matrix products correspond to composition since $(AB)\mathbf{x} = A(B\mathbf{x})$.]

EXAMPLE 4.34 Let $T_1: \mathcal{P}_4 \to \mathcal{P}_2$ be given by $T_1(p) = p''(x)$, and let $T_2: \mathcal{P}_2 \to \mathcal{P}_1$ be given by $T_2(q) = xq(1)$. For example, $T_1(x^4 - x^2 + 1) = 12x^2 - 2$ and $T_2(12x^2 - 2) = 10x$. Technically, the sum $T_1 + T_2$ is not defined since T_1 and T_2 have different domains and ranges; but the composition $L = T_2 \circ T_1$ is defined. In particular, $L: \mathcal{P}_4 \to \mathcal{P}_1$ and $L(p) = T_2(T_1(p)) = xp''(1)$. [This situation resembles the matrix situation in which an $(m \times n)$ matrix and an $(n \times r)$ matrix cannot be added unless $m = n$ and $n = r$, but the product can be formed.]

If we alter the domain and range designations of T_2 such that $T_2: \mathscr{P}_4 \to \mathscr{P}_2$, we can form the sum $T_1 + T_2$. We again define the operation of T_2 as $T_2(p) = xp(1)$. Then for $S = T_1 + T_2$, $S: \mathscr{P}_4 \to \mathscr{P}_2$ and $S(p) = p''(x) + xp(1)$. Since $\mathscr{P}_2$ is a subspace of $\mathscr{P}_4$, we can designate the ranges of T_1 and T_2 to be in $\mathscr{P}_4$; that is, $T_1 : \mathscr{P}_4 \to \mathscr{P}_4$ and $T_2 : \mathscr{P}_4 \to \mathscr{P}_4$. Under this designation both compositions $T_1 \circ T_2$ and $T_2 \circ T_1$ are defined. Then we have

$$(T_2 \circ T_1) = T_2(T_1(p)) = xp''(1)$$
$$(T_1 \circ T_2) = T_1(T_2(p)) = \theta(x) \qquad \text{(the zero polynomial).}$$

Therefore just as we saw with matrix multiplication, we should expect that $T_2 \circ T_1$ and $T_1 \circ T_2$ are different transformations. (In fact $T_1 \circ T_2$ is the zero transformation

---

To introduce the inverse of a linear transformation, let us first suppose X and Y are any sets, and f is a function $f: X \to Y$; and suppose $\mathscr{R}(f)$ denotes the range of f where $\mathscr{R}(f) \subseteq Y$. If $\mathscr{R}(f) = Y$, we say that f is a function from X ***onto*** Y. If f is one to one, we recall that we can define an ***inverse*** function, f^{-1}, where $f^{-1}: \mathscr{R}(f) \to X$ and f^{-1} is defined by $f^{-1}(y) = x$ if and only if $f(x) = y$. Similarly if $T: U \to V$ is a linear transformation, then T^{-1} can be defined on $\mathscr{R}(T)$ whenever T is one to one. [T is frequently called ***invertible*** when T is one to one and $V = \mathscr{R}(T)$.]

Theorem 4.16

Let U and V be vector spaces, and let $T: U \to V$ be a linear transformation. If T is one to one, then

1. $T^{-1}: \mathscr{R}(T) \to U$ is a linear transformation;
2. T^{-1} is one to one, and $(T^{-1})^{-1} = T$; and
3. $T^{-1} \circ T = I_U$ where I_U is the identity transformation on U.

Proof. For part (1) we need to show that the function $T^{-1}: \mathscr{R}(T) \to U$ satisfies Definition 4.8. Suppose $\mathbf{v}_1$ and $\mathbf{v}_2$ are vectors in $\mathscr{R}(T)$. Then there are vectors $\mathbf{u}_1$ and $\mathbf{u}_2$ in U such that $T(\mathbf{u}_1) = \mathbf{v}_1$ and $T(\mathbf{u}_2) = \mathbf{v}_2$. Equivalently

$$T^{-1}(\mathbf{v}_1) = \mathbf{u}_1 \qquad \text{and} \qquad T^{-1}(\mathbf{v}_2) = \mathbf{u}_2. \tag{4.22}$$

Furthermore $\mathbf{v}_1 + \mathbf{v}_2 = T(\mathbf{u}_1) + T(\mathbf{u}_2) = T(\mathbf{u}_1 + \mathbf{u}_2)$ means $T^{-1}(\mathbf{v}_1 + \mathbf{v}_2) = \mathbf{u}_1 + \mathbf{u}_2$. Using $T^{-1}(\mathbf{v}_1 + \mathbf{v}_2) = \mathbf{u}_1 + \mathbf{u}_2$ and (4.22), we see that $T^{-1}(\mathbf{v}_1 + \mathbf{v}_2) = T^{-1}(\mathbf{v}_1) + T^{-1}(\mathbf{v}_2)$. Next showing that $T^{-1}(c\mathbf{v}) = cT^{-1}(\mathbf{v})$ for all $\mathbf{v}$ in $\mathscr{R}(T)$ is equally easy and is left to the exercises as are parts (2) and (3). ■

EXAMPLE 4.35 Recall from Example 4.33 that $T: \mathscr{P}_4 \to \mathscr{P}_2$ where $T(p) = p''(x)$ was not a one-to-one transformation. We now see that we can change the domain of T to make this operator one to one. For example consider the subspace U of $C[0, 1]$ given by

$$U = \{f(x): f(x) = a \sin x + b \cos x + cx^2\},$$

and suppose $T: U \to C[0, 1]$ is given by $T(f) = f''(x)$. Now a basis for U is $\{\sin x, \cos x, x^2\}$; so a spanning set for $\mathscr{R}(T)$ is obtained by applying T to each vector in the basis for U. In particular $\mathscr{R}(T) = \text{Sp}\{-\sin x, -\cos x, 2\}$; and since these vectors are linearly independent in $\mathscr{R}(T)$, it follows that T is one to one (see Theorem 4.15). It is easy to obtain T^{-1} where $T^{-1}: \mathscr{R}(T) \to U$; and if $h(x) = -q \sin x - r \cos x + 2s$ is in $\mathscr{R}(T)$, then $T^{-1}(h) = q \sin x + r \cos x + sx^2$.

As might be guessed from the corresponding theorems for nonsingular matrices, other properties of invertible transformations can be established. For instance if S and T are invertible, then so is $S \circ T$; and we can show that $(S \circ T)^{-1} = T^{-1} \circ S^{-1}$. Also a companion result for part (3) in Theorem 4.16 is that if I_R denotes the identity transformation of $\mathscr{R}(T)$, then $T \circ T^{-1} = I_R$. As final example if $\mathbf{b}$ is in $\mathscr{R}(T)$ and if T is invertible, then $\mathbf{x} = T^{-1}(\mathbf{b})$ is the unique solution of $T(\mathbf{x}) = \mathbf{b}$.

Invertible transformations can serve also to introduce the quality of being isomorphic. In particular if U and V are vector spaces, and if $T: U \to V$ is one to one and onto, then U and V are said to be ***isomorphic*** vector spaces and T is called an isomorphism. If two vector spaces are isomorphic, they may be regarded as being indistinguishable (or equivalent) algebraically even though their elements may be different. It is easy to show that if U is isomorphic to V, and if V is isomorphic to W, then U and W are also isomorphic. In this respect we can show that all real n-dimensional vector spaces are isomorphic, and hence every real n-dimensional vector space has (essentially) the same algebraic character as R^n.

Theorem 4.17

If U is a real n-dimensional vector space, then U and R^n are isomorphic.

Proof. To prove this theorem, we need only exhibit the isomorphism; and a coordinate system on U will provide the means. Let $B = \{\mathbf{u}_1, \mathbf{u}_2, \ldots, \mathbf{u}_n\}$ be a basis for U, and let $T: U \to R^n$ be the linear transformation defined by

$$T(\mathbf{u}) = [\mathbf{u}]_B.$$

Since B is a basis, $\boldsymbol{\theta}_U$ is the only vector in $\mathscr{N}(T)$; and therefore T is one to one. Furthermore $T(\mathbf{u}_i)$ is the vector $\mathbf{e}_i$ in R^n; so $\mathscr{R}(T) = \text{Sp}\{T(\mathbf{u}_1), T(\mathbf{u}_2), \ldots, T(\mathbf{u}_n)\} = \text{Sp}\{\mathbf{e}_1, \mathbf{e}_2, \ldots, \mathbf{e}_n\} = R^n$. Hence T is one to one and onto. ■

To explain what we mean by the n-dimensional vector space U having the same algebraic character as R^n, we note that if $\mathbf{u} = a_1\mathbf{u}_1 + a_2\mathbf{u}_2 + \ldots + a_n\mathbf{u}_n$ and $\mathbf{w} = b_1\mathbf{u}_1 + b_2\mathbf{u}_2 + \ldots + b_n\mathbf{u}_n$, then the addition of $\mathbf{u}$ and $\mathbf{w}$ and the scalar multiplication $a\mathbf{u}$ are essentially the same as the addition of $[\mathbf{u}]_B$ and $[\mathbf{w}]_B$ and the scalar multiplication $a[\mathbf{u}]_B$ in R^n, respectively. That is, $[\mathbf{u} + \mathbf{w}]_B = [\mathbf{u}]_B + [\mathbf{w}]_B$, and $[a\mathbf{u}]_B = a[\mathbf{u}]_B$. (When we add $\mathbf{u}$ and $\mathbf{v}$ "coordinatewise" using the addition operation in U, we can think of this procedure as being the same as adding $[\mathbf{u}]_B$ and $[\mathbf{v}]_B$ in R^n.) As a specific example let $p(x) = 2 - 3x + x^2$ and $q(x) = -3 - x - 2x^2$ in $\mathscr{P}_2$. With the basis $B = \{1, x, x^2\}$ then

$$p(x)+q(x) = -1-4x-x^2 \qquad \text{while}$$

$$[p]_B+[q]_B = \begin{bmatrix} 2 \\ -3 \\ 1 \end{bmatrix} + \begin{bmatrix} -3 \\ -1 \\ -2 \end{bmatrix} = \begin{bmatrix} -1 \\ -4 \\ -1 \end{bmatrix} = [p+q]_B.$$

With the basis $Q = \{x-1, x+2, x^2-1\}$, $p(x) = -3(x-1)+(x^2-1)$ and $q(x) = (x-1)-2(x+2)-2(x^2-1)$; so $p(x)+q(x) = -2(x-1)-2(x+2)-(x^2-1)$ while

$$[p]_Q+[q]_Q = \begin{bmatrix} -3 \\ 0 \\ 1 \end{bmatrix} + \begin{bmatrix} 1 \\ -2 \\ -2 \end{bmatrix} = \begin{bmatrix} -2 \\ -2 \\ -1 \end{bmatrix} = [p+q]_Q.$$

An immediate corollary of Theorem 4.17 follows.

Corollary

If U and V are real n-dimensional vector spaces, then U and V are isomorphic.

EXAMPLE 4.36 Since $\dim(\mathscr{P}_n) = n+1$, it follows from Theorem 4.17 that $\mathscr{P}_n$ is isomorphic to R^{n+1}.

EXAMPLE 4.37 Let U and V be vector spaces, and let

$$L(U, V) = \{T: T \text{ is a linear transformation from } U \text{ to } V\}.$$

[$L(U, V)$ is the set of all possible linear transformations from U to V.] If T_1 and T_2 are in $L(U, V)$, then $T_1 + T_2$ and aT_1 are also linear transformations from U to V; that is, $T_1 + T_2$ and aT_1 belong to $L(U, V)$. Therefore the two closure conditions in Definition 4.1 are verified. It is easy to see that the other eight conditions are also valid for $L(U, V)$, and hence $L(U, V)$ is a vector space whose elements are linear transformations. This result is analogous to the result that the set of all $(m \times n)$ matrices is a vector space. In fact in Section 4.9 we show that if $\dim(U) = n$ and $\dim(V) = m$, then $L(U, V)$ is isomorphic to the vector space of all $(m \times n)$ matrices.

4.8 EXERCISES

1. Suppose that $S: U \to V$ and $T: U \to V$ are linear transformations. Prove that $Q = S + T$ is a linear transformation from U to V.
2. Suppose that $T: U \to V$ is a linear transformation, and suppose that a is a scalar. Prove that $Q = aT$ is a linear transformation from U to V.
3. Let $T: \mathscr{P}_3 \to \mathscr{P}_4$ be defined by $T(p) = q$ where $q(x) = (x+2)p(x)$, and let $S: \mathscr{P}_3 \to \mathscr{P}_4$ be defined by $S(p) = p'(0)$. Calculate the transformation $T+S$. Let $H: \mathscr{P}_4 \to \mathscr{P}_3$ be given by $H(p) = r$ where $r(x) = p'(x) + p(0)$. Calculate $H \circ T$ and $S \circ (H \circ T)$. Verify that T is one to one. Identify $\mathscr{R}(T)$ and exhibit T^{-1} where $T^{-1}: \mathscr{R}(T) \to \mathscr{P}_3$.

4. Find a basis for $\mathscr{R}(T)$ where T is as in Exercise 3. Can T^{-1} be extended in a meaningful way to all of $\mathscr{P}_4$?
5. Find $\mathscr{N}(H)$, $\mathscr{R}(H)$; and calculate the nullity and rank of H where H is defined in Exercise 3.
6. The functions e^x, e^{2x}, e^{3x} are linearly independent in $C[0, 1]$. Let V be the subspace of $C[0, 1]$ defined by $V = \mathrm{Sp}\{e^x, e^{2x}, e^{3x}\}$, and let $T: V \to V$ be given by $T(p) = p'(x)$. Show T is invertible, and calculate $T^{-1}(e^x)$, $T^{-1}(e^{2x})$, $T^{-1}(e^{3x})$. What is $T^{-1}(ae^x + be^{2x} + ce^{3x})$?
7. Let $T: \mathscr{P}_2 \to \mathscr{P}_2$ be given by $T(p) = p'(x)$. Show that T is not one to one, and T is not onto.
8. Suppose $S: V \to U$ and $T: U \to W$ are linear transformations. Show that $T \circ S$ is a linear transformation where $T \circ S: V \to W$.
9. Suppose $S: V \to U$ and $T: U \to W$, and suppose S is one to one and T is one to one. Prove that $T \circ S$ is one to one where $T \circ S: V \to W$. If $\mathscr{R}(S) = U$ and $\mathscr{R}(T) = W$, show that $T \circ S$ is invertible and that $(T \circ S)^{-1} = S^{-1} \circ T^{-1}$.
10. Recall that U and V are isomorphic if there is a linear transformation T where $T: U \to V$, T is one to one, and $V = \mathscr{R}(T)$. Prove that if U is isomorphic to V, and V is isomorphic to W, then U is isomorphic to W.
11. A set of transformations $L(U, V)$ is defined in Example 4.37. Prove that $L(U, V)$ is a vector space.
12. Exhibit an isomorphism between $\mathscr{P}_3$ and the set of all real (2×2) matrices. Exhibit an isomorphism between $\mathscr{P}_2$ and the set of all real (2×2) symmetric matrices.
13. Complete the proof of Theorem 4.16.

4.9 MATRIX REPRESENTATIONS FOR LINEAR TRANSFORMATIONS

In Example 4.37 we mentioned that the vector space $L(U, V)$, the set of all linear transformations from U to V, is isomorphic to the vector space of all $(m \times n)$ matrices when $\dim(U) = n$ and $\dim(V) = m$. While we have not proved this yet, the result should not be totally unexpected. Moreover the existence of such an isomorphism implies that there is a sort of equivalence between linear transformations and matrices.

Suppose that $T: U \to V$ is a linear transformation where $\dim(U) = n$, $\dim(V) = m$. Let $B = \{\mathbf{u}_1, \mathbf{u}_2, \ldots, \mathbf{u}_n\}$ be a basis for U, and let $C = \{\boldsymbol{v}_1, \boldsymbol{v}_2, \ldots, \boldsymbol{v}_m\}$ be a basis for V. As we know, the action of T on U is completely determined by the action of T on the basis B. So to describe T, let us represent the vectors $T(\mathbf{u}_1)$, $T(\mathbf{u}_2), \ldots, T(\mathbf{u}_n)$ in terms of the basis C for V:

basis

$$\begin{aligned} T(\mathbf{u}_1) &= q_{11}\boldsymbol{v}_1 + q_{21}\boldsymbol{v}_2 + \ldots + q_{m1}\boldsymbol{v}_m \\ T(\mathbf{u}_2) &= q_{12}\boldsymbol{v}_1 + q_{22}\boldsymbol{v}_2 + \ldots + q_{m2}\boldsymbol{v}_m \\ &\vdots \\ T(\mathbf{u}_n) &= q_{1n}\boldsymbol{v}_1 + q_{2n}\boldsymbol{v}_2 + \ldots + q_{mn}\boldsymbol{v}_m. \end{aligned} \tag{4.23}$$

In terms of the coordinate system defined on V by C, we see that (4.23) means

$$[T(\mathbf{u}_1)]_C = \begin{bmatrix} q_{11} \\ q_{21} \\ \vdots \\ q_{m1} \end{bmatrix}, \quad [T(\mathbf{u}_2)]_C = \begin{bmatrix} q_{12} \\ q_{22} \\ \vdots \\ q_{m2} \end{bmatrix}, \ldots, \quad [T(\mathbf{u}_n)]_C = \begin{bmatrix} q_{1n} \\ q_{2n} \\ \vdots \\ q_{mn} \end{bmatrix}.$$

Next if $\mathbf{u}$ is a vector in U, $\mathbf{u} = a_1\mathbf{u}_1 + a_2\mathbf{u}_2 + \ldots + a_n\mathbf{u}_n$, then

$$T(\mathbf{u}) = a_1T(\mathbf{u}_1) + a_2T(\mathbf{u}_2) + \ldots + a_nT(\mathbf{u}_n); \tag{4.24a}$$

and we note that

$$[T(\mathbf{u})]_C = a_1[T(\mathbf{u}_1)]_C + a_2[T(\mathbf{u}_2)]_C + \ldots + a_n[T(\mathbf{u}_n)]_C. \tag{4.24b}$$

Thus with respect to the coordinate system on V if (4.24a) holds, then

$$[T(\mathbf{u})]_C = a_1\mathbf{Q}_1 + a_2\mathbf{Q}_2 + \ldots + a_n\mathbf{Q}_n \tag{4.25}$$

where

$$\mathbf{Q}_i = [T(\mathbf{u}_i)]_C = \begin{bmatrix} q_{1i} \\ q_{2i} \\ \vdots \\ q_{mi} \end{bmatrix}, \quad 1 \leqslant i \leqslant n$$

Now the vectors $\mathbf{Q}_1, \mathbf{Q}_2, \ldots, \mathbf{Q}_n$ in (4.25) are in R^m; so (4.25) is just the product $Q\mathbf{a}$ where Q is the $(m \times n)$ matrix $Q = [\mathbf{Q}_1, \mathbf{Q}_2, \ldots, \mathbf{Q}_n]$ and $\mathbf{a}$ is the coordinate vector for $\mathbf{u}$,

$$\mathbf{a} = [\mathbf{u}]_B = \begin{bmatrix} a_1 \\ a_2 \\ \vdots \\ a_n \end{bmatrix}.$$

In particular we can write (4.25) precisely as

$$[T(\mathbf{u})]_C = Q[\mathbf{u}]_B. \tag{4.26a}$$

From (4.26a) it is clear that we can find the vector $\mathbf{v}$ such that $\mathbf{v} = T(\mathbf{u})$ by simply multiplying Q times the coordinate vector $[\mathbf{u}]_B$ and then reading off the coordinates of $\mathbf{v}$ with respect to C.

If $T: U \to V$ is any linear transformation where $\dim(U) = n$ and $\dim(V) = m$, then we have shown that there is an $(m \times n)$ matrix Q that "models" this transformation:

$$Q = [[T(\mathbf{u}_1)]_C, [T(\mathbf{u}_2)]_C, \ldots, [T(\mathbf{u}_n)]_C]. \tag{4.26b}$$

If we change the basis for U and/or the basis for V, the matrix Q may also change. However for fixed bases B and C it is easy to show that Q is unique. That is, if

$[T(\mathbf{u})]_C = Q_1[\mathbf{u}]_B$ and $[T(\mathbf{u})]_C = Q_2[\mathbf{u}]_B$ for all $\mathbf{u}$ in U, then $Q_1 = Q_2$ (see Exercise 4, Section 4.9). In this regard we call the matrix Q given in (4.26b) the ***matrix representation*** for T with respect to the bases B and C. An interesting special case of (4.26a) is provided when T is the identity transformation on U; that is $T: U \to U$ and $T(\mathbf{u}) = \mathbf{u}$ for all $\mathbf{u}$ in U. Then (4.26a) becomes $[\mathbf{u}]_C = Q[\mathbf{u}]_B$ and this relates the coordinates of $\mathbf{u}$ with respect to B to the coordinates of $\mathbf{u}$ with respect to C. (See Theorem 4.18.)

EXAMPLE 4.38 Consider the differential operator given by $T(f) = x^2 f'' - 2f' + xf$. In this example we will regard T as a linear transformation $T: \mathscr{P}_2 \to \mathscr{P}_3$ and find its representation as a (4×3) matrix [note that $\dim(\mathscr{P}_2) = 3$, $\dim(\mathscr{P}_3) = 4$]. We choose the natural bases $B = \{1, x, x^2\}$ and $C = \{1, x, x^2, x^3\}$ for $\mathscr{P}_2$ and $\mathscr{P}_3$, respectively. So to construct the (4×3) matrix Q that represents T, we need to find the coordinate vectors of $T(1)$, $T(x)$, and $T(x^2)$ with respect to C. We calculate that $T(1) = x$, $T(x) = x^2 - 2$, $T(x^2) = x^3 + 2x^2 - 4x$; so the coordinate vectors of $T(1)$, $T(x)$, $T(x^2)$ are

$$[T(1)]_C = \begin{bmatrix} 0 \\ 1 \\ 0 \\ 0 \end{bmatrix}, \qquad [T(x)]_C = \begin{bmatrix} -2 \\ 0 \\ 1 \\ 0 \end{bmatrix}, \qquad [T(x^2)]_C = \begin{bmatrix} 0 \\ -4 \\ 2 \\ 1 \end{bmatrix}.$$

Thus the matrix representation for T is the (4×3) matrix

$$Q = \begin{bmatrix} 0 & -2 & 0 \\ 1 & 0 & -4 \\ 0 & 1 & 2 \\ 0 & 0 & 1 \end{bmatrix}.$$

As an illustration, let $p(x) = x^2 - 3x + 1$. Then $T(p) = 2x^2 - 2(2x - 3) + (x^3 - 3x^2 + x) = x^3 - x^2 - 3x + 6$. To see the operation of (4.26a), note that $[p]_B$ and $Q[p]_B$ are given by

$$[p]_B = \begin{bmatrix} 1 \\ -3 \\ 1 \end{bmatrix} \quad \text{and} \quad Q[p]_B = \begin{bmatrix} 6 \\ -3 \\ -1 \\ 1 \end{bmatrix} = [T(p)]_C$$

Thus $T(p)$ is the vector in $\mathscr{P}_3$ whose coordinates with respect to the basis C are given by $Q[p]_B$. [With $Q[p]_B$ as listed above, it follows that $T(p) = 6 - 3x - x^2 + x^3$, which agrees with the direct calculation.]

EXAMPLE 4.39 Let A be an $(m \times n)$ matrix, and consider the linear transformation $T: R^n \to R^m$ defined by $T(\mathbf{x}) = A\mathbf{x}$. With respect to the natural bases for R^n and R^m the matrix representation of T is A.

EXAMPLE 4.40 Let $T: U \to V$ and $S: V \to W$ be linear transformations, and suppose

$\dim(U) = n$, $\dim(V) = m$, $\dim(W) = k$. Let matrix representations of T and S be Q and P, respectively [Q is $(m \times n)$ and P is $(k \times m)$]. Then the matrix representation of $S \circ T$ is PQ.

EXAMPLE 4.41 Let T_1 and T_2 be linear transformations where $T_1: U \rightarrow V$, $T_2: U \rightarrow V$, $\dim(U) = n$, $\dim(V) = m$. Suppose Q_1 and Q_2 are matrix representations for T_1 and T_2, respectively. Then the matrix representation for $T_1 + T_2$ is $Q_1 + Q_2$, and the matrix representation for aT_1 is aQ_1.

EXAMPLE 4.42 Let U and V be vector spaces with $\dim(U) = n$ and $\dim(V) = m$. Furthermore let R_{mn} denote the vector space of $(m \times n)$ matrices. Fix bases B and C for U and V, and define a function $\psi: L(U, V) \rightarrow R_{mn}$ where for T in $L(U, V)$, $\psi(T)$ is the matrix representation for T. It is not hard to show that ψ is an isomorphism. In particular ψ is a well-defined function since each T in $L(U, V)$ has a unique matrix representation. From Example 4.41 it also follows that ψ is a linear transformation; so to show that ψ is an isomorphism, we need show only that ψ is one to one and onto. Thus (see Exercise 15, Section 4.9) $L(U, V)$ and R_{mn} are isomorphic vector spaces.

When V is an n-dimensional vector space, then $L(V, V)$ is isomorphic to the vector space of $(n \times n)$ matrices. For $T: V \rightarrow V$ we usually choose the same basis B for both the domain and the range of T, and we refer to the representation as the matrix of T with respect to B. The following results should not be surprising.

EXAMPLE 4.43 Suppose that $T: V \rightarrow V$ is a linear transformation where $\dim(V) = n$, and let Q be the matrix representation for T with respect to a basis B. If T is invertible, then Q is nonsingular; and furthermore the matrix representation for T^{-1} is Q^{-1}. The matrix representation for the identity transformation on V, I_V, is the $(n \times n)$ identity matrix I. The matrix representation for the zero transformation on V is the $(n \times n)$ zero matrix. (Observe that the identity and the zero transformations always have the same matrix representations, regardless of what basis we choose for V. Thus changing the basis for V may change the matrix representation for T or may leave the representation unchanged.)

4.9 EXERCISES

1. Find the matrix representations for T, S, and H in Exercise 3, Section 4.8, with respect to the natural bases for $\mathscr{P}_3$ and $\mathscr{P}_4$.
2. Calculate the matrix representations for $H \circ T$ and $S \circ (H \circ T)$ where T, S, and H are as in Exercise 1. Relate these representations to those in Exercise 1 (see Example 4.40).
3. Find the matrix representations for T and T^{-1} where T is as in Exercise 6, Section 4.8. How are these two representations related? (See Example 4.43.)
4. Let U and V be finite-dimensional vector spaces, and let $T: U \rightarrow V$ be a linear

transformation. Let B be a basis for U, and C a basis for V, and prove that the matrix representation for T with respect to these bases is unique. (Hint: Suppose that $[T(\mathbf{u})]_C = Q_1[\mathbf{u}]_B$ and $[T(\mathbf{u})]_C = Q_2[\mathbf{u}]_B$ for all $\mathbf{u}$ in U. What if $\mathbf{u}$ is a basis vector?)

* 5. Let $S: \mathscr{P}_2 \to \mathscr{P}_3$ be given by $S(p) = x^3p'' - x^2p' + 3p$. Find the matrix representation of S with respect to the natural bases $B = \{1, x, x^2\}$ for $\mathscr{P}_2$ and $C = \{1, x, x^2, x^3\}$ for $\mathscr{P}_3$.

6. Let S be the transformation in Exercise 5, let the basis for $\mathscr{P}_2$ be $Q = \{x+1, x+2, x^2\}$, and let the basis for $\mathscr{P}_3$ be $C = \{1, x, x^2, x^3\}$. Find the representation for S.

* 7. Let $T: \mathscr{P}_2 \to R^3$ be given by

$$T(p) = \begin{bmatrix} p(0) \\ 3p'(1) \\ p'(1) + p''(0) \end{bmatrix}.$$

Find the representation of T with respect to the natural bases for $\mathscr{P}_2$ and R^3.

8. Find the representation for the transformation in Exercise 7 with respect to the natural basis for $\mathscr{P}_2$ and the basis $\{\mathbf{u}_1, \mathbf{u}_2, \mathbf{u}_3\}$ for R^3 where

$$\mathbf{u}_1 = \begin{bmatrix} 1 \\ 0 \\ 1 \end{bmatrix} \quad \mathbf{u}_2 = \begin{bmatrix} 0 \\ 1 \\ 1 \end{bmatrix} \quad \mathbf{u}_3 = \begin{bmatrix} 1 \\ 1 \\ 1 \end{bmatrix}.$$

9. Let $T: V \to V$ be a linear transformation where $B = \{\mathbf{v}_1, \mathbf{v}_2, \mathbf{v}_3, \mathbf{v}_4\}$ is a basis for V. Find matrix representation of T with respect to V if $T(\mathbf{v}_1) = \mathbf{v}_2$, $T(\mathbf{v}_2) = \mathbf{v}_3$, $T(\mathbf{v}_3) = \mathbf{v}_1 + \mathbf{v}_2$ and $T(\mathbf{v}_4) = \mathbf{v}_1 + 3\mathbf{v}_4$.

10. Let $T: R^2 \to R^3$ be given by $T(\mathbf{x}) = A\mathbf{x}$ where

$$A = \begin{bmatrix} 1 & 2 & 1 \\ 3 & 0 & 4 \end{bmatrix}.$$

Find the representation of T with respect to the natural bases for R^2 and R^3.

11. Prove the assertion in Example 4.39 (a special case is in Exercise 10).

12. Prove the assertion in Example 4.40 (a special case is in Exercise 2).

13. Let $S: \mathscr{P}_2 \to \mathscr{P}_3$ be the transformation in Exercise 5, and let $T: \mathscr{P}_2 \to \mathscr{P}_3$ be the transformation in Example 4.38. Let $L: \mathscr{P}_2 \to \mathscr{P}_3$ be the transformation $L = S + T$, and find the representation of L with respect to the natural bases for $\mathscr{P}_2$ and $\mathscr{P}_3$.

14. Prove the assertion in Example 4.41 (a special case is in Exercise 13).

15. Prove the assertion in Example 4.42. [Note that if $S: U \to V$ and $T: U \to V$ are transformations, then $S = T$ if $S(\mathbf{u}) = T(\mathbf{u})$ for all $\mathbf{u}$ in U. To show that ψ is onto, choose any $(m \times n)$ matrix A, and construct a linear transformation L from the action of A on $\{\mathbf{e}_1, \mathbf{e}_2, \ldots, \mathbf{e}_n\}$.]

16. Prove the assertions in Example 4.43.

4.10 CHANGE OF BASIS AND DIAGONALIZATION

In Section 4.9 we saw that a linear transformation in $L(U, V)$ could be represented as an $(m \times n)$ matrix when $\dim(U) = n$, $\dim(V) = m$. A consequence of this representation is that the properties of transformations can be studied by examining their corresponding matrix representations. Moreover we have a great deal of machinery in place for matrix theory; so matrices will provide a suitable analytical and computational framework for studying a linear transformation. To simplify matters somewhat, we consider only transformations from V to V where $\dim(V) = n$. So let $T: V \to V$ be a linear transformation, and suppose Q is the matrix representation for T with respect to a basis B; that is, if $\mathbf{w} = T(\mathbf{u})$, then $[\mathbf{w}]_B = Q[\mathbf{u}]_B$. As we know, when we change the basis B for V, we may well change the matrix representation for T. If we are interested in the properties of T, then it is reasonable to search for a basis for V that makes the matrix representation for T as simple as possible. Finding such a basis is the subject of this section.

A particularly nice matrix to deal with computationally is a diagonal matrix. If $T: V \to V$ is a linear transformation whose matrix representation with respect to B is a diagonal matrix

$$D = \begin{bmatrix} d_1 & 0 & 0 & \dots & 0 \\ 0 & d_2 & 0 & \dots & 0 \\ 0 & 0 & d_3 & \dots & 0 \\ \vdots & & & & \vdots \\ 0 & 0 & 0 & \dots & d_n \end{bmatrix},$$

then it is easy to analyze the action of T on V. In particular suppose $B = \{\mathbf{v}_1, \mathbf{v}_2, \dots, \mathbf{v}_n\}$, and suppose D is the matrix representation for T; so $[T(\mathbf{u})]_B = D[\mathbf{u}]_B$ for all $\mathbf{u}$ in V. If we note that $[\mathbf{v}_i]_B = \mathbf{e}_i$, then it follows from $[T(\mathbf{v}_i)]_B = D[\mathbf{v}_i]_B$ that $[T(\mathbf{v}_i)]_B = D\mathbf{e}_i$, or $[T(\mathbf{v}_i)]_B = d_i\mathbf{e}_i$. From interpreting this last equality in terms of the coordinate system on V, it follows that

$$T(\mathbf{v}_i) = d_i\mathbf{v}_i.$$

In terms of the action of T on V suppose $\mathbf{u}$ is a vector in V:

$$\mathbf{u} = a_1\mathbf{v}_1 + a_2\mathbf{v}_2 + \dots + a_n\mathbf{v}_n.$$

We immediately see that

$$T(\mathbf{u}) = a_1T(\mathbf{v}_1) + a_2T(\mathbf{v}_2) + \dots + a_nT(\mathbf{v}_n);$$

and since $T(\mathbf{v}_i) = d_i\mathbf{v}_i$, we have

$$T(\mathbf{u}) = a_1(d_1\mathbf{v}_1) + a_2(d_2\mathbf{v}_2) + \dots + a_n(d_n\mathbf{v}_n).$$

If T is a linear transformation that has a matrix representation that is diagonal, then T is called ***diagonalizable***. Unfortunately not every linear transformation is diagonalizable. Equivalently if $T: V \to V$ is a linear transformation, it may be that no matter what basis we choose for V, we never obtain a matrix representation for T that is diagonal.

In order to see how the matrix representations for T vary as the basis for V is changed, we state Theorem 4.18 below. We wish to emphasize that Theorem 4.18 has several purposes. First of all as the statement of the theorem indicates, we can use Theorem 4.18 to relate the coordinates of any vector $\mathbf{v}$ with respect to one basis B to the coordinates of $\boldsymbol{v}$ with respect to another basis C. Second we will use Theorem 4.18 as a tool to prove that all matrix representations for a transformation T are similar. That is, as we vary the basis for V, we produce matrix representations for T that are similar matrices. Since we know how to determine whether a matrix is similar to a diagonal matrix, we will be able to say when T is diagonalizable.

Theorem 4.18 (change of basis)

Let V be a vector space, and let $B = \{\mathbf{u}_1, \mathbf{u}_2, \ldots, \mathbf{u}_n\}$ and $C = \{\mathbf{w}_1, \mathbf{w}_2, \ldots, \mathbf{w}_n\}$ be bases for V. Then there is a nonsingular $(n \times n)$ matrix P such that for all $\boldsymbol{v}$ in V

$$[\boldsymbol{v}]_C = P[\boldsymbol{v}]_B. \tag{4.27}$$

Moreover the matrix P is given by $P = [\mathbf{P}_1, \mathbf{P}_2, \ldots, \mathbf{P}_n]$ where the ith column of P is $\mathbf{P}_i = [\mathbf{u}_i]_C$, $1 \leqslant i \leqslant n$.

The matrix P in (4.27) is called the ***transition*** matrix and shows how the coordinate vector of $\boldsymbol{v}$ with respect to C is related to the coordinate vector of $\boldsymbol{v}$ with respect to B for any $\boldsymbol{v}$ in V. The proof of Theorem 4.18 is easy and is based on (4.26); see the remarks following (4.26a) and (4.26b). Once it is shown that P is nonsingular, then we also have the relationship

$$[\boldsymbol{v}]_B = P^{-1}[\boldsymbol{v}]_C.$$

EXAMPLE 4.44 We have seen before that two bases for $\mathscr{P}_2$ are $B = \{1, x, x^2\}$ and $C = \{1, x+1, (x+1)^2\}$. To construct the transition matrix changing coordinates with respect to B into coordinates with respect to C, we follow Theorem 4.18 and determine the coordinates of 1, x, and x^2 in terms of 1, $x+1$, and $(x+1)^2$. This determination is easy, and we find

$$\begin{aligned} 1 &= 1 \\ x &= (x+1) - 1 \\ x^2 &= (x+1)^2 - 2(x+1) + 1. \end{aligned}$$

Thus with respect to C the coordinate vectors of B are

$$[1]_C = \begin{bmatrix} 1 \\ 0 \\ 0 \end{bmatrix}, \qquad [x]_C = \begin{bmatrix} -1 \\ 1 \\ 0 \end{bmatrix}, \qquad [x^2]_C = \begin{bmatrix} 1 \\ -2 \\ 1 \end{bmatrix}.$$

The transition matrix P is therefore

$$P = \begin{bmatrix} 1 & -1 & 1 \\ 0 & 1 & -2 \\ 0 & 0 & 1 \end{bmatrix}.$$

In particular any polynomial $q(x) = a_0 + a_1x + a_2x^2$ can be expressed in terms of 1, $x+1$, and $(x+1)^2$ by forming $[q]_C = P[q]_B$. Forming this, we find

$$[q]_C = \begin{bmatrix} a_0 - a_1 + a_2 \\ a_1 - 2a_2 \\ a_2 \end{bmatrix}.$$

So with respect to C we can write $q(x)$ as $q(x) = (a_0 - a_1 + a_2) + (a_1 - 2a_2)(x+1) + a_2(x+1)^2$ [a result which we can verify directly by multiplying out the new expression for $q(x)$].

Now let $T: V \to V$ be a linear transformation, and suppose Q_1 is the matrix representation for T with respect to a basis B; that is, $[T(\mathbf{u})]_B = Q_1[\mathbf{u}]_B$. We are interested in seeing how changing the basis for V will change the representation for T. To see this, let C be a new basis for V, and let Q_2 be the matrix representation for T with respect to C; so $[T(\mathbf{u})]_C = Q_2[\mathbf{u}]_C$. Next let P be the transition matrix from C to B so that $[\boldsymbol{v}]_B = P[\boldsymbol{v}]_C$ for all $\boldsymbol{v}$ in V. Collecting these details, let us suppose that $T(\mathbf{u}) = \mathbf{w}$. Then

$$[\mathbf{w}]_B = Q_1[\mathbf{u}]_B \qquad [\mathbf{w}]_C = Q_2[\mathbf{u}]_C$$
$$[\mathbf{w}]_B = P[\mathbf{w}]_C \qquad [\mathbf{u}]_B = P[\mathbf{u}]_C.$$

Our objective is to find a relation between the matrix representations Q_1 and Q_2 by using the four equations above. We first note that $[\mathbf{w}]_B = Q_1[\mathbf{u}]_B$ and $[\mathbf{w}]_B = P[\mathbf{w}]_C$, and so $P[\mathbf{w}]_C = Q_1[\mathbf{u}]_B$. Since $[\mathbf{u}]_B = P[\mathbf{u}]_C$, this expression becomes $P[\mathbf{w}]_C = Q_1 P[\mathbf{u}]_C$, or

$$[\mathbf{w}]_C = P^{-1}Q_1 P[\mathbf{u}]_C.$$

Now since the matrix representation is unique, we know that $Q_2 = P^{-1}Q_1P$. In other words Q_1 and Q_2 are similar whenever Q_1 and Q_2 are matrix representations for the same linear transformation T. Moreover the similarity transformation $Q_2 = P^{-1}Q_1P$ uses the transition matrix P, which transforms coordinates with respect to C to coordinates with respect to B.

This result suggests the following idea: if we are given a linear transformation $T: V \to V$ and the matrix representation, Q_1, for T with respect to a given basis B, then we should look for a "simple" matrix R (diagonal if possible) that is similar to Q_1, $R = S^{-1}Q_1S$. In this case we can use S^{-1} as a transition matrix to obtain a new basis C for V where $[\mathbf{u}]_C = S^{-1}[\mathbf{u}]_B$. With respect to the basis C, T will have the matrix representation R where $R = S^{-1}Q_1S$.

Given the transition matrix S^{-1}, it is an easy matter to find the actual basis vectors of C. In particular suppose that $B = \{\mathbf{u}_1, \mathbf{u}_2, \ldots, \mathbf{u}_n\}$ is the given basis for V, and we wish to find the vectors in $C = \{\boldsymbol{v}_1, \boldsymbol{v}_2, \ldots, \boldsymbol{v}_n\}$. Since $[\mathbf{u}]_C = S^{-1}[\mathbf{u}]_B$ for all $\mathbf{u}$ in V, we know that $S[\mathbf{u}]_C = [\mathbf{u}]_B$. Moreover with respect to C, $[\boldsymbol{v}_i]_C = \mathbf{e}_i$. So from $S[\boldsymbol{v}_i]_C = [\boldsymbol{v}_i]_B$ we obtain

$$S\mathbf{e}_i = [\boldsymbol{v}_i]_B, \quad 1 \leqslant i \leqslant n. \tag{4.28}$$

But if $S = [\mathbf{S}_1, \mathbf{S}_2, \ldots, \mathbf{S}_n]$, then $S\mathbf{e}_i = \mathbf{S}_i$ so that (4.28) tells us that the coordinate vector of $\boldsymbol{v}_i$ with respect to the known basis B is the ith column of S.

EXAMPLE 4.45 Consider the differential operator $T: \mathscr{P}_2 \to \mathscr{P}_2$ given by $T(p) = x^2p'' + (2x-1)p' + 3p$. With respect to the basis $B = \{1, x, x^2\}$, T has the matrix representation

$$Q_1 = \begin{bmatrix} 3 & -1 & 0 \\ 0 & 5 & -2 \\ 0 & 0 & 9 \end{bmatrix}.$$

Since Q_1 is triangular, we see that the eigenvalues are 3, 5, and 9; and since Q_1 has distinct eigenvalues, Q_1 can be diagonalized where (see Theorem 3.9) the matrix S of eigenvectors will diagonalize Q_1.

We calculate the eigenvectors $\mathbf{u}_1, \mathbf{u}_2, \mathbf{u}_3$ for Q_1 and form $S = [\mathbf{u}_1, \mathbf{u}_2, \mathbf{u}_3]$, which yields

$$S = \begin{bmatrix} 1 & 1 & 1 \\ 0 & -2 & -6 \\ 0 & 0 & 12 \end{bmatrix}.$$

In this case it follows that

$$S^{-1}Q_1S = \begin{bmatrix} 3 & 0 & 0 \\ 0 & 5 & 0 \\ 0 & 0 & 9 \end{bmatrix} = R.$$

In view of our remarks above, R is the matrix representation for T with respect to the basis $C = \{v_1, v_2, v_3\}$ where $[v_i]_B = \mathbf{S}_i$, or

$$[v_1]_B = \begin{bmatrix} 1 \\ 0 \\ 0 \end{bmatrix}, \qquad [v_2]_B = \begin{bmatrix} 1 \\ -2 \\ 0 \end{bmatrix}, \qquad [v_3]_B = \begin{bmatrix} 1 \\ -6 \\ 12 \end{bmatrix}.$$

Therefore the basis C is given precisely as $C = \{1, 1-2x, 1-6x+12x^2\}$. Moreover it is easy to see that $T(v_1) = 3v_1$, $T(v_2) = 5v_2$, $T(v_3) = 9v_3$ where $v_1 = 1$, $v_2 = 1-2x$, $v_3 = 1-6x+12x^2$.

4.10 EXERCISES

1. Prove Theorem 4.18 for $n = 2$ by verifying that $[\mathbf{v}]_C = P[\mathbf{v}]_B$ when $P = [\mathbf{P}_1, \mathbf{P}_2]$, $\mathbf{P}_1 = [\mathbf{u}_1]_C$, $\mathbf{P}_2 = [\mathbf{u}_2]_C$. Start with $\mathbf{v} = a_1\mathbf{u}_1 + a_2\mathbf{u}_2$, and express $\mathbf{u}_1$ and $\mathbf{u}_2$ in terms of $\mathbf{w}_1$ and $\mathbf{w}_2$ as $\mathbf{u}_1 = p_{11}\mathbf{w}_1 + p_{21}\mathbf{w}_2$, $\mathbf{u}_2 = p_{12}\mathbf{w}_1 + p_{22}\mathbf{w}_2$.
2. Find the transition matrix for R^2 when $B = \{\mathbf{u}_1, \mathbf{u}_2\}$, $C = \{\mathbf{w}_1, \mathbf{w}_2\}$,
$$\mathbf{w}_1 = \begin{bmatrix} 2 \\ 1 \end{bmatrix}, \quad \mathbf{w}_2 = \begin{bmatrix} 1 \\ 2 \end{bmatrix}, \quad \mathbf{u}_1 = \begin{bmatrix} 1 \\ 1 \end{bmatrix}, \quad \mathbf{u}_2 = \begin{bmatrix} 3 \\ 1 \end{bmatrix}.$$
3. Repeat Exercise 2 for the basis vectors
$$\mathbf{w}_1 = \begin{bmatrix} 4 \\ 3 \end{bmatrix}, \quad \mathbf{w}_2 = \begin{bmatrix} 2 \\ 3 \end{bmatrix}, \quad \mathbf{u}_1 = \begin{bmatrix} 4 \\ 1 \end{bmatrix}, \quad \mathbf{u}_2 = \begin{bmatrix} 2 \\ 1 \end{bmatrix}.$$

4. Let B be the natural basis for R^2, and let C be the basis in Exercise 2. Find the transition matrix, and represent the following vectors in terms of C.

$$\mathbf{a} = \begin{bmatrix} 4 \\ 2 \end{bmatrix} \quad \mathbf{b} = \begin{bmatrix} -2 \\ 0 \end{bmatrix} \quad \mathbf{c} = \begin{bmatrix} 9 \\ 5 \end{bmatrix} \quad \mathbf{d} = \begin{bmatrix} -5 \\ -1 \end{bmatrix}$$

5. Repeat Exercise 4 when C is the basis in Exercise 3. Represent the following in terms of C.

$$\mathbf{a} = \begin{bmatrix} 2 \\ 0 \end{bmatrix} \quad \mathbf{b} = \begin{bmatrix} 6 \\ -1 \end{bmatrix} \quad \mathbf{c} = \begin{bmatrix} 8 \\ 3 \end{bmatrix} \quad \mathbf{d} = \begin{bmatrix} 6 \\ 2 \end{bmatrix}$$

6. Using (4.26a) and (4.26b), prove Theorem 4.18.

7. Let $B = \{1, x, x^2, x^3\}$ and $C = \{x, x+1, x^2-2x, x^3+3\}$ be bases for $\mathscr{P}_3$. Find the transition matrix and use it to represent the following in terms of C.

$$p(x) = x^2 - 7x + 2 \qquad q(x) = x^3 + 9x - 1 \qquad r(x) = x^3 - 2x^2 + 6$$

8. Represent the following quadratic polynomials in the form $a_0 + a_1x + a_2x(x-1)$ by constructing the appropriate transition matrix.

$$p(x) = x^2 + 5x - 3 \qquad q(x) = 2x^2 - 6x + 8 \qquad r(x) = x^2 - 5$$

9. Let T: $\mathscr{P}_2 \to \mathscr{P}_2$ be given by $T(p) = xp'' + (x+1)p' + p$. Verify that T is diagonalizable, and find a basis C for which the matrix representation for T is diagonal. Let P be the transition matrix between $B = \{1, x, x^2\}$ and C, and use P to calculate $T(p)$ for the following.

a) $p(x) = x^2 + 7x - 8$ b) $p(x) = x^2 + 5$ c) $p(x) = 2x^2 - 3x + 4$

10. Repeat Exercise 9 for the transformation T: $\mathscr{P}_3 \to \mathscr{P}_3$ given by $T(p) = x^2p'' + (x-1)p' + p$. Calculate $T(p)$ for the following.

a) $p(x) = x^3 + 3x$ b) $p(x) = x^2 - 3x + 2$ c) $p(x) = 2x^3 - 7x + 8$

5

DETERMINANTS

5.1 INTRODUCTION

Determinants have played a major role in the historical development of matrix theory, and they possess a number of properties that are theoretically pleasing. For example in terms of linear algebra, determinants can be used to characterize nonsingular matrices, express solutions of nonsingular systems $A\mathbf{x} = \mathbf{b}$, and calculate the dimension of subspaces. In analysis, determinants are used to express vector cross products, to express the conversion factor (the Jacobian) when a change of variables is needed to evaluate a multiple integral, as a convenient test (the Wronskian) for linear independence of sets of functions, etc. We will explore the theory and some of the applications of determinants in this chapter.

5.2 COFACTOR EXPANSIONS OF DETERMINANTS

If A is an $(n \times n)$ matrix, the determinant of A, denoted $\det(A)$, is a number that we associate with A. Determinants are usually defined either in terms of "cofactors" or in terms of "permutations," and we elect to use the cofactor definition here. We begin with the definition of $\det(A)$ when A is a (2×2) matrix.

DEFINITION 5.1:

Let $A = (a_{ij})$ be a (2×2) matrix. The ***determinant*** of A is given by

$$\det(A) = a_{11}a_{22} - a_{12}a_{21}. \tag{5.1}$$

For notational purposes the determinant is often expressed using vertical bars:

$$\det(A) = \begin{vmatrix} a_{11} & a_{12} \\ a_{21} & a_{22} \end{vmatrix}. \tag{5.2}$$

EXAMPLE 5.1 For the matrices

$$A = \begin{bmatrix} 1 & 2 \\ -1 & 3 \end{bmatrix}, \quad B = \begin{bmatrix} 4 & 1 \\ 2 & 1 \end{bmatrix}, \quad \text{and } C = \begin{bmatrix} 3 & 4 \\ 6 & 8 \end{bmatrix},$$

we see that det $(A) = 5$, det $(B) = 2$, det $(C) = 0$. In the notation of (5.2)

$$\begin{vmatrix} 1 & 1 \\ 4 & 3 \end{vmatrix} = -1 \quad \text{and} \quad \begin{vmatrix} 2 & 2 \\ 3 & 5 \end{vmatrix} = 4.$$

We now define the determinant of an $(n \times n)$ matrix as a "weighted sum" of $((n-1) \times (n-1))$ determinants. It is convenient first to make a preliminary definition. Let $A = (a_{ij})$ be an $(n \times n)$ matrix, and let M_{rs} denote the $((n-1) \times (n-1))$ matrix obtained by deleting the rth row and sth column from A. The ***cofactor*** of the entry a_{rs} in A is the number A_{rs} defined by

$$A_{rs} = (-1)^{r+s} \det(M_{rs}).$$

* cofactor

[The number $\det(M_{rs})$ is called the ***minor*** of a_{rs}.]

minor

EXAMPLE 5.2 Let A be the (3×3) matrix

$$A = \begin{bmatrix} 1 & -1 & 2 \\ 2 & 3 & -3 \\ 4 & 5 & 1 \end{bmatrix}.$$

We calculate

$$A_{11} = \begin{vmatrix} 3 & -3 \\ 5 & 1 \end{vmatrix} = 18, \qquad A_{23} = -\begin{vmatrix} 1 & -1 \\ 4 & 5 \end{vmatrix} = -9, \qquad \text{and}$$

$$A_{32} = -\begin{vmatrix} 1 & 2 \\ 2 & -3 \end{vmatrix} = 7.$$

cofactors

We use cofactors in our definition of the determinant.

DEFINITION 5.2.

Let $A = (a_{ij})$ be an $(n \times n)$ matrix. Then the determinant of A is

$$\det(A) = a_{11}A_{11} + a_{12}A_{12} + \ldots + a_{1n}A_{1n}$$

where A_{1j} is the cofactor of a_{1j}, $1 \leqslant j \leqslant n$.

Determinants are defined only for *square* matrices. Note also the inductive nature of the definition. For example if A is (3×3), then det $(A) = a_{11}A_{11} + a_{12}A_{12} + a_{13}A_{13}$; and the cofactors A_{11}, A_{12}, A_{13} can be evaluated from Definition 5.1. Similarly the determinant of a (4×4) matrix is the sum of four (3×3) determinants where each (3×3) determinant is in turn the sum of three (2×2) determinants.

EXAMPLE 5.3 As a simple illustration of Definition 5.2, consider the (3×3) determinant

$$\begin{vmatrix} 3 & 2 & 1 \\ 2 & 1 & -3 \\ 4 & 0 & 1 \end{vmatrix} = 3\begin{vmatrix} 1 & -3 \\ 0 & 1 \end{vmatrix} - 2\begin{vmatrix} 2 & -3 \\ 4 & 1 \end{vmatrix} + \begin{vmatrix} 2 & 1 \\ 4 & 0 \end{vmatrix} = -29.$$

If A is (4×4), then $\det(A) = a_{11}A_{11} + a_{12}A_{12} + a_{13}A_{13} + a_{14}A_{14}$. In evaluating the cofactors A_{11}, A_{12}, A_{13}, and A_{14}, we need to evaluate four (3×3) determinants where each such evaluation is also done using Definition 5.2. For example

$$\det(A) = \begin{vmatrix} 1 & 2 & 0 & 2 \\ -1 & 2 & 3 & 1 \\ -3 & 2 & -1 & 0 \\ 2 & -3 & -2 & 1 \end{vmatrix} = A_{11} + 2A_{12} + 0A_{13} + 2A_{14}$$

where

$$A_{11} = \begin{vmatrix} 2 & 3 & 1 \\ 2 & -1 & 0 \\ -3 & -2 & 1 \end{vmatrix} = 2\begin{vmatrix} -1 & 0 \\ -2 & 1 \end{vmatrix} - 3\begin{vmatrix} 2 & 0 \\ -3 & 1 \end{vmatrix} + \begin{vmatrix} 2 & -1 \\ -3 & -2 \end{vmatrix} = -15$$

$$A_{12} = -\begin{vmatrix} -1 & 3 & 1 \\ -3 & -1 & 0 \\ 2 & -2 & 1 \end{vmatrix} = -\left(-\begin{vmatrix} -1 & 0 \\ -2 & 1 \end{vmatrix} - 3\begin{vmatrix} -3 & 0 \\ 2 & 1 \end{vmatrix} + \begin{vmatrix} -3 & -1 \\ 2 & -2 \end{vmatrix}\right)$$
$$= -18.$$

Similarly $A_{13} = 9$ and $A_{14} = -6$; so we calculate $\det(A) = 1(-15) + 2(-18) + 0(9) + 2(-6) = -63$.

Note in Example 5.3 that the calculation of the (4×4) determinant was simplified because of the zero entry in the $(1, 3)$ position. Clearly if we had some procedure for creating zero entries, we could simplify the computation of determinants since the cofactor of a zero entry need not be calculated. We will develop such simplifications in the next section.

EXAMPLE 5.4 Consider the determinant of the (4×4) lower-triangular matrix T:

$$\det(T) = \begin{vmatrix} 3 & 0 & 0 & 0 \\ 1 & 2 & 0 & 0 \\ 2 & 3 & 2 & 0 \\ 1 & 4 & 5 & 1 \end{vmatrix}.$$

Clearly det (T) is easy to calculate because of the zero entries:

$$\det(T) = 3\begin{vmatrix} 2 & 0 & 0 \\ 3 & 2 & 0 \\ 4 & 5 & 1 \end{vmatrix} = 3\cdot 2\begin{vmatrix} 2 & 0 \\ 5 & 1 \end{vmatrix} = 12.$$

Thus det (T) is the product of the diagonal entries (see Theorem 5.1 in Section 5.3).

5.2 EXERCISES

1. Calculate the determinants of the following.

$$A = \begin{bmatrix} 1 & 3 \\ 2 & 1 \end{bmatrix} \quad B = \begin{bmatrix} 6 & 7 \\ 7 & 3 \end{bmatrix} \quad C = \begin{bmatrix} 2 & 4 \\ 4 & 8 \end{bmatrix} \quad D = \begin{bmatrix} 1 & 3 \\ 0 & 2 \end{bmatrix}$$

2. Calculate the determinants of the following.

$$A = \begin{bmatrix} 4 & 3 \\ 1 & 7 \end{bmatrix} \quad B = \begin{bmatrix} 2 & -1 \\ 1 & 1 \end{bmatrix} \quad C = \begin{bmatrix} 4 & 1 \\ -2 & 1 \end{bmatrix} \quad D = \begin{bmatrix} 1 & 3 \\ 2 & 6 \end{bmatrix}$$

3. Calculate the cofactors A_{11}, A_{21}, A_{23}, and A_{33} for the following.

$$\text{a) } A = \begin{bmatrix} 1 & 2 & 1 \\ 0 & 1 & 3 \\ 2 & 1 & 1 \end{bmatrix} \quad \text{b) } A = \begin{bmatrix} 1 & 4 & 0 \\ 1 & 0 & 2 \\ 3 & 1 & 2 \end{bmatrix} \quad \text{c) } A = \begin{bmatrix} 2 & -1 & 3 \\ -1 & 2 & 2 \\ 3 & 2 & 1 \end{bmatrix}$$

4. Repeat Exercise 3 for the following.

$$\text{a) } A = \begin{bmatrix} 1 & 1 & 1 \\ 1 & 1 & 2 \\ 2 & 1 & 1 \end{bmatrix} \quad \text{b) } A = \begin{bmatrix} -1 & 1 & -1 \\ 2 & 1 & 0 \\ 0 & 1 & 3 \end{bmatrix} \quad \text{c) } A = \begin{bmatrix} 1 & 2 & 1 \\ 4 & 3 & 1 \\ 0 & 0 & 2 \end{bmatrix}$$

5. Calculate the determinants of the matrices in Exercise 3.

6. Calculate the determinants of the matrices in Exercise 4.

7. Calculate the determinants of the following.

$$A = \begin{bmatrix} 2 & 1 & -1 & 2 \\ 3 & 0 & 0 & 1 \\ 2 & 1 & 2 & 0 \\ 3 & 1 & 1 & 2 \end{bmatrix} \quad B = \begin{bmatrix} 1 & -1 & 1 & 2 \\ 1 & 0 & 1 & 3 \\ 0 & 0 & 2 & 4 \\ 1 & 1 & -1 & 1 \end{bmatrix}$$

8. Verify that det $(A) = 0$ when

$$A = \begin{bmatrix} 0 & a_{12} & a_{13} \\ 0 & a_{22} & a_{23} \\ 0 & a_{32} & a_{33} \end{bmatrix}.$$

9. Use the result of Exercise 8 to prove that if $U = (u_{ij})$ is a (4×4) upper-triangular matrix, then $\det(U) = u_{11}u_{22}u_{33}u_{44}$.

10. Prove that if $T = (t_{ij})$ is an $(n \times n)$ lower-triangular matrix where $2 \leqslant n \leqslant 4$, then $\det(T) = t_{11} \ldots t_{nn}$.

11. Calculate the determinants of the following.

$$A = \begin{bmatrix} 1 & 2 & 1 & 3 \\ 0 & 4 & 1 & 2 \\ 0 & 0 & 3 & 1 \\ 0 & 0 & 0 & 2 \end{bmatrix} \qquad B = \begin{bmatrix} 2 & 0 & 0 & 0 \\ 3 & 1 & 0 & 0 \\ 4 & 2 & 3 & 0 \\ 1 & 1 & 2 & 1 \end{bmatrix}$$

12. Verify that $\det(A^{\mathrm{T}}) = \det(A)$ for the matrices in Exercise 3. (This statement is true for all square matrices as we prove in Section 5.5.)

13. An $(n \times n)$ symmetric matrix A is called ***positive definite*** if $\mathbf{x}^{\mathrm{T}}A\mathbf{x} > 0$ for all $\mathbf{x}$ in R^n, $\mathbf{x} \neq \boldsymbol{\theta}$. Let A be a (2×2) symmetric matrix. Prove the following.

a) If A is positive definite, then $a_{11} > 0$ and $\det(A) > 0$.
b) If $a_{11} > 0$ and $\det(A) > 0$, then A is positive definite.
[Hint: For part (a), $\mathbf{x}^{\mathrm{T}}A\mathbf{x} = a_{11}$ when $\mathbf{x} = \mathbf{e}_1$. To show $\det(A) > 0$, let $\mathbf{x}$ be a vector in R^2 with components u and v. Calculate the number $\mathbf{x}^{\mathrm{T}}A\mathbf{x}$, multiply it by a_{11}, and collect terms; show that $a_{11}(\mathbf{x}^{\mathrm{T}}A\mathbf{x})$ has the form $p^2 + v^2 \det(A)$.]

14. Use induction to prove that the determinant of the $(n \times n)$ identity matrix is 1 for $n = 2, 3, \ldots$.

5.3 COLUMN OPERATIONS AND DETERMINANTS

In this section we will show how certain column operations simplify the calculation of determinants. In addition the properties we develop will be used later to demonstrate some of the connections between determinant theory and linear algebra. We will use three elementary column operations, which are analogous to the elementary row operations defined in Chapter 1. For a matrix A the elementary column operations are as follows.

1. Interchange two columns of A.
2. Multiply a column of A by a scalar c, $c \neq 0$.
3. Add a scalar multiple of one column of A to another column of A.

From Chapter 1 we know that row operations can be used to reduce a square matrix A to an upper-triangular matrix (that is, we know A can be reduced to echelon form, and a square matrix in echelon form is upper triangular). Similarly it is easy to show that column operations can be used to reduce a square matrix to lower-triangular form; we will show some examples of this and make a formal statement in the next section. One reason for reducing an $(n \times n)$ matrix A to a lower-triangular matrix T is that $\det(T)$ is trivial to evaluate (see Theorem 5.1). Then if we can calculate the effect that column operations have on the determinant, we can relate $\det(A)$ to $\det(T)$.

Theorem 5.1

If $T = (t_{ij})$ is an $(n \times n)$ lower-triangular matrix, then

$$\det(T) = t_{11}t_{22} \ldots t_{nn}. \tag{5.3}$$

det of singular matrix is 0

Proof. If T is a (2×2) lower-triangular matrix, then

$$\det(T) = \begin{vmatrix} t_{11} & 0 \\ t_{21} & t_{22} \end{vmatrix} = t_{11}t_{22}.$$

Proceeding inductively, suppose the theorem is true for any $(k \times k)$ lower-triangular matrix where $2 \leqslant k \leqslant n-1$. If T is an $(n \times n)$ lower-triangular matrix, then

$$\det(T) = \begin{vmatrix} t_{11} & 0 & 0 & \dots & 0 \\ t_{21} & t_{22} & 0 & \dots & 0 \\ \vdots & & & & \\ t_{n1} & t_{n2} & t_{n3} & \dots & t_{nn} \end{vmatrix} = t_{11}T_{11}.$$

Clearly T_{11} is the determinant of an $((n-1) \times (n-1))$ lower-triangular matrix; so $T_{11} = t_{22}t_{33} \dots t_{nn}$. Thus $\det(T) = t_{11}t_{22} \dots t_{nn}$, and the theorem is proved. ■

Before continuing, we want to make several comments about the organization of the remaining material in this chapter. First of all, several of the proofs are tedious and obscure when the most general case is considered, primarily because of problems in notation. For this reason we will occasionally prove only a restricted version of a determinant property when it is clear how notational modifications can be made to treat the general case. For example (see Theorem 5.2) it is easy to show that the sign of a determinant is changed when the first two columns are interchanged. The proof that the sign changes when the ith and jth columns are interchanged is exactly analogous, but the required notation is quite cumbersome.

With respect to the theoretical consistency of what follows, we state a theorem that we will prove in Section 5.5.

Theorem

If A is an $(n \times n)$ matrix, then $\det(A^{\mathrm{T}}) = \det(A)$.

In this section we will establish several properties concerning column operations and determinants and then prove the theorem above using only the properties of column operations. However since row operations on A are the same as column operations on A^{T}, it will follow from the theorem above that any property of a column operation is valid also for the corresponding row operation. For the sake of brevity we state the results of this section in terms of both columns and rows although the properties of row operations will not actually be established until $\det(A^{\mathrm{T}}) = \det(A)$ is proved.

To begin our analysis of the effect of a column operation on the determinant, we prove that a column interchange will change only the sign of the determinant.

Theorem 5.2

Let $A = [\mathbf{A}_1, \mathbf{A}_2, \mathbf{A}_3, \dots, \mathbf{A}_n]$ be an $(n \times n)$ matrix. If B is obtained from A by interchanging two columns (or rows) of A, then $\det(B) = -\det(A)$.

Proof. Because of the cumbersome notation, we prove this theorem only for the case that the first and second columns of A are interchanged. In particular we use induction to prove that $\det(B) = -\det(A)$ when $B = [\mathbf{A}_2, \mathbf{A}_1, \mathbf{A}_3, \ldots, \mathbf{A}_n]$.

If A is a (2×2) matrix and $B = [\mathbf{A}_2, \mathbf{A}_1]$, then

$$\det(B) = \begin{vmatrix} a_{12} & a_{11} \\ a_{22} & a_{21} \end{vmatrix} = a_{12}a_{21} - a_{11}a_{22} = -\det(A).$$

Inductively let us suppose that the result is correct for all $(k \times k)$ matrices, $2 \leqslant k \leqslant n-1$, and let $A = (a_{ij})$ be an $(n \times n)$ matrix. For this case

$$\det(B) = \begin{vmatrix} a_{12} & a_{11} & a_{13} & \cdots & a_{1n} \\ a_{22} & a_{21} & a_{23} & \cdots & a_{2n} \\ \vdots & & & & \vdots \\ a_{n2} & a_{n1} & a_{n3} & \cdots & a_{nn} \end{vmatrix} = a_{12} \begin{vmatrix} a_{21} & a_{23} & \cdots & a_{2n} \\ a_{31} & a_{33} & \cdots & a_{3n} \\ \vdots & & & \vdots \\ a_{n1} & a_{n3} & \cdots & a_{nn} \end{vmatrix}$$

$$-a_{11} \begin{vmatrix} a_{22} & a_{23} & \cdots & a_{2n} \\ a_{32} & a_{33} & \cdots & a_{3n} \\ \vdots & & & \vdots \\ a_{n2} & a_{n3} & \cdots & a_{nn} \end{vmatrix} + a_{13} \begin{vmatrix} a_{22} & a_{21} & a_{24} & \cdots & a_{2n} \\ a_{32} & a_{31} & a_{34} & & a_{3n} \\ \vdots & & & & \vdots \\ a_{n2} & a_{n1} & a_{n4} & \cdots & a_{nn} \end{vmatrix}$$

$$- \cdots + (-1)^{n+1} a_{1n} \begin{vmatrix} a_{22} & a_{21} & a_{23} & \cdots & a_{2,n-1} \\ a_{32} & a_{31} & a_{33} & \cdots & a_{3,n-1} \\ \vdots & & & & \vdots \\ a_{n2} & a_{n1} & a_{n3} & \cdots & a_{n,n-1} \end{vmatrix}.$$

The first two $((n-1) \times (n-1))$ determinants above are the same as in the cofactor expansion of $\det(A)$ *except* their signs have been reversed. The remaining $((n-1) \times (n-1))$ determinants above are the same as in the cofactor expansion of $\det(A)$ *except* that their first two columns are interchanged. By the induction hypothesis, changing the first two columns in each of these back to the original order changes their signs. These two observations show that $\det(B) = -\det(A)$. ■

EXAMPLE 5.5 Consider the three matrices.

$$A = \begin{bmatrix} 1 & 3 & 1 \\ 2 & 0 & 4 \\ 1 & 2 & 3 \end{bmatrix}, \quad B = \begin{bmatrix} 3 & 1 & 1 \\ 0 & 2 & 4 \\ 2 & 1 & 3 \end{bmatrix}, \quad C = \begin{bmatrix} 1 & 1 & 3 \\ 2 & 4 & 0 \\ 1 & 3 & 2 \end{bmatrix}.$$

If we write A in column form as $A = [\mathbf{A}_1, \mathbf{A}_2, \mathbf{A}_3]$, then $B = [\mathbf{A}_2, \mathbf{A}_1, \mathbf{A}_3]$ and $C = [\mathbf{A}_1, \mathbf{A}_3, \mathbf{A}_2]$. Theorem 5.2 asserts that $\det(B) = -\det(A)$, and $\det(C) = -\det(A)$; and this assertion is easily verified $[\det(A) = -10, \det(B) = \det(C) = 10]$.

By performing a sequence of column interchanges, we can produce any rearrangement of columns that we wish; and Theorem 5.2 can be used to find the determinant of the end result. For example if $A = [\mathbf{A}_1, \mathbf{A}_2, \mathbf{A}_3, \mathbf{A}_4]$ is a (4×4) matrix, and $B = [\mathbf{A}_4, \mathbf{A}_3, \mathbf{A}_1, \mathbf{A}_2]$, then we can relate $\det(B)$ to $\det(A)$ as follows: form $B_1 = [\mathbf{A}_4, \mathbf{A}_2, \mathbf{A}_3, \mathbf{A}_1]$; then form $B_2 = [\mathbf{A}_4, \mathbf{A}_3, \mathbf{A}_2, \mathbf{A}_1]$; and then form B by interchanging the last two columns of B_2. In this sequence $\det(B_1) = -\det(A)$, $\det(B_2) = -\det(B_1)$, and $\det(B) = -\det(B_2)$. Thus $\det(B_2) = \det(A)$, and $\det(B) = -\det(A)$.

Our next theorem shows that multiplying all entries in a column of A by a scalar c has the effect of multiplying the determinant by c.

Theorem 5.3

If A is an $(n \times n)$ matrix, and if B is the $(n \times n)$ matrix resulting from multiplying the kth column (or row) of A by a scalar c, then $\det(B) = c\det(A)$.

Proof. Again we use induction and for simplicity assume that the first column of A is multiplied by c. When A is (2×2), we have

$$\det(B) = \begin{vmatrix} ca_{11} & a_{12} \\ ca_{21} & a_{22} \end{vmatrix} = ca_{11}a_{22} - ca_{12}a_{21} = c\det(A).$$

Now suppose the theorem is valid for all k where $2 \leqslant k \leqslant n-1$. If A is $(n \times n)$, then

$$\det(B) = \begin{vmatrix} ca_{11} & a_{12} & \cdots & a_{1n} \\ ca_{21} & a_{22} & \cdots & a_{2n} \\ \vdots & & & \vdots \\ ca_{n1} & a_{n2} & \cdots & a_{nn} \end{vmatrix} = ca_{11}\begin{vmatrix} a_{22} & a_{23} & \cdots & a_{2n} \\ a_{32} & a_{33} & \cdots & a_{3n} \\ \vdots & & & \vdots \\ a_{n2} & a_{n3} & \cdots & a_{nn} \end{vmatrix}$$

$$-a_{12}\begin{vmatrix} ca_{21} & a_{23} & \cdots & a_{2n} \\ ca_{31} & a_{33} & \cdots & a_{3n} \\ \vdots & & & \vdots \\ ca_{n1} & a_{n3} & \cdots & a_{nn} \end{vmatrix} + \cdots$$

$$+(-1)^{n+1}a_{1n}\begin{vmatrix} ca_{21} & a_{22} & \cdots & a_{2,n-1} \\ ca_{31} & a_{32} & \cdots & a_{3,n-1} \\ \vdots & & & \vdots \\ ca_{n1} & a_{n2} & \cdots & a_{n,n-1} \end{vmatrix}.$$

Applying the induction hypothesis to the last $n-1$ determinants will show that $\det(B) = c\det(A)$. (We emphasize that the theorem is valid for $c = 0$.) ■

EXAMPLE 5.6 The (3×3) matrix A in Example 5.5 has a common factor of 2 in the second row. Thus $\det(A) = 2 \det(E)$ where

$$E = \begin{bmatrix} 1 & 3 & 1 \\ 1 & 0 & 2 \\ 1 & 2 & 3 \end{bmatrix} \times 2$$

So far we have analyzed the effect of two elementary column operations: column interchanges and multiplication of a column by a scalar. We now wish to show that the addition of a constant multiple of one column to another column does not change the determinant. We need several preliminary results to prove this.

Theorem 5.4

If A, B, and C are $(n \times n)$ matrices that are equal except that the kth column (or row) of A is equal to the sum of the kth columns (or rows) of B and C, then $\det(A) = \det(B) + \det(C)$.

Proof. We prove this statement for the first column; and we want to show that $\det(A) = \det(B) + \det(C)$ when $B = [\mathbf{B}_1, \mathbf{A}_2, \ldots, \mathbf{A}_n]$, $C = [\mathbf{C}_1, \mathbf{A}_2, \ldots, \mathbf{A}_n]$, and $A = [\mathbf{B}_1 + \mathbf{C}_1, \mathbf{A}_2, \ldots, \mathbf{A}_n]$. We use induction and leave the (2×2) case to the exercises (Exercise 4, Section 5.3).

Assuming the result true for $(k \times k)$ matrices where $2 \leqslant k \leqslant n-1$, we have in the $(n \times n)$ case

$$\det(A) = \begin{vmatrix} (b_{11}+c_{11}) & a_{12} & \cdots & a_{1n} \\ (b_{21}+c_{21}) & a_{22} & \cdots & a_{2n} \\ \vdots & & & \vdots \\ (b_{n1}+c_{n1}) & a_{n2} & \cdots & a_{nn} \end{vmatrix} = (b_{11}+c_{11}) \begin{vmatrix} a_{22} & a_{23} & \cdots & a_{2n} \\ a_{32} & a_{33} & \cdots & a_{3n} \\ \vdots & & & \vdots \\ a_{n2} & a_{n3} & \cdots & a_{nn} \end{vmatrix}$$

$$-a_{12} \begin{vmatrix} (b_{21}+c_{21}) & a_{23} & \cdots & a_{2n} \\ (b_{31}+c_{31}) & a_{33} & \cdots & a_{3n} \\ \vdots & & & \vdots \\ (b_{n1}+c_{n1}) & a_{n3} & \cdots & a_{nn} \end{vmatrix} + \cdots$$

$$+(-1)^{1+n} a_{1n} \begin{vmatrix} (b_{21}+c_{21}) & a_{22} & \cdots & a_{2,n-1} \\ (b_{31}+c_{31}) & a_{32} & \cdots & a_{3,n-1} \\ \vdots & & & \vdots \\ (b_{n1}+c_{n1}) & a_{n2} & \cdots & a_{n,n-1} \end{vmatrix}.$$

The induction assumption allows us to express the cofactors A_{1j} as $A_{1j} = B_{1j} + C_{1j}$ for $2 \leqslant j \leqslant n$, and the theorem is proved when we regroup the terms. ∎

EXAMPLE 5.7 Consider the matrices A, B, and C:

$$A=\begin{bmatrix}1&3&2\\0&4&7\\2&1&8\end{bmatrix},\qquad B=\begin{bmatrix}1&1&2\\0&2&7\\2&0&8\end{bmatrix},\qquad C=\begin{bmatrix}1&2&2\\0&2&7\\2&1&8\end{bmatrix}.$$

The second columns of B and C sum to the second column of A; so $\det(A)=\det(B)+\det(C)$ as is easily checked.

Theorem 5.5

Let A be an $(n\times n)$ matrix. If the jth column (or row) of A is a multiple of the kth column (or row) of A, then $\det(A)=0$.

Proof. Let $A=[\mathbf{A}_1,\mathbf{A}_2,\ldots,\mathbf{A}_j,\ldots,\mathbf{A}_k,\ldots,\mathbf{A}_n]$, and suppose that $\mathbf{A}_j=c\mathbf{A}_k$. Define B to be the matrix $B=[\mathbf{A}_1,\mathbf{A}_2,\ldots,\mathbf{A}_k,\ldots,\mathbf{A}_k,\ldots,\mathbf{A}_n]$, and observe that $\det(A)=c\det(B)$. Now if we interchange the jth and kth columns of B, the matrix B remains the same; but the determinant changes sign (Theorem 5.2). This $[\det(B)=-\det(B)]$ can happen only if $\det(B)=0$; and since $\det(A)=c\det(B)$, we must have $\det(A)=0$. ■

Two special cases of Theorem 5.5 are particularly interesting. If A has two identical columns ($c=1$ in the proof above), or if A has a zero column ($c=0$ in the proof), then $\det(A)=0$.

Theorem 5.4 and 5.5 can be used to analyze the effect of the last elementary column operation.

Theorem 5.6

If A is an $(n\times n)$ matrix, and if a multiple of the kth column (or row) is added to the jth column (or row), then the determinant is not changed.

Proof. Let $A=[\mathbf{A}_1,\mathbf{A}_2,\ldots,\mathbf{A}_j,\ldots,\mathbf{A}_k,\ldots,\mathbf{A}_n]$, and let $B=[\mathbf{A}_1,\mathbf{A}_2,\ldots,\mathbf{A}_j+c\mathbf{A}_k,\ldots,\mathbf{A}_k,\ldots,\mathbf{A}_n]$. By Theorem 5.5 $\det(B)=\det(A)+\det(Q)$ where $Q=[\mathbf{A}_1,\mathbf{A}_2,\ldots,c\mathbf{A}_k,\ldots,\mathbf{A}_k,\ldots,\mathbf{A}_n]$. By Theorem 5.6 $\det(Q)=0$; so $\det(B)=\det(A)$, and the theorem is proved. ■

As shown in the examples that follow, we can use elementary column operations to introduce zero entries into the first row of a matrix A. The analysis of how these operations affect the determinant will allow us to relate this effect back to $\det(A)$.

EXAMPLE 5.8 Consider the (4×4) matrix A from Example 5.3:

$$A=\begin{bmatrix}1&2&0&2\\-1&2&3&1\\-3&2&-1&0\\2&-3&-2&1\end{bmatrix}.$$

In Example 5.3 a laborious cofactor expansion showed that $\det(A) = -63$. In column form, $A = [\mathbf{A}_1, \mathbf{A}_2, \mathbf{A}_3, \mathbf{A}_4]$; and it is clear that we can introduce a zero into the (1, 2) position by replacing $\mathbf{A}_2$ by $\mathbf{A}_2 - 2\mathbf{A}_1$. Similarly replacing $\mathbf{A}_4$ by $\mathbf{A}_4 - 2\mathbf{A}_1$ zeros the (1, 4) entry. Moreover by Theorem 5.6 the determinant is unchanged. The details are

$$\det(A) = \begin{vmatrix} 1 & 2 & 0 & 2 \\ -1 & 2 & 3 & 1 \\ -3 & 2 & -1 & 0 \\ 2 & -3 & -2 & 1 \end{vmatrix} = \begin{vmatrix} 1 & 0 & 0 & 2 \\ -1 & 4 & 3 & 1 \\ -3 & 8 & -1 & 0 \\ 2 & -7 & -2 & 1 \end{vmatrix}$$

$$= \begin{vmatrix} 1 & 0 & 0 & 0 \\ -1 & 4 & 3 & 3 \\ -3 & 8 & -1 & 6 \\ 2 & -7 & -2 & -3 \end{vmatrix}.$$

Thus it follows that $\det(A)$ is given by

$$\det(A) = \begin{vmatrix} 4 & 3 & 3 \\ 8 & -1 & 6 \\ -7 & -2 & -3 \end{vmatrix}.$$

We now wish to create zeros in the (1, 2) and (1, 3) positions of this (3×3) determinant. To avoid using fractions, we multiply the second and third columns by 4 (using Theorem 5.3), and then add a multiple of -3 times column one to columns two and three:

$$\det(A) = \begin{vmatrix} 4 & 3 & 3 \\ 8 & -1 & 6 \\ -7 & -2 & -3 \end{vmatrix} = \frac{1}{16}\begin{vmatrix} 4 & 12 & 12 \\ 8 & -4 & 24 \\ -7 & -8 & -12 \end{vmatrix} = \frac{1}{16}\begin{vmatrix} 4 & 0 & 0 \\ 8 & -28 & 0 \\ -7 & 13 & 9 \end{vmatrix}.$$

Thus we again find $\det(A) = -63$.

As in Gauss elimination, column interchanges are sometimes desirable and serve to keep order in the computation. Consider

$$\det(A) = \begin{vmatrix} 0 & 1 & 3 & 1 \\ 1 & -2 & -2 & 2 \\ 3 & 4 & 2 & -2 \\ 4 & 3 & -1 & 1 \end{vmatrix} = -\begin{vmatrix} 1 & 0 & 3 & 1 \\ -2 & 1 & -2 & 2 \\ 4 & 3 & 2 & -2 \\ 3 & 4 & -1 & 1 \end{vmatrix}.$$

Use column one to introduce zeros along the first row:

$$\det(A) = -\begin{vmatrix} 1 & 0 & 0 & 0 \\ -2 & 1 & 4 & 4 \\ 4 & 3 & -10 & -6 \\ 3 & 4 & -10 & -2 \end{vmatrix} = -\begin{vmatrix} 1 & 4 & 4 \\ 3 & -10 & -6 \\ 4 & -10 & -2 \end{vmatrix}.$$

Again column one can be used to introduce zeros:

$$\det(A) = -\begin{vmatrix} 1 & 0 & 0 \\ 3 & -22 & -18 \\ 4 & -26 & -18 \end{vmatrix} = -\begin{vmatrix} -22 & -18 \\ -26 & -18 \end{vmatrix};$$

and we calculate the (2×2) determinant to find $\det(A) = 72$.

5.3 EXERCISES

1. Use elementary column operations to zero the last two entries in the first row, and then calculate the determinant of the original matrix.

a) $\begin{vmatrix} 1 & 2 & 1 \\ 2 & 0 & 1 \\ 1 & -1 & 1 \end{vmatrix}$ b) $\begin{vmatrix} 2 & 4 & -2 \\ 0 & 2 & 3 \\ 1 & 1 & 2 \end{vmatrix}$ c) $\begin{vmatrix} 0 & 1 & 2 \\ 3 & 1 & 2 \\ 2 & 0 & 3 \end{vmatrix}$

2. Repeat Exercise 1 for the following.

a) $\begin{vmatrix} 2 & 2 & 4 \\ 1 & 0 & 1 \\ 2 & 1 & 2 \end{vmatrix}$ b) $\begin{vmatrix} 0 & 1 & 3 \\ 2 & 1 & 2 \\ 1 & 1 & 2 \end{vmatrix}$ c) $\begin{vmatrix} 1 & 1 & 1 \\ 2 & 1 & 2 \\ 3 & 0 & 2 \end{vmatrix}$

3. Let $U = (u_{ij})$ be an $(n \times n)$ upper-triangular matrix. Use the observation made after Theorem 5.5 to prove that $\det(U) = u_{11}U_{11}$. Use induction to prove that $\det(U) = u_{11}u_{22} \ldots u_{nn}$.

4. Verify that Theorem 5.4 is valid for $n = 2$.

5. Use only column interchanges to produce a triangular determinant, and evaluate the determinant of the original matrix.

a) $\begin{vmatrix} 1 & 0 & 0 & 0 \\ 2 & 0 & 0 & 3 \\ 1 & 1 & 0 & 1 \\ 1 & 4 & 2 & 2 \end{vmatrix}$ b) $\begin{vmatrix} 0 & 0 & 2 & 0 \\ 0 & 0 & 1 & 3 \\ 0 & 4 & 1 & 3 \\ 2 & 1 & 5 & 6 \end{vmatrix}$ c) $\begin{vmatrix} 0 & 1 & 0 & 0 \\ 0 & 2 & 0 & 3 \\ 2 & 1 & 0 & 6 \\ 3 & 2 & 2 & 4 \end{vmatrix}$

6. Prove that if A is $(n \times n)$, then $\det(cA) = c^n \det(A)$. (Hint: Write A in column form and recall the definition of cA.)

7. Use column operations to zero the (1, 2), (1, 3), (1, 4), (2, 3), and (2, 4) positions; and evaluate the original determinant.

a) $\begin{vmatrix} 1 & 2 & 0 & 3 \\ 2 & 5 & 1 & 1 \\ 2 & 0 & 4 & 3 \\ 0 & 1 & 6 & 2 \end{vmatrix}$ b) $\begin{vmatrix} 2 & 4 & -2 & -2 \\ 1 & 3 & 1 & 2 \\ 1 & 3 & 1 & 3 \\ -1 & 2 & 1 & 2 \end{vmatrix}$ c) $\begin{vmatrix} 1 & 1 & 2 & 1 \\ 0 & 1 & 4 & 1 \\ 2 & 1 & 3 & 0 \\ 2 & 2 & 1 & 2 \end{vmatrix}$

8. Find a (2×2) matrix A and a (2×2) matrix B where $\det(A + B)$ is not equal to $\det(A) + \det(B)$.

9. Let $y = mx + b$ be the equation of the line through the points (x_1, y_1) and (x_2, y_2) in the plane. Show that the equation is given also by

$$\begin{vmatrix} x & y & 1 \\ x_1 & y_1 & 1 \\ x_2 & y_2 & 1 \end{vmatrix} = 0.$$

10. Let (x_1, y_1), (x_2, y_2), (x_3, y_3) be the vertices of a triangle in the plane where these vertices are numbered clockwise. Prove that the area of the triangle is given by

$$\text{area} = \frac{1}{2}\begin{vmatrix} x_1 & y_1 & 1 \\ x_2 & y_2 & 1 \\ x_3 & y_3 & 1 \end{vmatrix}.$$

11. Let $\mathbf{x}$ and $\mathbf{y}$ be vectors in R^3, and let $A = I + \mathbf{x}\mathbf{y}^{\mathrm{T}}$. Show that $\det(A) = 1 + \mathbf{y}^{\mathrm{T}}\mathbf{x}$. (Hint: If $B = \mathbf{x}\mathbf{y}^{\mathrm{T}}$, $B = [\mathbf{B}_1, \mathbf{B}_2, \mathbf{B}_3]$, then $A = [\mathbf{B}_1 + \mathbf{e}_1, \mathbf{B}_2 + \mathbf{e}_2, \mathbf{B}_3 + \mathbf{e}_3]$. Therefore $\det(A) = \det[\mathbf{B}_1, \mathbf{B}_2 + \mathbf{e}_2, \mathbf{B}_3 + \mathbf{e}_3] + \det[\mathbf{e}_1, \mathbf{B}_2 + \mathbf{e}_2, \mathbf{B}_3 + \mathbf{e}_3]$. Use Theorems 5.4 and 5.5 to show that the first determinant is equal to $\det[\mathbf{B}_1, \mathbf{e}_2, \mathbf{B}_3 + \mathbf{e}_3]$, etc.)

12. Use column operations to prove that

$$\begin{vmatrix} 1 & a & a^2 \\ 1 & b & b^2 \\ 1 & c & c^2 \end{vmatrix} = (b-a)(c-a)(c-b).$$

13. As in Exercise 12, evaluate

$$\begin{vmatrix} 1 & a & a^2 & a^3 \\ 1 & b & b^2 & b^3 \\ 1 & c & c^2 & c^3 \\ 1 & d & d^2 & d^3 \end{vmatrix}.$$

(The determinants in Exercises 12 and 13 are called ***Vandermonde*** determinants.)

5.4 NONSINGULAR MATRICES AND CRAMER'S RULE

In Section 5.3 we saw how to calculate the effect that a column operation has on the determinant of a matrix. In this section we will use this information to analyze the relationships between determinants, nonsingular matrices, and solutions of systems $A\mathbf{x} = \mathbf{b}$. We begin with a special case of a result we need; that is, column operations can be used to reduce a square matrix to lower-triangular form.

EXAMPLE 5.9 Consider the (3×3) matrix

$$A = \begin{bmatrix} 1 & 2 & -3 \\ 3 & 6 & 1 \\ -2 & 1 & 0 \end{bmatrix}.$$

If we add a multiple of -2 times column one to column two, and a multiple of 3 times column one to column three, we obtain

$$B = \begin{bmatrix} 1 & 0 & 0 \\ 3 & 0 & 10 \\ -2 & 5 & -6 \end{bmatrix}$$

Interchanging columns two and three in B, we arrive at a lower-triangular matrix T:

$$T = \begin{bmatrix} 1 & 0 & 0 \\ 3 & 10 & 0 \\ -2 & 6 & 5 \end{bmatrix}.$$

[From the results of Section 5.3, $\det(A) = \det(B)$ and $\det(T) = -\det(B)$; so $\det(A) = -50$.]

Just as row operations can be used to reduce a square matrix A to upper-triangular form, it is easy to prove that column operations can be used to reduce A to lower-triangular form.

Theorem 5.7

If A is an $(n \times n)$ matrix, then elementary column operations can be used to reduce A to a lower-triangular matrix T. Furthermore T is nonsingular if and only if A is nonsingular.

Proof. Since the proof of the first part parallels the proof of Theorem 1.1, we give only an outline. Let A be the matrix

$$A = \begin{bmatrix} a_{11} & a_{12} & \cdots & a_{1n} \\ a_{21} & a_{22} & \cdots & a_{2n} \\ \vdots & & & \vdots \\ a_{n1} & a_{n2} & \cdots & a_{nn} \end{bmatrix}.$$

If $a_{11} \neq 0$, then adding a multiple of $-a_{1k}/a_{11}$ times column one to column k will zero the $(1, k)$ entry, $2 \leqslant k \leqslant n$. If $a_{11} = 0$ but $a_{1j} \neq 0$ for some j, $2 \leqslant j \leqslant n$, then we can interchange columns one and j and then zero the $(1, k)$ entries as above for $2 \leqslant k \leqslant n$. Thus by a sequence of column operations we can reduce A to a matrix of the form

$$B = \begin{bmatrix} b_{11} & 0 & 0 & \cdots & 0 \\ b_{21} & b_{22} & b_{23} & \cdots & b_{2n} \\ \vdots & & & & \vdots \\ b_{n1} & b_{n2} & b_{n3} & \cdots & b_{nn} \end{bmatrix}.$$

(Note : If all the entries in the first row of A were zero, we would already have the form of B.)

If $b_{22} \neq 0$, we can use column operations to zero the $(2, k)$ entries of B where $3 \leqslant k \leqslant n$. If $b_{22} = 0$ but $b_{2j} \neq 0$ for some j, $3 \leqslant j \leqslant n$, a column interchange can be made; and the $(2, k)$ entries can be zeroed for $3 \leqslant k \leqslant n$. Continuing in this way, we can use a sequence of column operations to zero all the entries above the main diagonal; and we produce a lower-triangular matrix T.

To complete the proof requires only a simple demonstration that the elementary column operations used to triangularize A preserve singularity and nonsingularity. That is, suppose Q is any $(n \times n)$ matrix, and suppose P is an $(n \times n)$ matrix obtained from Q by interchanging columns or by adding a multiple of the jth column of Q to the kth column of Q. Then it can be shown (Exercise 6, Section 5.4) that Q is nonsingular if and only if P is nonsingular.

Given Exercise 6, Section 5.4, if A is nonsingular, then each step of the triangularization process must produce a nonsingular matrix; and hence the end result, T, is also nonsingular. If we work backwards from T being nonsingular, all the intermediate matrices leading up to T must have been nonsingular as well. ■

Theorem 5.7 can be used to make a connection between singular matrices and their determinants.

Theorem 5.8

If A is an $(n \times n)$ matrix, then A is singular if and only if $\det(A) = 0$.

Proof. By Theorem 5.7 elementary column operations can reduce A to a lower-triangular matrix T, and moreover A is singular if and only if T is singular. From the results of Section 5.4, $|\det(A)| = |\det(T)|$, and $\det(T)$ is the product of the diagonal entries of T. By Exercise 10, Section 1.7, T is singular if and only if one of the diagonal entries of T is zero; so by Theorem 5.1, T is singular if and only if $\det(T) = 0$. Putting these results together means that if A is singular, then T is singular, and $0 = \det(T) = \det(A)$. Conversely if $\det(A) = 0$, then $\det(T) = 0$, and T must be singular. But if T is singular, then so is A; and the proof is complete. ■

A major result in determinant theory is Cramer's rule, which gives a formula for the solution of any system $A\mathbf{x} = \mathbf{b}$ when A is nonsingular.

Theorem 5.9 (Cramer's rule)

Let $A = [\mathbf{A}_1, \mathbf{A}_2, \ldots, \mathbf{A}_n]$ be a nonsingular $(n \times n)$ matrix, and let $\mathbf{b}$ be any vector in R^n. For each i, $1 \leqslant i \leqslant n$, let B_i be the matrix $B_i = [\mathbf{A}_1, \ldots, \mathbf{A}_{i-1}, \mathbf{b}, \mathbf{A}_{i+1}, \ldots, \mathbf{A}_n]$. Then the ith component, x_i, of the solution of $A\mathbf{x} = \mathbf{b}$ is given by

$$x_i = \frac{\det(B_i)}{\det(A)}. \tag{5.4}$$

Proof. Observe that (5.4) is defined since $\det(A) \neq 0$ when A is nonsingular. We give the proof only for $i = 1$ in order to keep the notation simple.

Now since A is nonsingular, the system $A\mathbf{x} = \mathbf{b}$ has a unique solution. Thus there are unique values $x_1, x_2, \ldots, x_n$ such that

$$x_1\mathbf{A}_1 + x_2\mathbf{A}_2 + \ldots + x_n\mathbf{A}_n = \mathbf{b}. \tag{5.5}$$

Using the properties of determinants and (5.5), we have

$$\begin{aligned} x_1 \det(A) &= \det[x_1\mathbf{A}_1, \mathbf{A}_2, \ldots, \mathbf{A}_n] \\ &= \det[\mathbf{b} - x_2\mathbf{A}_2 - \ldots - x_n\mathbf{A}_n, \mathbf{A}_2, \ldots, \mathbf{A}_n] \\ &= \det[\mathbf{b}, \mathbf{A}_2, \ldots, \mathbf{A}_n] - x_2 \det[\mathbf{A}_2, \mathbf{A}_2, \ldots, \mathbf{A}_n] - \ldots \\ &\quad - x_n \det[\mathbf{A}_n, \mathbf{A}_2, \ldots, \mathbf{A}_n]. \end{aligned}$$

By Theorem 5.5 the last $n-1$ determinants are zero; so we have

$$x_1 \det(A) = \det[\mathbf{b}, \mathbf{A}_2, \ldots, \mathbf{A}_n];$$

and this equality verifies (5.4) for $i = 1$. ■

EXAMPLE 5.10 Consider the system

$$\begin{aligned} 3x_1 + 2x_2 &= 4 \\ 5x_1 + 4x_2 &= 6. \end{aligned}$$

To solve this system by Cramer's rule, we write the system as $A\mathbf{x} = \mathbf{b}$, and we form $B_1 = [\mathbf{b}, \mathbf{A}_2]$ and $B_2 = [\mathbf{A}_1, \mathbf{b}]$:

$$A = \begin{bmatrix} 3 & 2 \\ 5 & 4 \end{bmatrix}, \quad B_1 = \begin{bmatrix} 4 & 2 \\ 6 & 4 \end{bmatrix}, \quad B_2 = \begin{bmatrix} 3 & 4 \\ 5 & 6 \end{bmatrix}.$$

We calculate $\det(A) = 2$, $\det(B_1) = 4$, and $\det(B_2) = -2$. Thus from (5.4) the solution is

$$x_1 = \frac{4}{2} = 2 \quad \text{and} \quad x_2 = \frac{-2}{2} = -1.$$

EXAMPLE 5.11 We use Cramer's rule to solve

$$\begin{aligned} x_1 - x_2 + x_3 &= 0 \\ x_1 + x_2 - 2x_3 &= 1 \\ x_1 + 2x_2 + x_3 &= 6. \end{aligned}$$

Writing the system as $A\mathbf{x} = \mathbf{b}$, we have

$$A = \begin{bmatrix} 1 & -1 & 1 \\ 1 & 1 & -2 \\ 1 & 2 & 1 \end{bmatrix}, \quad B_1 = \begin{bmatrix} 0 & -1 & 1 \\ 1 & 1 & -2 \\ 6 & 2 & 1 \end{bmatrix},$$

$$B_2 = \begin{bmatrix} 1 & 0 & 1 \\ 1 & 1 & -2 \\ 1 & 6 & 1 \end{bmatrix}, \quad B_3 = \begin{bmatrix} 1 & -1 & 0 \\ 1 & 1 & 1 \\ 1 & 2 & 6 \end{bmatrix}.$$

A calculation shows that $\det(A) = 9$, $\det(B_1) = 9$, $\det(B_2) = 18$, and $\det(B_3) = 9$. Thus by (5.4) the solution is

$$x_1 = \frac{9}{9} = 1, \qquad x_2 = \frac{18}{9} = 2, \qquad x_3 = \frac{9}{9} = 1.$$

As a computational tool, Cramer's rule is rarely competitive with Gauss elimination. However as a theoretical tool, Cramer's rule is valuable; and we will use it to prove the next theorem. The theorem below is perhaps surprising because $\det(A+B)$ is not usually equal to $\det(A)+\det(B)$ (see Exercise 8, Section 5.3).

Theorem 5.10

If A and B are $(n \times n)$ matrices, then $\det(AB) = \det(A)\det(B)$.

Proof. We dispose of the easy case first. If either A or B is singular, then AB is singular by Theorem 1.13. Calling on Theorem 5.8, we see that $\det(AB) = 0 = \det(A)\det(B)$. Thus we may as well assume that both A and B are nonsingular.

The remainder of the proof is rather involved although each step is elementary. Our technique is to use induction and to show that for $i = 1, 2, \ldots, n$, there is a nonsingular matrix Q such that

$$\frac{\det(AB)}{\det(B)} = \frac{\det(AQ)}{\det(Q)}$$

where Q has the form $Q = [\mathbf{e}_1, \mathbf{e}_2, \ldots, \mathbf{e}_i, \mathbf{Q}_{i+1}, \ldots, \mathbf{Q}_n]$ (when $i = n$, the theorem is proved).

We first write A and B in column form: $A = [\mathbf{A}_1, \mathbf{A}_2, \ldots, \mathbf{A}_n]$, $B = [\mathbf{B}_1, \mathbf{B}_2, \ldots, \mathbf{B}_n]$. Since B is nonsingular, we know there is a vector $\mathbf{x}$ such that $B\mathbf{x} = \mathbf{e}_1$. Let x_k be any nonzero component of $\mathbf{x}$, and define the matrix C by $C = [\mathbf{B}_1, \ldots, \mathbf{B}_{k-1}, \mathbf{e}_1, \mathbf{B}_{k+1}, \ldots, \mathbf{B}_n]$. By Cramer's rule

$$x_k = \frac{\det(C)}{\det(B)};$$

and since $x_k \neq 0$, we know that C is nonsingular. Also for use below we observe that $AC = [A\mathbf{B}_1, \ldots, A\mathbf{B}_{k-1}, \mathbf{A}_1, A\mathbf{B}_{k+1}, \ldots, A\mathbf{B}_n]$. Now since $B\mathbf{x} = \mathbf{e}_1$, it follows that $AB\mathbf{x} = \mathbf{A}_1$; and from Cramer's rule we obtain another expression for x_k:

$$x_k = \frac{\det(AC)}{\det(AB)}.$$

Given these two expressions, we can assert that

$$\frac{\det(AB)}{\det(B)} = \frac{\det(AC)}{\det(C)}. \tag{5.6}$$

If we interchange columns one and k in C and call the result D, then $\det(D) = -\det(C)$. Furthermore interchanging columns one and k of AC produces the matrix AD; so $\det(AD) = -\det(AC)$. From this result and (5.6)

$$\frac{\det(AB)}{\det(B)} = \frac{\det(AD)}{\det(D)} \tag{5.7}$$

where D is nonsingular and has the form $D = [\mathbf{e}_1, \mathbf{D}_2, \ldots, \mathbf{D}_n]$.

We continue inductiviely: suppose $1 \leqslant j \leqslant n-1$ and suppose

$$\frac{\det(AB)}{\det(B)} = \frac{\det(AQ)}{\det(Q)} \tag{5.8}$$

where Q is nonsingular and has the form $Q = [\mathbf{e}_1, \mathbf{e}_2, \ldots, \mathbf{e}_j, \mathbf{Q}_{j+1}, \ldots, \mathbf{Q}_n]$. (Note that $AQ = [\mathbf{A}_1, \mathbf{A}_2, \ldots, \mathbf{A}_j, A\mathbf{Q}_{j+1}, \ldots, A\mathbf{Q}_n]$.) There is a vector $\mathbf{y}$ such that $Q\mathbf{y} = \mathbf{e}_{j+1}$; and it is easy to show (Exercise 7, Section 5.4) that some component y_r of $\mathbf{y}$ must be nonzero where $j+1 \leqslant r \leqslant n$. Furthermore $AQ\mathbf{y} = \mathbf{A}_{j+1}$; so Cramer's rule can be used to give two expressions for y_r. As before, equating these leads to

$$\frac{\det(AQ)}{\det(Q)} = \frac{\det[\mathbf{A}_1, \ldots, \mathbf{A}_j, A\mathbf{Q}_{j+1}, \ldots, \mathbf{A}_{j+1}, \ldots, A\mathbf{Q}_n]}{\det[\mathbf{e}_1, \ldots, \mathbf{e}_j, \mathbf{Q}_{j+1}, \ldots, \mathbf{e}_{j+1}, \ldots, \mathbf{Q}_n]}. \tag{5.9}$$

In (5.9) the vectors $\mathbf{A}_{j+1}$ and $\mathbf{e}_{j+1}$ are in the rth column. Thus interchanging columns $j+1$ and r, and then calling on (5.8), we have

$$\frac{\det(AB)}{\det(B)} = \frac{\det(AP)}{\det(P)}$$

where P is nonsingular and has the form $P = [\mathbf{e}_1, \mathbf{e}_2, \ldots, \mathbf{e}_{j+1}, \mathbf{P}_{j+2}, \ldots, \mathbf{P}_n]$.

Continuing the process, we eventually arrive at

$$\frac{\det(AB)}{\det(B)} = \frac{\det(AI)}{\det(I)} \tag{5.10}$$

where I is the identity matrix. Then since $\det(I) = 1$, the theorem is proved. ■

5.4 EXERCISES

1. Use column operations to reduce the matrices to lower-triangular form, and evaluate the determinant of the original matrix.

$$A = \begin{bmatrix} 0 & 1 & 3 \\ 1 & 2 & 1 \\ 3 & 4 & 1 \end{bmatrix} \qquad B = \begin{bmatrix} 1 & 2 & 1 \\ 2 & 4 & 3 \\ 2 & 1 & 3 \end{bmatrix} \qquad C = \begin{bmatrix} 2 & 2 & 4 \\ 1 & 3 & 4 \\ -1 & 2 & 1 \end{bmatrix}$$

2. Use Cramer's rule to solve these systems.

a) $\begin{aligned} x_1 + x_2 &= 3 \\ x_1 - x_2 &= -1 \end{aligned}$ b) $\begin{aligned} x_1 + 3x_2 &= 4 \\ x_1 - x_2 &= 0 \end{aligned}$

3. Use Cramer's rule to solve these systems.

a)
$$\begin{aligned} x_1 - 2x_2 + x_3 &= -1 \\ x_1 \quad\quad + x_3 &= 3 \\ x_1 - x_2 \quad\quad &= 0 \end{aligned}$$

b)
$$\begin{aligned} x_1 + x_2 + x_3 &= 2 \\ x_1 + 2x_2 + x_3 &= 2 \\ x_1 + 3x_2 - x_3 &= -4 \end{aligned}$$

4. Use Cramer's rule to solve these systems.

a)
$$\begin{aligned} x_1 + x_2 + x_3 - x_4 &= 2 \\ x_2 - x_3 + x_4 &= 1 \\ x_3 - x_4 &= 0 \\ x_3 + 2x_4 &= 3 \end{aligned}$$

b)
$$\begin{aligned} 2x_1 - x_2 + x_3 &= 3 \\ x_1 + x_2 \quad\quad &= 3 \\ x_2 - x_3 &= 1 \end{aligned}$$

5. a) If A is nonsingular, prove that $\det(A^{-1}) = 1/\det(A)$.
   b) If $A^2 = I$, prove that $|\det(A)| = 1$.

6. Let $Q = [\mathbf{Q}_1, \mathbf{Q}_2, \ldots, \mathbf{Q}_n]$ be an $(n \times n)$ matrix, and suppose Q is nonsingular. Let P be the matrix obtained from Q by a column interchange.
   a) Prove that Q is singular if and only if P is singular. (Hint: If $Q\mathbf{x} = \boldsymbol{\theta}$, then $x_1\mathbf{Q}_1 + \ldots + x_n\mathbf{Q}_n = \boldsymbol{\theta}$. What vector $\mathbf{y}$ satisfies $P\mathbf{y} = \boldsymbol{\theta}$?)

   Suppose P is obtained from Q by adding a multiple of c times column j to column k. That is, $P = [\mathbf{Q}_1, \mathbf{Q}_2, \ldots, \mathbf{Q}_j, \ldots, \mathbf{Q}_k + c\mathbf{Q}_j, \ldots, \mathbf{Q}_n]$.
   b) Prove that if Q is nonsingular, then P is nonsingular. (Hint: Suppose a linear combination of the columns of P add to $\boldsymbol{\theta}$. Use the fact that Q has linearly independent columns to deduce that P has as well.)
   (c) Prove that if P is nonsingular, then Q is nonsingular. [Hint: Use a column operation on P to recover Q; then apply part (b).]

7. As in the proof of Theorem 5.10, suppose Q is a nonsingular $(n \times n)$ matrix of the form $Q = [\mathbf{e}_1, \mathbf{e}_2, \ldots, \mathbf{e}_j, \mathbf{Q}_{j+1}, \ldots, \mathbf{Q}_n]$. Let $\mathbf{y}$ be a vector in R^n such that $Q\mathbf{y} = \mathbf{e}_{j+1}$. Prove that some component y_r of $\mathbf{y}$ must be nonzero where $j+1 \leqslant r \leqslant n$. (Hint: Suppose $Q\mathbf{y} = \mathbf{e}_{j+1}$ and $y_k = 0$ for $j+1 \leqslant k \leqslant n$.)

8. We know that AB and BA are not usually equal. However show that if A and B are $(n \times n)$, then $\det(AB) = \det(BA)$. (Caution: Do not divide by zero.)

9. Suppose S is a nonsingular $(n \times n)$ matrix, and suppose A and B are $(n \times n)$ matrices such that $SAS^{-1} = B$. Prove that $\det(A) = \det(B)$.

10. Suppose A is $(n \times n)$, and $A^2 = A$. What is $\det(A)$?

11. If $\det(A) = 3$, what is $\det(A^5)$?

12. Let A and C be square matrices, and let Q be a matrix of the form

$$Q = \begin{bmatrix} A & \mathcal{O} \\ B & C \end{bmatrix}.$$

Convince yourself that $\det(Q) = \det(A)\det(C)$. (Hint: Reduce C to lower-triangular form with column operations; then reduce A.)

13. Verify the result in Exercise 12 for the matrix

$$Q = \begin{bmatrix} 1 & 2 & 0 & 0 & 0 \\ 2 & 1 & 0 & 0 & 0 \\ 3 & 5 & 1 & 2 & 2 \\ 7 & 2 & 3 & 5 & 1 \\ 1 & 8 & 1 & 4 & 1 \end{bmatrix}.$$

5.5 APPLICATIONS OF DETERMINANTS; INVERSES AND WRONSKIANS

Now that we have $\det(AB) = \det(A)\det(B)$, we are ready to prove that $\det(A^{\mathrm{T}}) = \det(A)$ and to establish some other useful properties of determinants. First however we need two preliminary results. One of these is a companion to Theorem 5.1: if U is an $(n \times n)$ upper-triangular matrix, then $\det(U) = u_{11}u_{22} \ldots u_{nn}$. This result is easy to prove (see Exercise 3, Section 5.3). The other result we require is also easy to prove (Exercises 4–6, Section 5.5) and shows that we can consider the usual row reduction of a matrix A to echelon form the same as forming QA where Q is a special sort of matrix.

Theorem 5.11

Let A be an $(n \times n)$ matrix. Then there is a nonsingular $(n \times n)$ matrix Q such that $QA = U$ where U is upper triangular. Moreover $\det(Q^{\mathrm{T}}) = \det(Q)$.

With these preliminaries, it is easy to prove Theorem 5.12.

Theorem 5.12

If A is an $(n \times n)$ matrix, then $\det(A^{\mathrm{T}}) = \det(A)$.

Proof. By Theorem 5.11 there is an $(n \times n)$ matrix Q such that $QA = U$ where U is upper triangular and $\det(Q^{\mathrm{T}}) = \det(Q)$. Given $QA = U$, it follows that

$$A^{\mathrm{T}}Q^{\mathrm{T}} = U^{\mathrm{T}}$$

where U^{T} is a lower-triangular matrix.

Applying Theorem 5.10 to $QA = U$ and $A^{\mathrm{T}}Q^{\mathrm{T}} = U^{\mathrm{T}}$, we have

$$\det(Q)\det(A) = \det(U) \qquad \text{and} \qquad \det(A^{\mathrm{T}})\det(Q^{\mathrm{T}}) = \det(U^{\mathrm{T}}).$$

However since U and U^{T} are triangular matrices with the same main-diagonal entries, $\det(U) = \det(U^{\mathrm{T}})$. Thus we can conclude that $\det(A) = \det(A^{\mathrm{T}})$; note that $\det(Q) = \det(Q^{\mathrm{T}})$ and $\det(Q) \neq 0$ by Theorem 5.11. ■

At this point we know that Theorems 5.2–5.6 in Section 5.3 are valid for rows as well as for columns. In particular we can use row operations to reduce a matrix A to a triangular matrix T and conclude that $\det(A) = \det(T)$.

EXAMPLE 5.12 We return to the (4×4) matrix A in Example 5.3 where $\det(A) = -63$:

$$\det(A) = \begin{vmatrix} 1 & 2 & 0 & 2 \\ -1 & 2 & 3 & 1 \\ -3 & 2 & -1 & 0 \\ 2 & -3 & -2 & 1 \end{vmatrix}.$$

By using row operations, we can reduce det(A) to

$$\det(A) = \begin{vmatrix} 1 & 2 & 0 & 2 \\ 0 & 4 & 3 & 3 \\ 0 & 8 & -1 & 6 \\ 0 & -7 & -2 & -3 \end{vmatrix}. \tag{5.11}$$

Now we switch rows two and three, and then switch columns two and three in order to get the number -1 into the pivot position. Following this switch, we zero the (2, 3) and (2, 4) positions with row operations; and we find

$$\det(A) = \begin{vmatrix} 1 & 0 & 2 & 2 \\ 0 & -1 & 8 & 6 \\ 0 & 3 & 4 & 3 \\ 0 & -2 & -7 & -3 \end{vmatrix}$$

$$= \begin{vmatrix} 1 & 0 & 2 & 2 \\ 0 & -1 & 8 & 6 \\ 0 & 0 & 28 & 21 \\ 0 & 0 & -23 & -15 \end{vmatrix}.$$

[The sign of the first determinant above is the same as det(A) since the first determinant is the result of two interchanges.] A quick calculation shows that the last determinant has the value -63.

The next theorem shows that we can evaluate det(A) by using an expansion along any row or any column we choose. Computationally this ability is useful when some row or column contains a number of zero entries.

Theorem 5.13

Let $A = (a_{ij})$ be an $(n \times n)$ matrix. Then

$$\det(A) = a_{i1}A_{i1} + a_{i2}A_{i2} + \ldots + a_{in}A_{in} \tag{5.12a}$$

$$\det(A) = a_{1j}A_{1j} + a_{2j}A_{2j} + \ldots + a_{nj}A_{nj}. \tag{5.12b}$$

Proof. We establish only (5.12a), which is an expansion of det(A) along the ith row. Expansion of det(A) along the jth column in (5.12b) is proved the same way.

Form a matrix B from A in the following manner: interchange row i first with row $i-1$ and then with row $i-2$; continue until row i is the top row of B. In other words bring row i to the top and push the other rows down so that they retain their same relative ordering. This procedure requires $i-1$ interchanges; so det$(A) = (-1)^{i-1}$ det (B). An inspection shows that the cofactors $B_{11}, B_{12}, \ldots, B_{1n}$ are also related to the cofactors $A_{i1}, A_{i2}, \ldots, A_{in}$ by $B_{1k} = (-1)^{i-1} A_{ik}$. To see this relationship, one need only observe that if M is the minor of the $(1, k)$ entry of B, then M is the minor of the (i, k) entry of A. Therefore $B_{1k} = (-1)^{k+1} M$ and $A_{ik} = (-1)^{i+k} M$, which shows that $B_{1k} = (-1)^{i-1} A_{ik}$. With this equality and Definition 5.2

$$\begin{aligned}\det(B) &= b_{11}B_{11} + b_{12}B_{12} + \dots + b_{1n}B_{1n} \\ &= a_{i1}B_{11} + a_{i2}B_{12} + \dots + a_{in}B_{1n} \\ &= (-1)^{i-1}(a_{i1}A_{i1} + a_{i2}A_{i2} + \dots + a_{in}A_{in}).\end{aligned}$$

Since $\det(A) = (-1)^{i-1}\det(B)$, formula (5.12a) is proved. ■

We next show how determinants can be used to give a formula for the inverse of a nonsingular matrix. We first prove a lemma, which is similar in appearance to Theorem 5.13. In words the lemma says that the sum of the products of entries from the ith row with cofactors from the kth row is zero when $i \neq k$ (and by Theorem 5.13 this sum is the determinant when $i = k$).

Lemma

If A is an $(n \times n)$ matrix and if $i \neq k$, then $a_{i1}A_{k1} + a_{i2}A_{k2} + \dots + a_{in}A_{kn} = 0$.

Proof. For i and k given, $i \neq k$, let B be the $(n \times n)$ matrix obtained from A by deleting the kth row of A and replacing it by the ith row of A; that is, B has two equal rows, the ith and kth, and B is the same as A for all rows but the kth.

In this event it is clear that $\det(B) = 0$, that the cofactor B_{kj} is equal to A_{kj}, and that the entry b_{kj} is equal to a_{ij}. Putting these together gives

$$\begin{aligned}0 = \det(B) &= b_{k1}B_{k1} + b_{k2}B_{k2} + \dots + b_{kn}B_{kn} \\ &= a_{i1}A_{k1} + a_{i2}A_{k2} + \dots + a_{in}A_{kn};\end{aligned}$$

thus the lemma is proved. ■

This lemma can be used to derive a formula for A^{-1}. In particular let A be an $(n \times n)$ matrix, and let C denote the "matrix of cofactors"; $C = (c_{ij})$ is $(n \times n)$, and $c_{ij} = A_{ij}$. The ***adjoint*** matrix of A, denoted $\text{Adj}(A)$ is equal to C^{T}. With these preliminaries we prove Theorem 5.14.

Theorem 5.14

If A is an $(n \times n)$ nonsingular matrix, then

$$A^{-1} = \frac{1}{\det(A)}\text{Adj}(A).$$

Proof. Let $B = (b_{ij})$ be the matrix product of A and $\text{Adj}(A)$. Then the ijth entry of

B is

$$b_{ij} = a_{i1}A_{j1} + a_{i2}A_{j2} + \ldots + a_{in}A_{jn};$$

and by the lemma and Theorem 5.13, $b_{ij} = 0$ when $i \neq j$ while $b_{ii} = \det(A)$. Therefore B is equal to a multiple of det (A) times I, and the theorem is proved. ■

EXAMPLE 5.13 Let A be the matrix

$$A = \begin{bmatrix} 1 & -1 & 2 \\ 2 & 1 & -3 \\ 4 & 1 & 1 \end{bmatrix}.$$

We calculate the nine required cofactors and find

$$\begin{array}{lll} A_{11} = 4 & A_{12} = -14 & A_{13} = -2 \\ A_{21} = 3 & A_{22} = -\ 7 & A_{23} = -5 \\ A_{31} = 1 & A_{32} = \ \ \ 7 & A_{33} = \ \ \ 3. \end{array}$$

The adjoint matrix (the transpose of the cofactor matrix) is

$$\operatorname{Adj}(A) = \begin{bmatrix} 4 & 3 & 1 \\ -14 & -7 & 7 \\ -2 & -5 & 3 \end{bmatrix}.$$

A multiplication shows that the product of A and $\operatorname{Adj}(A)$ is

$$\begin{bmatrix} 14 & 0 & 0 \\ 0 & 14 & 0 \\ 0 & 0 & 14 \end{bmatrix};$$

so $A^{-1} = (1/14)\operatorname{Adj}(A)$ where of course $\det(A) = 14$.

As a final application of determinant theory, we develop a simple test for linear independence of a set of functions. Suppose $f_0(x), f_1(x), \ldots, f_n(x)$ are real-valued functions defined on an interval $[a, b]$. If there exist scalars $a_0, a_1, \ldots, a_n$ (not all of which are zero) such that

$$a_0 f_0(x) + a_1 f_1(x) + \ldots + a_n f_n(x) = 0 \tag{5.13}$$

for all x in $[a, b]$, then $\{f_0(x), f_1(x), \ldots, f_n(x)\}$ is a linearly dependent set of functions (see Section 4.4). If the only scalars for which (5.13) holds for all x in $[a, b]$ are $a_0 = a_1 = \ldots = a_n = 0$, then the set is linearly independent.

A test for linear independence can be formulated from (5.13) as follows. If $a_0, a_1, \ldots, a_n$ are scalars satisfying (5.13), and if the functions $f_i(x)$ are sufficiently differentiable, then we can differentiate both sides of the identity (5.13) and have $a_1 f_0^{(i)}(x) + a_1 f_1^{(i)}(x) + \dots + a_n f_n^{(i)}(x) = 0,\ 1 \leqslant i \leqslant n$. In matrix terms, these equations are

$$\begin{bmatrix} f_0(x) & f_1(x) & \dots & f_n(x) \\ f'_0(x) & f'_1(x) & \dots & f'_n(x) \\ \vdots & & & \vdots \\ f_0^{(n)}(x) & f_1^{(n)}(x) & \dots & f_n^{(n)}(x) \end{bmatrix} \begin{bmatrix} a_0 \\ a_1 \\ \vdots \\ a_n \end{bmatrix} = \begin{bmatrix} 0 \\ 0 \\ \vdots \\ 0 \end{bmatrix}. \qquad (5.14)$$

If we denote the coefficient matrix in (5.14) as $W(x)$, then $\det(W(x))$ is called the *Wronskian* for $\{f_0(x), f_1(x), \dots, f_n(x)\}$. If there is a point x_0 in $[a, b]$ such that det $(W(x_0)) \neq 0$, then the matrix $W(x)$ is nonsingular at $x = x_0$; and the implication is that $a_0 = a_1 = \dots = a_n = 0$. In summary, if the Wronskian is nonzero at any point in $[a, b]$, then $\{f_0(x), f_1(x), \dots, f_n(x)\}$ is a linearly independent set of functions. Note, however, that det $(W(x)) = 0$ for all x in $[a, b]$ does not imply linear dependence (see Example 5.14).

EXAMPLE 5.14 Let $F_1 = \{x, \cos x, \sin x\}$ and $F_2 = \{\sin^2 x, |\sin x| \sin x\}$ for $-1 \leqslant x \leqslant 1$. The respective Wronskians are

$$w_1(x) = \begin{vmatrix} x & \cos x & \sin x \\ 1 & -\sin x & \cos x \\ 0 & -\cos x & -\sin x \end{vmatrix} = x$$

and

$$w_2(x) = \begin{vmatrix} \sin^2 x & |\sin x| \sin x \\ \sin 2x & |\sin 2x \end{vmatrix} = 0.$$

Since $w_1(x) \neq 0$ for $x \neq 0$, F_1 is linearly independent. Even though $w_2(x) = 0$ for all x in $[-1, 1]$, F_2 is also linearly independent, for if $a_1 \sin^2 x + a_2 |\sin x| \sin x = 0$, then at $x = 1$, $a_1 + a_2 = 0$; and at $x = -1$, $a_1 - a_2 = 0$; so $a_1 = a_2 = 0$.

5.5 EXERCISES

1. Use row operations to reduce the determinants to upper-triangular form, and evaluate the original determinant.

a) $\begin{vmatrix} 1 & 2 & 1 \\ 2 & 3 & 2 \\ -1 & 4 & 1 \end{vmatrix}$ b) $\begin{vmatrix} 0 & 3 & 1 \\ 1 & 2 & 1 \\ 2 & -2 & 2 \end{vmatrix}$ c) $\begin{vmatrix} 1 & 0 & 1 \\ 0 & 2 & 4 \\ 3 & 2 & 1 \end{vmatrix}$

2. Repeat Exercise 1 for the determinants in Exercise 1, Section 5.3.
3. Let P be the (3×3) matrix $P = [\mathbf{e}_2, \mathbf{e}_1, \mathbf{e}_3]$.
   a) If A is a (3×3) matrix, verify that forming PA amounts to interchanging rows one and two of A.
   b) Verify that P is symmetric.
4. (This problem is a generalization of Exercise 3.) Let P be the $(n \times n)$ matrix obtained from the $(n \times n)$ identity I by interchanging column i and column j of I.
   a) If $\mathbf{x}$ is a vector in R^n, prove that forming $P\mathbf{x}$ amounts to interchanging the ith and jth components of $\mathbf{x}$. (Hint: Write P in column form and express $P\mathbf{x}$ as a linear combination of the columns of P.)
   b) If A is an $(n \times n)$ matrix, prove that forming PA amounts to interchanging rows i and j of A. [Hint: Use part (a).]
   c) Prove that P is symmetric by showing that $p_{rs} = p_{sr}$ for $1 \leqslant r, s \leqslant n$.
5. Let L be the $(n \times n)$ lower-triangular matrix defined as follows: the main-diagonal entries of L are all 1; and all the other entries of L are zero except the jkth entry, which is the number c where $j > k$. That is, L is the identity matrix except for one entry below the main diagonal—the jkth entry, which is equal to c. If A is an $(n \times n)$ matrix, prove that forming LA amounts to adding a multiple of c times row k to row j. (Hint: Determine the rsth entry of LA by using the definition of matrix multiplication.)
6. Exercises 4 and 5 show that we can use a product PA to interchange rows of A, and a product LA to add a multiple of one row to another. Therefore A can be row reduced to an upper-triangular matrix U by applying a sequence of "elementary matrices" of the form P or L. That is, there are matrices $Q_1, Q_2, \ldots, Q_r$ such that

$$Q_r Q_{r-1} \cdots Q_2 Q_1 A = U$$

   where each Q_m is a matrix of the form P or L. Let $Q = Q_r Q_{r-1} \cdots Q_2 Q_1$, and prove that $\det(Q^{\mathrm{T}}) = \det(Q)$.
7. Let A be the matrix

$$A = \begin{bmatrix} 0 & 1 & 3 \\ 1 & 2 & 4 \\ 2 & 2 & 1 \end{bmatrix}.$$

   Find a matrix P_1 (as in Exercise 4) and two matrices L_1 and L_2 (as in Exercise 5) such that $L_2 L_1 P_1 A = U$ where U is upper triangular. Calculate the product $Q = L_2 L_1 P_1$, and verify that $\det(Q) = \det(Q^{\mathrm{T}})$.
8. An $(n \times n)$ matrix A is called ***skew symmetric*** if $A^{\mathrm{T}} = -A$. Prove that if A is skew symmetric, then $\det(A) = (-1)^n \det(A)$. If n is odd, prove also that A is must be singular.
9. Calculate the adjoint matrix for these matrices.

$$A = \begin{bmatrix} 1 & 0 & 1 \\ 2 & 1 & 2 \\ 1 & 1 & 2 \end{bmatrix} \qquad B = \begin{bmatrix} 2 & 1 & 0 \\ 3 & 0 & 1 \\ 0 & 1 & 1 \end{bmatrix} \qquad C = \begin{bmatrix} 1 & 1 & 1 \\ 1 & 2 & 2 \\ 1 & 3 & 1 \end{bmatrix}$$

10. Prove that if A is singular, then the product of A and Adj (A) is the zero matrix. (Hint: Consider the proof of Theorem 5.14.)
11. Use Theorem 5.14 to calculate the inverse of each matrix in Exercise 9.

12. Determine whether the following sets of functions are linearly independent on the interval $[-1, 1]$.

$G_1 = \{1, x, x^2\}$ $\quad G_2 = \{e^x, e^{2x}, e^{3x}\}$ $\quad G_3 = \{1, \cos^2 x, \sin^2 x\}$

$G_4 = \{1, \cos x, \cos 2x\}$ $\quad G_5 = \{x^2, x|x|\}$ $\quad G_6 = \{x^2, 1 + x^2, 2 - x^2\}$

6
NUMERICAL METHODS IN LINEAR ALGEBRA

†6.1 COMPUTER ARITHMETIC AND ROUNDOFF

The advent of the computer has greatly increased the range of problems in which matrix theory and linear algebra are applicable. Problems that are too large and complex to consider solving by hand are now solved routinely in seconds when the proper techniques are coded and executed on the computer. As computer technology advances, we can expect that problems too large for today's machines will become routine tomorrow. However, every computer has computational limitations that must be considered; and two of these important limitations form the subject of this chapter.

1. Efficient algorithms are necessary; even the largest computer may be unable to perform the number of calculations required by a poorly conceived algorithm.
2. Computer arithmetic is inherently inaccurate; every arithmetic operation in the computer is a potential source of error.

In essence our goal is to design computer programs that solve mathematical problems in a reasonable amount of time with a reasonable standard of accuracy.

In this section we will consider some of the limitations to accuracy. In particular we will describe ***roundoff error***, which loosely means any error introduced into a computation when performing arithmetic operations restricted to a finite number of decimal places. We will explain roundoff error in more detail later; for now we want to observe only that the presence of roundoff error means that arithmetic operations in the machine obey hardly any of the rules for arithmetic in the real number system. For example in the real number system we expect that

$$\begin{aligned} a+b &= b+a \\ a+(b+c) &= (a+b)+c \\ ab &= ba \\ a(bc) &= (ab)c \\ a(b+c) &= ab+ac. \end{aligned} \tag{6.1}$$

† This section contains background material on computer arithmetic. A casual reading of Section 6.1 is all that is required for the later sections, for only indirect reference will be made to Section 6.1.

In any digital device from the smallest hand calculator to the largest computer, the associative and distributive laws in (6.1) do not hold. For example if we form $a+(b+c)$ in a machine and then form $(a+b)+c$, we should not necessarily expect the same result.

EXAMPLE 6.1 Different machines produce different sorts of roundoff errors. As an example, consider the three numbers

$$a = \frac{3.}{7.} \qquad b = \frac{5.}{9.} \qquad c = \frac{7.}{11.}.$$

On an HP-45 calculator the products $a(bc)$ and $(ab)c$ are different while $a(b+c)$ is equal to $ab+ac$. By contrast when the same operations were coded in FORTRAN and run on an IBM 370/158 (using one of the several FORTRAN compilers), the opposite result was produced. That is, $(ab)c$ and $a(bc)$ were equal while $a(b+c)$ was not equal to $ab+ac$. With regard to hand calculators we note that some of them display fewer digits than the calculators use internally. So for example to see the difference between $a(bc)$ and $(ab)c$ for various numbers a, b, and c, it may be necessary to calculate $r = (ab)c$, store r in memory, then calculate $s = a(bc)$, and form

$$q = \frac{1}{r-s}.$$

An example should serve to illustrate how roundoff errors occur when number representations are limited to a finite number of places. For instance suppose the restriction is that every number a must be represented in the form

$$a = \pm .a_1a_2a_3a_4a_5 \times 10^c \tag{6.2}$$

where a_i is an integer with $0 \leqslant a_i < 10$ for $1 \leqslant i \leqslant 5$, and where $a_1 \neq 0$ unless $a = 0$. For example if $a = 2147.6$ and $f = 16.211$, then we will represent a and f as

$$a = .21476 \times 10^4 \qquad f = .16211 \times 10^2.$$

(These restrictions are exactly like those under which a computer or hand calculator must operate.)

In order to add a and f, we must first align the decimal places; so we express f in the form $f = .0016211 \times 10^4$. We can now form the sum $s = a+f$ and obtain

$$s = a+f = .2163811 \times 10^4.$$

Since we are restricted to representing s in the form (6.2), we will have to discard two of the seven digits in s and represent $s = a+f$ as

$$a+f = .21638 \times 10^4.$$

This representation is obviously an error (a roundoff error), and any further operations we carry out using s will be contaminated by this error. Multiplying a

times f produces an error also since

$$af = .348147436 \times 10^5,$$

and we will have to store af as $.34814 \times 10^5$ or as $.34815 \times 10^5$ (depending on the rounding rule we choose to use). Again a roundoff error has been committed and will contaminate all further calculations.

We next illustrate a particularly disastrous kind of roundoff error. Suppose $a = .21476 \times 10^4$, and suppose $q = .13129 \times 10^{-2}$. To form $a+q$, we must again align the decimal places; and we write q as

$$q = .00000013129 \times 10^4.$$

Therefore $a+q$ is given by

$$a+q = .21476013129 \times 10^4,$$

and the sum $a+q$ is stored as

$$a+q = .21476 \times 10^4.$$

In effect we have said that $a+q = a$; and while this statement is not a bad approximation, it is an error. On philosophical grounds it is a very bad error—we have lost all the information contained in q. In general we should be careful when there is a possibility of adding numbers of widely varying magnitudes since accuracy may be degraded. There are many other potential (some quite subtle) sources of computational error, but it would be inappropriate to discuss them in this brief introduction. We will mention some of the principal strategies of error control for numerical procedures in linear algebra in subsequent sections.

The principles illustrated above carry over to the computer; only the details are different. At the machine level a number can be thought of as being represented by a finite string of 0's and 1's. The arithmetic operations are actually carried out (most usually) in base 2, base 8, base 10, or base 16; and the base dictates the actual arrangement of the 0's and 1's in the string. To be specific, suppose the base of the arithmetic system is b. Then any real number a is represented in the form

$$a = \pm .a_1 a_2 \ldots a_m \times b^c \tag{6.3}$$

where each a_i is an integer, $0 \leq a_i < b$, and where $a_1 \neq 0$ unless $a = 0$. In (6.3), m is fixed and determined by the "word length" of the computer. [Many hand calculators use $b = 10$ and a value of m on the order of 10 to 12. The IBM 360 and 370 computers use $b = 16$ (hexadecimal arithmetic) and $m = 6$.]

The IBM 360 and 370 computers serve as convenient examples to illustrate some of the details of (6.3). In these computers a "real" number is represented as a string of thirty-two 0's and 1's as in Fig. 6.1. In Fig. 6.1, the compartments labeled Byte 1, Byte 2, Byte 3, and Byte 4 each represent a string of eight 0's and 1's. (The 0's and 1's represent binary digits or "bits," and there are eight bits to a byte.) Bytes 2, 3, and 4 each consist of two strings of four 0's and 1's—these are $h_1, h_2, \ldots, h_6$ in Fig. 6.1. A real number a is then represented internally in the form

$$a = \pm .h_1h_2h_3h_4h_5h_6 \times 16^c \tag{6.4}$$

where each h_i is a hexadecimal digit, an integer between 0 and 15. These hexadecimal digits are each represented in base 2 (as 0's and 1's) according to Table 6.1. In this scheme A stands for 10, B for 11, etc. For example the number $d = .B3C \times 16^2$ is the same as

$$d = \left(\frac{11}{16} + \frac{3}{16^2} + \frac{12}{16^3}\right) \times 16^2.$$

Thus $d = 11 \times 16 + 3 + 12/16 = 176 + 3 + .75$, or $d = 179.75$ in base 10. In the context of Fig. 6.1 the number d has the form $d = .B3C000 \times 16^2$; so Byte 2 would contain h_1h_2 or 10110011, Byte 3 would have h_3h_4 or 11000000, and Byte 4 is 00000000.

Byte 1	Byte 2		Byte 3		Byte 4	
	h_1	h_2	h_3	h_4	h_5	h_6

Figure 6.1

TABLE 6.1

0	0000	8	1000
1	0001	9	1001
2	0010	A	1010
3	0011	B	1011
4	0100	C	1100
5	0101	D	1101
6	0110	E	1110
7	0111	F	1111

In the hexadecimal representation (6.4) if $a \geqslant 0$, then 0 occupies the first location of Byte 1 while 1 is present if $a < 0$. The remaining 7 locations of Byte 1 store the exponent c. One of the vagaries of hexadecimal arithmetic is that certain nice decimal numbers, such as .1, have infinite hexadecimal expansions. That is, if $(.h_1h_2h_3 \ldots)_{16}$ represents

$$\frac{h_1}{16} + \frac{h_2}{16^2} + \frac{h_3}{16^3} + \ldots,$$

then 1/10 has the form $(.1999999 \ldots)_{16}$.

The representation (6.3) is for "real" variables; "integer" variables are treated somewhat differently. In addition different software systems on the same computer may use different rules of arithmetic.

6.1 EXERCISES

1. The following binary (base-2) numbers are expressed as in (6.3). Convert them to their decimal (base-10) equivalents.

$$a = .101 \times 2^3 \qquad b = .11011 \times 2^3 \qquad c = .10011 \times 2^6$$

2. Convert the octal (base-8) numbers to their decimal equivalents.

$$a = .1314 \times 8^3 \qquad b = .7113 \times 8^2 \qquad c = .624 \times 8^4$$

3. Convert the hexadecimal numbers to their decimal equivalents.

$$a = .B29 \times 16^3 \qquad b = .91C \times 16^2 \qquad c = .333 \times 16^2$$

4. Does every hexadecimal number with a finite expansion (such as those in Exercise 3) have a finite decimal expansion?
5. Convert the following decimal numbers to base 2 in the form of (6.3).

$$a = 14 \qquad b = 29.5 \qquad c = 6.75 \qquad d = 134.875$$

6. Convert the decimal numbers in Exercise 5 to base 8.
7. Convert the decimal numbers in Exercise 5 to base 16.
8. Give an example of a decimal number that has a finite decimal expansion but an infinite expansion in base 8.
9. Prove that the hexadecimal expansion of the decimal number 1/10 is .19999 . . . . That is, show that

$$1/10 = 1/16 + \sum_{n=2}^{\infty} 9/16^n.$$

(Hint: The infinite series is a geometric series.)

10. Verify that $(ab)c \neq a(bc)$ on a hand calculator where $a = 1./7.$, $b = 1./9.$, $c = 1./11.$
11. Suppose we are using a machine that does base-10 arithmetic and carries five places [in (6.3), $b = 10$ and $m = 5$]. Suppose that the result of any arithmetic operation is truncated to five places. (For example if $a + b = .314263 \times 10^2$, then $a + b$ is stored as $.31426 \times 10^2$. If $a + b = .314269 \times 10^2$, then $a + b$ is also stored as $.31426 \times 10^2$.) Given the machine numbers

$$a = .20211 \times 10^2 \qquad c = .50000 \times 10^1$$
$$b = .31323 \times 10^{-1} \qquad d = .60000 \times 10^{-4},$$

calculate what the machine will store as the result of performing $a + b$, ab, $a(b + 3c)$, $c + (d + d)$, and $(c + d) + d$.

6.2 GAUSS ELIMINATION

Besides being easy to understand and serving as a valuable theoretical tool, Gauss elimination is one of the most popular computational procedures for solving a linear system $A\mathbf{x} = \mathbf{b}$. In this section we will discuss some of the aspects that are related to implementing Gauss elimination on the computer. In order not to obscure the main points, we will restrict our discussion to systems $A\mathbf{x} = \mathbf{b}$ where A is square. The basic operations in Gauss elimination or reduction to echelon form are row interchanges and the addition of a multiple of one row to another. Clearly the only source of computational errors is that of adding one row to another; so we analyze this basic operation first.

To begin, consider the $(n \times n)$ system

$$\begin{array}{llllll} a_{11}x_1 & +a_{12}x_2 & + \dots + & a_{1n}x_n & = & b_1 \\ a_{21}x_1 & +a_{22}x_2 & + \dots + & a_{2n}x_n & = & b_2 \\ \vdots & & & \vdots & & \vdots \\ a_{n1}x_1 & +a_{n2}x_2 & + \dots + & a_{nn}x_n & = & b_n. \end{array} \tag{6.5a}$$

As we know if $a_{11} \neq 0$, we can eliminate x_1 from the kth equation by multiplying row one by $-a_{k1}/a_{11}$ and adding the result to row k for $2 \leqslant k \leqslant n$. But as we saw in the last section if we add a large number to a small one, we essentially erase much of the information contained in the small number. Similarly if $-a_{k1}/a_{11}$ is very large, then much of the information in equation k is lost when we add a multiple of $-a_{k1}/a_{11}$ times the first row to row k. On the other hand if $-a_{k1}/a_{11}$ is quite small (relative to the entries in the kth row), then we will disturb the entries in the kth row only slightly; that is, the information contained in the kth equation will be left very nearly intact.

In summary, roundoff-error considerations suggest that if we replace row k by a "large" multiple of row one, then we have in effect replaced row k by a multiple of row one. Alternatively if we replace row k by a "small" multiple of row one, then we have left row k reasonably intact. Since the multiplier for row one that is used to eliminate row k is $-a_{k1}/a_{11}$, it follows that we would like a_{k1}/a_{11} to be as small as possible in absolute value. This observation leads to a "pivoting strategy."

1. Search column one for the largest entry, say a_{r1}; that is,

$$|a_{r1}| \geqslant |a_{i1}|, \quad 1 \leqslant i \leqslant n.$$

2. Interchange row one and row r in (6.5a).

By interchanging row one and row r, we obtain an equivalent system:

$$\begin{array}{l} a_{r1}x_1 + a_{r2}x_2 + \dots + a_{rn}x_n = b_r \\ a_{21}x_1 + a_{22}x_2 + \dots + a_{2n}x_n = b_2 \\ \vdots \qquad\qquad\qquad\quad \vdots \quad\; \vdots\,; \\ a_{n1}x_1 + a_{n2}x_2 + \dots + a_{nn}x_n = b_n \end{array} \tag{6.5b}$$

and we are assured that when we eliminate x_1 from row k by multiplying the first row of (6.5b) by $-a_{k1}/a_{r1}$, the multiplier $-a_{k1}/a_{r1}$ is as small as possible.

EXAMPLE 6.2 As an illustration of the observations above, consider the (2×2) system

$$\begin{aligned} -.001x_1 + x_2 &= 1 \\ x_1 + x_2 &= 2. \end{aligned} \tag{6.6a}$$

To eliminate x_1 from row two, we multiply row one by 1000, add the result to row two, and obtain an equivalent system:

$$\begin{aligned} -.001\,x_1 + x_2 &= 1 \\ 1001\,x_2 &= 1002. \end{aligned}$$

If we were working with a machine that carries only three places, the coefficients in the last equation would be rounded to three places:

$$\begin{aligned} -.001\,x_1 + x_2 &= 1 \\ 1000\,x_2 &= 1000. \end{aligned} \tag{6.6b}$$

Solving (6.6b), we would have $x_2 = 1$ and $x_1 = 0$; and this result is certainly not a solution of (6.6a), either mathematically or even in our hypothetical three-place machine. In effect we have erased the information in the second equation of (6.6a) by adding a multiple of 1000 times row one to row two—in our three-place machine $x_1 = 0$ and $x_2 = 1$ satisfy equation one but not equation two of (6.6a); the information in equation two has been lost in (6.6b).

Suppose now that we pivot, interchanging equations one and two in (6.6a), and obtain

$$\begin{aligned} x_1 + x_2 &= 2 \\ -.001x_1 + x_2 &= 1. \end{aligned} \tag{6.7a}$$

If we eliminate x_1 in equation two by multiplying equation one by .001 and adding the result to equation two, we obtain

$$\begin{aligned} x_1 + x_2 &= 2 \\ 1.001x_2 &= 1.002, \end{aligned}$$

which becomes in our three-place machine

$$\begin{aligned} x_1 + x_2 &= 2 \\ x_2 &= 1. \end{aligned} \tag{6.7b}$$

Solving (6.7b), we get $x_2 = 1$ and $x_1 = 1$; and either mathematically or in our machine $x_1 = 1$ and $x_2 = 1$ is a better solution of (6.7a) than our first solution was. In fact in the machine $x_1 = 1$ and $x_2 = 1$ solve (6.7a) "exactly." The mathematical solution is of course

$$x_1 = \frac{1000}{1001} = 1 - \frac{1}{1001} \qquad x_2 = \frac{1002}{1001} = 1 + \frac{1}{1001};$$

so our second "machine solution" is not too far off.

The pivoting strategy described above is called ***partial pivoting***. In detail we interchange equation r and equation one in (6.5a) where $|a_{r1}| \geqslant |a_{i1}|$, $1 \leqslant i \leqslant n$; and then we eliminate x_1 from equations 2, 3, . . . , n. The result is a system of the form

$$\begin{aligned} a'_{11}x_1 + a'_{12}x_2 + \dots + a'_{1n}x_n &= b'_1 \\ a'_{22}x_2 + \dots + a'_{2n}x_n &= b'_2 \\ \vdots \qquad\qquad \vdots \quad &\quad \vdots \\ a'_{n2}x_2 + \dots + a'_{nn}x_n &= b'_n. \end{aligned}$$

Given the system above, we search column two, rows 2, 3, . . . , n for an entry a'_{j2} where $|a'_{j2}| \geqslant |a'_{i2}|, 2 \leqslant i \leqslant n$. We then interchange rows j and two; and we eliminate x_2 in rows 3, 4, . . . , n by using the multiplier $-a'_{k2}/a'_{j2}$ to eliminate x_2 in row k. Having done this procedure, we move to column three and search for the largest entry in rows 3, 4, . . . , n, etc.

Instead of partial pivoting, we might consider ***total pivoting***, which requires more of a search. That is, in (6.5a) we could ask for the largest entry a_{rs} where $|a_{rs}| \geqslant |a_{ij}|$, $1 \leqslant i \leqslant n$, $1 \leqslant j \leqslant n$. Having a_{rs}, we could interchange row one and row r, and then interchange column one and column s in order to move a_{rs} to the (1, 1) position. While this strategy is desirable, it is expensive and requires n^2 comparisons to find the largest entry, a_{rs}. In many Gauss elimination programs partial pivoting is used rather than total pivoting.

The other computational aspect of Gauss elimination that bears mentioning is that the procedure requires on the order of $n^3/3$ multiplies and adds to solve an $(n \times n)$ system. To elaborate, we note that one crude measure of the "efficiency" of an algorithm is an ***operations count,*** a count of the number of arithmetic operations that must be executed when the algorithm is used. In the exercises the reader is asked to show that given an $(n \times n)$ system $A\mathbf{x} = \mathbf{b}$, then about $n^3/3$ multiplies and adds are necessary to reduce the augmented matrix $[A \,\vdots\, \mathbf{b}]$ to echelon form. Given the echelon form, it then takes about n^2 multiplies and adds to solve the system. The number n^2 tells us why it is not efficient to solve $A\mathbf{x} = \mathbf{b}$ by calculating A^{-1} and then forming $A^{-1}\mathbf{b}$. To see why, we need only note that forming the product $A^{-1}\mathbf{b}$ when A^{-1} is $(n \times n)$ takes n^2 multiplies and adds, the same number of operations we perform when we solve a system that is in echelon form. Thus the only way that forming $A^{-1}\mathbf{b}$ could be competitive with simply solving $A\mathbf{x} = \mathbf{b}$ directly is if fewer operations were required to obtain A^{-1} than to reduce $[A \,\vdots\, \mathbf{b}]$ to echelon form. Since finding A^{-1} is most easily done by reducing the $(n \times 2n)$ matrix $[A \,\vdots\, I]$ to echelon form, it is almost obvious that we need to do more work to find A^{-1} than to reduce $[A \,\vdots\, \mathbf{b}]$ (a count shows that more than twice as much work is required to find A^{-1} than to solve $A\mathbf{x} = \mathbf{b}$). In addition more than efficiency is at stake—a procedure that uses twice as many arithmetic operations has the potential of making twice as many errors. We can expect that $A^{-1}\mathbf{b}$ is not so accurate a solution to $A\mathbf{x} = \mathbf{b}$ as is the solution we calculate directly.

In Fig. 6.2 we have listed a FORTRAN program, Subroutine GAUSS, which implements Gauss elimination with partial pivoting. Our purpose in this listing is to provide the interested reader with an easy to understand and easy to use computational tool. This program, as well as others we include later was written with an emphasis on simplicity and readability with a minimum of user options. In the interest of clarity we included only enough comment statements to highlight the main computational segments of the program. The following is a description of the parameters that are not explained in the comments.

Subroutine GAUSS uses Gauss elimination with partial pivoting to solve $A\mathbf{x} = \mathbf{b}$. The required inputs are the following.

A	An (N × N) matrix.
B	An (N × 1) vector.
N	The size of A.
MAXDIM	The declared dimension of the array containing A in the calling program.

The outputs from Subroutine GAUSS are the following.

X	The machine solution of $A\mathbf{x} = \mathbf{b}$.
IERROR	A flag set equal to 2 if Gauss elimination cannot proceed because of a zero pivot. IERROR is set equal to 1 if Gauss elimination is successful.
RNORM	This number is set equal to the norm of the residual vector, $\mathbf{b} - A\mathbf{x}$, where $\mathbf{x}$ is the machine solution.

In brief, GAUSS sets up the augmented matrix $[A \mid \mathbf{b}]$ and stores this in array AUG. AUG is reduced to echelon form; and if no zero diagonal entries are found, the solution X is calculated. As a crude error test, RNORM $= \|\mathbf{b} - A\mathbf{x}\|$ is calculated and returned to the calling program.

```
      SUBROUTINE GAUSS(A,B,X,N,MAXDIM,IERROR,RNORM)
      DIMENSION A(MAXDIM,MAXDIM),B(MAXDIM),X(MAXDIM)
      DIMENSION AUG(50,51)
      NM1=N-1
      NP1=N+1
C
C   SET UP THE AUGMENTED MATRIX FOR AX=B.
C
         DO 2 I=1,N
            DO 1 J=1,N
            AUG(I,J)=A(I,J)
    1       CONTINUE
         AUG(I,NP1)=B(I)
    2    CONTINUE
C
C   THE OUTER LOOP USES ELEMENTARY ROW OPERATIONS TO TRANSFORM
C   THE AUGMENTED MATRIX TO ECHELON FORM.
C
         DO 8 I=1,NM1
C
C   SEARCH FOR THE LARGEST ENTRY IN COLUMN I, ROWS I THROUGH N.
C   IPIVOT IS THE ROW INDEX OF THE LARGEST ENTRY.
C
         PIVOT=0.
            DO 3 J=I,N
            TEMP=ABS(AUG(J,I))
            IF(PIVCT.GE.TEMP)  GO TO 3
            PIVOT=TEMP
            IPIVOT=J
```

Figure 6.2

```
    3       CONTINUE
         IF(PIVOT.EQ.0.)  GO TO 13
         IF(IPIVOT.EQ.I)  GO TO  5
C
C   INTERCHANGE ROW I AND ROW IPIVOT.
C
            DO 4 K=I,NP1
            TEMP=AUG(I,K)
            AUG(I,K)=AUG(IPIVOT,K)
            AUG(IPIVOT,K)=TEMP
    4       CONTINUE
C
C   ZERO ENTRIES (I+1,I), (I+2,I),...,(N,I) IN THE AUGMENTED MATRIX.
C
    5   IP1=I+1
            DO 7 K=IP1,N
            Q=-AUG(K,I)/AUG(I,I)
            AUG(K,I)=0.
              DO 6 J=IP1,NP1
              AUG(K,J)=Q*AUG(I,J)+AUG(K,J)
    6         CONTINUE
    7       CONTINUE
    8     CONTINUE
         IF(AUG(N,N).EQ.0.)  GO TO 13
C
C   BACKSOLVE TO OBTAIN A SOLUTION TO AX=B.
C
        X(N)=AUG(N,NP1)/AUG(N,N)
          DO 10 K=1,NM1
          Q=0.
            DO 9 J=1,K
            Q=Q+AUG(N-K,NP1-J)*X(NP1-J)
    9       CONTINUE
          X(N-K)=(AUG(N-K,NP1)-Q)/AUG(N-K,N-K)
   10     CONTINUE




C
C   CALCULATE THE NORM OF THE RESIDUAL VECTOR, B-AX.
C   SET IERROR=1 AND RETURN.
C
        RSQ=0.
          DO 12 I=1,N
          Q=0.
            DO 11 J=1,N
            Q=Q+A(I,J)*X(J)
   11       CONTINUE
          RESI=B(I)-Q
          RMAG=ABS(RESI)
          RSQ=RSQ+RMAG**2
   12     CONTINUE
        RNORM=SQRT(RSQ)
        IERROR=1
        RETURN
C
C   ABNORMAL RETURN --- REDUCTION TO ECHELON FORM PRODUCES A ZERO
C   ENTRY ON THE DIAGONAL.  THE MATRIX A MAY BE SINGULAR.
C
   13 IERROR=2
        RETURN
        END
```

Figure 6.2 *(continued)*

An example of a simple program that uses Subroutine GAUSS to solve a linear system is given in Fig. 6.3. As a guide to interpreting the output of a linear-equation solver such as GAUSS, we first observe that the machine may not recognize a coefficient matrix A as being singular when it is indeed singular. For example with the inputs

$$\mathrm{N} = 2, \qquad \mathrm{A} = \begin{bmatrix} 1 & 4 \\ 3 & 12 \end{bmatrix}, \qquad \mathrm{B} = \begin{bmatrix} 4 \\ 5 \end{bmatrix},$$

the following was output*:

$$\mathrm{IERROR} = 1,\ \mathrm{RNORM} = .117047\mathrm{E}02,\ \mathrm{X} = \begin{bmatrix} -.420336\mathrm{E}17 \\ .105084\mathrm{E}17 \end{bmatrix}.$$

Here the size of $\|\mathbf{b} - A\mathbf{x}\|$ gives a clue that the machine solution is suspect.

```
      DIMENSION A(20,20),B(20),X(20)
      MAXDIM=20
    1 READ 100,N
      IF(N.LE.1)  STOP
         DO 2 I=1,N
         READ 101,(A(I,J),J=1,N),B(I)
    2    CONTINUE
      CALL GAUSS(A,B,X,N,MAXDIM,IERROR,RNORM)
      PRINT 102,IERROR
      IF(IERROR.EQ.2)  GO TO 1
      PRINT 103,RNORM
      PRINT 104,(X(I),I=1,N)
  100 FORMAT(I2)
  101 FORMAT(21F4.0)
  102 FORMAT(8H IERROR=,I3)
  103 FORMAT(7H RNORM=,E20.6)
  104 FORMAT(1H ,6E16.6)
      GO TO 1
      END
```

Figure 6.3

Runs with nonsingular matrices are much more satisfactory. For example with input of

$$\mathrm{N} = 4, \qquad \mathrm{A} = \begin{bmatrix} 5 & 7 & 6 & 5 \\ 7 & 10 & 8 & 7 \\ 6 & 8 & 10 & 9 \\ 5 & 7 & 9 & 10 \end{bmatrix} \qquad \mathrm{B} = \begin{bmatrix} 23 \\ 32 \\ 33 \\ 31 \end{bmatrix}, \tag{6.8}$$

the output was

$$\mathrm{IERROR} = 1,\ \mathrm{RNORM} = .502430\mathrm{E}{-14},\ \mathrm{X} = \begin{bmatrix} .100000\mathrm{E}01 \\ .100000\mathrm{E}01 \\ .100000\mathrm{E}01 \\ .100000\mathrm{E}01 \end{bmatrix};$$

and the machine solution is correct to as many places as listed.

---

* This output was from an IBM 370/158, using double precision arithmetic (single precision on an IBM 360 or 370 is not suited for most scientific computation).

As a final note of caution, suppose $\mathbf{x}_m$ denotes the machine solution to $A\mathbf{x} = \mathbf{b}$ and $\mathbf{x}_t$ denotes the actual (mathematical) solution. We would like some reasonable estimate of how far $\mathbf{x}_m$ is from $\mathbf{x}_t$, but unfortunately this estimate is hard to obtain when we do not know $\mathbf{x}_t$. The number RNORM serves as a rough guide; and if RNORM $= \|\mathbf{b} - A\mathbf{x}_m\|$ is not small, then we suspect that the matrix A is badly behaved in some fashion. But even if RNORM is small, we may have to be careful. To see why, suppose $\mathbf{r} = \mathbf{b} - A\mathbf{x}_m$, and suppose the true solution is $\mathbf{x}_t$ so that $\mathbf{b} = A\mathbf{x}_t$. In this case

$$\mathbf{r} = \mathbf{b} - A\mathbf{x}_m = A\mathbf{x}_t - A\mathbf{x}_m;$$

and if A is nonsingular, we can write $\mathbf{r} = A\mathbf{x}_t - A\mathbf{x}_m = A(\mathbf{x}_t - \mathbf{x}_m)$ as

$$A^{-1}\mathbf{r} = \mathbf{x}_t - \mathbf{x}_m.$$

From this expression we see that even if $\mathbf{r}$ is "small," $\mathbf{x}_t - \mathbf{x}_m$ may be "large" if A^{-1} is "large."

6.2 EXERCISES

1. Write a program that uses Subroutine GAUSS to calculate A^{-1} where A is $(n \times n)$. To do this, set up an $(n \times n)$ array AINV(I, J) to hold A^{-1} where the jth column of AINV is the result of solving $A\mathbf{x} = \mathbf{e}_j$ by GAUSS. Test your program on the matrix A in (6.8); it is known that

$$A^{-1} = \begin{bmatrix} 68 & -41 & -17 & 10 \\ -41 & 25 & 10 & -6 \\ -17 & 10 & 5 & -3 \\ 10 & -6 & -3 & 2 \end{bmatrix}.$$

2. Use Subroutine GAUSS to solve the system $A\mathbf{x} = \mathbf{b}$ given in (6.8). Use your *machine* version of A^{-1}, found by the program in Exercise 1, to calculate $A^{-1}\mathbf{b}$; and compare the two answers by calculating the norm of the residual vector.

3. Use GAUSS to solve $A\mathbf{x} = \mathbf{b}$ where A, $\mathbf{b}$, and the solution $\mathbf{x}_t$ are

$$A = \begin{bmatrix} 1 & -2 & 3 & 1 \\ -2 & 1 & -2 & -1 \\ 3 & -2 & 1 & 5 \\ 1 & -1 & 5 & 3 \end{bmatrix}, \quad \mathbf{b} = \begin{bmatrix} 3 \\ -4 \\ 7 \\ 8 \end{bmatrix}, \quad \mathbf{x}_t = \begin{bmatrix} 1 \\ 1 \\ 1 \\ 1 \end{bmatrix}.$$

4. Small changes in the right-hand side of $A\mathbf{x} = \mathbf{b}$ may lead to relatively large changes in the solution. As an illustration, solve $A\mathbf{x} = \mathbf{b}$ where A is the matrix in (6.8) and $\mathbf{b}$ is

$$\text{a) } \mathbf{b} = \begin{bmatrix} 23.01 \\ 31.99 \\ 32.99 \\ 31.01 \end{bmatrix}, \quad \text{b) } \mathbf{b} = \begin{bmatrix} 23.1 \\ 31.9 \\ 32.9 \\ 31.1 \end{bmatrix}.$$

5. Let A_1 and A_2 be the matrices obtained from A in (6.8) by replacing the (1, 1) entry of A by 5.01 and 4.99, respectively. Using the program in Exercise 1, calculate A_1^{-1} and A_2^{-1}. Compare your answers with A^{-1}.

6. FORTRAN has provisions for complex arithmetic, and it is easy to modify Subroutine GAUSS to solve systems with complex coefficients. The required changes are these (see the listing in Fig. 6.2).

   a) Replace the two dimension declarations by

```
COMPLEX A(MAXDIM,MAXDIM),B(MAXDIM),X(MAXDIM)
COMPLEX AUG(50,51)
COMPLEX CTEMP,Q,RESI,CABS.
```

   b) In the row-interchange loop, change TEMP to CTEMP so that the loop reads

```
DO 4 K=I,NP1
CTEMP=AUG(I,K)
AUG(I,K)=AUG(IPIVOT,K)
AUG(IPIVOT,K)=CTEMP
CONTINUE.
```

   c) Change the divide check statement just before the backsolving segment to

```
IF(CABS(AUG(N,N)).EQ.0.)   GC TO 13.
```

In any program that calls GAUSS, the arrays A, B, and X must also be declared complex. There are two ways to make a complex assignment. For example if S is declared to be complex, and we want to assign $3+4i$ to S, we can write S = (3., 4.) or we can write S = CMPLX(3., 4.). To read in a complex matrix A, we can read the real and imaginary parts of A into two real arrays U and V. In particular if $a_{jk} = u_{jk} + iv_{jk}$, then the statement

A(J, K) = CMPLX(U(J, K), V(J, K))

will set up the matrix A. Finally all complex-valued functions and variables must be declared complex. Thus the functions CABS, CSQRT, CMPLX, CSIN, etc. must be declared to be of complex type in the beginning.

Modify GAUSS to do complex arithmetic, and use it to solve the systems in Exercise 5, Section 3.5.

7. If Q is an $(n \times n)$ matrix and $\mathbf{b}$ is in R^n, verify that forming the product $Q\mathbf{b}$ requires n^2 multiplies.

8. If T is an $(n \times n)$ upper-triangular matrix and $\mathbf{b}$ is a vector in R^n, verify that it requires $1+2+3+\ldots+n = n(n+1)/2$ multiplies and/or divides to solve $T\mathbf{x} = \mathbf{b}$. (Hint: Write out the system $T\mathbf{x} = \mathbf{b}$, and note that it takes one divide to find x_n, two multiplies and/or divides to obtain x_{n-1}, etc. Observe that Exercises 7 and 8 show that calculating $A^{-1}\mathbf{b}$ requires more effort than solving $T\mathbf{x} = \mathbf{b}$.)

9. Let $[A \mid \mathbf{b}]$ be the augmented matrix for the $(n \times n)$ system $A\mathbf{x} = \mathbf{b}$. To zero the $(r, 1)$ entry of $[A \mid \mathbf{b}]$ with a row operation requires the formation of a multiple $m = a_{r1}/a_{11}$ (one divide) and the replacement of the (r, j) entry of $[A \mid \mathbf{b}]$ by $a_{rj} - ma_{1j}$ for $j = 2, 3, \ldots, n+1$. Thus $n+1$ multiplies and/or divides are required for each row in the first step of reducing $[A \mid \mathbf{b}]$ to echelon form; a total of $(n-1)(n+1)$ multiplies and/or divides is needed.

   a) Verify that $(n-2)n$ multiplies and/or divides are necessary to zero the $(3, 2), (4, 2), \ldots, (n, 2)$ entries.

b) Verify that $(n-i)(n-i+2)$ multiplies and/or adds are required at the ith stage of Gauss elimination (zeroing the entries below the main diagonal in column i).

c) The process of Gauss elimination stops after column $n-1$; so a total of

$$S = (n-1)(n+1) + (n-2)n + \ldots + (n-i)(n-i+2) + \ldots + 1\cdot 3$$

multiplies and/or divides is required. Evaluate this sum by expressing it as

$$S = \sum_{i=1}^{n-1} (n-i)(n-i+2).$$

6.3 THE POWER METHOD FOR EIGENVALUES

Finding the eigenvalues of an $(n \times n)$ matrix A is a harder computational problem than is solving $A\mathbf{x} = \mathbf{b}$. In particular as we commented in Chapter 3, we cannot usually expect to find the eigenvalues of an $(n \times n)$ matrix in a finite number of steps when $n \geqslant 5$ (since this process amounts to solving a polynomial equation). Given the development of Chapter 3, one might feel that the following is a reasonable computational scheme for finding the eigenvalues of A.

1. Reduce A to Hessenberg form, H.
2. Find the characteristic polynomial, $p(t)$, for H.
3. Find the roots of $p(t) = 0$.

Unfortunately the approach above may well lead to severe errors, and it is relatively easy to see why. In particular when A is reduced to Hessenberg form, roundoff errors will occur; and the Hessenberg matrix found by the machine will not be quite what it should be (if exact arithmetic were used). As the example below shows, even though two matrices are almost the same, it is possible for their eigenvalues to be substantially different.

EXAMPLE 6.3 This example is due to Forsythe and is an extreme instance of two nearly equal matrices with different eigenvalues. Let H and $H+E$ be $(n \times n)$ Hessenberg matrices where

$$H = \begin{bmatrix} 1 & 0 & 0 & \ldots & 0 & 0 \\ 1 & 1 & 0 & \ldots & 0 & 0 \\ 0 & 1 & 1 & \ldots & 0 & 0 \\ \vdots & & & & & \vdots \\ 0 & 0 & 0 & \ldots & 1 & 0 \\ 0 & 0 & 0 & \ldots & 1 & 1 \end{bmatrix} \qquad H+E = \begin{bmatrix} 1 & 0 & 0 & \ldots & 0 & \varepsilon \\ 1 & 1 & 0 & \ldots & 0 & 0 \\ 0 & 1 & 1 & \ldots & 0 & 0 \\ \vdots & & & & & \\ 0 & 0 & 0 & \ldots & 1 & 0 \\ 0 & 0 & 0 & \ldots & 1 & 1 \end{bmatrix}.$$

Thus H consists entirely of 0's except for the main diagonal and the subdiagonal, which contain 1's. The matrix $H+E$ is equal to H except for the entry ε in the $(1, n)$ position.

It can be shown (Exercise 3, Section 6.3) that the characteristic polynomials of H and $H+E$ are

$$p(t) = (1-t)^n \qquad \text{and} \qquad q(t) = (1-t)^n + (-1)^{n+1}\varepsilon, \quad \text{respectively.}$$

To see how the eigenvalues might differ, suppose $n = 10$ and $\varepsilon = 2^{-10}$. We see that $q(t) = 0$ means $(1-t)^{10} = 2^{-10}$; and one solution to this equation is $1-t = 2^{-1}$, or $t = .5$ (a change in H of amount 2^{-10} produces a 50 percent change in one eigenvalue of H). Not every perturbation of entries in H will lead to such a dramatic change in the eigenvalues. For example if the (1, 1) entry of H is changed to $1+\varepsilon$, then an eigenvalue of H has been perturbed only by amount ε.

Computational experience, as well as examples such as the one above, suggests that no matter how carefully we carry out Hessenberg reduction on A, the inevitable roundoff errors will produce a Hessenberg matrix in the machine whose eigenvalues are not the same as those of A (and may differ substantially from the eigenvalues of A). To overcome this difficulty, we employ a device that is relatively common in numerical methods, that of "refinement." Briefly our approach will be to reduce A to Hessenberg form H, find the eigenvalues of H, and then regard these eigenvalues of H as *estimates* to the eigenvalues of A. We next apply the inverse power method (as described in the next section) to the original matrix A and refine (or correct) these estimates by iteration. Before the inverse power method is described, it is convenient to consider a closely related algorithm, the power method.

The power method is an iterative procedure that can be used to estimate the "dominant" eigenvalue of a matrix. To begin, suppose A is an $(n \times n)$ matrix, and suppose A has eigenvalues $\lambda_1, \lambda_2, \ldots, \lambda_n$ with corresponding eigenvectors $\mathbf{u}_1, \mathbf{u}_2, \ldots, \mathbf{u}_n$; so $A\mathbf{u}_j = \lambda_j\mathbf{u}_j$, $1 \leqslant j \leqslant n$. (Of course we do not know the eigenvalues or the eigenvectors; the objective of the power method is to estimate an eigenvalue accurately.) For simplicity we will put a rather severe restriction on the eigenvectors; we will assume that $\{\mathbf{u}_1, \mathbf{u}_2, \ldots, \mathbf{u}_n\}$ is linearly independent. [Recall that the eigenvectors may have complex components; so linear independence means that the only scalars, either real or complex, that satisfy $b_1\mathbf{u}_1 + b_2\mathbf{u}_2 + \ldots + b_n\mathbf{u}_n = \boldsymbol{\theta}$ are the scalars $b_1 = b_2 = \ldots = b_n = 0$. Moreover linear independence means that any $(n \times 1)$ vector $\mathbf{v}$ where $\mathbf{v}$ may have real or complex components can be expressed as a linear combination of $\mathbf{u}_1, \mathbf{u}_2, \ldots, \mathbf{u}_n$.] Given this assumption, let us choose some initial vector $\mathbf{v}_0$ where $\mathbf{v}_0 \neq \boldsymbol{\theta}$. By our linear independence assumption we know that $\mathbf{v}_0$ can be expressed in the form

$$\mathbf{v}_0 = a_1\mathbf{u}_1 + a_2\mathbf{u}_2 + \ldots + a_n\mathbf{u}_n. \tag{6.9a}$$

[Again we do not know the eigenvectors; so we do not know the coefficients $a_1, a_2, \ldots, a_n$ in (6.9a); all we know is that $\mathbf{v}_0$ can be expressed in the form above. A typical initial vector $\mathbf{v}_0$ is one that has each component equal to 1.] If we form the sequence $\mathbf{v}_k = A\mathbf{v}_{k-1}$ for $k = 1, 2, \ldots$, then as we saw in Chapter 3, each vector $\mathbf{v}_k$ in the calculated sequence has the form

$$\mathbf{v}_k = a_1\lambda_1^k\mathbf{u}_1 + a_2\lambda_2^k\mathbf{u}_2 + \ldots + a_n\lambda_n^k\mathbf{u}_n. \tag{6.9b}$$

Now suppose the eigenvalues are ordered so that $|\lambda_1| \geqslant |\lambda_2| \geqslant |\lambda_3| \geqslant \ldots \geqslant |\lambda_n|$, and let us write (6.9b) as

$$\mathbf{v}_k = \lambda_1^k\left(a_1\mathbf{u}_1 + a_2\left(\frac{\lambda_2}{\lambda_1}\right)^k\mathbf{u}_2 + \ldots + a_n\left(\frac{\lambda_n}{\lambda_1}\right)^k\mathbf{u}_n\right). \tag{6.10}$$

If $|\lambda_1| > |\lambda_2|$, then the terms $(\lambda_i/\lambda_1)^k$ are small for large k where $2 \leqslant i \leqslant n$. Thus from (6.10) we expect (if $a_1 \neq 0$) that

$$\mathbf{v}_k \simeq \lambda_1^k a_1\mathbf{u}_1. \tag{6.11}$$

(That is, we expect that $\mathbf{v}_k$ is nearly a multiple of $\mathbf{u}_1$. Even though we don't know a_1, λ_1, or $\mathbf{u}_1$, we do have the calculated vector $\mathbf{v}_k$. Under the assumptions that $a_1 \neq 0$ and $|\lambda_1| > |\lambda_2|$, we do know that the vectors $\mathbf{v}_0, \mathbf{v}_1, \mathbf{v}_2, \ldots$ are aligning themselves along $\mathbf{u}_1$.) To obtain an estimate to λ_1, we utilize two successive calculated vectors $\mathbf{v}_k$ and $\mathbf{v}_{k+1}$ where as above, we expect that

$$\mathbf{v}_{k+1} \simeq \lambda_1^{k+1}a_1\mathbf{u}_1 \qquad \text{or} \qquad \mathbf{v}_{k+1} \simeq \lambda_1\mathbf{v}_k.$$

Now if we form the quotient $\beta_k = \mathbf{w}^{\mathrm{T}}\mathbf{v}_{k+1}/\mathbf{w}^{\mathrm{T}}\mathbf{v}_k$ where $\mathbf{w}$ is any vector such that $\mathbf{w}^{\mathrm{T}}\mathbf{u}_1 \neq 0$, we have

$$\beta_k = \frac{\mathbf{w}^{\mathrm{T}}\mathbf{v}_{k+1}}{\mathbf{w}^{\mathrm{T}}\mathbf{v}_k} \simeq \frac{\lambda_1^{k+1}a_1\mathbf{w}^{\mathrm{T}}\mathbf{u}_1}{\lambda_1^k a_1\mathbf{w}^{\mathrm{T}}\mathbf{u}_1} = \lambda_1. \tag{6.12a}$$

The approximation in (6.12a) is the essence of the power method. With respect to (6.12a) we note that to the extent that (6.11) is valid, a reasonable choice for $\mathbf{w}$ is the vector $\mathbf{v}_k$ itself; and this choice leads to the approximation

$$\beta_k = \frac{\mathbf{v}_k^{\mathrm{T}}\mathbf{v}_{k+1}}{\mathbf{v}_k^{\mathrm{T}}\mathbf{v}_k} \simeq \lambda_1. \tag{6.12b}$$

The left-hand side of (6.12b) is called the ***Rayleigh quotient***; and it can be shown that if $a_1 \neq 0$ in (6.9) and if $|\lambda_1| > |\lambda_2|$, then

$$\lim_{k\to\infty}\frac{\mathbf{v}_k^{\mathrm{T}}\mathbf{v}_{k+1}}{\mathbf{v}_k^{\mathrm{T}}\mathbf{v}_k} = \lambda_1.$$

In summary the power method proceeds as follows.

1. Guess an initial vector $\mathbf{v}_0$.
2. Form the sequence $\mathbf{v}_k = A\mathbf{v}_{k-1}$, $k = 1, 2, \ldots$.
3. For each k calculate the Rayleigh quotient in (6.12b).

The output of the power method is in part the sequence of numbers $\beta_0, \beta_1, \beta_2, \ldots$ in (6.12b), which, one hopes, converges to the dominant eigenvalue of A. (By dominant we of course mean the eigenvalue that is largest in absolute value.)

EXAMPLE 6.4 Let A be the matrix

$$\begin{bmatrix} 1 & -1 & 2 \\ -2 & 0 & 5 \\ 6 & -3 & 6 \end{bmatrix}.$$

It is easy to verify that the eigenvalues of A are $\lambda_1 = 5$, $\lambda_2 = 3$, and $\lambda_3 = -1$; and corresponding eigenvectors are

$$\mathbf{u}_1 = \begin{bmatrix} 5 \\ 16 \\ 18 \end{bmatrix} \quad \mathbf{u}_2 = \begin{bmatrix} 1 \\ 6 \\ 4 \end{bmatrix} \quad \mathbf{u}_3 = \begin{bmatrix} -1 \\ -2 \\ 0 \end{bmatrix}.$$

With the initial vector $\boldsymbol{v}_0$ given below, we generate

$$\boldsymbol{v}_0 = \begin{bmatrix} 0 \\ 1 \\ 3 \end{bmatrix}, \quad \boldsymbol{v}_1 = \begin{bmatrix} 5 \\ 15 \\ 15 \end{bmatrix}, \quad \boldsymbol{v}_2 = \begin{bmatrix} 20 \\ 65 \\ 75 \end{bmatrix}, \quad \boldsymbol{v}_3 = \begin{bmatrix} 105 \\ 335 \\ 375 \end{bmatrix}.$$

The corresponding Rayleigh quotients are

$$\frac{\boldsymbol{v}_0^{\mathrm{T}}\boldsymbol{v}_1}{\boldsymbol{v}_0^{\mathrm{T}}\boldsymbol{v}_0} = \frac{60}{10} = 6 \qquad \frac{\boldsymbol{v}_1^{\mathrm{T}}\boldsymbol{v}_2}{\boldsymbol{v}_1^{\mathrm{T}}\boldsymbol{v}_1} = \frac{2200}{475} \simeq 4.63 \qquad \frac{\boldsymbol{v}_2^{\mathrm{T}}\boldsymbol{v}_3}{\boldsymbol{v}_2^{\mathrm{T}}\boldsymbol{v}_2} = \frac{52000}{10250} \simeq 5.07,$$

and they appear to be converging to the dominant eigenvalue $\lambda_1 = 5$. To illustrate the point that the vectors $\boldsymbol{v}_k$ are aligning themselves in the direction of $\mathbf{u}_1$, we observe that

$$\boldsymbol{v}_1 = \begin{bmatrix} 5 \\ 15 \\ 15 \end{bmatrix} \quad 4\boldsymbol{v}_2 = \begin{bmatrix} 5 \\ 16.25 \\ 18.75 \end{bmatrix} \quad 21\boldsymbol{v}_3 = \begin{bmatrix} 5 \\ 15.95 \\ 17.86 \end{bmatrix} \quad \text{(approximately).}$$

The vector $\boldsymbol{v}_3$ for example is close to being a scalar multiple of $\mathbf{u}_1$.

As a more realistic illustration, we used a computer program that implements the power method and produced the estimates listed in Table 6.2. The program used "scaling," which we define shortly, and Rayleigh quotients to produce the sequence of estimates to λ_1.

Several issues related to the power method have not been resolved. First if the power method finds just the dominant eigenvalue, how do we find the other $n-1$ eigenvalues of A and how do we find the eigenvectors? The answer to this is that the power method is not suited to find the other eigenvalues, and we will use the inverse power method to find them if we have reasonably good estimates. The eigenvectors are easily found when we use "scaling" as described later. Second if we choose an initial vector $\boldsymbol{v}_0$ at random, what if $a_1 = 0$ in (6.9)? This is more of a problem in theory than it is in practice. However a thorough discussion would lead us too far afield, and the interested reader can consult any good numerical analysis book. Third what if A does not have a single dominant eigenvalue; what if $|\lambda_1| = |\lambda_2|$? This situation, it develops, is a real flaw in the power method. To see the dimensions of this problem, let us consider (6.10) in a bit more detail. First of all if $\lambda_1 = \lambda_2$ and if $|\lambda_2| > |\lambda_3| \geqslant \ldots \geqslant |\lambda_n|$, then (6.10) leads to a slightly different approximation in (6.11):

$$\boldsymbol{v}_k \simeq \lambda_1^k (a_1 \mathbf{u}_1 + a_2 \mathbf{u}_2).$$

However for large k we still expect that [as in (6.12a)]

$$\beta_k = \frac{\mathbf{w}^{\mathrm{T}} \boldsymbol{v}_{k+1}}{\mathbf{w}^{\mathrm{T}} \boldsymbol{v}_k} \simeq \frac{\lambda_1^{k+1} \mathbf{w}^{\mathrm{T}} (a_1 \mathbf{u}_1 + a_2 \mathbf{u}_2)}{\lambda_1^k \mathbf{w}^{\mathrm{T}} (a_1 \mathbf{u}_1 + a_2 \mathbf{u}_2)} = \lambda_1 .$$

In general it is not hard to show that a multiple dominant eigenvalue will not affect the convergence of the power method. However if $\lambda_1 = -\lambda_2$, then we do have problems. In this case, (6.10) shows that

$$\boldsymbol{v}_k \simeq \lambda_1^k (a_1 \mathbf{u}_1 + (-1)^k a_2 \mathbf{u}_2),$$

and the quotients $\beta_k = \mathbf{w}^{\mathrm{T}} \boldsymbol{v}_{k+1} / \mathbf{w}^{\mathrm{T}} \boldsymbol{v}_k$ will exhibit an oscillatory behavior and will not converge. Furthermore if A is a real matrix and if λ_1 is complex, then we know that $\overline{\lambda}_1$ is also an eigenvalue of A. In this event $|\lambda_1| = |\overline{\lambda}_1|$; so it can be expected that the power method will have difficulty whenever the dominant eigenvalue of A is complex. This difficulty can be overcome, but overcoming it requires quite an effort. We will see that the inverse power method does not suffer from this same problem.

TABLE 6.2

k	β_k		k	β_k	
1	0.4666666E	01	14	0.5002243E	01
2	0.7127658E	01	15	0.5001345E	01
3	0.5770836E	01	16	0.5000809E	01
4	0.5460129E	01	17	0.5000484E	01
5	0.5245014E	01	18	0.5000289E	01
6	0.5142666E	01	19	0.5000171E	01
7	0.5083027E	01	20	0.5000103E	01
8	0.5049140E	01	21	0.5000067E	01
9	0.5029204E	01	22	0.5000038E	01
10	0.5017425E	01	23	0.5000021E	01
11	0.5010424E	01	24	0.5000011E	01
12	0.5006239E	01	25	0.5000006E	01
13	0.5003741E	01	26	0.5000005E	01

A thoughtful reader with some computational experience may have noted another potential problem. If $|\lambda_1| > 1$, then the components of $\boldsymbol{v}_k$ are growing rapidly with k, and overflow threatens. Similarly if $|\lambda_1| < 1$, we have the potential for underflow. (Every digital device has a "largest" and "smallest" number that it can represent; the IBM 360 and 370 computers cannot represent positive numbers larger than about $.73 \times 10^{76}$ or positive numbers smaller than $.53 \times 10^{-78}$. Attempting to calculate a number whose magnitude is outside this range may cause a program to cease execution.) This problem is easily remedied by *scaling* the vectors $\mathbf{v}_k$. In particular suppose at the kth stage that $\|\mathbf{v}_k\| = 1$. We then calculate

$$\mathbf{z}_{k+1} = A\mathbf{v}_k$$

and form the Rayleigh quotient

$$\beta_k = \mathbf{v}_k^{\mathrm{T}}\mathbf{z}_{k+1} \simeq \lambda_1.$$

(Note that $\mathbf{v}_k^{\mathrm{T}}\mathbf{v}_k = 1$ since $\|\mathbf{v}_k\| = 1$.) Next we define $\mathbf{v}_{k+1}$ by

$$\mathbf{v}_{k+1} = \mathbf{z}_{k+1}/\|\mathbf{z}_{k+1}\|.$$

Since $\|\mathbf{v}_{k+1}\| = 1$, the $(k+1)$st stage proceeds just as the kth stage above. It is not hard to prove that this scaling process does not affect the convergence of the power method [this statement is nearly obvious from (6.9b) since scaling $\mathbf{v}_k$ is the same as adjusting the coefficients $a_1, a_2, \ldots, a_n$]. Furthermore as we expect from (6.11), it can be shown that the vectors $\mathbf{v}_k$ converge to an eigenvector of A corresponding to λ_1 when scaling is employed and when the power method itself is converging.

We want to make one final observation about the power method. If $|\lambda_1| > |\lambda_2|$, then it should also be expected from (6.10) that the rate of convergence of the power method is governed by the size of λ_2/λ_1. If the ratio is small, convergence is rapid since (6.11) is now a very good approximation. To quantify these statements, suppose the power method with scaling is employed; and suppose $\beta_k = \mathbf{v}_k^{\mathrm{T}}A\mathbf{v}_k$ denotes the kth Rayleigh quotient. It is not too hard to show that

$$\lim_{k\to\infty} \frac{\beta_{k+1} - \lambda_1}{\beta_k - \lambda_1} = \frac{\lambda_2}{\lambda_1}.$$

That is, $\beta_{k+1} - \lambda_1 \simeq (\lambda_2/\lambda_1)(\beta_k - \lambda_1)$, and this result can be interpreted to mean that the error $\beta_k - \lambda_1$ is decreased by about a factor of λ_2/λ_1 with each step of the power method.

EXAMPLE 6.5 As we mentioned at the end of Example 6.4, scaling was employed in the program that produced the results in Table 6.2. The initial vector $\mathbf{v}_0$ was chosen as

$$\mathbf{v}_0 = \frac{1}{\sqrt{3}}\begin{bmatrix} 1 \\ 1 \\ 1 \end{bmatrix};$$

so $\|\mathbf{v}_0\| = 1$. When scaling is used, the vectors $\mathbf{v}_k$ are converging to $\mathbf{u}_1/\|\mathbf{u}_1\|$ where (to six places)

$$\frac{\mathbf{u}_1}{\|\mathbf{u}_1\|} = \begin{bmatrix} .203279 \\ .650493 \\ .731804 \end{bmatrix}.$$

Some of the vectors $\boldsymbol{v}_k$ produced in Example 6.4 were

$$\boldsymbol{v}_6 = \begin{bmatrix} .205821 \\ .642018 \\ .738546 \end{bmatrix}, \quad \boldsymbol{v}_{12} = \begin{bmatrix} .203392 \\ .650114 \\ .732109 \end{bmatrix}, \quad \boldsymbol{v}_{18} = \begin{bmatrix} .203284 \\ .650475 \\ .731818 \end{bmatrix},$$

$$\boldsymbol{v}_{24} = \begin{bmatrix} .203279 \\ .650492 \\ .731805 \end{bmatrix}.$$

Finally to demonstrate the rate of convergence, we note that $\lambda_2/\lambda_1 = .6$. If we let $\beta_k = \boldsymbol{v}_k A \boldsymbol{v}_k$ be the kth estimate to λ_1, then a check of Table 6.2 will show that

$$\beta_{k+1} - 5 \simeq .6(\beta_k - 5).$$

Equivalently the $(k+1)$st error is approximately .6 times the kth error.

6.3 EXERCISES

1. For the matrix A in Example 6.4 and for the starting vector

$$\boldsymbol{v}_0 = \begin{bmatrix} 1 \\ 1 \\ 1 \end{bmatrix},$$

calculate $\boldsymbol{v}_1, \boldsymbol{v}_2, \boldsymbol{v}_3, \boldsymbol{v}_4, \boldsymbol{v}_5$ where $\boldsymbol{v}_{i+1} = A\boldsymbol{v}_i$. Form the Rayleigh quotients as in (6.12b) to estimate the dominant eigenvalue of A.

2. Let A be the (2×2) matrix

$$A = \begin{bmatrix} 3 & -4 \\ 2 & -3 \end{bmatrix}.$$

a) Verify that A has eigenvalues $\lambda = 1$ and $\lambda = -1$.
b) For $\boldsymbol{v}_0 = \mathbf{e}_1$, calculate $\boldsymbol{v}_1, \boldsymbol{v}_2, \boldsymbol{v}_3, \boldsymbol{v}_4$, and $\boldsymbol{v}_5$.
c) Form the Rayleigh quotients and verify that the power method cannot work for A.

3. Verify that the characteristic polynomial of $H+E$ in Example 6.3 is $q(t) = (1-t)^n + (-1)^{n+1}\varepsilon$. [Hint: Calculate $\det(H + E - tI)$.]

4. Suppose A is an $(n \times n)$ matrix with eigenvalues $\lambda_1, \lambda_2, \ldots, \lambda_n$ where $|\lambda_1| > |\lambda_2| \geqslant \ldots \geqslant |\lambda_n|$; and suppose $\boldsymbol{v}_0$ is given by (6.9a) where $a_1 \neq 0$. Suppose $\mathbf{w}$ is a fixed vector in R^n such that $\mathbf{w}^T\mathbf{u}_1 \neq 0$; and suppose for $k = 0, 1, \ldots$ that

$$\beta_k = \frac{\mathbf{w}^T \boldsymbol{v}_{k+1}}{\mathbf{w}^T \boldsymbol{v}_k}$$

where $\boldsymbol{v}_k = A\boldsymbol{v}_{k-1}$, $k = 1, 2, \ldots$. [Thus $\boldsymbol{v}_k$ has the form of (6.10).] Prove that

$$\lim_{k \to \infty} \beta_k = \lambda_1.$$

5. Using the same hypotheses as Exercise 4, prove that

$$\lim_{k\to\infty} \frac{\beta_{k+1}-\lambda_1}{\beta_k-\lambda_1} = \frac{\lambda_2}{\lambda_1}$$

when $a_2 \neq 0$ and $|\lambda_2| > |\lambda_3|$.

6. If A is an $(n \times n)$ symmetric matrix, we can assume that A has n linearly independent eigenvectors $\mathbf{u}_1, \mathbf{u}_2, \ldots, \mathbf{u}_n$ where $\mathbf{u}_i^T\mathbf{u}_j = 0$ when $i \neq j$ and $\mathbf{u}_i^T\mathbf{u}_i = 1$, $1 \leqslant i, j \leqslant n$. Calculate the Rayleigh quotient (6.12b) under this assumption, and convince yourself that convergence should be

$$\lim_{k\to\infty} \frac{\beta_{k+1}-\lambda_1}{\beta_k-\lambda_1} = \left(\frac{\lambda_2}{\lambda_1}\right)^2.$$

7. Let A be an $(n \times n)$ symmetric matrix and let $\mathbf{x}$ be any nonzero vector in R^n. As in Exercise 5, we can assume that $\mathbf{x} = a_1\mathbf{u}_1 + a_2\mathbf{u}_2 + \ldots + a_n\mathbf{u}_n$ where the eigenvectors $\mathbf{u}_1, \mathbf{u}_2, \ldots, \mathbf{u}_n$ are orthonormal.
   a) Calculate $\mathbf{x}^T\mathbf{x}$.
   b) Calculate $\mathbf{x}^T A\mathbf{x}$.
   c) Suppose the eigenvalues of A are $\lambda_1, \lambda_2, \ldots, \lambda_n$ where $a \leqslant \lambda_i \leqslant b$, $1 \leqslant i \leqslant n$. Prove that

$$a \leqslant \frac{\mathbf{x}^T A\mathbf{x}}{\mathbf{x}^T\mathbf{x}} \leqslant b.$$

6.4 THE INVERSE POWER METHOD FOR THE EIGENVALUE PROBLEM

The inverse power method is nothing more than the power method applied to the matrix $(A-\alpha I)^{-1}$. If α is a reasonably good estimate to an eigenvalue λ of A, then several steps of the inverse power method will give a very accurate estimate to λ and a corresponding eigenvector. To develop the theoretical basis of the inverse power method, we suppose α is any scalar that is not an eigenvalue of A, and suppose $A\mathbf{u} = \lambda\mathbf{u}$, $\mathbf{u} \neq \boldsymbol{\theta}$. Since $A\mathbf{u} = \lambda\mathbf{u}$, it is clear that

$$(A-\alpha I)\mathbf{u} = (\lambda-\alpha)\mathbf{u}.$$

Next since α is not an eigenvalue of A, $A-\alpha I$ is nonsingular; and we can write

$$(A-\alpha I)^{-1}\mathbf{u} = \frac{1}{\lambda-\alpha}\mathbf{u}. \tag{6.13}$$

Clearly (6.13) says that $1/(\lambda-\alpha)$ is an eigenvalue of $(A-\alpha I)^{-1}$ and that $\mathbf{u}$ is a corresponding eigenvector.

Suppose A has eigenvalues $\lambda_1, \lambda_2, \ldots, \lambda_n$; and suppose α_1 is a good estimate to λ_1. By (6.13) the eigenvalues of $(A-\alpha_1 I)^{-1}$ are $\mu_1, \mu_2, \ldots, \mu_n$ where

$$\mu_1 = \frac{1}{\lambda_1-\alpha_1},\ \mu_2 = \frac{1}{\lambda_2-\alpha_1},\ \ldots,\ \mu_n = \frac{1}{\lambda_n-\alpha_1}. \tag{6.14}$$

If $\alpha_1 \simeq \lambda_1$, then we can expect that μ_1 is the dominant eigenvalue of $(A-\alpha_1 I)^{-1}$; that is, if the eigenvalues $\lambda_1, \lambda_2, \ldots, \lambda_n$ are reasonably separated, and if α_1 is a good estimate to λ_1, then we can anticipate that $|\mu_1| >> |\mu_i|$ for $2 \leqslant i \leqslant n$. From Section 6.3 we know that if $|\mu_1| > |\mu_i|$ for $2 \leqslant i \leqslant n$, then the power method applied to $(A-\alpha_1 I)^{-1}$ will converge to μ_1; and the rate of convergence will be governed by the largest number among

$$\frac{\mu_2}{\mu_1}, \frac{\mu_3}{\mu_1}, \ldots, \frac{\mu_n}{\mu_1}.$$

Furthermore we would expect rapid convergence when α_1 is close to λ_1 since $|\mu_1|$ is then quite large with respect to $|\mu_2|, |\mu_3|, \ldots, |\mu_n|$.

As an example to fix these ideas, suppose A is a (3×3) matrix with eigenvalues $\lambda_1 = 3$, $\lambda_2 = -3$, and $\lambda_3 = 1$. Since $|\lambda_1| = |\lambda_2|$ and $\lambda_1 \neq \lambda_2$, the power method will not converge when applied to A. However for $\alpha_1 = 2.9$ the matrix $(A-\alpha_1 I)^{-1}$ has approximate eigenvalues

$$\mu_1 = 10. \qquad \mu_2 = -.169 \qquad \mu_3 = -.526,$$

and the power method applied to $(A-\alpha_1 I)^{-1}$ will converge to $\mu_1 = 10$. Moreover convergence will be extremely rapid, governed by the ratio $|\mu_3/\mu_1| = 1/19 \simeq .0526$. That is, the errors decrease by about a factor of $1/19$ per step, and so k steps will decrease the initial error by a factor of $(1/19)^k$. If $k = 3$ for instance, then $(1/19)^3 \simeq .000146$.

In general if A is $(n \times n)$ and has eigenvalues $\lambda_1, \lambda_2, \ldots, \lambda_n$, and if we have good estimates $\alpha_1, \alpha_2, \ldots, \alpha_n$ to $\lambda_1, \lambda_2, \ldots, \lambda_n$, then we can expect that the power method applied to $(A-\alpha_j I)^{-1}$ will converge to μ_j where

$$\mu_j = \frac{1}{\lambda_j - \alpha_j}.$$

Having μ_j, we can then recover λ_j from

$$\lambda_j = \frac{1}{\mu_j} + \alpha_j.$$

We wish to emphasize the eigenvalue "refinement" aspect of this procedure. When we transform A to a Hessenberg matrix H, we do not expect (in the presence of roundoff error) that the eigenvalues of H and A will be exactly the same. However if the eigenvalues of H are $\alpha_1, \alpha_2, \ldots, \alpha_n$, then we can regard them as estimates to the eigenvalues of A. The estimates $\alpha_1, \alpha_2, \ldots, \alpha_n$ can be quickly corrected to $\lambda_1, \lambda_2, \ldots, \lambda_n$ when the power method is applied to $(A-\alpha_j I)^{-1}$ for $1 \leqslant j \leqslant n$. (The matrix A of course has not been contaminated by the roundoff errors made in Hessenberg reduction.) As we have tried to point out above if the estimates α_j are good relative to the separation of $\lambda_1, \lambda_2, \ldots, \lambda_n$, then the inverse power method will converge since the eigenvalue $\mu_j = 1/(\lambda_j - \alpha_j)$ will dominate all the other eigenvalues of $(A-\alpha_j I)^{-1}$.

As another example, suppose A is a (6×6) real matrix with eigenvalues $\lambda_1, \lambda_2, \ldots, \lambda_6$. Suppose further that λ_1 is complex, $\lambda_1 = a + ib$; suppose $\lambda_2 = \lambda_1$, $\lambda_3 = \lambda_4 = \bar{\lambda}_1$; and suppose λ_5 and λ_6 are real where the ordering is

$$|\lambda_1| = |\lambda_2| = |\lambda_3| = |\lambda_4| > |\lambda_5| > |\lambda_6|.$$

This hypothetical situation is diagramed in Fig. 6.4 and represents a case that the power method cannot deal with. [In Fig. 6.4 we think of a complex number $z = x + iy$ as being "identified" with the coordinates (x, y).] Given the situation illustrated in Fig. 6.4, suppose we have estimates α_1 to λ_1, α_3 to λ_3, α_5 to λ_5, and α_6 to λ_6 where α_j is in the circle centered at λ_j for $j = 1, 3, 5, 6$. Clearly the power method applied to $(A - \alpha_j I)^{-1}$ will converge to $1/(\lambda_j - \alpha_j)$, $j = 1, 3, 5, 6$. [Recall that a multiple eigenvalue, such as $1/(\lambda_1 - \alpha_1)$ or $1/(\lambda_3 - \alpha_3)$, does not affect the convergence of the power method.]

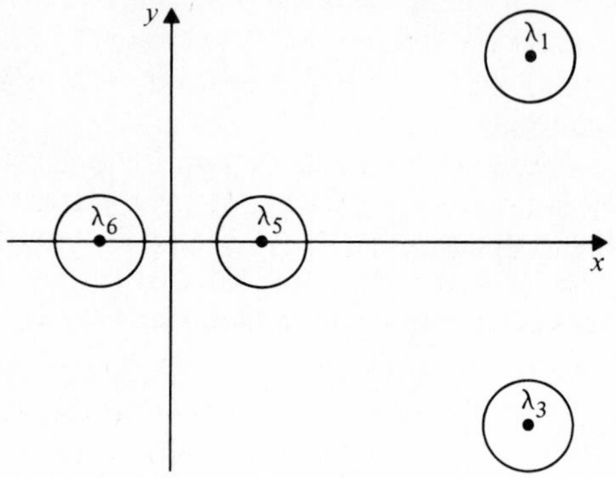

Fig. 6.4 A distribution of eigenvalues in the complex plane.

One remaining point must be made about the inverse power method: since the inverse power method amounts to applying the power method to $(A - \alpha I)^{-1}$, we might feel that we should calculate the sequence

$$\mathbf{v}_k = (A - \alpha I)^{-1}\mathbf{v}_{k-1}, \; k = 1, 2, \ldots,$$

where $\mathbf{v}_0$ is some starting vector. However since an inverse matrix should not usually be calculated, we rewrite this iteration as

$$(A - \alpha I)\mathbf{v}_k = \mathbf{v}_{k-1}, \; k = 1, 2, \ldots. \tag{6.15}$$

Thus given $\mathbf{v}_0$, we solve $(A - \alpha I)\mathbf{x} = \mathbf{v}_0$ to find $\mathbf{v}_1$, and then solve $(A - \alpha I)\mathbf{x} = \mathbf{v}_1$ to find $\mathbf{v}_2$, etc. In an algorithmic form the inverse power method with scaling proceeds as follows. Let $\mathbf{v}_0$ be a starting vector where $\|\mathbf{v}_0\| = 1$ and for $k = 0, 1, 2, \ldots$.

1. Find $\mathbf{z}_{k+1}$ such that $(A - \alpha I)\mathbf{z}_{k+1} = \mathbf{v}_k$.
2. Set $\beta_k = \mathbf{z}_{k+1}^T \mathbf{v}_k$. (6.16)
3. Let $\eta_{k+1} = \|\mathbf{z}_{k+1}\|$.
4. Set $\mathbf{v}_{k+1} = \mathbf{z}_{k+1}/\eta_{k+1}$, and return to step (1).

If we terminate this iteration after r steps with β_r and $\mathbf{v}_r$, then

$$\lambda = \frac{1}{\beta_r} + \alpha$$

is an estimate to the eigenvalue of A that is nearest to α, and $\boldsymbol{v}_r$ is an estimate to a corresponding eigenvector. If A is $(n \times n)$, then a reasonable starting vector $\boldsymbol{v}_0$ of norm 1 is

$$\boldsymbol{v}_0 = \frac{1}{\sqrt{n}} \begin{bmatrix} 1 \\ 1 \\ 1 \\ \vdots \\ 1 \end{bmatrix}.$$

As a final observation, we recall that our discussion of the power method in Section 6.3 presupposed that A had a set of n linearly independent eigenvectors—in (6.9a) we supposed that $\boldsymbol{v}_0$ could be expressed as a linear combination of the eigenvectors of A. In Exercises 6 and 7, Section 6.4, the reader is asked to show that this assumption is not necessary by using the fact that every $(n \times n)$ matrix A has a set of n linearly independent generalized eigenvectors. Thus the power method always converges in theory when $|\lambda_1| > |\lambda_2| \geqslant |\lambda_3| \geqslant \ldots \geqslant |\lambda_n|$ and when $a_1 \neq 0$ in (6.9a). In practice the restriction $a_1 \neq 0$ can be ignored, and the inverse power method satisfies the first restriction when sufficiently good estimates are available.

Figure 6.5 lists a FORTRAN program, Subroutine INVPOW, that implements the inverse power method as outlined in (6.16). The required inputs are the following.

A	An (N × N) matrix.
ALPHA	An estimate to an eigenvalue of A.
N	The size of A.
TOL	A convergence criterion. Control will return to the calling program when $\lvert\beta_{k+1} - \beta_k\rvert \leqslant$ TOL.
MAXITR	The maximum number of iterations to be executed in INVPOW.
MAXDIM	The declared dimension of the array containing A in the calling program.

The outputs from Subroutine INVPOW are the following.

EIGNVL	The machine estimate of the eigenvalue of A that is nearest ALPHA.
EIGNVC	The machine estimate of an eigenvector of A corresponding to EIGNVL.
IFLAG	A flag set equal to 1 if the convergence test is met. IFLAG is set equal to 2 or 3 to signal an abnormal return.

Figure 6.6 gives an example of a simple program that uses Subroutine INVPOW to find the eigenvalue of A that is nearest to α. As an example, we input the matrix

$$A = \begin{bmatrix} 4 & -5 & 0 & 3 \\ 0 & 4 & -3 & -5 \\ 5 & -3 & 4 & 0 \\ 3 & 0 & 5 & 4 \end{bmatrix},$$

```
      SUBROUTINE INVPOW(A,EIGNVC,ALPHA,EIGNVL,N,TOL,MAXITR,MAXDIM,IFLAG)
      DIMENSION A(MAXDIM,MAXDIM),EIGNVC(MAXDIM)
      DIMENSION AWORK(50,50),OLDVCT(50),REFVCT(50)
C
C  INITIALIZE THE ITERATION COUNTER AND SET UP THE WORKING
C  MATRIX AWORK=A-ALPHA*I.
C
      ITR=0
        DO 2 I=1,N
          DO 1 J=1,N
          AWCRK(I,J)=A(I,J)
    1     CONTINUE
        AWORK(I,I)=AWORK(I,I)-ALPHA
    2   CONTINUE
C
C  CHOOSE A STARTING VECTOR OF NORM 1.
C
      AN=N
      Q=1./SQRT(AN)
        DO 3 I=1,N
        OLDVCT(I)=Q
    3   CONTINUE
C
C  CARRY OUT THE INVERSE POWER METHOD ITERATION.  REFVAL AND REFVCT
C  ARE REFINED ESTIMATES TO AN EIGENVALUE OF (A-ALPHA*I)**(-1)
C  AND A CORRESPONDING EIGENVECTOR.
C
      OLDVAL=ALPHA
    4 ITR=ITR+1
      CALL GAUSS(AWORK,OLDVCT,REFVCT,N,50,IERROR,RNORM)
      IF(IERROR.EQ.2)  GO TO 10
C
C  CALCULATE THE RAYLEIGH QUOTIENT AND THE NORM OF REFVCT.
C
      RAYQTN=0.

      SCALSQ=0.
        DO 5 I=1,N
        RAYQTN=RAYQTN+OLDVCT(I)*REFVCT(I)
        SCALSQ=SCALSQ+REFVCT(I)**2
    5   CONTINUE
C
C  CALCULATE A REFINED ESTIMATE OF AN EIGENVALUE OF A AND
C  SCALE REFVCT TO HAVE NORM 1.
C
      REFVAL=(1./RAYQTN)+ALPHA
      SCALE=SQRT(SCALSQ)
        DO 6 I=1,N
        REFVCT(I)=REFVCT(I)/SCALE
    6   CONTINUE
C
C  TEST FOR CONVERGENCE AND EXCESSIVE ITERATIONS.
C
      IF(ABS(OLDVAL-REFVAL).LE.TCL)  GO TO 8
      IF(ITR.GE.MAXITR)  GO TO 11
      OLDVAL=REFVAL
        DO 7 I=1,N
        OLDVCT(I)=REFVCT(I)
    7   CONTINUE
      GO TO 4
C
C  SUCCESSFUL RETURN.
C
    8 IFLAG=1
      EIGNVL=REFVAL
        DO 9 I=1,N
        EIGNVC(I)=REFVCT(I)
    9   CONTINUE
      RETURN
C
C  ABNORMAL RETURN --- AWORK MAY BE SINGULAR.
C
   10 IFLAG=2
      RETURN
C
C  ABNORMAL RETURN --- LIMIT ON NUMBER OF ITERATIONS IS EXCEEDED.
C
   11 IFLAG=3
      RETURN
      END
```

Figure 6.5

which has eigenvalues

$$\lambda_1 = 12, \qquad \lambda_2 = 1+5i, \qquad \lambda_3 = 1-5i, \qquad \lambda_4 = 2$$

and corresponding eigenvectors

$$\mathbf{u}_1 = \begin{bmatrix} 1 \\ -1 \\ 1 \\ 1 \end{bmatrix}, \quad \mathbf{u}_2 = \begin{bmatrix} 1 \\ -i \\ -i \\ -1 \end{bmatrix}, \quad \mathbf{u}_3 = \begin{bmatrix} 1 \\ i \\ i \\ -1 \end{bmatrix}, \quad \mathbf{u}_4 = \begin{bmatrix} 1 \\ 1 \\ -1 \\ 1 \end{bmatrix}.$$

We input $\alpha = 2.1$ as an estimate to λ_4, and Subroutine INVPOW returned $\lambda = .20000\text{E } 01$ and an estimated eigenvector

$$\begin{bmatrix} -.50000 \\ -.50000 \\ .50000 \\ -.50000 \end{bmatrix}.$$

```
      DIMENSION A(20,20),EIGNVC(20)
      MAXDIM=20
    1 READ 100,N
      IF(N.LT.2)  STOP
        DO 2 I=1,N
        READ 101,(A(I,J),J=1,N)
    2   CONTINUE
      READ 101,ALPHA
      TOL=.1E-06
      MAXITR=10
      CALL INVPOW(A,EIGNVC,ALPHA,EIGNVL,N,TOL,MAXITR,MAXDIM,IFLAG)
      PRINT 102,IFLAG,EIGNVL
      PRINT 103,(EIGNVC(K),K=1,N)
      GO TO 1
  100 FORMAT(I2)
  101 FORMAT(10F4.0)
  102 FORMAT(1H0,6H FLAG=,I3,5X,36HREFINED ESTIMATE OF AN EIGENVALUE IS,
     1E15.5)
  103 FORMAT(1H0,30HA CORRESPONDING EIGENVECTOR IS,/,6E18.5)
      END
```

Figure 6.6

6.4 EXERCISES

1. Use Subroutine INVPOW to refine the given eigenvalue estimates of A where A is the (4×4) matrix at the end of this section. Include a print statement in INVPOW after the convergence test to print the array OLDVCT and the constant OLDVAL. Use the following value for α.

   a) $\alpha = 2.1$ b) $\alpha = 2.01$ c) $\alpha = 12.1$ d) $\alpha = 12.01$

2. Use INVPOW to refine the eigenvalue estimates of

$$A = \begin{bmatrix} 33 & 16 & 72 \\ -24 & -10 & -57 \\ -8 & -4 & -17 \end{bmatrix}.$$

a) $\alpha = 1.1$ b) $\alpha = 1.4$ c) $\alpha = 1.9$ d) $\alpha = 3.02$
(The eigenvalues of A are $\lambda = 1$, $\lambda = 2$, $\lambda = 3$.)

3. The eigenvalues of

$$A = \begin{bmatrix} 6 & 4 & 4 & 1 \\ 4 & 6 & 1 & 4 \\ 4 & 1 & 6 & 4 \\ 1 & 4 & 4 & 6 \end{bmatrix}$$

are $\lambda_1 = 15$, $\lambda_2 = \lambda_3 = 5$, $\lambda_4 = -1$. An eigenvector corresponding to λ_1 is

$$\mathbf{u}_1 = \begin{bmatrix} .5 \\ .5 \\ .5 \\ .5 \end{bmatrix}.$$

Since INVPOW uses $\mathbf{u}_1$ as its starting vector, we can expect this "textbook" eigenvalue problem to cause the subroutine difficulty. Try estimates of $\alpha = 5.1$ and $\alpha = -.9$ in INVPOW. If the program converges to $\lambda = 15$, modify the starting vector in INVPOW by setting the initial vector, OLDVCT, equal to a unit vector $\mathbf{e}_j$ (this can be done in the loop that terminates at the statement 3 CONTINUE). Run the program again with the estimates above.

4. The eigenvalues of

$$A = \begin{bmatrix} 2 & 1 & 3 & 4 \\ 1 & -3 & 1 & 5 \\ 3 & 1 & 6 & -2 \\ 4 & 5 & -2 & -1 \end{bmatrix}$$

are $\lambda = 7.9329$, $\lambda = -8.0286$, $\lambda = 5.6689$, and $\lambda = -1.5732$ to the place given. Run INVPOW with $\alpha = 8$, $\alpha = -8$, $\alpha = 5.5$, and $\alpha = -1.6$.

5. Write a calling program for INVPOW that checks the output by calculating $\|A\mathbf{u} - \lambda\mathbf{u}\|$ where λ and $\mathbf{u}$ are the eigenvalue and eigenvector estimates returned by INVPOW.

6. In Chapter 3 we saw that any $(n \times n)$ matrix A has a set of n linearly independent eigenvectors and generalized eigenvectors. As a simple example, suppose A is a (3×3) matrix with two distinct eigenvalues, λ and γ, with corresponding eigenvectors $\mathbf{u}$ and $\mathbf{v}$. Thus

$$A\mathbf{u} = \lambda\mathbf{u}, \qquad \mathbf{u} \neq \boldsymbol{\theta};$$
$$A\mathbf{v} = \gamma\mathbf{v}, \qquad \mathbf{v} \neq \boldsymbol{\theta}.$$

Suppose $\mathbf{q}$ is a generalized eigenvector corresponding to λ so that

$$(A - \lambda I)\mathbf{q} = \mathbf{u}, \quad \mathbf{q} \neq \boldsymbol{\theta}.$$

Prove that $A^k\mathbf{q} = \lambda^k\mathbf{q} + k\lambda^{k-1}\mathbf{u}$, $k = 1, 2, \ldots$.

7. With the same hypotheses as in Exercise 6, suppose that $|\lambda| > |\gamma|$; and suppose $\mathbf{x}_0$ is a starting vector such that

$$\mathbf{x}_0 = a_1\mathbf{q} + a_2\mathbf{u} + a_3\mathbf{v}.$$

Suppose $\mathbf{x}_k = A\mathbf{x}_{k-1}$, $k = 1, 2, \ldots$; and suppose $\mathbf{w}$ is a fixed vector such that either $\mathbf{w}^T\mathbf{q} \neq 0$ or $\mathbf{w}^T\mathbf{u} \neq 0$. Prove that

$$\lim_{k \to \infty} \frac{\mathbf{w}^T\mathbf{x}_{k+1}}{\mathbf{w}^T\mathbf{x}_k} = \lambda.$$

8. Modify Subroutine INVPOW so that complex estimates α can be employed. Using the complex version of GAUSS, test your modification on the matrix in Exercise 1 with $\alpha = 1.1 + 4.9i$ as an estimate.

6.5 HOUSEHOLDER TRANSFORMATIONS AND REDUCTION TO HESSENBERG FORM

In Chapter 3 we saw how to use elementary similarity transformations to reduce a matrix A to Hessenberg form. For actual numerical calculations a different sort of transformation procedure is usually used—the Householder transformation. One reason for this choice is that Householder transformations preserve symmetric matrices. Furthermore Householder transformations can be used in other applications, such as finding least-squares solutions to inconsistent systems.

To define the Householder transformation, we suppose $\mathbf{u}$ is any vector in R^n, $\mathbf{u} \neq \boldsymbol{\theta}$. We first note that $\mathbf{u}\mathbf{u}^{\mathrm{T}}$ is an $(n \times n)$ matrix; so the following $(n \times n)$ matrix Q can be defined:

$$Q = I - \frac{2}{\|\mathbf{u}\|^2}\mathbf{u}\mathbf{u}^{\mathrm{T}} \tag{6.17}$$

where I is the $(n \times n)$ identity. It is easy to show that Q is a symmetric matrix (Exercise 4, Section 6.5), and we now show that Q is an *orthogonal* matrix (a matrix B is orthogonal if $B^{\mathrm{T}}B = I$). Showing that Q is orthogonal is easy; and since $Q^{\mathrm{T}} = Q$, we need only calculate

$$Q^{\mathrm{T}}Q = QQ = \left(I - \frac{2}{\|\mathbf{u}\|^2}\mathbf{u}\mathbf{u}^{\mathrm{T}}\right)\left(I - \frac{2}{\|\mathbf{u}\|^2}\mathbf{u}\mathbf{u}^{\mathrm{T}}\right)$$

$$= I - \frac{4}{\|\mathbf{u}\|^2}\mathbf{u}\mathbf{u}^{\mathrm{T}} + \frac{4}{\|\mathbf{u}\|^4}(\mathbf{u}\mathbf{u}^{\mathrm{T}})(\mathbf{u}\mathbf{u}^{\mathrm{T}}).$$

But the matrix product $(\mathbf{u}\mathbf{u}^{\mathrm{T}})(\mathbf{u}\mathbf{u}^{\mathrm{T}})$ can be associated as $\mathbf{u}(\mathbf{u}^{\mathrm{T}}\mathbf{u})\mathbf{u}^{\mathrm{T}} = \mathbf{u}\|\mathbf{u}\|^2\mathbf{u}^{\mathrm{T}} = \|\mathbf{u}\|^2\mathbf{u}\mathbf{u}^{\mathrm{T}}$; and so we see that $Q^{\mathrm{T}}Q = I$.

Since $Q^{-1} = Q^{\mathrm{T}} = Q$, we know that QAQ is a similarity transformation. In order to show how to reduce A to Hessenberg form using a sequence of Householder transformations, we first note how a product $Q\boldsymbol{v}$ would be formed in practice where $\boldsymbol{v}$ is any vector in R^n. If we write Q as $Q = I - b\mathbf{u}\mathbf{u}^{\mathrm{T}}$ where $b = 2/\|\mathbf{u}\|^2$, then $Q\boldsymbol{v}$ is given as

$$Q\boldsymbol{v} = \boldsymbol{v} - b(\mathbf{u}\mathbf{u}^{\mathrm{T}})\boldsymbol{v}. \tag{6.18}$$

But $(\mathbf{u}\mathbf{u}^{\mathrm{T}})\boldsymbol{v} = \mathbf{u}(\mathbf{u}^{\mathrm{T}}\boldsymbol{v})$; and we observe that $\mathbf{u}^{\mathrm{T}}\boldsymbol{v}$ is a scalar, say $\mathbf{u}^{\mathrm{T}}\boldsymbol{v} = a$. It follows that (6.18) is the same as

$$Q\boldsymbol{v} = \boldsymbol{v} - (ab)\mathbf{u} \tag{6.19}$$

where $ab = (2\mathbf{u}^{\mathrm{T}}\boldsymbol{v})/\|\mathbf{u}\|^2$. Equation (6.19) is significant in terms of computation, for it says that to form the product $Q\boldsymbol{v}$ it is necessary to calculate only one scalar product, $a = \mathbf{u}^{\mathrm{T}}\boldsymbol{v}$, and then to perform a vector subtraction. In fact we can calculate $Q\boldsymbol{v}$ without ever forming the matrix Q.

A Householder transformation is normally used, as is seen shortly, to zero certain specified entries in some vector $\boldsymbol{v}$. To illustrate how this zeroing is done, suppose $\boldsymbol{v}$ is in R^n:

$$\boldsymbol{v} = \begin{bmatrix} v_1 \\ v_2 \\ \vdots \\ v_{k-1} \\ v_k \\ v_{k+1} \\ \vdots \\ v_n \end{bmatrix}; \tag{6.20}$$

and suppose we want to find a Householder transformation Q so that when we form $\mathbf{w} = Q\boldsymbol{v}$, the result $\mathbf{w}$ satisfies these conditions.

1. The first $k-1$ components of $\mathbf{w}$ are $v_1, v_2, \ldots, v_{k-1}$.
2. The last $n-k$ components of $\mathbf{w}$ are all zero.

That is, if $\mathbf{w} = Q\boldsymbol{v}$, then we want $\mathbf{w}$ to have the form

$$\mathbf{w} = \begin{bmatrix} v_1 \\ v_2 \\ \vdots \\ v_{k-1} \\ \sigma \\ 0 \\ 0 \\ \vdots \\ 0 \end{bmatrix}. \tag{6.21}$$

By using (6.19) it is easy to derive a vector $\mathbf{u}$ so that Q has the desired properties. In short we specify $\mathbf{u}$ according to these conditions.

1. $u_1 = u_2 = \ldots = u_{k-1} = 0$.
2. $u_{k+1} = v_{k+1}, u_{k+2} = v_{k+2}, \ldots, u_n = v_n$.
3. $u_k = v_k - s$ where $s = \pm\sqrt{v_k^2 + v_{k+1}^2 + \ldots + v_n^2}$.

Thus given $\boldsymbol{v}$ as in (6.20), we define $\mathbf{u}$ to be

$$\mathbf{u} = \begin{bmatrix} 0 \\ 0 \\ \vdots \\ 0 \\ v_k - s \\ v_{k+1} \\ \vdots \\ v_n \end{bmatrix} \tag{6.22}$$

We will soon remove the ambiguity of the sign of s in condition (3). For now, we calculate $Q\boldsymbol{v}$ according to (6.19). First we note that

$$\begin{aligned} \|\mathbf{u}\|^2 &= (v_k - s)^2 + v_{k+1}^2 + \dots + v_n^2 \\ &= 2s^2 - 2v_k s \end{aligned}$$

and $\mathbf{u}^{\mathrm{T}}\boldsymbol{v} = s^2 - v_k s$. Thus with $ab = (2\mathbf{u}^{\mathrm{T}}\boldsymbol{v})/\|\mathbf{u}\|^2$ we have that $ab = 1$. From (6.19) it follows that $Q\boldsymbol{v}$ has the desired form of (6.21):

$$Q\boldsymbol{v} = \boldsymbol{v} - \mathbf{u} = \begin{bmatrix} v_1 \\ v_2 \\ \vdots \\ v_{k-1} \\ s \\ 0 \\ 0 \\ \vdots \\ 0 \end{bmatrix}. \tag{6.23}$$

EXAMPLE 6.6 Suppose $\boldsymbol{v}$ is the vector

$$\boldsymbol{v} = \begin{bmatrix} 3 \\ 4 \\ 2 \\ 1 \\ 2 \end{bmatrix},$$

and suppose we want an orthogonal matrix Q such that forming $Q\boldsymbol{v}$ zeros the last three entries in $\boldsymbol{v}$. We must calculate $\mathbf{u}$ as in (6.22) with $k = 2$, and we first find $s = \pm\sqrt{4^2+2^2+1^2+2^2} = \pm 5$. If we choose s to be -5, then $\mathbf{u}$ is the vector

$$\mathbf{u} = \begin{bmatrix} 0 \\ 9 \\ 2 \\ 1 \\ 2 \end{bmatrix}.$$

With this choice, Q is the (5×5) matrix

$$Q = I - \frac{1}{45}\mathbf{u}\mathbf{u}^{\mathrm{T}}.$$

A quick calculation shows that $Q\boldsymbol{v} = \boldsymbol{v} - \mathbf{u}$, or

$$Q\boldsymbol{v} = \begin{bmatrix} 3 \\ -5 \\ 0 \\ 0 \\ 0 \end{bmatrix}.$$

Note especially that we did not form Q in order to evaluate $Q\boldsymbol{v}$. In fact as we noted in (6.19) if $\mathbf{x}$ is any vector in R^5, then $Q\mathbf{x}$ is given simply as

$$Q\mathbf{x} = \mathbf{x} - \frac{\mathbf{u}^{\mathrm{T}}\mathbf{x}}{45}\mathbf{u}.$$

For example if

$$\mathbf{x} = \begin{bmatrix} 3 \\ 1 \\ 1 \\ 2 \\ 1 \end{bmatrix},$$

then $Q\mathbf{x} = \mathbf{x} - \frac{1}{3}\mathbf{u}$.

As we have observed, condition (3), which defines the number s in (6.22), is ambiguous; we can choose s to be either positive or negative; and in either case the product $Q\boldsymbol{v}$ will have the form of (6.23); the last $n-k$ entries in $\boldsymbol{v}$ are zero; the first $k-1$ entries are left unchanged. From roundoff-error considerations we want to choose s so that $\|\mathbf{u}\|^2$ is as large as possible where as we have seen, $\|\mathbf{u}\|^2 = 2s^2 - 2v_k s$. Clearly $\|\mathbf{u}\|^2$ is as large as possible if v_k and s have opposite signs, and this observation provides the rule for choosing s as either positive or negative. (In the example above we chose s negative since the pivot element, $v_2 = 4$, was positive. If we had selected $s = 5$, then $\|\mathbf{u}\|^2$ would have been 10 instead of 90.) The reason for the rule is found in (6.19)—if $\mathbf{x}$ is any vector, then $Q\mathbf{x}$ is given by

$$Q\mathbf{x} = \mathbf{x} - (ab)\mathbf{u} \tag{6.24}$$

where $ab = (2\mathbf{u}^T\mathbf{x})/\|\mathbf{u}\|^2$. By choosing $\|\mathbf{u}\|^2$ as large as possible, we expect to disturb $\mathbf{x}$ as little as possible when we form $Q\mathbf{x}$ for most vectors $\mathbf{x}$.

We are now prepared to show how Householder transformations can be used to reduce an $(n \times n)$ matrix A to Hessenberg form. We first note that the product QA is very easy to calculate despite the fact that we have not actually formed Q. In particular if $A = [\mathbf{A}_1, \mathbf{A}_2, \ldots, \mathbf{A}_n]$, then QA is given by

$$QA = [Q\mathbf{A}_1, Q\mathbf{A}_2, \ldots, Q\mathbf{A}_n]. \tag{6.25}$$

We calculate the columns of QA as in (6.24); that is, $Q\mathbf{A}_j = \mathbf{A}_j - m_j\mathbf{u}$ where $m_j = (2\mathbf{u}^T\mathbf{A}_j)/\|\mathbf{u}\|^2$ for $1 \leqslant j \leqslant n$. When B is $(n \times n)$, it is similarly easy to form the product BQ by subtracting multiples of $\mathbf{u}^T$ from the rows of B. For verification it is convenient to set $C = B^T$ and consider

$$(BQ)^T = Q^TB^T = QC = [Q\mathbf{C}_1, Q\mathbf{C}_2, \ldots, Q\mathbf{C}_n].$$

As in (6.25), $Q\mathbf{C}_j = \mathbf{C}_j - p_j\mathbf{u}$ where $p_j = (2\mathbf{u}^T\mathbf{C}_j)/\|\mathbf{u}\|^2$. However the columns of QC are the rows of $(QC)^T = C^TQ^T = BQ$. From above, the jth row of $(QC)^T$ is $\mathbf{C}_j^T - p_j\mathbf{u}^T$, and thus the jth row of BQ is constructed by the rule

$$j\text{th row of } BQ = j\text{th row of } B - p_j\mathbf{u}^T. \tag{6.26}$$

Rather than giving the general algorithm for Hessenberg reduction using Householder transformations, we illustrate the process with a (4×4) matrix A. Let A be given by

$$A = \begin{bmatrix} a_{11} & a_{12} & a_{13} & a_{14} \\ a_{21} & a_{22} & a_{23} & a_{24} \\ a_{31} & a_{32} & a_{33} & a_{34} \\ a_{41} & a_{42} & a_{43} & a_{44} \end{bmatrix}.$$

Writing A in column form, $A = [\mathbf{A}_1, \mathbf{A}_2, \mathbf{A}_3, \mathbf{A}_4]$, we choose a Householder transformation Q_1, which zeros the lst two entries of $\mathbf{A}_1$; that is, $\mathbf{u}$, Q_1, and $Q_1\mathbf{A}_1$ have the form

$$\mathbf{u} = \begin{bmatrix} 0 \\ a_{21} - s \\ a_{31} \\ a_{41} \end{bmatrix}, \quad Q_1 = I - \frac{2}{\|\mathbf{u}\|^2}\mathbf{u}\mathbf{u}^T, \quad Q_1\mathbf{A}_1 = \begin{bmatrix} a_{11} \\ s \\ 0 \\ 0 \end{bmatrix}$$

where $s = \pm\sqrt{a_{21}^2 + a_{31}^2 + a_{41}^2}$, and where the sign of s is chosen so that $sa_{21} \leqslant 0$. (Note that if $a_{21} = a_{31} = a_{41} = 0$, then the first step of Hessenberg reduction is not necessary.)

After we have Q_1, the matrix $Q_1 A$ has the form

$$Q_1 A = \begin{bmatrix} a_{11} & a'_{12} & a'_{13} & a'_{14} \\ s & a'_{22} & a'_{23} & a'_{24} \\ 0 & a'_{32} & a'_{33} & a'_{34} \\ 0 & a'_{42} & a'_{43} & a'_{44} \end{bmatrix}.$$

As seen above, forming $(Q_1 A)Q_1$ amounts to subtracting multiples of $\mathbf{u}^T$ from the rows of $Q_1 A$. Since the first component of $\mathbf{u}$ is zero, forming $Q_1 A Q_1$ does not disturb the first column of $Q_1 A$. Thus the first columns of $Q_1 A$ and $Q_1 A Q_1$ agree, and the (3, 1) and (4, 1) entries of $Q_1 A Q_1$ are zero. $Q_1 A Q_1$ is a matrix that is similar to A and has the form

$$Q_1 A Q_1 = B = \begin{bmatrix} b_{11} & b_{12} & b_{13} & b_{14} \\ b_{21} & b_{22} & b_{23} & b_{24} \\ 0 & b_{32} & b_{33} & b_{34} \\ 0 & b_{42} & b_{43} & b_{44} \end{bmatrix}.$$

If $b_{42} = 0$, then we are done; B is in Hessenberg form. Otherwise we choose a vector $\mathbf{w}$ to zero the last entry of $\mathbf{B}_2$ where $B = [\mathbf{B}_1, \mathbf{B}_2, \mathbf{B}_3, \mathbf{B}_4]$. In detail we have

$$\mathbf{w} = \begin{bmatrix} 0 \\ 0 \\ b_{32} - s \\ b_{42} \end{bmatrix}, \qquad Q_2 = I - \frac{2}{\|\mathbf{w}\|^2}\mathbf{w}\mathbf{w}^T, \quad Q_2 \mathbf{B}_2 = \begin{bmatrix} b_{12} \\ b_{22} \\ s \\ 0 \end{bmatrix}$$

where $s = \pm\sqrt{b_{32}^2 + b_{42}^2}$ with $sb_{32} \leqslant 0$. We now calculate $Q_2 B = [Q_2 \mathbf{B}_1, Q_2 \mathbf{B}_2, Q_2 \mathbf{B}_3, Q_2 \mathbf{B}_4]$; we observe that $Q_2 \mathbf{B}_1 = \mathbf{B}_1$ since

$$Q_2 \mathbf{B}_1 = \mathbf{B}_1 - \frac{2\mathbf{w}^T \mathbf{B}_1}{\|\mathbf{w}\|^2}\mathbf{w};$$

and clearly $\mathbf{w}^T \mathbf{B}_1 = 0$. Thus $Q_2 B$ has zeros in the (3, 1), (4, 1), and (4, 2) positions. As above, forming $(Q_2 B)Q_2$ does not disturb the first two columns of $Q_2 B$ since forming $(Q_2 B)Q_2$ amounts to subtracting multiples of $\mathbf{w}^T$ from the rows of $Q_2 B$.

The net result of what we have done is to obtain a matrix $Q_2 Q_1 A Q_1 Q_2$, which is similar to A and which is in Hessenberg form. One striking feature of Householder transformations is that if A is symmetric, then so is QAQ [since $(QAQ)^T = Q^T A^T Q^T = QAQ$]. Thus if A is symmetric, and if A is reduced to Hessenberg form by a sequence of Householder transformations, then the resulting Hessenberg matrix is also symmetric. Since a Hessenberg matrix has zeros below the subdiagonal, a symmetric Hessenberg matrix also has zeros above the superdiagonal. For instance a symmetric (5 × 5) Hessenberg matrix has the form

$$H = \begin{bmatrix} \times & \times & 0 & 0 & 0 \\ \times & \times & \times & 0 & 0 \\ 0 & \times & \times & \times & 0 \\ 0 & 0 & \times & \times & \times \\ 0 & 0 & 0 & \times & \times \end{bmatrix}.$$

[Such a matrix is called ***tridiagonal*** and has many nice features. For example if the (3, 2) entry of H is zero, then so is the (2, 3) entry; and H is block diagonal, a very good form for the eigenvalue problem (see Exercise 5, Section 6.5).]

```
      SUBROUTINE SIMTRN(A,H,N,MAXDIM)
      DIMENSION A(MAXDIM,MAXDIM),H(MAXDIM,MAXDIM)
      DIMENSION U(50)
      NM2=N-2
C
C  INITIALIZE H--H WILL CONTAIN THE MACHINE ESTIMATE TO A HESSENBERG
C  MATRIX SIMILAR TO A WHEN CONTROL IS RETURNED TO THE CALLING PROGRAM.
C
      DO 2 I=1,N
        DO 1 J=1,N
        H(I,J)=A(I,J)
    1   CONTINUE
    2 CONTINUE
C
C  USE HOUSEHOLDER TRANSFORMATIONS TO ZERO ENTRIES BELOW THE
C  SUBDIAGONAL IN COLUMN J, J=1,2,...,N-2.
C
      DO 6 J=1,NM2
      JP1=J+1
      JP2=J+2
C
C  CALCULATE THE VECTOR U IN EQUATION (6.22), WHERE K=J+1.
C
      SSQ=0.
        DO 3 I=JP1,N
        SSQ=SSQ+H(I,J)**2
    3   CONTINUE
      IF(SSQ.EQ.0.)  GO TO 6
      S=SQRT(SSQ)
      IF(H(JP1,J).GE.0.)  S=-S
      B=1./(SSQ-H(JP1,J)*S)
        DO 4 L=1,J
        U(L)=0.
    4   CONTINUE
      U(JP1)=H(JP1,J)-S
      H(JP1,J)=S
        DO 5 L=JP2,N
        U(L)=H(L,J)
        H(L,J)=0.
    5   CONTINUE
C
C  CALCULATE QHQ, WHERE Q IS THE HOUSEHOLDER TRANSFORMATION
C  DEFINED BY THE VECTOR U.
C
      CALL PRODCT(H,U,B,JP1,N,MAXDIM)
    6 CONTINUE
      RETURN
      END
```

Fig. 6.7 Subroutine SIMTRN.

Subroutine SIMTRN, listed in Fig. 6.7, uses Householder transformations to produce a Hessenberg matrix H, which is similar to A. Subroutine SIMTRN proceeds in stages in the obvious fashion. That is, given A, SIMTRN calculates a Householder transformation Q_1 such that $H_1 = Q_1 A Q_1$ has zeros in the (3, 1), (4, 1), . . . , (N, 1) positions. Given H_1, SIMTRN finds Q_2 so that $H_2 = Q_2 H_1 Q_2$ has zeros in the (4, 2), (5, 2), . . . , (N, 2) positions, etc. For clarity SIMTRN calls Subroutine PRODCT to calculate $Q_i H_{i-1} Q_i$. In addition Subroutine PRODCT calls Subroutine SCPROD to calculate the scalar products $\mathbf{u}^T\boldsymbol{v}$ required by (6.25) and (6.26). The necessary inputs are these.

A	An (N × N) matrix.
N	The size of A.
MAXDIM	The declared dimension of the array containing A in the calling program.

The output from Subroutine SIMTRN is this.

H	A machine estimate to an (N × N) Hessenberg matrix that is similar to A.

The two subroutines called by SIMTRN and a simple program that uses SIMTRN to transform A to Hessenberg form are listed in Figs. 6.8 and 6.9.

EXAMPLE 6.7 The (4 × 4) symmetric matrix

$$A = \begin{bmatrix} 6 & 4 & 4 & 1 \\ 4 & 6 & 1 & 4 \\ 4 & 1 & 6 & 4 \\ 1 & 4 & 4 & 6 \end{bmatrix}$$

has eigenvalues $\lambda_1 = 15$, $\lambda_2 = \lambda_3 = 5$, and $\lambda_4 = -1$. Corresponding eigenvectors are

$$\mathbf{u}_1 = \begin{bmatrix} 1 \\ 1 \\ 1 \\ 1 \end{bmatrix}, \qquad \mathbf{u}_2 = \mathbf{u}_3 = \begin{bmatrix} -1 \\ -1 \\ 1 \\ 1 \end{bmatrix}, \qquad \mathbf{u}_4 = \begin{bmatrix} 1 \\ -1 \\ -1 \\ 1 \end{bmatrix}.$$

This matrix was read in by the program in Fig. 6.9 and produced this output:

$$\text{H} = \begin{bmatrix} 0.6000\text{E } 01 & -0.5745\text{E } 01 & -0.2271\text{E}{-17} & 0.2225\text{E}{-15} \\ -0.5745\text{E } 01 & 0.8909\text{E } 01 & -0.5143\text{E } 01 & -0.6661\text{E}{-15} \\ 0.0 & -0.5143\text{E } 01 & 0.4091\text{E } 01 & 0.1998\text{E}{-14} \\ 0.0 & 0.0 & 0.4441\text{E}{-15} & 0.5000\text{E } 01 \end{bmatrix}$$

Although the input matrix A is symmetric, the output H is not quite symmetric and reflects the effect of roundoff error. (We expected H to be tridiagonal, but h_{13}, h_{14}, and h_{24} are not quite zero. The entries below the subdiagonal were set to zero by SIMTRN.) Interpreting the small entries as being zero, the machine approximation to a Hessenberg matrix that is similar to A is to four places

$$H = \begin{bmatrix} 6.000 & -5.745 & 0 & 0 \\ -5.745 & 8.909 & -5.143 & 0 \\ 0 & -5.143 & 4.091 & 0 \\ 0 & 0 & 0 & 5.000 \end{bmatrix}$$

Note that H is block diagonal, and we see that $\lambda = 5.000$ is probably one of the eigenvalues of A. We will return to this example in the next section.

```
      SUBROUTINE PRODCT(H,U,B,JP1,N,MAXDIM)
      DIMENSION H(MAXDIM,MAXDIM)
      DIMENSION U(50),V(50)
C
C   CALCULATE QH.
C
        DO 3 K=JP1,N
          DO 1 I=1,N
          V(I)=H(I,K)
    1     CONTINUE
        CALL SCPROD(U,V,UDOTV,N)
          DO 2 I=JP1,N
          H(I,K)=H(I,K)-UDOTV*B*U(I)
    2     CONTINUE
    3   CONTINUE
C
C   CALCULATE (QH)Q.
C
        DO 6 K=1,N
          DO 4 I=1,N
          V(I)=H(K,I)
    4     CONTINUE
        CALL SCPROD(U,V,UDOTV,N)
          DO 5 I=JP1,N
          H(K,I)=H(K,I)-UDOTV*B*U(I)
    5     CONTINUE
    6   CONTINUE
      RETURN
      END
      SUBROUTINE SCPROD(U,V,UDOTV,N)
      DIMENSION U(50),V(50)
C
C   THIS SUBPROGRAM CALCULATES UDOTV, WHERE UDOTV IS THE
C   SCALAR PRODUCT OF TWO N-DIMENSIONAL VECTORS, U AND V.
C
      UDOTV=0.
        DO 1 I=1,N
        UDOTV=UDOTV+U(I)*V(I)
    1   CONTINUE
      RETURN
      END
```

Fig. 6.8 Subroutines PRODCT and SCPROD.

```
      DIMENSION A(20,20),H(20,20)
      MAXDIM=20
    1 READ 100,N
      IF(N.LT.3) STOP
        DO 2 I=1,N
        READ 101,(A(I,J),J=1,N)
```

Figure 6.9

```
    2   CONTINUE
        CALL SIMTRN(A,H,N,MAXDIM)
        DO 3 I=1,N
        PRINT 102,(H(I,J),J=1,N)
    3   CONTINUE
        GO TO 1
  100 FORMAT(I2)
  101 FORMAT(10F4.0)
  102 FORMAT(1H0,8E12.4)
      END
```

Figure 6.9 *(continued)*

6.5 EXERCISES

1. Use Subroutine SIMTRN to reduce the following matrices to Hessenberg form.

$$A = \begin{bmatrix} 1 & 2 & 1 & 3 \\ 2 & 4 & 1 & 7 \\ 1 & 1 & 0 & 1 \\ 3 & 7 & 1 & 3 \end{bmatrix} \quad B = \begin{bmatrix} 2 & 0 & 1 & 0 \\ 0 & 2 & 3 & 1 \\ 1 & 3 & 5 & 2 \\ 0 & 1 & 2 & 1 \end{bmatrix} \quad C = \begin{bmatrix} 1 & 1 & 1 & 1 \\ 1 & 1 & 1 & 1 \\ 1 & 1 & 1 & 1 \\ 1 & 1 & 1 & 1 \end{bmatrix}$$

2. Use SIMTRN to reduce the (12×12) matrix A to Hessenberg form.

$$A = \begin{bmatrix} 12 & 11 & 10 & 9 & \dots & 2 & 1 \\ 11 & 11 & 10 & 9 & \dots & 2 & 1 \\ 10 & 10 & 10 & 9 & \dots & 2 & 1 \\ \vdots & & & & & & \vdots \\ 2 & 2 & 2 & 2 & & 2 & 1 \\ 1 & 1 & 1 & 1 & \dots & 1 & 1 \end{bmatrix}$$

3. Reduce the following matrices to Hessenberg form using SIMTRN.

$$A = \begin{bmatrix} 1 & 2 & 1 & 3 \\ 0 & 4 & 1 & 5 \\ 2 & 1 & 1 & 2 \\ 0 & 1 & 4 & 1 \end{bmatrix} \quad B = \begin{bmatrix} 3 & 6 & 4 & 2 \\ 1 & 5 & 3 & 0 \\ 2 & 1 & 8 & 4 \\ 1 & 3 & 9 & 6 \end{bmatrix}$$

4. Prove that the matrix Q in (6.17) is symmetric.
5. Suppose H is a block-diagonal matrix of the form

$$H = \begin{bmatrix} H_{11} & \mathcal{O} \\ \mathcal{O} & H_{22} \end{bmatrix}$$

where H_{11} and H_{22} are square. If λ is an eigenvalue of either H_{11} or H_{22}, then we know from Chapter 3 that λ is an eigenvalue of H. How would you construct an eigenvector for H corresponding to λ? [Hint: See (3.73).]

6. Use results from Chapter 3 to prove that the eigenvalues of a standard symmetric Hessenberg matrix all have algebraic multiplicity one. (Hint: Recall that a symmetric matrix is diagonalizable.)

7. Suppose A is symmetric and has an eigenvalue λ of multiplicity 2. If A is reduced to Hessenberg form by Householder transformations, convince yourself that $QAQ = H$ must have the block-diagonal form of Exercise 5 where λ is an eigenvalue of H_{11} and of H_{22}.
8. Write a subroutine that calculates the characteristic polynomial of a standard Hessenberg matrix H. Test your program on the Hessenberg matrices obtained in Exercises 1, 2, and 3.
9. Most computing facilities have a routine that finds the roots of $p(t) = 0$ where $p(t)$ is a polynomial. Using SIMTRN, the subroutine from Exercise 8, and the root finder, write a program that estimates the eigenvalues of the matrices in Exercises 1, 2, and 3. Then use INVPOW to refine these estimates in the *original* matrices. Check your results by calculating $\|A\mathbf{u} - \lambda\mathbf{u}\|$ where λ and $\mathbf{u}$ are the returned eigenvalue and eigenvector. [Caution: Any procedure that estimates eigenvalues by finding the roots of the characteristic equation has a number of computational pitfalls. In particular while the zeros of the calculated characteristic polynomial are refined to be accurate eigenvalues by INVPOW (and these answers can be checked directly), we cannot be certain that we have found all the eigenvalues unless we obtain n distinct eigenvalues when A is $(n \times n)$.]

6.6 ESTIMATING THE EIGENVALUES OF HESSENBERG MATRICES

Given an $(n \times)$ matrix A, we have shown how Householder transformations can be used to reduce A to a Hessenberg matrix H. Now we come to the hard part—estimating the eigenvalues of H. Once we obtain estimates $\alpha_1, \alpha_2, \ldots, \alpha_n$ to the eigenvalues of H, we can refine them with the inverse power method and in the process find the corresponding eigenvectors for A. If we are confronted with the problem of finding the eigenvalues of H, our first impulse is to calculate the characteristic polynomial, $p(t)$, for H and then find the roots of $p(t) = 0$. This impulse has both reasonable and unreasonable aspects in terms of computation. The bad aspect is that because of roundoff errors, we cannot expect to be able to find the coefficients of $p(t)$ exactly. Suppose for example that the characteristic polynomial of H is $p(t)$, but our machine calculations produce a slightly different polynomial, $q(t)$:

$$\begin{aligned} p(t) &= t^n + a_{n-1}t^{n-1} + \ldots + a_1 t + a_0 \\ q(t) &= t^n + b_{n-1}t^{n-1} + \ldots + b_1 t + b_0. \end{aligned} \tag{6.27}$$

It is an unfortunate fact that even though the coefficients of $p(t)$ and $q(t)$ might be quite close, the roots of $p(t) = 0$ and $q(t) = 0$ may not be so close. We already have an illustration at hand in Example 6.3 where

$$p(t) = (1-t)^{10}$$

$$q(t) = (1-t)^{10} - 2^{-10}.$$

In terms of (6.27) the coefficients of $p(t)$ and $q(t)$ satisfy

$$b_9 = a_9, \qquad b_8 = a_8, \qquad \ldots, \qquad b_1 = a_1, \qquad b_0 = a_0 - 2^{-10};$$

so $p(t)$ and $q(t)$ are polynomials with nearly equal coefficients. However $t = 1/2$ and $t = 3/2$ are roots of $q(t) = 0$ while $t = 1$ is the only root of $p(t) = 0$. [The other eight roots of $q(t) = 0$ are complex. Of course $p(1/2) = p(3/2) = 2^{-10}$. which is quite small; but still $t = 1/2$ and $t = 3/2$ are not roots of $p(t) = 0$.]

The problem of estimating eigenvalues of Hessenberg matrices can be split in terms of computational difficulty into two classes—the eigenvalue problem is relatively easy when the Hessenberg matrix is symmetric, but the problem may be hard for a nonsymmetric matrix. We discuss the easy case first. Suppose H is an $(n \times n)$ symmetric Hessenberg matrix. Then H is tridiagonal and has the form

$$H = \begin{bmatrix} d_1 & b_1 & 0 & 0 & \cdots & 0 & 0 \\ b_1 & d_2 & b_2 & 0 & \cdots & 0 & 0 \\ 0 & b_2 & d_3 & b_3 & \cdots & 0 & 0 \\ 0 & 0 & b_3 & d_4 & \cdots & 0 & 0 \\ \vdots & & & & & & \vdots \\ 0 & 0 & 0 & 0 & \cdots & b_{n-1} & d_n \end{bmatrix}.$$

Suppose we define the sequence of polynomials

$$\begin{aligned} p_0(t) &= 1 \\ p_1(t) &= d_1 - t \\ p_2(t) &= (d_2 - t)p_1(t) - b_1^2 p_0(t) \\ &\vdots \\ p_i(t) &= (d_i - t)p_{i-1}(t) - b_{i-1}^2 p_{i-2}(t) \\ &\vdots \\ p_n(t) &= (d_n - t)p_{n-1}(t) - b_{n-1}^2 p_{n-2}(t). \end{aligned} \tag{6.28}$$

It is an easy exercise to show (Exercise 3, Section 6.6) that $p_n(t)$ is the characteristic polynomial for H.

If the subdiagonal entries $b_1, b_2, \ldots, b_{n-1}$ are all nonzero, then a very effective algorithm due to Givens can be used to isolate the roots of $p_n(t) = 0$. The algorithm proceeds as follows.

1. Let c be some real number.
2. Calculate the numbers $p_0(c), p_1(c), \ldots, p_n(c)$.
3. Let $N(c)$ be the number of agreements in sign in the sequence $p_0(c), p_1(c), \ldots, p_n(c)$.
4. $N(c)$ is equal to the number of roots of $p_n(t) = 0$ that are in the interval $[c, \infty)$.

In the event that $p_k(c) = 0$ for some k, we take the sign of $p_k(c)$ to be that of $p_{k-1}(c)$. [Note that two successive terms in (6.28) cannot both vanish at $t = c$ unless $p_i(c) = 0$ for all i, $0 \leqslant i \leqslant n$. Thus if $p_k(c) = 0$, then $p_{k-1}(c) \neq 0$; and $p_{k-1}(c)$ has a well-defined sign.]

It is not too hard to show that assertion (4) in the algorithm is true, but it is not appropriate to include the proof here. As an example if $n = 4$ and if $\{p_0(c), p_1(c), p_2(c), p_3(c), p_4(c)\}$ has the sign pattern

$$\{+, +, +, -, -\},$$

then $N(c) = 3$; there are 3 roots of $P_4(t) = 0$ in the interval $[c, \infty)$. Similarly the pattern $\{+, -, +, -, -\}$ leads to $N(c) = 1$, and the pattern $\{+, -, 0, +, +\}$ gives $N(c) = 2$.

To use the Givens algorithm for computational purposes, we would first determine an interval $[a, b]$ that contains all the roots of $p_n(t) = 0$; that is, $[a, b]$ contains all the eigenvalues of H. (Since H is symmetric, all the eigenvalues of H are real.) Next let c be the mid-point of $[a, b]$. If $N(c) > N(b)$, then there is at least one eigenvalue in $[c, b]$. Let d be the mid-point of $[c, b]$. If $N(d) > N(b)$, there is at least one eigenvalue in $[d, b]$; on the other hand if $N(d) = N(b)$, then any eigenvalue in $[c, b]$ must be in $[c, d]$. In this fashion by repeatedly halving and testing subintervals we can determine a small subinterval $[r, s]$ that contains $N(r) - N(s)$ eigenvalues of H. The remaining eigenvalues of H must be in $[a, r]$, and the bisection procedure described above can be applied to the interval $[a, r]$.

This process can be terminated when we have determined k small subintervals, $I_1, I_2, \ldots, I_k$, whose union contains all the eigenvalues of H. The mid-point of an interval, I_j, will serve as an estimate, α_j, to an eigenvalue of H. This estimate, α_j, can then be passed to the inverse power method and corrected to be a good estimate of an eigenvalue of A.

Several points must be made about the practical implementation of the Givens algorithm. First of all, we need not demand that the intervals I_j be terribly small since the mid-point of I_j will be used only as an initial guess for an eigenvalue of A. Second we note that the polynomials $p_i(t)$ in (6.28) are not actually calculated since all we need is the value of $p_i(c)$ for various numbers c, $i = 0, 1, \ldots, n$. To calculate $\{p_0(c), p_1(c), \ldots, p_n(c)\}$, we merely compute the numbers

$$\begin{aligned} P_0 &= 1 \\ P_1 &= d_1 - c \\ P_i &= (d_i - c)P_{i-1} - b_{i-1}^2 P_{i-2} \\ &\vdots \end{aligned}$$

for $i = 2, 3, \ldots, n$; and we note that the number P_k is in fact $p_k(c)$. Finally to start the algorithm, we need an initial interval $[a, b]$, which we know contains all the eigenvalues of H. Now if $C = (c_{ij})$ is an $(n \times n)$ real matrix, then it can be shown (see Exercise 4, Section 6.6) that every eigenvalue λ of C satisfies the inequality

$$|\lambda| \leqslant \|C\|$$

where

$$\|C\| = \left(\sum_{i=1}^{n} \sum_{j=1}^{n} c_{ij}^2 \right)^{1/2}. \tag{6.29}$$

Thus for the symmetric Hessenberg matrix H we know that every eigenvalue of H is in the interval $[-\|H\|, \|H\|]$ where

$$\|H\| = \left(\sum_{i=1}^{n} d_i^2 + 2\sum_{i=1}^{n-1} b_i^2\right)^{1/2}.$$

Subroutine GIVENS listed in Fig. 6.10 applies the Givens alogrithm to a symmetric Hessenberg matrix H. The required inputs are the following.

- D — An $(N \times 1)$ array containing the diagonal entries of H.
- B — An $((N-1) \times 1)$ array containing the subdiagonal entries of H. These entries are all presumed to be nonzero.
- RADIUS — A scalar defining the desired radius of a subinterval containing an eigenvalue of H.
- N — The size of H.
- MAXDIM — The declared dimension of the array containing H in the calling program.

The outputs from Subroutine GIVENS are the following.

- ESTEIG — An array containing machine estimates to the eigenvalues of H. If ESTEIG $(k) = \alpha_k$, then there is at least one eigenvalue of H in the interval $[\alpha_k - r, \alpha_k + r]$ where $r =$ RADIUS.
- NEIG — An array of integers. If NEIG(K) $= j$, there are j eigenvalues of H in $[\alpha_k - r, \ \alpha_k + r]$.
- LSTEIG — An integer set equal to the number of subintervals found.

In summary

$$\text{NEIG}(1) + \text{NEIG}(2) + \ldots + \text{NEIG}(\text{LSTEIG}) = \text{N},$$

and NEIG(K) eigenvalues of H are between ESTEIG(K) − RADIUS and ESTEIG(K) + RADIUS. Subroutine GIVENS calls a subprogram, Subroutine SGNCTR, which evaluates the quantity $N(c)$ required by the Givens algorithm; this subprogram is also listed in Fig. 6.10.

Figure 6.11 lists a simple program that employs Subroutine GIVENS. The program reads in a *symmetric* matrix A, uses SIMTRN to reduce A to a symmetric Hessenberg matrix, and then calls GIVENS to isolate the eigenvalues of H. When the program in Fig. 6.11 was given as input the (4×4) matrix A from Example 6.7, the program found the following:

Number of eigenvalues within .1 of 15.02 is 1
Number of eigenvalues within .1 of 5.022 is 2
Number of eigenvalues within .1 of −.9585 is 1.

As we noted in Example 6.7, the eigenvalues of A are $\lambda_1 = 15$, $\lambda_2 = \lambda_3 = 5$, and $\lambda_4 = 1$. The combination of Householder transformations and the Givens algorithm has worked well on this example.

```
      SUBROUTINE GIVENS(D,B,ESTEIG,NEIG,RADIUS,N,MAXDIM,LSTEIG)
C
C  INITIALIZATION PHASE--K, IS THE INDEX OF A SUBINTERVAL OF LENGTH
C  2*RADIUS WHICH CONTAINS NEIG(K) EIGENVALUES OF H.  ALL EIGENVALUES OF
C  H ARE BETWEEN A1=-HNORM AND A2=HNORM.  NTOGO IS THE NUMBER OF
C  EIGENVALUES YET TO BE FOUND AT STAGE K.  SGNCTR IS A SUBROUTINE THAT
C  CALCULATES THE NUMBER OF SIGN AGREEMENTS, N(C).
C
      DIMENSION D(MAXDIM),B(MAXDIM),ESTEIG(MAXDIM),NEIG(MAXDIM)
      NM1=N-1
      K=1
      NTOGO=N
      HNRMSQ=0.
        DO 1  I=1,NM1
        HNRMSQ=HNRMSQ+2.*B(I)**2+D(I)**2
    1   CONTINUE
      HNRMSQ=HNRMSQ+D(N)**2
      HNORM=SQRT(HNRMSQ)
      A1=-HNORM
      A2=HNORM
      CALL SGNCTR(D,B,A2,N,NOFA2,MAXDIM)
C
C  BASIC LOOP--BEGINS WITH AN INTERVAL (A1,A2) AND DETERMINES A
C  SUBINTERVAL OF LENGTH 2*RADIUS CONTAINING NEIG(K) EIGENVALUES OF
C  H, FOR K=1,2,...,LSTEIG.  NBISCT IS THE NUMBER OF BISECTIONS
C  NECESSARY TO PRODUCE THE K-TH SUBINTERVAL.
C
    2 NBISCT=ALOG(2.*(A2-A1)/RADIUS)/ALOG(2.)
      ICTR=0
    3 EVALPT=(A1+A2)/2.
      CALL SGNCTR(D,B,EVALPT,N,NOFEPT,MAXDIM)
      IF(NOFEPT.EQ.NOFA2)  GO TO 4
      A1=EVALPT
      GO TO 5
    4 A2=EVALPT
    5 ICTR=ICTR+1
      IF(ICTR.LT.NBISCT)  GO TO 3
C
C  SET K-TH ESTIMATE OF AN EIGENVALUE EQUAL TO MID-POINT OF
C  K-TH SUBINTERVAL, CALCULATE NEIG(K) AND RETURN IF ALL THE
C  EIGENVALUES OF H HAVE BEEN ISCLATED.
C
      ESTEIG(K)=(A1+A2)/2.
      CALL SGNCTR(D,B,A1,N,NOFA1,MAXDIM)
      CALL SGNCTR(D,B,A2,N,NOFA2,MAXDIM)
      NEIG(K)=NOFA1-NOFA2
      LSTEIG=K
      NTOGO=NTOGO-NEIG(K)
      IF(NTOGO.LE.0)  RETURN
      K=K+1
      A2=A1
      NOFA2=NOFA1
      A1=-HNCRM
      GO TO 2
      END
      SUBROUTINE SGNCTR(D,B,EVALPT,N,NOFEPT,MAXDIM)
      DIMENSION D(MAXDIM),B(MAXDIM)
C
C  THIS SUBPROGRAM CALCULATES THE NUMBER OF SIGN AGREEMENTS, N(C),
C  AT C=EVALPT.  THE NUMBER N(EVALPT) IS RETURNED AS NOFEPT.
C
      NOFEPT=0
      PIM2=1.
      PIM1=D(1)-EVALPT
      IF(PIM2*PIM1.GE.0.)  NOFEPT=1
        DO 1  I=2,N
        PI=(D(I)-EVALPT)*PIM1-B(I-1)**2*PIM2
        IF(PI*PIM1.GT.0.)  NOFEPT=NOFEPT+1
        IF(PI.EQ.0.)  NOFEPT=NOFEPT+1
        PIM2=PIM1
        PIM1=PI
    1   CONTINUE
      RETURN
      END
```

Fig. 6.10 Subroutine GIVENS and SGNCTR.

```
      DIMENSION A(20,20),H(20,20),D(20),B(20),ESTEIG(20),NEIG(20)
      MAXDIM=20
    1 READ 100,N
      IF(N.LT.3)  STOP
        DO 2 I=1,N
        READ 101,(A(I,J),J=1,N)
    2   CONTINUE
      CALL SIMTRN(A,H,N,MAXDIM)
        DO 3 I=1,N
        PRINT 102,(H(I,J),J=1,N)
    3   CONTINUE
      NM1=N-1
        DO 4 I=1,NM1
        D(I)=H(I,I)
        B(I)=H(I+1,I)
        IF(B(I).EQ.0.)  GO TO 1
    4   CONTINUE
      D(N)=H(N,N)
      RADIUS=.1
      CALL GIVENS(D,B,ESTEIG,NEIG,RADIUS,N,MAXDIM,LSTEIG)
        DO 5 I=1,LSTEIG
        PRINT 103,RADIUS,ESTEIG(I),NEIG(I)
    5   CONTINUE
      GO TO 1
  100 FORMAT(I2)
  101 FORMAT(10F4.0)
  102 FORMAT(1H0,8E12.4)
  103 FORMAT(1H0,28HNUMBER OF EIGENVALUES WITHIN,F8.4,2X,2HOF,
     1E16.4,4H  IS,I3)
      END
```

Fig. 6.11 A program to isolate the eigenvalues of a symmetric matrix A.

The inverse power method can be incorporated into the program in Fig. 6.11 to produce a complete package for finding the eigenvalues and eigenvectors of a symmetric matrix. Such a program is listed in Fig. 6.12.

EXAMPLE 6.8 As a simple test of the program in Fig. 6.12, we input the matrix

$$A = \begin{bmatrix} 2 & 4 & -6 \\ 4 & 2 & -6 \\ -6 & -6 & -15 \end{bmatrix},$$

which has eigenvalues $\lambda_1 = -18$, $\lambda_2 = 9$, $\lambda_3 = -2$, and corresponding eigenvectors

$$\mathbf{u}_1 = \begin{bmatrix} 1 \\ 1 \\ 4 \end{bmatrix}, \qquad \mathbf{u}_2 = \begin{bmatrix} 2 \\ 2 \\ -1 \end{bmatrix}, \qquad \mathbf{u}_3 = \begin{bmatrix} 1 \\ -1 \\ 0 \end{bmatrix}.$$

The program returned eigenvalues correct to the five places printed and eigenvectors (of length 1):

$$\mathbf{v}_1 = \begin{bmatrix} -.23570 \\ -.23570 \\ -.94281 \end{bmatrix}, \qquad \mathbf{v}_2 = \begin{bmatrix} .66667 \\ .66667 \\ -.33333 \end{bmatrix}, \qquad \mathbf{v}_3 = \begin{bmatrix} .70711 \\ -.70711 \\ -.14343\text{E-}10 \end{bmatrix}.$$

```
      DIMENSION A(20,20),H(20,20),D(20),B(20),ESTEIG(20),NEIG(20)
      DIMENSION EIGNVC(20)
      MAXDIM=20
    1 READ 100,N
      IF(N.LT.3)  STOP
        DO 2 I=1,N
        READ 101,(A(I,J),J=1,N)
    2   CONTINUE
      CALL SIMTRN(A,H,N,MAXDIM)
        DO 3 I=1,N
        PRINT 102,(H(I,J),J=1,N)
    3   CONTINUE
      NM1=N-1
        DO 4 I=1,NM1
        D(I)=H(I,I)
        B(I)=H(I+1,I)
        IF(B(I).EQ.0.)  GO TO 1
    4   CONTINUE
      D(N)=H(N,N)
      RADIUS=.1
      CALL GIVENS(D,B,ESTEIG,NEIG,RADIUS,N,MAXDIM,LSTEIG)
        DO 5 I=1,LSTEIG
        PRINT 103,RADIUS,ESTEIG(I),NEIG(I)
    5   CONTINUE
      TOL=.1E-06
      MAXITR=10
        DO 6 I=1,LSTEIG
        ALPHA=ESTEIG(I)
        CALL INVPOW(A,EIGNVC,ALPHA,EIGNVL,N,TOL,MAXITR,MAXDIM,IFLAG)
        PRINT 104,IFLAG,EIGNVL
        PRINT 105,(EIGNVC(K),K=1,N)
    6   CONTINUE
      GO TO 1
  100 FORMAT(I2)
  101 FORMAT(10F4.0)
  102 FORMAT(1H0,8E12.4)
  103 FORMAT(1H0,28HNUMBER OF EIGENVALUES WITHIN,F8.4,2X,2HOF,
     1E16.4,4H  IS,I3)
  104 FORMAT(1H0,6H FLAG=,I3,5X,36HREFINED ESTIMATE OF AN EIGENVALUE IS,
     1E15.5)
  105 FORMAT(1H0,30HA CORRESPONDING EIGENVECTOR IS,/,6E18.5)
      END
```

Fig. 6.12 A program to find the eigenvalues and eigenvectors of a symmetric matrix A.

EXAMPLE 6.9 It is possible (especially when running a contrived problem rather than a real-world problem) that an input symmetric matrix might have an eigenvector each of whose entries is 1. This sort of matrix might fool the inverse power method, and Subroutine INVPOW may have to be modified to accommodate this. For example when we input the matrix A from Example 6.7 to the program in Fig. 6.12, the program found estimates $\lambda_1 = 15.02$, $\lambda_2 = \lambda_3 = 5.022$, and $\lambda_4 = -.9585$ as we noted above. When these are passed to the inverse power method, convergence is to $\lambda = 15$ for each of these estimates because the eigenvector $\mathbf{u}_1$ corresponding to $\lambda_1 = 15$ is

$$\mathbf{u}_1 = \begin{bmatrix} 1 \\ 1 \\ 1 \\ 1 \end{bmatrix}.$$

To get the inverse power method to work properly, we changed the definition of the starting vector (OLDVCT in Subroutine INVPOW) to

$$\begin{bmatrix} 0 \\ 1/\sqrt{3} \\ 1/\sqrt{3} \\ 1/\sqrt{3} \end{bmatrix}.$$

With this modification all three eigenvectors and eigenvalues were found correctly. (Note that the output from the Givens algorithm alerted us that the inverse power method had been fooled.)

As we mentioned earlier, the eigenvalue problem for a nonsymmetric matrix may be quite difficult. If A is not symmetric, the currently accepted practice is to reduce A to Hessenberg form and then to employ some version of a procedure called the QR algorithm. Since the QR algorithm is rather hard to explain thoroughly, we refer the reader to a more advanced reference.

As a rather *ad hoc* procedure that can be used for a nonsymmetric matrix A, we can first transform A to a Hessenberg matrix H. Next using the ideas in Chapter 3, we can calculate the characteristic polynomial for H and call a polynomial root-finding routine to obtain estimates to the eigenvalues of H. (Any large computing center has a number of root-finding routines, which are relatively easy to use.) These estimated eigenvalues can then be passed to the inverse power method for correction. There is however one complication—if A is not symmetric, A may well have some complex eigenvalues. In this event Subroutine INVPOW and Subroutine GAUSS (which is called by INVPOW) will have to be modified so that they can do complex arithmetic. These modifications are relatively trivial and can be accomplished by changing several statements in GAUSS and INVPOW. In addition as we mentioned in Exercise 9, Section 6.5, we have no guarantee that we have found all the eigenvalues of A.

6.6 EXERCISES

1. Use the program in Fig. 6.12 to find the eigenvalues and the eigenvectors of the matrices in Exercises 1 and 2, Section 6.5.
2. It can be shown that the eigenvalue problem for symmetric matrices is well behaved in that small changes in a symmetric matrix produce proportionately small changes in the eigenvalues. To illustrate this, choose one of the matrices in Exercise 1, add .01 to the (1, 4) and (4, 1) entries, and calculate the eigenvalues and the eigenvectors of the new matrix.
3. Prove that $p_n(t)$ as defined in (6.28) is the characteristic polynomial of the associated symmetric tridiagonal matrix H.
4. Let $C = (c_{ij})$ be an $(n \times n)$ matrix and let λ be an eigenvalue of C. Prove that $|\lambda| \leqslant \|C\|$ where $\|C\|$ is given in (6.29). To prove this, suppose $C\mathbf{x} = \lambda\mathbf{x}$where $\|\mathbf{x}\| = 1$, and show

that $\|C\mathbf{x}\| = |\lambda|$. Use the Cauchy–Schwarz inequality and the properties of norms to show that

$$\begin{aligned}\|C\mathbf{x}\| &\leqslant |x_1|\,\|\mathbf{C}_1\| + |x_2|\,\|\mathbf{C}_2\| + \ldots + |x_n|\,\|\mathbf{C}_n\| \\ &\leqslant \|\mathbf{u}\|\,\|\boldsymbol{v}\|\end{aligned}$$

where

$$\mathbf{u} = \begin{bmatrix} |x_1| \\ |x_2| \\ \vdots \\ |x_n| \end{bmatrix} \qquad \boldsymbol{v} = \begin{bmatrix} \|\mathbf{C}_1\| \\ \|\mathbf{C}_2\| \\ \vdots \\ \|\mathbf{C}_n\| \end{bmatrix}.$$

Finally show that $\|\mathbf{u}\| = 1$ and $\|\boldsymbol{v}\| = \|C\|$.

5. (Hyman's method) Let H be a (4×4) standard Hessenberg matrix, and α a scalar; and consider

$$\begin{aligned}(h_{11}-\alpha)x_1 + h_{12}x_2 + h_{13}x_3 + h_{14}x_4 &= g(\alpha) \\ h_{21}x_1 + (h_{22}-\alpha)x_2 + h_{23}x_3 + h_{24}x_4 &= 0 \\ h_{32}x_2 + (h_{33}-\alpha)x_3 + h_{34}x_4 &= 0 \\ h_{43}x_3 + (h_{44}-\alpha)x_4 &= 0.\end{aligned}$$

In this system, set $x_4 = 1$; determine x_3, x_2, and x_1 from the last three equations; and then let $g(\alpha)$ be defined by the first equation. Prove that $g(\alpha) = 0$ if and only if α is an eigenvalue of H. [Note: One half of this proof is easy; if $g(\alpha) = 0$, then you have generated a nontrivial solution of $(H - \alpha I)\mathbf{x} = \boldsymbol{\theta}$. To complete the proof, show that if α is an eigenvalue of H, then x_4 can always be chosen to be equal to 1. Also note that this procedure can clearly be extended to $(n \times n)$ matrices H.]

6. If H is any matrix and λ is any real eigenvalue of H, then $-\|H\| \leqslant \lambda \leqslant \|H\|$ by Exercise 4. If H is a standard Hessenberg matrix, we can attempt to isolate the real eigenvalues of H by evaluating $g(\alpha)$ for various α in $[-\|H\|, \|H\|]$ where $g(\alpha)$ is as in Exercise 5. Write a subroutine that evaluates $g(\alpha)$ for various α, and test your program on

$$A = \begin{bmatrix} 15 & 11 & 6 & -9 & -15 \\ 1 & 3 & 9 & -3 & -8 \\ 7 & 6 & 6 & -3 & -11 \\ 7 & 7 & 5 & -3 & -11 \\ 17 & 12 & 5 & -10 & -16 \end{bmatrix}.$$

(Note: The matrix A has one real eigenvalue and of course must be reduced to Hessenberg form before Hyman's method can be implemented.)

APPENDIX: REVIEW OF GEOMETRIC VECTORS

Many physical quantities cannot be described mathematically by a single real number; an example is force, which has both magnitude and direction. Such physical quantities are called vectors, and we can represent them in two-space or three-space as directed line segments. Such a line segment, $\overrightarrow{AB}$, directed from A to B is called a ***geometric vector***. We note that a force vector is completely determined by its length and direction; so the initial point A in a representation $\overrightarrow{AB}$ is irrelevant. Similarly any two geometric vectors having the same length and direction are called ***equivalent*** and can be regarded as being equal. In Fig. A.1 the geometric vectors $\overrightarrow{OP}$, $\overrightarrow{AB}$, and $\overrightarrow{CD}$ are all equivalent.

Given an arbitrary geometric vector $\boldsymbol{v} = \overrightarrow{AB}$, there is a unique geometric vector $\overrightarrow{OP}$ (called the position vector) that is equivalent to $\boldsymbol{v}$ and has its initial point at the origin of the coordinate system (see Fig. A.1). If the coordinates of the terminal point P are (v_1, v_2), then we say that (v_1, v_2) are the ***components*** of $\boldsymbol{v}$; and we write this designation as

$$\boldsymbol{v} = (v_1, v_2). \tag{A.1}$$

Similarly for a geometric vector in three-space, we would write $\boldsymbol{v} = (v_1, v_2, v_3)$ where the components of $\boldsymbol{v}$ are the coordinates of the terminal point of its equivalent position vector. If $\boldsymbol{v} = \overrightarrow{AB}$ is any geometric vector, and if A and B have coordinates (a_1, a_2) and (b_1, b_2), respectively, then clearly $\overrightarrow{OP}$ is equivalent to $\overrightarrow{AB}$ where the point P has coordinates $(b_1 - a_1, b_2 - a_2)$; that is, the line segment $\overrightarrow{OP}$ has the same

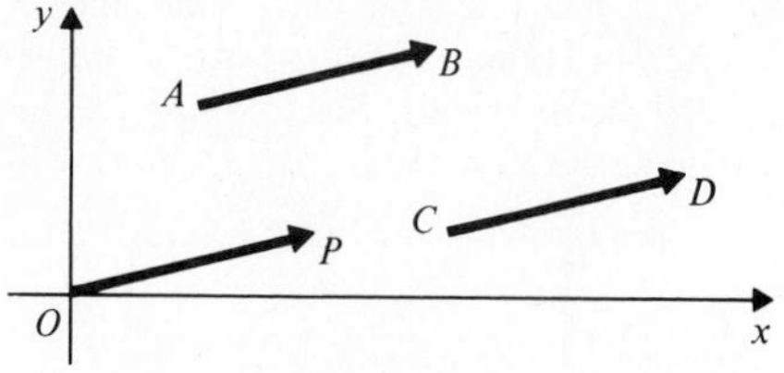

Fig. A.1 Equivalent vectors.

length and direction as $\overrightarrow{AB}$. Therefore the components of $\boldsymbol{v} = \overrightarrow{AB}$ are

$$\boldsymbol{v} = (b_1 - a_1, b_2 - a_2). \tag{A.2}$$

From (A.2) one sees how to find the components of any geometric vector $\boldsymbol{v} = \overrightarrow{AB}$. Conversely if we are given the components of a geometric vector $\boldsymbol{v}$ as in (A.1), then we can construct the equivalent geometric vector $\overrightarrow{AB}$ whose initial point is at A; the coordinates of the terminal point B are $(a_1 + v_1, a_2 + v_2)$.

If $\mathbf{u}$ and $\boldsymbol{v}$ are any two geometric vectors, then the sum $\mathbf{u} + \boldsymbol{v}$ is determined as follows. Position $\boldsymbol{v}$ so that the initial point of $\boldsymbol{v}$ coincides with the terminal point of $\mathbf{u}$. Then $\mathbf{u} + \boldsymbol{v}$ is the geometric vector represented by the line segment from the initial point of $\mathbf{u}$ to the terminal point of $\boldsymbol{v}$ (see Fig. A.2). The components of $\mathbf{u} + \boldsymbol{v}$ are easy to calculate given the components $\mathbf{u} = (u_1, u_2)$ and $\boldsymbol{v} = (v_1, v_2)$. In particular if the initial point of $\mathbf{u}$ is at $A = (a_1, a_2)$, then the terminal point of $\mathbf{u}$ is at $(a_1 + u_1, a_2 + u_2)$. Similarly the terminal point of $\boldsymbol{v}$ is at $(a_1 + u_1 + v_1, a_2 + u_2 + v_2)$; so the sum $\mathbf{u} + \boldsymbol{v}$ has its initial point at (a_1, a_2) and its terminal point at $(a_1 + u_1 + v_1, a_2 + u_2 + v_2)$. Thus by (A.2) the components of $\mathbf{u} + \boldsymbol{v}$ are $(u_1 + v_1, u_2 + v_2)$; see Fig. A.3. Similarly in three-space if $\mathbf{u} = (u_1, u_2, u_3)$ and $\boldsymbol{v} = (v_1, v_2, v_3)$, then $\mathbf{u} + \boldsymbol{v} = (u_1 + v_1, u_2 + v_2, u_3 + v_3)$.

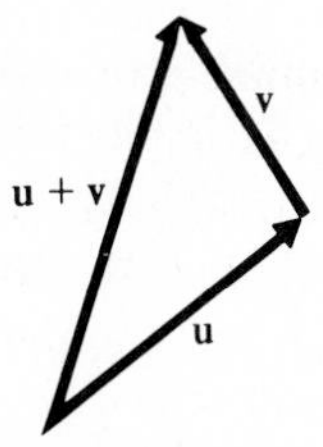

Figure A.2

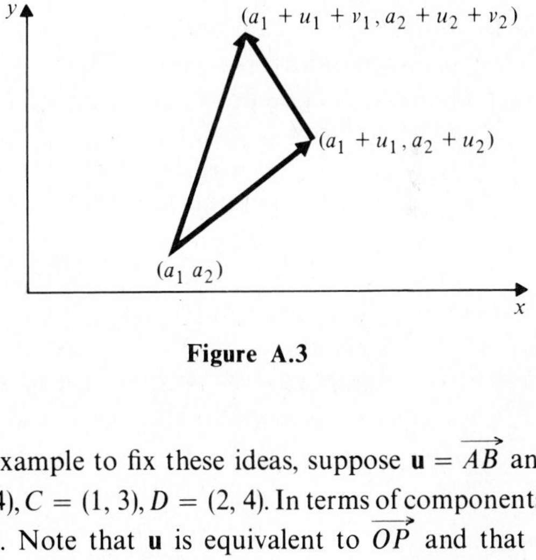

Figure A.3

EXAMPLE A.1 As a numerical example to fix these ideas, suppose $\mathbf{u} = \overrightarrow{AB}$ and $\boldsymbol{v} = \overrightarrow{CD}$ where $A = (7, 2)$, $B = (4, 4)$, $C = (1, 3)$, $D = (2, 4)$. In terms of components we have $\mathbf{u} = (-3, 2)$ and $\boldsymbol{v} = (1, 1)$. Note that $\mathbf{u}$ is equivalent to $\overrightarrow{OP}$ and that $\boldsymbol{v}$ is equivalent to $\overrightarrow{OQ}$ where $P = (-3, 2)$ and $Q = (1, 1)$; see Fig. A.4, in which it is clear that $\overrightarrow{OP}$ and $\overrightarrow{AB}$ have the same length and direction as do $\overrightarrow{OQ}$ and $\overrightarrow{CD}$. In Fig. A.5 the vector $\mathbf{u} + \boldsymbol{v}$ is illustrated. Since $\mathbf{u} = (-3, 2)$ and $\boldsymbol{v} = (1, 1)$, it follows that $\mathbf{u} + \boldsymbol{v} = (-2, 3)$, or $\mathbf{u} + \boldsymbol{v}$ is equivalent to $\overrightarrow{OR}$.

If $\boldsymbol{v} = (v_1, v_2)$ is a geometric vector, then the ***length*** or ***magnitude*** of $\boldsymbol{v}$ is defined to be $\|\boldsymbol{v}\| = \sqrt{v_1{}^2 + v_2{}^2}$. [If $\boldsymbol{v} = \overrightarrow{AB}$, then (A.2) says that $\|\boldsymbol{v}\|$ is just the length of the

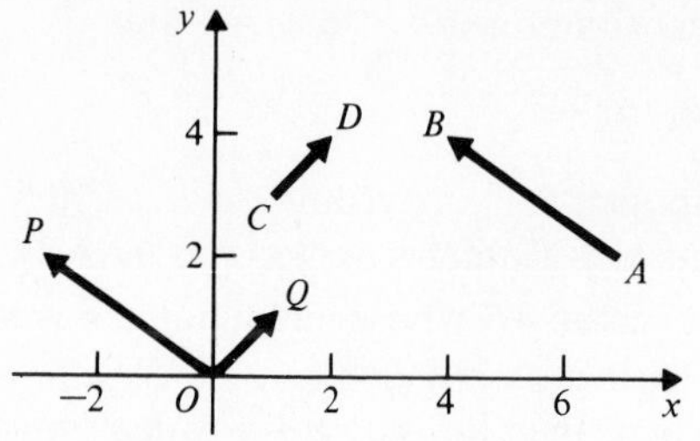

Figure A.4

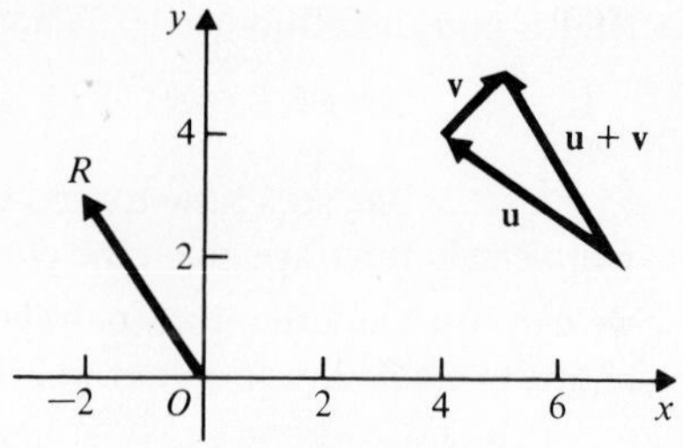

Figure A.5

line segment from A to B. Similarly if $\boldsymbol{v} = (v_1, v_2, v_3)$ is a geometric vector in three-space, then $\| \boldsymbol{v} \| = \sqrt{v_1{}^2 + v_2{}^2 + v_3{}^2}$.) If c is a scalar, we define $c\boldsymbol{v}$ to be the geometric vector whose magnitude is $|c| \, \| \boldsymbol{v} \|$ and whose direction is the same as that of $\boldsymbol{v}$ when $c > 0$, and is opposite that of $\boldsymbol{v}$ when $c < 0$. For $c = 0$ the vector $c\boldsymbol{v}$ is defined to be the zero vector $\boldsymbol{\theta}$ (the direction of $\boldsymbol{\theta}$ is not defined, but the magnitude of $\boldsymbol{\theta}$ is 0). If $\boldsymbol{v}$ has components (v_1, v_2), it is an easy exercise to show that $c\boldsymbol{v}$ has components (cv_1, cv_2). For the special case $c = -1$ we usually write $-\boldsymbol{v}$ instead of $(-1)\boldsymbol{v}$; and we observe that if $\mathbf{u} = (u_1, u_2)$ and $\boldsymbol{v} = (v_1, v_2)$, then $\mathbf{u} - \boldsymbol{v} = (u_1 - v_1, u_2 - v_2)$.

In two-space and three-space we define the angle between two vectors $\mathbf{u}$ and $\boldsymbol{v}$ in a natural way, using components. That is, if $\mathbf{u} = (u_1, u_2, u_3)$ and $\boldsymbol{v} = (v_1, v_2, v_3)$ are two vectors in three-space, then the angle φ between $\mathbf{u}$ and $\boldsymbol{v}$ is the angle AOB where $A = (u_1, u_2, u_3), O = (0, 0, 0), B = (v_1, v_2, v_3)$. (Note: If $\mathbf{u} = c\boldsymbol{v}, c > 0$, then $\varphi = 0$; if $c < 0$, then $\varphi = \pi$.) Figure A.6 illustrates the definition in two-space; the angle between $\mathbf{u}$ and $\boldsymbol{v}$ is φ where $\mathbf{u}$ is equivalent to $\overrightarrow{OA}$ and $\boldsymbol{v}$ is equivalent to $\overrightarrow{OB}$. For $\mathbf{u} = (u_1, u_2, u_3)$ and $\boldsymbol{v} = (v_1, v_2, v_3)$ we define the ***dot product*** of $\mathbf{u}$ and $\boldsymbol{v}$, denoted by $\mathbf{u} \cdot \boldsymbol{v}$, as

$$\mathbf{u} \cdot \boldsymbol{v} = u_1 v_1 + u_2 v_2 + u_3 v_3. \tag{A.3}$$

Similarly if $\mathbf{u} = (u_1, u_2)$ and $\boldsymbol{v} = (v_1, v_2)$ are in two-space, then $\mathbf{u} \cdot \boldsymbol{v} = u_1 v_1 + u_2 v_2$. There is a useful connection between $\mathbf{u} \cdot \boldsymbol{v}$ and the angle between $\mathbf{u}$ and $\boldsymbol{v}$.

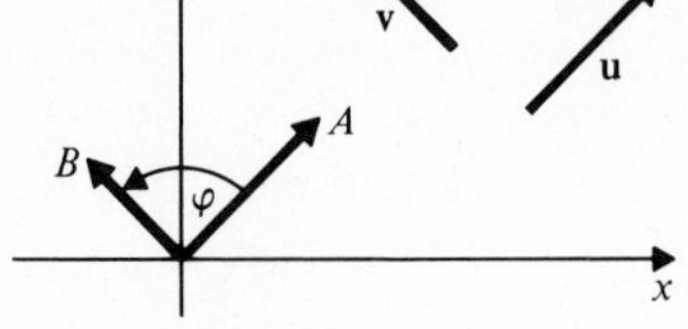

Figure A.6

Theorem A.1

If φ is the angle between two geometric vectors $\mathbf{u}$ and $\boldsymbol{v}$, then

$$\mathbf{u} \cdot \boldsymbol{v} = \|\mathbf{u}\| \, \|\boldsymbol{v}\| \cos \varphi.$$

Proof. If $\mathbf{u} = c\boldsymbol{v}$, then $\varphi = 0$ or $\varphi = \pi$, and the proof is immediate. For $0 < \varphi < \pi$, we can use the law of cosines in Fig. A.6 to deduce

$$\|\overrightarrow{AB}\|^2 = \|\overrightarrow{OA}\|^2 + \|\overrightarrow{OB}\|^2 - 2\|\overrightarrow{OA}\| \, \|\overrightarrow{OB}\| \cos \varphi. \tag{A.4}$$

With reference to Fig. A.6 we see that $\overrightarrow{AB}$ is equivalent to $\boldsymbol{v} - \mathbf{u}$; so if $\mathbf{u} = (u_1, u_2)$ and $\boldsymbol{v} = (v_1, v_2)$ then $\|\overrightarrow{AB}\|^2 = (v_1 - u_1)^2 + (v_2 - u_2)^2$. Similarly $\|\overrightarrow{OA}\|^2 = u_1{}^2 + u_2{}^2$ and $\|\overrightarrow{OB}\|^2 = v_1{}^2 + v_2{}^2$; so (A.4) reduces to

$$-2u_1v_1 - 2u_2v_2 = -2\|\overrightarrow{OA}\| \, \|\overrightarrow{OB}\| \cos \varphi.$$

Since $\|\overrightarrow{OA}\| = \|\mathbf{u}\|$ and $\|\overrightarrow{OB}\| = \|\boldsymbol{v}\|$, Theorem A.1 is proved in two-space and similarly follows in three-space. ■

From Theorem A.1 if $\mathbf{p}$ and $\mathbf{q}$ are *nonzero* geometric vectors, then $\mathbf{p}$ and $\mathbf{q}$ are perpendicular if and only if $\mathbf{p} \cdot \mathbf{q} = 0$. Given vectors $\mathbf{u}$ and $\boldsymbol{v}$, it is frequently necessary to express $\mathbf{u}$ as the sum of two vectors $\mathbf{r}$ and $\mathbf{s}$ where $\mathbf{r}$ is a scalar multiple of $\boldsymbol{v}$ and $\mathbf{s}$ is perpendicular to $\boldsymbol{v}$. That is, we want $\mathbf{u} = \mathbf{r} + \mathbf{s}$ where $\mathbf{r} = c\boldsymbol{v}$ and $\mathbf{s} \cdot \boldsymbol{v} = 0$ (see Fig. A.7). To determine $\mathbf{r}$ and $\mathbf{s}$, we set $\mathbf{u} = \mathbf{r} + \mathbf{s} = c\boldsymbol{v} + \mathbf{s}$; and we calculate $\mathbf{u} \cdot \boldsymbol{v} = c(\boldsymbol{v} \cdot \boldsymbol{v}) + (\mathbf{s} \cdot \boldsymbol{v}) = c\,\|\boldsymbol{v}\|^2$. Thus $\mathbf{r} = c\boldsymbol{v}$ is found from $c = \mathbf{u} \cdot \boldsymbol{v} / \|\boldsymbol{v}\|^2$; and once we have $\mathbf{r}$, $\mathbf{s}$ is given by $\mathbf{s} = \mathbf{u} - \mathbf{r}$.

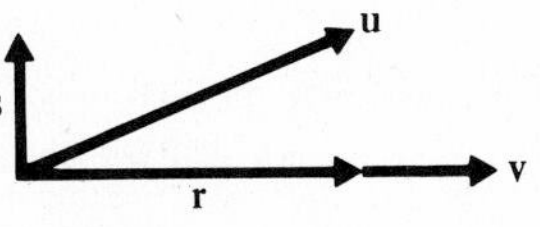

Figure A.7

If $\mathbf{u}$ and $\boldsymbol{v}$ are any two nonzero vectors in three-space, it is often important to be able to construct a vector $\mathbf{x}$ that is perpendicular to both $\mathbf{u}$ and $\boldsymbol{v}$. If $\mathbf{u} = (u_1, u_2, u_3)$ and $\boldsymbol{v} = (v_1, v_2, v_3)$, and if $\mathbf{u}$ is not a multiple of $\boldsymbol{v}$, then Theorem A.1 shows that we want a vector $\mathbf{x} = (x_1, x_2, x_3)$ such that $\mathbf{u} \cdot \mathbf{x} = 0$ and $\boldsymbol{v} \cdot \mathbf{x} = 0$. Writing these two conditions out, we see that $\mathbf{x}$ will be perpendicular to both $\mathbf{u}$ and $\boldsymbol{v}$ if

$$\begin{aligned} u_1x_1 + u_2x_2 + u_3x_3 &= 0. \\ v_1x_1 + v_2x_2 + v_3x_3 &= 0. \end{aligned} \tag{A.5}$$

Solving the system of equations in (A.5) shows that one vector $\mathbf{x}$ that satisfies $\mathbf{u} \cdot \mathbf{x} = 0$ and $\boldsymbol{v} \cdot \mathbf{x} = 0$ is

$$\mathbf{x} = (u_2v_3 - u_3v_2,\ u_3v_1 - u_1v_3,\ u_1v_2 - u_2v_1). \tag{A.6}$$

We define the ***cross product*** of $\mathbf{u}$ and $\boldsymbol{v}$, denoted by $\mathbf{u} \times \boldsymbol{v}$, to be the geometric vector $\mathbf{x}$ in (A.6). For example if $\mathbf{u} = (1, 2, 1)$ and $\boldsymbol{v} = (6, 0, 3)$, then

$$\mathbf{u} \times \boldsymbol{v} = (6, 3, -12);$$

and we see that $\mathbf{u} \cdot (\mathbf{u} \times \boldsymbol{v}) = 0$ and $\boldsymbol{v} \cdot (\mathbf{u} \times \boldsymbol{v}) = 0$. Theorem A.2 summarizes some of the properties of the cross product.

Theorem A.2

Let $\mathbf{u}$, $\boldsymbol{v}$, and $\mathbf{w}$ be geometric vectors in three-space.

1. $\mathbf{u} \times \boldsymbol{v} = -(\boldsymbol{v} \times \mathbf{u})$.
2. $(\mathbf{u} \times \boldsymbol{v}) \cdot \mathbf{w} = \mathbf{u} \cdot (\boldsymbol{v} \times \mathbf{w})$. (the "box product")
3. $\mathbf{u} \times (\boldsymbol{v} \times \mathbf{w}) = (\mathbf{u} \cdot \mathbf{w})\boldsymbol{v} - (\mathbf{u} \cdot \boldsymbol{v})\mathbf{w}$. (the "triple product")
4. $\|\mathbf{u} \times \boldsymbol{v}\|^2 = \|\mathbf{u}\|^2 \|\boldsymbol{v}\|^2 - (\mathbf{u} \cdot \boldsymbol{v})^2$.
5. $\mathbf{u} \cdot (\mathbf{u} \times \boldsymbol{v}) = \boldsymbol{v} \cdot (\mathbf{u} \times \boldsymbol{v}) = 0$.

It is conventional to let $\mathbf{i}$, $\mathbf{j}$, and $\mathbf{k}$ denote the standard unit vectors

$$\mathbf{i} = (1, 0, 0) \qquad \mathbf{j} = (0, 1, 0) \qquad \mathbf{k} = (0, 0, 1);$$

and we observe that if $\boldsymbol{v} = (v_1, v_2, v_3)$, then $\boldsymbol{v} = v_1\mathbf{i} + v_2\mathbf{j} + v_3\mathbf{k}$. These unit vectors can be used also to remember the form of $\mathbf{u} \times \boldsymbol{v}$. That is, if $\mathbf{u} = (u_1, u_2, u_3)$ and $\boldsymbol{v} = (v_1, v_2, v_3)$, then $\mathbf{u} \times \boldsymbol{v}$ can be represented symbolically as a (3×3) determinant:

$$\mathbf{u} \times \boldsymbol{v} = \begin{vmatrix} \mathbf{i} & \mathbf{j} & \mathbf{k} \\ u_1 & u_2 & u_3 \\ v_1 & v_2 & v_3 \end{vmatrix} = \mathbf{i}\begin{vmatrix} u_2 & u_3 \\ v_2 & v_3 \end{vmatrix} - \mathbf{j}\begin{vmatrix} u_1 & u_3 \\ v_1 & v_3 \end{vmatrix} + \mathbf{k}\begin{vmatrix} u_1 & u_2 \\ v_1 & v_2 \end{vmatrix}.$$

Finally if we use Theorem .1 and part (4) of Theorem A.2, we obtain the important formula

$$\|\mathbf{u} \times \boldsymbol{v}\| = \|\mathbf{u}\| \, \|\boldsymbol{v}\| \sin \varphi;$$

and so $\|\mathbf{u} \times \boldsymbol{v}\|$ is the area of the parallelogram determined by $\mathbf{u}$ and $\boldsymbol{v}$.

We conclude by showing how geometric vectors can be used to express lines and planes in three-space. Consider the line L in three-space that passes through the point $P_0 = (x_0, y_0, z_0)$ and is parallel to the nonzero geometric vector $\boldsymbol{v} = (v_1, v_2, v_3)$. If $P = (x, y, z)$ is an arbitrary point on L, then the geometric vector $\overrightarrow{PP_0}$ is parallel to $\boldsymbol{v}$, and so $\overrightarrow{PP_0} = t\boldsymbol{v}$. Componentwise this equation is

$$x - x_0 = tv_1, \qquad y - y_0 = tv_2, \qquad z - z_0 = tv_3. \tag{A.7}$$

Since the point P is on L if and only if (A.7) is satisfied for some value of t, (A.7) gives the parametric equations of the line L. If v_1, v_2, and v_3 are all nonzero, we can eliminate the parameter t in (A.7) and obtain the symmetric equations of the line L as

$$\frac{x - x_0}{v_1} = \frac{y - y_0}{v_2} = \frac{z - z_0}{v_3}. \tag{A.8}$$

[There are obvious modifications to (A.8) if any of v_1, v_2, or v_3 is zero.]

Finally we consider the equation of the plane Π that passes through the point $P_0 = (x_0, y_0, z_0)$ and has the nonzero geometric vector $\boldsymbol{v} = (v_1, v_2, v_3)$ perpendicular to it; that is, $\boldsymbol{v}$ is "normal" to Π. If $P = (x, y, z)$ is any point in Π, then the geometric vector $\overrightarrow{PP_0}$ is perpendicular to $\boldsymbol{v}$. Setting $\overrightarrow{PP_0} \cdot \boldsymbol{v} = 0$ yields

$$v_1(x - x_0) + v_2(y - y_0) + v_3(z - z_0) = 0,$$

which is called the "point-normal" equation of the plane Π.

ANSWERS TO ODD-NUMBERED EXERCISES*

1.1 EXERCISES

1. a) $x_1 = 1, x_2 = -3$ (b) $x_1 = (3x_2 + 5)/2$ (c) $x_1 = 0, x_2 = 2$ (d) inconsistent (e) $x_1 = 2 - x_2$ (f) $x_1 = x_3, x_2 = 2x_3 - 3$ (g) $x_1 = x_2 + 10, x_3 = -6$ (h) $x_1 = -x_2 + x_3 + 2$ (i) inconsistent (j) $x_1 = 1, x_2 = 2, x_3 = 2$ (k) inconsistent (l) $x_1 = 1, x_2 = -2, x_3 = 3$ (m) $x_1 = -2x_3 + 3, x_2 = 3x_3 - 2$ (n) $x_1 = 2x_3, x_2 = 1$ **3.** $x_1 = 2, x_2 = 1, x_3 = -2.$ **5.** a) $b = 16$ (b) $b = -9$; solutions are not unique. **7.** One form is $x_1 = 2x_3 + 3x_4 - 12, x_2 = 3x_3 + 4x_4 - 16$ while $x_3 = -4x_1 + 3x_2, x_4 = -2x_2 + 3x_1 + 4$ is another. **9.** a) $a = 2$ (b) $a = -1$ (c) any value of a except $a = 8$

1.2 EXERCISES

1. a) $\begin{bmatrix} 1 & 2 \\ 2 & -1 \end{bmatrix}$ and $\begin{bmatrix} 1 & 2 & -5 \\ 2 & -1 & 5 \end{bmatrix}$ (c) $\begin{bmatrix} 1 & 2 \\ 2 & 6 \end{bmatrix}$ and $\begin{bmatrix} 1 & 2 & 4 \\ 2 & 6 & 12 \end{bmatrix}$

(g) $\begin{bmatrix} 1 & -1 & 1 \\ 2 & -2 & 3 \end{bmatrix}$ and $\begin{bmatrix} 1 & -1 & 1 & 4 \\ 2 & -2 & 3 & 2 \end{bmatrix}$ (k) $\begin{bmatrix} 1 & 2 & 4 \\ 1 & 0 & 2 \\ 2 & 3 & 7 \end{bmatrix}$ and

$\begin{bmatrix} 1 & 2 & 4 & 1 \\ 1 & 0 & 2 & 0 \\ 2 & 3 & 7 & 0 \end{bmatrix}$ **3.** a) $A = \begin{bmatrix} 2 & 1 & 6 \\ 4 & 3 & 8 \end{bmatrix}$ is row equivalent to $\begin{bmatrix} 2 & 1 & 6 \\ 0 & 1 & -4 \end{bmatrix}$.

b) The matrix A represents the system

$$\begin{aligned} 2x_1 + x_2 &= 6 \\ 4x_1 + 3x_2 &= 8. \end{aligned}$$

c) $x_1 = 5, x_2 = -4$

---

* Many of the problems have answers that contain parameters or answers that can be written in a variety of forms. For problems of this sort we have presented one possible form of the answer. Your solution may have a different form and still be correct. You can frequently check your solution by inserting it in the original problem or by showing that two different forms for the answer are equivalent.

5. a) A and C are in echelon form. B and D are row equivalent to

$$\begin{bmatrix} 2 & -1 & 3 \\ 0 & 1 & 1 \\ 0 & 0 & -3 \\ 0 & 0 & 0 \end{bmatrix} \quad \text{and} \quad \begin{bmatrix} 1 & -3 & 4 & 1 & -3 \\ 0 & 1 & 2 & -2 & 0 \\ 0 & 0 & -4 & 3 & 5 \\ 0 & 0 & 0 & 1 & 15 \end{bmatrix}, \text{ respectively.}$$

(b) A and B correspond to inconsistent systems. For C: $x_1 = -5x_3 + 8$, $x_2 = (-x_3 + 3)/2$, $x_4 = 2$. For D: the solution is $x_1 = -28$, $x_2 = 10$, $x_3 = 10$, $x_4 = 15$.

7. The reductions can proceed as

a) $$\begin{bmatrix} 1 & -1 & -1 \\ 1 & 1 & 3 \end{bmatrix} \rightarrow \begin{bmatrix} 1 & -1 & -1 \\ 0 & 2 & 4 \end{bmatrix}; \quad \begin{matrix} x_1 = 1 \\ x_2 = 2 \end{matrix}$$

b) $$\begin{bmatrix} 1 & 1 & -1 & 2 \\ 2 & 0 & -1 & 1 \end{bmatrix} \rightarrow \begin{bmatrix} 1 & 1 & -1 & 2 \\ 0 & -2 & 1 & -3 \end{bmatrix}; \quad \begin{matrix} x_1 = (x_3 + 1)/2 \\ x_2 = (x_3 + 3)/2 \end{matrix}$$

c) $$\begin{bmatrix} 1 & 3 & -1 & 1 \\ 2 & 5 & 1 & 5 \\ 1 & 1 & 1 & 3 \end{bmatrix} \rightarrow \begin{bmatrix} 1 & 3 & -1 & 1 \\ 0 & -1 & 3 & 3 \\ 0 & -2 & 2 & 2 \end{bmatrix}$$

$$\rightarrow \begin{bmatrix} 1 & 3 & -1 & 1 \\ 0 & -1 & 3 & 3 \\ 0 & 0 & -4 & -4 \end{bmatrix}; \quad \begin{matrix} x_1 = 2 \\ x_2 = 0 \\ x_3 = 1 \end{matrix}$$

d) $$\begin{bmatrix} 1 & 1 & 2 & 6 \\ 3 & 4 & -1 & 5 \\ -1 & 1 & 1 & 2 \end{bmatrix} \rightarrow \begin{bmatrix} 1 & 1 & 2 & 6 \\ 0 & 1 & -7 & -13 \\ 0 & 2 & 3 & 8 \end{bmatrix}$$

$$\rightarrow \begin{bmatrix} 1 & 1 & 2 & 6 \\ 0 & 1 & -7 & -13 \\ 0 & 0 & 17 & 34 \end{bmatrix}; \quad \begin{matrix} x_1 = 1 \\ x_2 = 1 \\ x_3 = 2 \end{matrix}$$

e) $$\begin{bmatrix} 1 & 1 & -3 & -1 \\ 1 & 2 & -5 & -2 \\ -1 & -3 & 7 & 3 \end{bmatrix} \rightarrow \begin{bmatrix} 1 & 1 & -3 & -1 \\ 0 & 1 & -2 & -1 \\ 0 & -2 & 4 & 2 \end{bmatrix}$$

$$\rightarrow \begin{bmatrix} 1 & 1 & -3 & -1 \\ 0 & 1 & -2 & -1 \\ 0 & 0 & 0 & 0 \end{bmatrix}; \quad \begin{matrix} x_1 = x_3 \\ x_2 = 2x_3 - 1 \end{matrix}$$

f) $$\begin{bmatrix} 1 & 1 & 1 & 1 \\ 2 & 3 & 1 & 2 \\ 1 & -1 & 3 & 2 \end{bmatrix} \rightarrow \begin{bmatrix} 1 & 1 & 1 & 1 \\ 0 & 1 & -1 & 0 \\ 0 & -2 & 2 & 1 \end{bmatrix}$$

$$\rightarrow \begin{bmatrix} 1 & 1 & 1 & 1 \\ 0 & 1 & -1 & 0 \\ 0 & 0 & 0 & 1 \end{bmatrix}; \quad \text{inconsistent}$$

9. A, B, and D are row equivalent to

$$\begin{bmatrix} 1 & 0 & 3 & 0 \\ 0 & 2 & -2 & 0 \\ 0 & 0 & 0 & 1 \end{bmatrix}, \quad \begin{bmatrix} 2 & 0 & 0 \\ 0 & 1 & 0 \\ 0 & 0 & -3 \\ 0 & 0 & 0 \end{bmatrix}, \quad \begin{bmatrix} 1 & 0 & 0 & 0 & -28 \\ 0 & 1 & 0 & 0 & 10 \\ 0 & 0 & -4 & 0 & -40 \\ 0 & 0 & 0 & 1 & 15 \end{bmatrix},$$

respectively.

11. A and C represent inconsistent systems. B and D represent systems with infinitely many solutions. E and F represent systems with unique solutions.

13. All the systems are consistent. The solutions are (a) $x_1 = 0$, $x_2 = 0$ (b) $x_1 = 5x_3$, $x_2 = -3x_3$ (c) $x_1 = 0$, $x_2 = 0$, $x_3 = 0$ (d) $x_1 = 3x_4$, $x_2 = -2x_4$, $x_3 = 0$ (e) $x_1 = -x_2$, $x_3 = 0$ (f) $x_1 = -x_2 - 4x_5$, $x_3 = x_5$, $x_4 = -x_5$, x_2 and x_5 arbitrary.

1.3 EXERCISES

1. Systems (a) and (d) are homogeneous; (b), (c), and (d) have nontrivial solutions.

3. a) $x_1 = 0$, $x_2 = 1$, $x_3 = 1$ (b) $x_1 = 1$, $x_2 = 0$, $x_3 = 1$ (c) $x_1 = -x_2 - 1$, $x_3 = 1$ (d) $x_1 = -1$, $x_2 = 0$, $x_3 = 1$, $x_4 = 0$. For (a): $x_2 = 2$, $x_1 = 0$, $x_3 = 2$ is a solution. For (c): $x_2 = 2$, $x_1 = -2 - x_3$ is a solution. For (b) and (d): x_2 must be 0.

5. One matrix C can be found according to

$$\begin{bmatrix} 0 & 1 & 2 & 1 \\ 1 & -1 & 3 & 4 \\ 2 & -1 & 8 & 1 \end{bmatrix} \to \begin{bmatrix} 1 & -1 & 3 & 4 \\ 0 & 1 & 2 & 1 \\ 2 & -1 & 8 & 1 \end{bmatrix} \to \begin{bmatrix} 1 & -1 & 3 & 4 \\ 0 & 1 & 2 & 1 \\ 0 & 1 & 2 & -7 \end{bmatrix}$$

$$\to \begin{bmatrix} 1 & -1 & 3 & 4 \\ 0 & 1 & 2 & 1 \\ 0 & 0 & 0 & -8 \end{bmatrix}.$$

The system represented by B is not consistent.

7. a) No. (b) From part (a), the system can have a nontrivial solution only if $(3-\lambda)^2 - 4 = 0$. The only values of λ for which $\lambda^2 - 6\lambda + 5 = 0$ are $\lambda = 5$ and $\lambda = 1$. For $\lambda = 5$, the solution is $x_1 = x_2$; and for $\lambda = 1$, the solution is $x_1 = -x_2$.

1.4 EXERCISES

1. a) AB is (2×4); BA is not defined. (b) Neither AB nor BA is defined. (c) AB is not defined; BA is (6×7). (d) AB is (2×2); BA is (3×3). (e) AB is (3×1); BA is not defined. (f) $A(BC)$ and $(AB)C$ are both (2×4). (g) AB is (4×4); BA is (1×1).

3. a) $\begin{bmatrix} 5 & 16 \\ 5 & 18 \end{bmatrix}$ (b) $\begin{bmatrix} 4 & 11 \\ 6 & 19 \end{bmatrix}$ (c) $\begin{bmatrix} 50 & 11 \\ 16 & 10 \\ 3 & -2 \\ 28 & 4 \end{bmatrix}$ (d) $\begin{bmatrix} 11 \\ 13 \end{bmatrix}$ (e) $[6 \quad 20]$ (f) 14

(g) 66 (h) $\begin{bmatrix} 16 \\ 8 \\ -1 \\ 4 \end{bmatrix}$ (i) $\begin{bmatrix} 3 & 5 \\ 2 & 8 \end{bmatrix}$ (j) $\begin{bmatrix} 25 \\ 39 \end{bmatrix}$ (k) $\begin{bmatrix} 5 & 10 \\ 8 & 12 \\ 15 & 20 \\ 8 & 17 \end{bmatrix}$ (l) $\begin{bmatrix} 124 \\ 120 \\ 171 \\ 202 \end{bmatrix}$

5. $\mathbf{A}_1 = \begin{bmatrix} 2 \\ 1 \end{bmatrix}$ $\mathbf{A}_2 = \begin{bmatrix} 3 \\ 4 \end{bmatrix}$ $\mathbf{D}_1 = \begin{bmatrix} 2 \\ 2 \\ 1 \\ 1 \end{bmatrix}$ $\mathbf{D}_2 = \begin{bmatrix} 1 \\ 0 \\ -1 \\ 3 \end{bmatrix}$

$\mathbf{D}_3 = \begin{bmatrix} 3 \\ 0 \\ 1 \\ 1 \end{bmatrix}$ $\mathbf{D}_4 = \begin{bmatrix} 6 \\ 4 \\ -1 \\ 2 \end{bmatrix}$ $\mathbf{A}_1$ is in R^2, $\mathbf{D}_1$ is in R^4.

7. a) $\begin{bmatrix} 2 & -1 \\ 1 & 1 \end{bmatrix}\begin{bmatrix} x_1 \\ x_2 \end{bmatrix} = \begin{bmatrix} 3 \\ 3 \end{bmatrix}$; $\begin{bmatrix} 1 & -3 & 1 \\ 1 & -2 & 1 \\ 0 & 1 & -1 \end{bmatrix}\begin{bmatrix} x_1 \\ x_2 \\ x_3 \end{bmatrix} = \begin{bmatrix} 1 \\ 2 \\ 1 \end{bmatrix}$

(b) $x_1\begin{bmatrix} 2 \\ 1 \end{bmatrix} + x_2\begin{bmatrix} -1 \\ 1 \end{bmatrix} = \begin{bmatrix} 3 \\ 3 \end{bmatrix}$; $x_1\begin{bmatrix} 1 \\ 1 \\ 0 \end{bmatrix} + x_2\begin{bmatrix} -3 \\ -2 \\ 1 \end{bmatrix} + x_3\begin{bmatrix} 1 \\ 1 \\ -1 \end{bmatrix} = \begin{bmatrix} 1 \\ 2 \\ 1 \end{bmatrix}$

(c) $2\begin{bmatrix} 2 \\ 1 \end{bmatrix} + \begin{bmatrix} -1 \\ 1 \end{bmatrix} = \begin{bmatrix} 3 \\ 3 \end{bmatrix}$; $4\begin{bmatrix} 1 \\ 1 \\ 0 \end{bmatrix} + \begin{bmatrix} -3 \\ -2 \\ 1 \end{bmatrix} = \begin{bmatrix} 1 \\ 2 \\ 1 \end{bmatrix}$

9. Let $B = [\mathbf{B}_1, \mathbf{B}_2]$. If $AB = I$, then $[A\mathbf{B}_1, A\mathbf{B}_2] = I$ or $A\mathbf{B}_1 = \begin{bmatrix} 1 \\ 0 \end{bmatrix}$, $A\mathbf{B}_2 = \begin{bmatrix} 0 \\ 1 \end{bmatrix}$. Solving for $\mathbf{B}_1$ and $\mathbf{B}_2$ yields $B = \begin{bmatrix} 2 & -1 \\ -1 & 1 \end{bmatrix}$.

11. a) $Q = \begin{bmatrix} 1 & 3 \\ 2 & 2 \end{bmatrix}$ $B = \begin{bmatrix} -1 & 6 \\ 1 & 0 \end{bmatrix}$ (b) $Q = \begin{bmatrix} 0 & -1 & -1 \\ 1 & 0 & -1 \\ -1 & -1 & 2 \end{bmatrix}$ There is no (3×3) matrix B such that $AB = C$.

1.5 EXERCISES

1. $A^{\mathrm{T}} = \begin{bmatrix} 3 & 4 & 2 \\ 1 & 7 & 6 \end{bmatrix}$, $B^{\mathrm{T}} = \begin{bmatrix} 1 & 7 & 6 \\ 2 & 4 & 0 \\ 1 & 3 & 1 \end{bmatrix}$, $C^{\mathrm{T}} = \begin{bmatrix} 2 & 6 & 2 \\ 1 & 1 & 1 \\ 4 & 3 & 1 \\ 0 & 5 & 0 \end{bmatrix}$,

$D^{\mathrm{T}} = \begin{bmatrix} 2 & 1 \\ 1 & 4 \end{bmatrix}$, $E^{\mathrm{T}} = \begin{bmatrix} 3 & 2 \\ 6 & 3 \end{bmatrix}$, $F^{\mathrm{T}} = \begin{bmatrix} 1 & 6 & 2 \\ 6 & 0 & 4 \\ 2 & 4 & 8 \end{bmatrix}$

3. D and F are symmetric as is $A^{\mathrm{T}}A$.

5. a) $\mathbf{y}^T\mathbf{x} = 5$, $\mathbf{x}^T\mathbf{y} = 5$, $\mathbf{x}^T\mathbf{x} = 5$, $\mathbf{u}^T\mathbf{b} = 5$, $\mathbf{x}\mathbf{y}^T = \begin{bmatrix} -3 & -4 \\ 6 & 8 \end{bmatrix}$, $\quad \mathbf{y}\mathbf{x}^T = \begin{bmatrix} -3 & 6 \\ -4 & 8 \end{bmatrix}$

(b) $\|\mathbf{x}\| = \sqrt{5}$, $\|\mathbf{y}\| = 5$, $\|\mathbf{u}\| = \sqrt{14}$, $\|\mathbf{b}\| = \sqrt{6}$. (c) $\|A\mathbf{x}\| = \sqrt{201}$, $\|A\mathbf{x}-\mathbf{b}\| = \sqrt{149}$, $\|A\mathbf{y}-\mathbf{b}\| = \sqrt{2449}$. (d) $\mathbf{x}^TE\mathbf{x} = -1$, $\mathbf{x}^TD\mathbf{y} = 28$, $\mathbf{x}^TD\mathbf{x} = 14$.

7. a) $A^T = \begin{bmatrix} 1 & 0 \\ 3 & 1 \end{bmatrix}$; so $A = \begin{bmatrix} 1 & 3 \\ 0 & 1 \end{bmatrix}$. (b) $A^T = \begin{bmatrix} 5 & -3 \\ 11 & -7 \end{bmatrix}$; so $A = \begin{bmatrix} 5 & 11 \\ -3 & -7 \end{bmatrix}$.

(c) $\begin{bmatrix} 14 \\ 18 \end{bmatrix}$, $[6, 8]$, 132, $\sqrt{1348}$ **15.** When A and B are symmetric, $(AB)^T = B^TA^T = BA$. Thus AB is symmetric if and only if $AB = BA$.

1.6 EXERCISES

1. a) linearly independent b) linearly dependent, $\boldsymbol{v}_3 = 2\boldsymbol{v}_1$ c) linearly independent (d) linearly independent (e) linearly dependent, $\boldsymbol{v}_1 = \boldsymbol{v}_2 - \boldsymbol{v}_4$ (f) linearly dependent, $\boldsymbol{v}_3 = 2\boldsymbol{v}_1$ **3.** a) linearly independent (b) linearly independent (c) linearly dependent, $\mathbf{w}_4 = -(7\mathbf{w}_1 + 6\mathbf{w}_2 - 4\mathbf{w}_3)/5$, (d) linearly independent (e) linearly dependent, $\mathbf{w}_5 = 3\mathbf{w}_2 - \mathbf{w}_4$, (f) linearly dependent, $\mathbf{w}_5 = (7\mathbf{w}_1 + 21\mathbf{w}_2 - 4\mathbf{w}_3)/5$ **5.** a) Yes. (b) No.

1.7 EXERCISES

1. A and C are singular; B and D are nonsingular.

$$B^{-1} = \begin{bmatrix} 3 & -2 \\ -4 & 3 \end{bmatrix} \qquad D^{-1} = \begin{bmatrix} 11 & -1 & -2 \\ -15 & 1 & 3 \\ 5 & 0 & -1 \end{bmatrix}$$

3. a) $A^{-1} = \begin{bmatrix} -2 & 1 & -1 \\ -8 & 4 & -3 \\ 3 & -1 & 1 \end{bmatrix}$ (b) solution is $A^{-1}\begin{bmatrix} 1 \\ 3 \\ 1 \end{bmatrix} = \begin{bmatrix} 0 \\ 1 \\ 1 \end{bmatrix}$

11. $T_1^{-1} = \begin{bmatrix} 1 & -2 \\ 0 & 1 \end{bmatrix}$ $\quad T_2^{-1} = \begin{bmatrix} 1 & -3 & 5 \\ 0 & 1 & -2 \\ 0 & 0 & 1 \end{bmatrix}$ **13.** a) $p(t) = 5 - 3t + t^2$

b) $p(t) = -4 + 2t + t^2$ (c) $p(t) = -4 + t + 3t^2$

1.8 EXERCISES

1. The inverses are

$$\begin{bmatrix} 0 & 1 & 3 \\ 5 & 5 & 4 \\ 1 & 1 & 1 \end{bmatrix}, \quad \begin{bmatrix} -1/2 & -2/3 & -1/6 & 7/6 \\ 1 & 1/3 & 1/3 & -4/3 \\ 0 & -1/3 & -1/3 & 1/3 \\ -1/2 & 1 & 1/2 & 1/2 \end{bmatrix}, \quad \begin{bmatrix} 1 & 0 & 0 \\ -2 & 1 & 0 \\ 5 & -4 & 1 \end{bmatrix}.$$

3. The solutions are

$$\text{(a) } A^{-1}\begin{bmatrix} 8 \\ -11 \\ 4 \end{bmatrix} = \begin{bmatrix} 1 \\ 1 \\ 1 \end{bmatrix}, \quad \text{(b) } A^{-1}\begin{bmatrix} 7 \\ -9 \\ 3 \end{bmatrix} = \begin{bmatrix} 0 \\ 2 \\ 1 \end{bmatrix}, \quad \text{(c) } A^{-1}\begin{bmatrix} -6 \\ 8 \\ -2 \end{bmatrix} = \begin{bmatrix} 2 \\ 2 \\ 0 \end{bmatrix}.$$

5.

$$[A \mid \mathbf{b}_1, \mathbf{b}_2, \mathbf{b}_3, \mathbf{b}_4] = \begin{bmatrix} 1 & 3 & -4 & 1 & 2 & 4 & 0 \\ 2 & -1 & 3 & 1 & 3 & -1 & 0 \\ 8 & 3 & 1 & 5 & -4 & 5 & 0 \end{bmatrix} \text{ is row equivalent to}$$

$$\begin{bmatrix} 1 & 3 & -4 & 1 & 2 & 4 & 0 \\ 0 & -7 & 11 & -1 & -1 & -9 & 0 \\ 0 & 0 & 0 & 0 & -17 & 0 & 0 \end{bmatrix}.$$ Thus the solution to $A\mathbf{x} = \mathbf{b}_k$ is as follows: for $k = 1$, $x_1 = -(5x_3 - 4)/7$, $x_2 = (11x_3 + 1)/7$; for $k = 2$, the system is inconsistent; for $k = 3$, $x_1 = -(5x_3 - 1)/7$, $x_2 = (11x_3 + 9)/7$; for $k = 4$, $x_1 = -5x_3/7$, $x_2 = 11x_3/7$.

2.2 EXERCISES

1. Conditions (a), (b), (d), and (f) define subspaces; (c) and (e) do not.

3. a) $\begin{bmatrix} 1 \\ 0 \\ 2 \end{bmatrix}, \begin{bmatrix} 0 \\ 1 \\ -1 \end{bmatrix}$ (b) $\begin{bmatrix} 1 \\ 1 \\ 0 \end{bmatrix}, \begin{bmatrix} 0 \\ 1 \\ 1 \end{bmatrix}$ (d) $\begin{bmatrix} 0 \\ 1 \\ 0 \end{bmatrix}, \begin{bmatrix} 2 \\ 0 \\ 1 \end{bmatrix}$ (f) $\begin{bmatrix} 1 \\ 0 \\ 0 \end{bmatrix}, \begin{bmatrix} 0 \\ 0 \\ 1 \end{bmatrix}$

5. b) $\mathbf{u}_1 = \begin{bmatrix} 1 \\ 2 \\ 1 \end{bmatrix}$ (c) $A = \begin{bmatrix} 1 & -1 & 1 \\ 1 & 0 & -1 \end{bmatrix}$

7. a) $\begin{bmatrix} -2 \\ 1 \\ 1 \\ 0 \end{bmatrix}, \begin{bmatrix} -3 \\ -1 \\ 0 \\ 1 \end{bmatrix}$ (b) $\begin{bmatrix} 1 \\ -2 \\ 1 \\ 0 \\ 0 \end{bmatrix}, \begin{bmatrix} 2 \\ -3 \\ 0 \\ 1 \\ 0 \end{bmatrix}$ (c) $\begin{bmatrix} -3 \\ -2 \\ 1 \end{bmatrix}$

9. a) $3b_1 + b_2 - 5b_3 = 0$; $\begin{bmatrix} 1 \\ -3 \\ 0 \end{bmatrix}, \begin{bmatrix} 0 \\ 5 \\ 1 \end{bmatrix}$ (b) $b_1 + b_2 + b_3 - 2b_4 = 0$;

$\begin{bmatrix} 2 \\ 0 \\ 0 \\ 1 \end{bmatrix}, \begin{bmatrix} 0 \\ 2 \\ 0 \\ 1 \end{bmatrix}, \begin{bmatrix} 0 \\ 0 \\ 2 \\ 1 \end{bmatrix}$ (c) $-4b_1 + 5b_2 + 6b_3 = 0$; $\begin{bmatrix} 5 \\ 4 \\ 0 \end{bmatrix}, \begin{bmatrix} 6 \\ 0 \\ 4 \end{bmatrix}$

11. a) $\begin{bmatrix} -1 \\ 1 \\ 0 \end{bmatrix}, \begin{bmatrix} -1 \\ 0 \\ 1 \end{bmatrix}$ (b) $\begin{bmatrix} 1 \\ 0 \\ 1 \end{bmatrix}, \begin{bmatrix} 0 \\ 1 \\ 0 \end{bmatrix}$ (c) $\begin{bmatrix} 1 \\ 0 \\ 0 \end{bmatrix}, \begin{bmatrix} 0 \\ 0 \\ 1 \end{bmatrix}$ (d) $\begin{bmatrix} 1 \\ 2 \\ 0 \end{bmatrix}, \begin{bmatrix} 0 \\ 3 \\ 1 \end{bmatrix}$

13. a) $\begin{bmatrix}1\\-1\\1\\0\end{bmatrix}, \begin{bmatrix}1\\0\\0\\1\end{bmatrix}$ (b) $\begin{bmatrix}-1\\0\\1\\0\end{bmatrix}, \begin{bmatrix}3\\-2\\0\\1\end{bmatrix}$ (c) $\begin{bmatrix}-2\\1\\0\\0\end{bmatrix}, \begin{bmatrix}-1\\0\\1\\0\end{bmatrix}, \begin{bmatrix}0\\0\\0\\1\end{bmatrix}$

(d) $\begin{bmatrix}1\\1\\0\\0\end{bmatrix}, \begin{bmatrix}0\\0\\1\\0\end{bmatrix}, \begin{bmatrix}0\\0\\0\\1\end{bmatrix}$ (e) $\begin{bmatrix}0\\0\\1\\0\end{bmatrix}, \begin{bmatrix}0\\0\\0\\1\end{bmatrix}$

21. $\mathcal{N}(A) = \{\theta\}, \mathcal{R}(A) = R^n$

2.3 EXERCISES

1. a) $\begin{bmatrix}1\\0\\1\\0\end{bmatrix}, \begin{bmatrix}-1\\1\\0\\1\end{bmatrix}$ (b) $\begin{bmatrix}-1\\2\\1\\0\end{bmatrix}, \begin{bmatrix}-2\\1\\0\\1\end{bmatrix}$ (c) $\begin{bmatrix}1\\1\\0\\0\end{bmatrix}, \begin{bmatrix}-1\\0\\1\\0\end{bmatrix}, \begin{bmatrix}3\\0\\0\\1\end{bmatrix}$

(d) $\begin{bmatrix}2\\2\\1\\1\end{bmatrix}$ **3.** a) $\begin{bmatrix}1\\3\\1\end{bmatrix}, \begin{bmatrix}0\\-1\\-1\end{bmatrix}$ (b) $\begin{bmatrix}1\\1\\2\end{bmatrix}, \begin{bmatrix}0\\0\\1\end{bmatrix}$ (c) $\begin{bmatrix}1\\2\\2\\0\end{bmatrix}, \begin{bmatrix}0\\1\\-2\\1\end{bmatrix}$

5. a) $\begin{bmatrix}-1\\-1\\1\\0\end{bmatrix}, \begin{bmatrix}-1\\1\\0\\1\end{bmatrix}$ (b) $\begin{bmatrix}-1\\-1\\1\end{bmatrix}$ (c) $\begin{bmatrix}1\\-1\\1\\0\end{bmatrix}, \begin{bmatrix}-2\\1\\0\\1\end{bmatrix}$

7. a) $\begin{bmatrix}1\\2\end{bmatrix}$ (b) $\begin{bmatrix}1\\2\end{bmatrix}, \begin{bmatrix}0\\-3\end{bmatrix}$ (c) $\begin{bmatrix}1\\2\\-1\\3\end{bmatrix}, \begin{bmatrix}0\\1\\0\\0\end{bmatrix}, \begin{bmatrix}0\\0\\0\\5\end{bmatrix}$ d) $\begin{bmatrix}1\\2\\1\end{bmatrix}, \begin{bmatrix}0\\1\\-2\end{bmatrix}$

9. a) $\begin{bmatrix}1\\1\end{bmatrix}, \begin{bmatrix}1\\0\end{bmatrix}$ (b) $\begin{bmatrix}1\\1\end{bmatrix}$ (c) $\begin{bmatrix}1\\1\end{bmatrix}, \begin{bmatrix}1\\0\end{bmatrix}$

11. The vectors in (a), (b), and (e) are in $\mathcal{R}(A)$. (a) $\mathbf{x} = \begin{bmatrix}2\\1\\0\end{bmatrix} + c\begin{bmatrix}1\\1\\1\end{bmatrix}$ (b) $\mathbf{x} = \begin{bmatrix}1\\-1\\0\end{bmatrix}$ $+ c\begin{bmatrix}-1\\1\\1\end{bmatrix}$ (e) $\mathbf{x} = \begin{bmatrix}3\\-4\\0\end{bmatrix} + c\begin{bmatrix}-1\\1\\1\end{bmatrix}$

2.4 EXERCISES

1. $\dim(\mathcal{R}(A)) = 2$, $\dim(\mathcal{N}(A)) = 2$, $\dim(\mathcal{R}(B)) = 3$, $\dim(\mathcal{N}(B)) = 1$, $\dim(\mathcal{R}(C)) = 3$, $\dim(\mathcal{N}(C)) = 0$, $\dim(\mathcal{R}(D)) = 2$, $\dim(\mathcal{N}(D)) = 1$. Basis vectors are

$\mathcal{N}(A)$: $\begin{bmatrix} -1 \\ -1 \\ 1 \\ 0 \end{bmatrix}, \begin{bmatrix} 1 \\ -2 \\ 0 \\ 1 \end{bmatrix}$ $\quad \mathcal{R}(B)$: $\begin{bmatrix} 1 \\ 2 \\ 3 \end{bmatrix}, \begin{bmatrix} 0 \\ 1 \\ 1 \end{bmatrix}, \begin{bmatrix} 0 \\ 0 \\ -6 \end{bmatrix}$ $\quad \mathcal{N}(B)$: $\begin{bmatrix} 1 \\ -2 \\ 1 \\ 0 \end{bmatrix}$

$\mathcal{R}(D)$: $\begin{bmatrix} 1 \\ 3 \\ 1 \end{bmatrix}, \begin{bmatrix} 0 \\ 1 \\ 1 \end{bmatrix}$ $\quad$ 3. a) A basis is $\begin{bmatrix} 1 \\ 1 \\ -2 \end{bmatrix}, \begin{bmatrix} 0 \\ -1 \\ 1 \end{bmatrix}, \begin{bmatrix} 0 \\ 0 \\ 1 \end{bmatrix}$ and $\dim(W) = 3$.

(b) A basis is $\begin{bmatrix} 1 \\ 2 \\ -1 \\ 1 \end{bmatrix}, \begin{bmatrix} 0 \\ -5 \\ 4 \\ -1 \end{bmatrix}, \begin{bmatrix} 0 \\ 0 \\ -3 \\ 12 \end{bmatrix}$ and $\dim(W) = 3$.

5. c) From (a) the components of any vector in W are of the form $x_1 = a_1 + 3a_2$, $x_2 = 2a_1 + a_2$, $x_3 = -a_1 + 2a_2$. If $ax_1 + bx_2 + x_3 = 0$ is to hold for all vectors in W, then it must be that $a(a_1 + 3a_2) + b(2a_1 + a_2) + (-a_1 + 2a_2) = 0$. Collecting like terms gives $a = -1$ and $b = 1$; W represents the plane $-x_1 + x_2 + x_3 = 0$.

9. A basis for W is $\begin{bmatrix} -2 \\ 1 \\ 0 \\ 0 \end{bmatrix}, \begin{bmatrix} 3 \\ 0 \\ 1 \\ 0 \end{bmatrix}, \begin{bmatrix} 1 \\ 0 \\ 0 \\ 1 \end{bmatrix}$ and $\dim(W) = 3$. $\quad$ 11. a) 1 $\quad$ (b) 2

(c) 2 $\quad$ (d) 3

2.5 EXERCISES

1. $\mathbf{a} = \frac{2}{3}\mathbf{u}_1 - \frac{1}{2}\mathbf{u}_2 + \frac{1}{6}\mathbf{u}_3$, $\quad \mathbf{b} = \mathbf{u}_1 + \mathbf{u}_2$, $\quad \mathbf{c} = 3\mathbf{u}_1$, $\quad \mathbf{d} = \frac{4}{3}\mathbf{u}_1 + \frac{2}{6}\mathbf{u}_3$.

5. a) $\begin{bmatrix} 0 \\ 0 \\ 1 \\ 0 \end{bmatrix}, \begin{bmatrix} 1 \\ 1 \\ 0 \\ 1 \end{bmatrix}, \begin{bmatrix} 1 \\ -2 \\ 0 \\ 1 \end{bmatrix}$ $\quad$ (b) $\begin{bmatrix} 1 \\ 0 \\ 1 \\ 2 \end{bmatrix}, \begin{bmatrix} 1 \\ 1 \\ -1 \\ 0 \end{bmatrix}, \begin{bmatrix} 1 \\ -2 \\ -1 \\ 0 \end{bmatrix}$

7. Basis for $\mathcal{R}(A)$: $\begin{bmatrix} 1 \\ 2 \\ 1 \end{bmatrix}, \begin{bmatrix} 0 \\ 5 \\ 1 \end{bmatrix}$; orthogonal basis: $\begin{bmatrix} 1 \\ 2 \\ 1 \end{bmatrix}, \begin{bmatrix} -11 \\ 8 \\ -5 \end{bmatrix}$.

Basis for $\mathcal{N}(A)$: $\begin{bmatrix} -3 \\ -1 \\ 1 \\ 0 \end{bmatrix}, \begin{bmatrix} -1 \\ -3 \\ 0 \\ 1 \end{bmatrix}$; orthogonal basis: $\begin{bmatrix} -3 \\ -1 \\ 1 \\ 0 \end{bmatrix}, \begin{bmatrix} 7 \\ -27 \\ -6 \\ 11 \end{bmatrix}$.

2.6 EXERCISES

1. An orthogonal basis is $\begin{bmatrix} 1 \\ 0 \\ 2 \end{bmatrix}, \begin{bmatrix} 2 \\ 5 \\ -1 \end{bmatrix}$. (a) $\frac{1}{3}\begin{bmatrix} 4 \\ 1 \\ 7 \end{bmatrix}$ (b) $\frac{1}{3}\begin{bmatrix} 1 \\ 1 \\ 1 \end{bmatrix}$ (c) $\begin{bmatrix} 3 \\ 1 \\ 5 \end{bmatrix}$ (d) $\frac{1}{6}\begin{bmatrix} 8 \\ 11 \\ 5 \end{bmatrix}$ 3. a) $\mathbf{x}^* = \frac{1}{13}\begin{bmatrix} -5 \\ 7 \end{bmatrix}$ (b) $\mathbf{x}^* = \frac{1}{3}\begin{bmatrix} -8 \\ 5 \\ 0 \end{bmatrix} + a\begin{bmatrix} -5 \\ 3 \\ 1 \end{bmatrix}$ where a is arbitrary 5. a) $p(x) = (13x + 11)/10$ (b) $p(x) = (-7x + 5)/10$
7. a) $p(x) = (x^2 + 2)/3$ (b) $p(x) = (5x^2 - x - 1)/20$
9. $q(x) = (5 - 4\cos x - 2\sin x)/8$

3.2 EXERCISES

1. a) $M_{11} = 7$, $M_{21} = 7$, $M_{31} = -7$, $M_{32} = 14$, $M_{33} = 7$ (b) $M_{11} = 14$, $M_{21} = 6$, $M_{31} = 2$, $M_{32} = -26$, $M_{33} = -10$ (c) $M_{11} = -6$, $M_{21} = -1$, $M_{31} = 4$, $M_{32} = 1$, $M_{33} = 14$ 3. a) $2M_{11} + 3M_{21} + 5M_{31} = 0$, $5M_{31} + M_{32} + 3M_{33} = 0$ (b) $-3M_{11} - 4M_{21} + M_{31} = -64$, $M_{31} + M_{32} + 4M_{33} = -64$ (c) $4M_{11} + 7M_{21} + M_{31} = -27$, $M_{31} - 3M_{32} - 2M_{33} = -27$ 5. The (3×3) matrix A in Exercise 1 (a) is singular, and the eigenvectors are the nontrivial solutions of $A\mathbf{x} = \boldsymbol{\theta}$. The solution of $A\mathbf{x} = \boldsymbol{\theta}$ is $x_1 = -x_3$, $x_2 = -2x_3$; so the eigenvectors corresponding to $\lambda = 0$ are all of the form

$$\mathbf{x} = a\begin{bmatrix} -1 \\ -2 \\ 1 \end{bmatrix}, \qquad a \neq 0.$$

7. $\det(A) = 5$, $\det(B) = 0$, $\det(C) = 0$, $\det(D) = 6$

9. For B: $a\begin{bmatrix} 1 \\ 1 \end{bmatrix}$, $a \neq 0$. For C: $a\begin{bmatrix} 2 \\ 1 \\ -1 \end{bmatrix}$, $a \neq 0$.

15. a) A sequence of row interchanges can be used to put A in triangular form:

$$A = \begin{bmatrix} 0 & 2 & 1 & 3 \\ 0 & 0 & 0 & 1 \\ 2 & 3 & 4 & 1 \\ 0 & 0 & 1 & 1 \end{bmatrix} \to \begin{bmatrix} 2 & 3 & 4 & 1 \\ 0 & 0 & 0 & 1 \\ 0 & 2 & 1 & 3 \\ 0 & 0 & 1 & 1 \end{bmatrix} \to \begin{bmatrix} 2 & 3 & 4 & 1 \\ 0 & 2 & 1 & 3 \\ 0 & 0 & 0 & 1 \\ 0 & 0 & 1 & 1 \end{bmatrix} \to \begin{bmatrix} 2 & 3 & 4 & 1 \\ 0 & 2 & 1 & 3 \\ 0 & 0 & 1 & 1 \\ 0 & 0 & 0 & 1 \end{bmatrix} = B.$$

Each row interchange switches the sign of the determinant; so $\det(B) = -\det(A)$. Since $\det(B) = 4$, it follows that $\det(A) = -4$. (b) Row interchanges and the addition of multiples of one row to another will put A into a nearly triangular form:

$$A = \begin{bmatrix} 0 & 1 & 1 & 2 \\ 1 & 2 & 1 & 1 \\ 3 & 1 & 2 & 2 \\ -1 & 2 & 0 & 1 \end{bmatrix} \to \begin{bmatrix} 1 & 2 & 1 & 1 \\ 0 & 1 & 1 & 2 \\ 3 & 1 & 2 & 2 \\ -1 & 2 & 0 & 1 \end{bmatrix} \to \begin{bmatrix} 1 & 2 & 1 & 1 \\ 0 & 1 & 1 & 2 \\ 0 & -5 & -1 & -1 \\ 0 & 4 & 1 & 2 \end{bmatrix} \to \begin{bmatrix} 1 & 2 & 1 & 1 \\ 0 & 1 & 1 & 2 \\ 0 & 0 & 4 & 9 \\ 0 & 0 & -3 & -6 \end{bmatrix}$$
$= B.$

Expanding along the first column gives $\det(B) = 3$. Since $\det(B) = -\det(A)$, it follows that $\det(A) = -3$. (c) $\det(A) = -96$ (d) $\det(A) = 0$

3.3 EXERCISES

1. For A: $p(t) = t^2 - 4t + 3 = (t-3)(t-1)$; $\lambda = 3$, $\mathbf{x} = a\begin{bmatrix} 0 \\ 1 \end{bmatrix}$, $a \neq 0$; $\lambda = 1$, $\mathbf{x} = a\begin{bmatrix} 1 \\ -1 \end{bmatrix}$, $a \neq 0$. For B: $p(t) = t^2 - 4t + 3 = (t-3)(t-1)$; $\lambda = 3$, $\mathbf{x} = a\begin{bmatrix} 1 \\ -1 \end{bmatrix}$, $a \neq 0$; $\lambda = 1$, $\mathbf{x} = a\begin{bmatrix} 1 \\ 1 \end{bmatrix}$, $a \neq 0$. For C: $p(t) = t^2 - 4t + 4 = (t-2)^2$; $\lambda = 2$, $\mathbf{x} = a\begin{bmatrix} -1 \\ 1 \end{bmatrix}$, $a \neq 0$.

3. For A: $p(t) = -t^3 + t^2 + t - 1 = -(t-1)^2(t+1)$; $\lambda = 1$, $\mathbf{x} = a\begin{bmatrix} 1 \\ -3 \\ 2 \end{bmatrix}$, $a \neq 0$; $\lambda = -1$, $\mathbf{x} = a\begin{bmatrix} 1 \\ -1 \\ 2 \end{bmatrix}$, $a \neq 0$. For B: $p(t) = -t^3 - 2t^2 - t = -t(t+1)^2$; $\lambda = 0$, $\mathbf{x} = a\begin{bmatrix} 1 \\ -2 \\ 2 \end{bmatrix}$, $a \neq 0$; $\lambda = -1$, $\mathbf{x} = a\begin{bmatrix} 1 \\ -1 \\ 2 \end{bmatrix}$, $a \neq 0$. For C: $p(t) = -t^3 + 2t^2 + t - 2 = -(t-2)(t-1)(t+1)$; $\lambda = 2$, $\mathbf{x} = a\begin{bmatrix} 1 \\ -1 \\ 2 \end{bmatrix}$, $a \neq 0$; $\lambda = 1$, $\mathbf{x} = a\begin{bmatrix} 3 \\ -1 \\ 7 \end{bmatrix}$, $a \neq 0$; $\lambda = -1$, $\mathbf{x} = a\begin{bmatrix} 1 \\ 2 \\ 2 \end{bmatrix}$, $a \neq 0$. For D: $p(t) = -t^3 + 3t^2 - 3t + 1 = -(t-1)^3$; $\lambda = 1$, $\mathbf{x} = a\begin{bmatrix} 4 \\ 8 \\ 0 \end{bmatrix} + b\begin{bmatrix} -3 \\ 0 \\ 8 \end{bmatrix}$, $|a| + |b| > 0$. For E: $p(t) = -t^3 + 6t^2 - 12t + 8 = -(t-2)^3$; $\lambda = 2$, $\mathbf{x} = a\begin{bmatrix} 1 \\ 0 \\ 0 \end{bmatrix} + b\begin{bmatrix} 0 \\ 1 \\ -1 \end{bmatrix}$, $|a| + |b| > 0$.

5. For A: $\lambda = -1$, $\mathbf{x} = a\begin{bmatrix} 1 \\ -1 \\ -1 \\ 1 \end{bmatrix}$; $\lambda = 5$, $\mathbf{x} = a\begin{bmatrix} 0 \\ -1 \\ 1 \\ 0 \end{bmatrix} + b\begin{bmatrix} -1 \\ 0 \\ 0 \\ 1 \end{bmatrix}$; $\lambda = 15$, $\mathbf{x} = a\begin{bmatrix} 1 \\ 1 \\ 1 \\ 1 \end{bmatrix}$. For B: $\lambda = 1$, $\mathbf{x} = a\begin{bmatrix} -1 \\ 1 \\ 0 \\ 0 \end{bmatrix}$; $\lambda = 2$, $\mathbf{x} = a\begin{bmatrix} 0 \\ 0 \\ -1 \\ 1 \end{bmatrix}$; $\lambda = 5$, $\mathbf{x} = a\begin{bmatrix} -1 \\ -1 \\ 2 \\ 2 \end{bmatrix}$; $\lambda = 10$, $\mathbf{x} = a\begin{bmatrix} 2 \\ 2 \\ 1 \\ 1 \end{bmatrix}$. For C: $\lambda = 2$,

$$\mathbf{x} = a\begin{bmatrix}1\\-1\\0\\0\end{bmatrix} + b\begin{bmatrix}1\\0\\-1\\0\end{bmatrix} + c\begin{bmatrix}1\\0\\0\\-1\end{bmatrix}; \lambda = -2, \mathbf{x} = a\begin{bmatrix}1\\1\\1\\1\end{bmatrix}$$

13. a) The vectors generated from $\mathbf{x}_0$ are $\mathbf{x}_1 = \begin{bmatrix}1\\-7\\1\end{bmatrix}$, $\mathbf{x}_2 = \begin{bmatrix}9\\-7\\17\end{bmatrix}$, $\mathbf{x}_3 = \begin{bmatrix}17\\-23\\33\end{bmatrix}$, $\mathbf{x}_4 = \begin{bmatrix}41\\-39\\81\end{bmatrix}$, $\mathbf{x}_5 = \begin{bmatrix}81\\-87\\161\end{bmatrix}$. The estimates to the dominant eigenvalue are $\beta_0 = -\frac{5}{3} \simeq -1.667$, $\beta_1 = \frac{75}{51} \simeq 1.471$, $\beta_2 = \frac{875}{419} \simeq 2.088$, $\beta_3 = \frac{4267}{1907} \simeq 2.238$, $\beta_4 = \frac{19755}{9763} \simeq 2.023$.

3.4 EXERCISES

3. a) The general solution is $\mathbf{x}_k = b_1 3^k \begin{bmatrix}1\\-1\end{bmatrix} + b_2\begin{bmatrix}1\\1\end{bmatrix}$; and choosing $b_1 = 1, b_2 = 2$ will satisfy the starting condition. (b) The general solution is $\mathbf{x}_k = b_1 2^k \begin{bmatrix}1\\-1\\2\end{bmatrix} + b_2\begin{bmatrix}3\\-1\\7\end{bmatrix} + b_3(-1)^k\begin{bmatrix}1\\2\\2\end{bmatrix}$; and $b_1 = 2$, $b_2 = 2$, $b_3 = -5$ will satisfy the starting condition. **5.** a) $s_0 = 6, t_0 = 5$ (b) $s_0 = 4, t_0 = 2$ **7.** a) The general solution is $\mathbf{x}(t) = b_1 e^t\begin{bmatrix}1\\-2\end{bmatrix} + b_2 e^{5t}\begin{bmatrix}1\\2\end{bmatrix}$ or $\begin{aligned}u(t) &= b_1 e^t + b_2 e^{5t}\\ v(t) &= -2b_1 e^t + 2b_2 e^{5t}.\end{aligned}$ Choosing $b_1 = 1$, $b_2 = 1$ will satisfy the initial conditions. (b) The general solution is $\mathbf{x}(t) = b_1 e^t\begin{bmatrix}0\\1\\0\end{bmatrix} + b_2 e^{2t}\begin{bmatrix}-1\\2\\2\end{bmatrix} + b_3 e^{3t}\begin{bmatrix}-1\\1\\1\end{bmatrix}$ or $\begin{aligned}u(t) &= -b_2 e^{2t} - b_3 e^{3t}\\ v(t) &= b_1 e^t + 2b_2 e^{2t} + b_3 e^{3t}\\ w(t) &= 2b_2 e^{2t} + b_3 e^{3t}.\end{aligned}$

Choosing $b_1 = 1, b_2 = -1, b_3 = 2$ will satisfy the initial conditions.

9. a) $x_1(t) = e^{2t}\begin{bmatrix}-a\\a\end{bmatrix}$, a arbitrary (b) $\mathbf{y}_0 = \begin{bmatrix}a\\0\end{bmatrix}$ The general solution is

$$\mathbf{x}(t) = b_1 e^{2t}\begin{bmatrix}-1\\1\end{bmatrix} + b_2\left(te^{2t}\begin{bmatrix}-1\\1\end{bmatrix} + e^2 t\begin{bmatrix}1\\0\end{bmatrix}\right) \text{ or } \begin{aligned}u(t) &= e^{2t}(-a - bt + b)\\ v(t)\, &e^{2t}(a + bt).\end{aligned}$$

3.5 EXERCISES

1. a) $3+2i$ (b) $7-3i$ (c) 6 (d) $-4i$ (e) 17 (f) $10-11i$ (g) $\frac{10}{17}-\frac{11}{17}i$ (h) $\frac{8}{169}+\frac{53}{169}i$ (i) $6+4i$ (j) $\frac{2}{13}-\frac{3}{13}i$ **3.** For A: $\lambda = 4+2i$, $\mathbf{x} = a\begin{bmatrix}4\\-1+i\end{bmatrix}$; $\lambda = 4-2i$, $\mathbf{x} = a\begin{bmatrix}4\\-1-i\end{bmatrix}$. For B: $\lambda = 2i$, $\mathbf{x} = a\begin{bmatrix}2\\-1+i\end{bmatrix}$; $\lambda = -2i$,

$\mathbf{x} = a\begin{bmatrix} 2 \\ -1-i \end{bmatrix}$. For C: $\lambda = 2$, $\mathbf{x} = a\begin{bmatrix} 0 \\ 1 \\ -1 \end{bmatrix}$; $\lambda = 2+i$, $\mathbf{x} = a\begin{bmatrix} -5 \\ 2+i \\ -5 \end{bmatrix}$; $\lambda = 2-i$,

$\mathbf{x} = a\begin{bmatrix} -5 \\ 2-i \\ -5 \end{bmatrix}$. **5.** a) $x_1 = 2-i$, $x_2 = 3-2i$ (b) $x_1 = i$, $x_2 = 2$

15. a) $\mathbf{w}(t) = e^{(1+3i)t}\begin{bmatrix} i \\ 1 \end{bmatrix}$; $\mathbf{u}(t) = e^t\begin{bmatrix} -\sin 3t \\ \cos 3t \end{bmatrix}$, $\mathbf{v}(t) = e^t\begin{bmatrix} \cos 3t \\ \sin 3t \end{bmatrix}$

(b) $\mathbf{w}(t) = e^{(2+5i)t}\begin{bmatrix} 13 \\ -1+5i \end{bmatrix}$; $\mathbf{u}(t) = \begin{bmatrix} 13e^{2t}\cos 5t \\ -e^{2t}(\cos 5t + 5\sin 5t) \end{bmatrix}$,

$\mathbf{v}(t) = \begin{bmatrix} 13e^{2t}\sin 5t \\ e^{2t}(5\cos 5t - \sin 5t) \end{bmatrix}$ (c) The eigenvalue $\lambda = -1$ gives

$\mathbf{x}_1(t) = e^{-t}\begin{bmatrix} 1 \\ -1 \\ 2 \end{bmatrix}$ while $\lambda = 1+i$ gives $\mathbf{w}(t) = e^{(1+i)t}\begin{bmatrix} 1 \\ -i \\ 3 \end{bmatrix}$;

$\mathbf{u}(t) = \begin{bmatrix} e^t\cos t \\ e^t\sin t \\ 3e^t\cos t \end{bmatrix}$, $\mathbf{v}(t) = \begin{bmatrix} e^t\sin t \\ -e^t\cos t \\ 3e^t\sin t \end{bmatrix}$

3.6 EXERCISES

1. $S = \begin{bmatrix} -1 & 0 & 0 \\ 1 & 4 & 0 \\ 0 & -1 & 1 \end{bmatrix}$

5. A and B are not diagonalizable; C can be diagonalized by

$$S = \begin{bmatrix} -3 & 1 & -1 \\ 1 & 2 & 1 \\ -7 & 2 & -2 \end{bmatrix}.$$

7. If $R^{-1}BR = E$, then $(R^{-1}BR)^{10} = E^{10}$ or $B^{10} = RE^{10}R^{-1}$ where

$$R = \begin{bmatrix} 1 & 1 & 1 \\ -1 & 3 & -1 \\ 3 & 2 & 2 \end{bmatrix}, \quad E^{10} = \begin{bmatrix} 1 & 0 & 0 \\ 0 & 1024 & 0 \\ 0 & 0 & 1024 \end{bmatrix},$$

$$R^{-1} = \frac{1}{4}\begin{bmatrix} -8 & 0 & 4 \\ 1 & 1 & 0 \\ 11 & -1 & -4 \end{bmatrix}.$$

11. a) $[\mathbf{e}_1, \mathbf{e}_2, \mathbf{e}_3]$, $[\mathbf{e}_1, \mathbf{e}_3, \mathbf{e}_2]$, $[\mathbf{e}_2, \mathbf{e}_1, \mathbf{e}_3]$, $[\mathbf{e}_2, \mathbf{e}_3, \mathbf{e}_1]$, $[\mathbf{e}_3, \mathbf{e}_1, \mathbf{e}_2]$, $[\mathbf{e}_3, \mathbf{e}_2, \mathbf{e}_1]$ (b) There are $n!$

15. a) The permutation matrix P is $\begin{bmatrix} 1 & 0 & 0 \\ 0 & 0 & 1 \\ 0 & 1 & 0 \end{bmatrix}$

3.7 EXERCISES

3. a) $H = \begin{bmatrix} -7 & -8 & -3 \\ 8 & 9 & 3 \\ 0 & 1 & 1 \end{bmatrix}$, $\quad Q = \begin{bmatrix} 1 & 0 & 0 \\ 0 & 1 & 0 \\ 0 & -4 & 1 \end{bmatrix}$

(b) $H = \begin{bmatrix} -6 & 31 & -14 \\ -1 & 6 & -2 \\ 0 & 2 & 1 \end{bmatrix}$, $\quad Q = \begin{bmatrix} 1 & 0 & 0 \\ 0 & 1 & 0 \\ 0 & 2 & 1 \end{bmatrix}$

(c) $H = \begin{bmatrix} 1 & 1 & 3 \\ 1 & 3 & 1 \\ 0 & 4 & 2 \end{bmatrix}$, $\quad Q = \begin{bmatrix} 1 & 0 & 0 \\ 0 & 0 & 1 \\ 0 & 1 & 0 \end{bmatrix}$

5. a) $H = \begin{bmatrix} 1 & -3 & -1 & -1 \\ -1 & -1 & -1 & -1 \\ 0 & 0 & 2 & 0 \\ 0 & 0 & 0 & 2 \end{bmatrix}$ $\quad$ (b) $H = \begin{bmatrix} 6 & 33 & 8 & 4 \\ 1 & 38 & 8 & 4 \\ 0 & -120 & -25 & -15 \\ 0 & 0 & 0 & 5 \end{bmatrix}$

(c) $H = \begin{bmatrix} 4 & 2 & 3 & 2 \\ 2 & 0 & 1 & -1 \\ 0 & 6 & 2 & 1 \\ 0 & 0 & -2 & 1 \end{bmatrix}$ $\quad$ **7.** a) $H = \begin{bmatrix} 1 & 2 & 1 & 1 \\ 2 & 1 & 1 & 1 \\ 0 & 0 & 2 & 4 \\ 0 & 0 & -2 & -2 \end{bmatrix}$

(b) $H = \begin{bmatrix} -2 & 0 & -4 & 1 \\ -1 & 1 & -4 & 1 \\ 0 & 1 & -3 & 1 \\ 0 & 0 & -8 & 3 \end{bmatrix}$ $\quad$ (c) $H = \begin{bmatrix} 2 & -1 & 0 & -1 \\ -1 & 2 & -2 & 1 \\ 0 & 0 & -3 & 6 \\ 0 & 0 & -5 & 10 \end{bmatrix}$

3.8 EXERCISES

1. a) $p(t) = t^3 - 2t^2 - t + 2$ $\quad$ (b) $p(t) = t^3 - t^2 - t + 1$ $\quad$ (c) $p(t) = t^3 - 3t^2 + t$

3. a) The only eigenvalue of B_{11} is $\lambda = 2$; $\lambda = 1$ and $\lambda = 3$ are the eigenvalues of B_{22}. An eigenvector is $\lambda = 2$, $\mathbf{x} = \begin{bmatrix} 1 \\ -1 \\ 0 \\ 0 \end{bmatrix}$; $\quad \lambda = 1$, $\mathbf{x} = \begin{bmatrix} -9 \\ 5 \\ 1 \\ 1 \end{bmatrix}$; $\quad \lambda = 3$, $\mathbf{x} = \begin{bmatrix} 3 \\ -9 \\ 1 \\ -1 \end{bmatrix}$. (b) The eigenvalues of B_{11} are $\lambda = 0, \lambda = 2$. The eigenvalues of B_{22} are λ

$= 3$, $\lambda = 4$. An eigenvector is $\lambda = 0$, $\mathbf{x} = \begin{bmatrix} 1 \\ -1 \\ 0 \\ 0 \end{bmatrix}$; $\lambda = 2$, $\mathbf{x} = \begin{bmatrix} 1 \\ 1 \\ 0 \\ 0 \end{bmatrix}$;

$\lambda = 3$, $\mathbf{x} = \begin{bmatrix} 0 \\ -1 \\ 1 \\ -1 \end{bmatrix}$; $\lambda = 4$, $\mathbf{x} = \begin{bmatrix} 3 \\ 5 \\ 0 \\ 4 \end{bmatrix}$. (c) The eigenvalues of the (3×3) matrix B_{11} are $\lambda = 0$, $\lambda = -1$. The eigenvalue of B_{22} is $\lambda = 2$. An eigenvector is $\lambda = -1$,

$\mathbf{x} = \begin{bmatrix} 2 \\ 0 \\ -1 \\ 0 \end{bmatrix}$; $\lambda = 0$, $\mathbf{x} = \begin{bmatrix} -1 \\ 1 \\ 1 \\ 0 \end{bmatrix}$; $\lambda = 2$, $\mathbf{x} = \begin{bmatrix} 1 \\ 15 \\ 1 \\ 6 \end{bmatrix}$.

3.9 EXERCISES

1. $q(A) = \begin{bmatrix} -1 & 0 \\ 0 & -1 \end{bmatrix}$; $q(B) = \begin{bmatrix} 0 & 0 \\ 0 & 0 \end{bmatrix}$; $q(C) = \begin{bmatrix} 15 & -2 & 14 \\ 5 & -2 & 10 \\ -1 & -4 & 6 \end{bmatrix}$

3. a) $q(t) = (t^3 + t - 1)p(t) + t + 2$ (b) $q(B) = B + 2I = \begin{bmatrix} 4 & -1 \\ -1 & 4 \end{bmatrix}$

3.10 EXERCISES

1. a) $p(t) = (t-2)^2$; $\mathbf{v}_1 = \begin{bmatrix} 1 \\ -1 \end{bmatrix}$, $\mathbf{v}_2 = \begin{bmatrix} 1 \\ -2 \end{bmatrix}$ (b) $p(t) = t(t+1)^2$; for $\lambda = -1$,

$\mathbf{v}_1 = \begin{bmatrix} -2 \\ 0 \\ 1 \end{bmatrix}$, $\mathbf{v}_2 = \begin{bmatrix} 0 \\ 1 \\ 1 \end{bmatrix}$; for $\lambda = 0$, $\mathbf{v}_1 = \begin{bmatrix} -1 \\ 1 \\ 1 \end{bmatrix}$

(c) $p(t) = (t-1)^2(t+1)$; for $\lambda = 1$,

$\mathbf{v}_1 = \begin{bmatrix} -2 \\ 0 \\ 1 \end{bmatrix}$, $\mathbf{v}_2 = \begin{bmatrix} 5/2 \\ 1/2 \\ 0 \end{bmatrix}$; for $\lambda = -1$, $\mathbf{v}_1 = \begin{bmatrix} -9 \\ -1 \\ 1 \end{bmatrix}$ **3.** a) $QAQ^{-1} = H$ where

$$Q = \begin{bmatrix} 1 & 0 & 0 \\ 0 & 1 & 0 \\ 0 & 3 & 1 \end{bmatrix}, \quad H = \begin{bmatrix} 8 & -69 & 21 \\ 1 & -10 & 3 \\ 0 & -4 & 1 \end{bmatrix}, \quad Q^{-1} = \begin{bmatrix} 1 & 0 & 0 \\ 0 & 1 & 0 \\ 0 & -3 & 1 \end{bmatrix}.$$

The characteristic polynomial for H is $p(t) = (t+1)^2(t-1)$; and the eigenvectors and generalized eigenvectors are

$$\lambda = -1,\ \mathbf{v}_1 = \begin{bmatrix} 3 \\ 1 \\ 2 \end{bmatrix},\qquad \mathbf{v}_2 = \begin{bmatrix} -7/2 \\ -1/2 \\ 0 \end{bmatrix};\qquad \lambda = 1,\ \mathbf{w}_1 = \begin{bmatrix} -3 \\ 0 \\ 1 \end{bmatrix}.$$

The general solution of $\mathbf{y}' = H\mathbf{y}$ is $\mathbf{y}(t) = c_1 e^{-t}\mathbf{v}_1 + c_2 e^{-t}(\mathbf{v}_2 + t\mathbf{v}_1) + c_3 e^t \mathbf{w}_1$; and the initial condition $\mathbf{y}(0) = Q\mathbf{x}_0$ can be met by choosing $c_1 = 0$, $c_2 = 2$, $c_3 = -2$. Finally $\mathbf{x}(t) = Q^{-1}\mathbf{y}(t)$, or

$$\mathbf{x}(t) = \begin{bmatrix} e^{-t}(6t-7) \quad +6e^t \\ e^{-t}(2t-1) \\ e^{-t}(-2t+3) - 2e^t \end{bmatrix}.$$

(b) $QAQ^{-1} = H$ where

$$Q = \begin{bmatrix} 1 & 0 & 0 \\ 0 & 1 & 0 \\ 0 & 3 & 1 \end{bmatrix},\qquad H = \begin{bmatrix} 2 & 4 & -1 \\ -3 & -4 & 1 \\ 0 & 3 & -1 \end{bmatrix},\qquad Q^{-1} = \begin{bmatrix} 1 & 0 & 0 \\ 0 & 1 & 0 \\ 0 & -3 & 1 \end{bmatrix}.$$

The characteristic polynomial is $p(t) = (t+1)^3$; and

$$\mathbf{v}_1 = \begin{bmatrix} 1 \\ 0 \\ 3 \end{bmatrix},\qquad \mathbf{v}_2 = \begin{bmatrix} 0 \\ 1 \\ 3 \end{bmatrix},\qquad \mathbf{v}_3 = \begin{bmatrix} 0 \\ 1 \\ 4 \end{bmatrix}.$$

The general solution of $\mathbf{y}' = H\mathbf{y}$ is $\mathbf{y}(t) = e^{-t}[c_1\mathbf{v}_1 + c_2(\mathbf{v}_2 + t\mathbf{v}_1) + c_3(\mathbf{v}_3 + t\mathbf{v}_2 + \frac{t^2}{2}\mathbf{v}_1)]$; and the initial condition $\mathbf{y}(0) = Q\mathbf{x}_0$ is met with $c_1 = -1$, $c_2 = -5$, $c_3 = 4$. Finally $Q^{-1}\mathbf{y}(t)$ solves $\mathbf{x}' = A\mathbf{x}$, $\mathbf{x}(0) = \mathbf{x}_0$. (c) $QAQ^{-1} = H$ where

$$Q = \begin{bmatrix} 1 & 0 & 0 \\ 0 & 1 & 0 \\ 0 & 3 & 1 \end{bmatrix},\qquad H = \begin{bmatrix} 1 & 4 & -1 \\ -3 & -5 & 1 \\ 0 & 3 & -2 \end{bmatrix},\qquad Q^{-1} = \begin{bmatrix} 1 & 0 & 0 \\ 0 & 1 & 0 \\ 0 & -3 & 1 \end{bmatrix}.$$

The characteristic polynomial is $p(t) = (t+2)^3$; and

$$\mathbf{v}_1 = \begin{bmatrix} 1 \\ 0 \\ 3 \end{bmatrix},\qquad \mathbf{v}_2 = \begin{bmatrix} 0 \\ 1 \\ 3 \end{bmatrix},\qquad \mathbf{v}_3 = \begin{bmatrix} 0 \\ 1 \\ 4 \end{bmatrix}.$$

The general solution of $\mathbf{y}' = H\mathbf{y}$ is $\mathbf{y}(t) = e^{-2t}[c_1\mathbf{v}_1 + c_2(\mathbf{v}_2 + t\mathbf{v}_1) + c_3(\mathbf{v}_3 + t\mathbf{v}_2 + \frac{t^2}{2}\mathbf{v}_1)]$; and the initial condition $\mathbf{y}(0) = Q\mathbf{x}_0$ is met with $c_1 = -1$, $c_2 = -5$, $c_3 = 4$. Finally $Q^{-1}\mathbf{y}(t)$ solves $\mathbf{x}' = A\mathbf{x}$, $\mathbf{x}(0) = \mathbf{x}_0$.

5. a) Eigenvectors and generalized eigenvectors for H_{22} are $\lambda = 1$, $\mathbf{w} = \begin{bmatrix} 0 \\ 0 \\ 1 \end{bmatrix}$;

$$\lambda = 2,\ \mathbf{v}_1 = \begin{bmatrix} 0 \\ 1 \\ 1 \end{bmatrix},\qquad \mathbf{v}_2 = \begin{bmatrix} 1 \\ 1 \\ 0 \end{bmatrix};\ \text{and for } H_{11},\quad \lambda = 1,\ \mathbf{u} = \begin{bmatrix} 0 \\ 1 \end{bmatrix};\quad \lambda = 2,\ \mathbf{r} = \begin{bmatrix} 1 \\ 1 \end{bmatrix}.$$

For $\lambda = 1$ the system (3.73) is inconsistent when $\boldsymbol{v} = \mathbf{w}$; but (3.75) is consistent for $\mathbf{u}_m = \mathbf{u}$ and $\alpha = -1$. An eigenvector and generalized eigenvector for $\lambda = 1$ is [see (3.77) and (3.78)]

$$\mathbf{z}_1 = \begin{bmatrix} 0 \\ -1 \\ 0 \\ 0 \\ 0 \end{bmatrix}, \quad \mathbf{z}_2 = \begin{bmatrix} -1 \\ 0 \\ 0 \\ 0 \\ 1 \end{bmatrix}.$$

For $\lambda = 2$ the system (3.73) is inconsistent when $\boldsymbol{v} = \boldsymbol{v}_1$; but (3.75) is consistent for $\mathbf{u}_m = \mathbf{r}$ and $\alpha = 1$.
Thus an eigenvector and a generalized eigenvector of order 2 corresponding to $\lambda = 2$ are

$$\mathbf{w}_1 = \begin{bmatrix} 1 \\ 1 \\ 0 \\ 0 \\ 0 \end{bmatrix}. \quad \mathbf{w}_2 = \begin{bmatrix} 1 \\ 0 \\ 0 \\ 1 \\ 1 \end{bmatrix}.$$

To get the last generalized eigenvector, consider (3.76) with $\mathbf{z} = \mathbf{w}_2$; that is, $\mathbf{y} = \boldsymbol{v}_1$, $\boldsymbol{v} = \boldsymbol{v}_2$, $\mathbf{x}_1 = \mathbf{e}_1$, and $\mathbf{u}_m = \mathbf{u}$. This system is consistent with $\alpha = 0$, and a solution is

$$\mathbf{w}_3 = \begin{bmatrix} -1 \\ 0 \\ 1 \\ 1 \\ 0 \end{bmatrix}.$$

Note that $(H - 2I)\mathbf{w}_3 = \mathbf{w}_2$; so $\mathbf{w}_3$ is a generalized eigenvector of order 3. Clearly $\{\mathbf{z}_1, \mathbf{z}_2, \mathbf{w}_1, \mathbf{w}_2, \mathbf{w}_3\}$ is a linearly independent set of vectors.

(b) Eigenvectors and generalized eigenvectors for H_{22} are $\lambda = 1$, $\mathbf{w} = \begin{bmatrix} 1 \\ -1 \end{bmatrix}$; $\lambda = 2$, $\boldsymbol{v} = \begin{bmatrix} 0 \\ 1 \end{bmatrix}$; and for H_{11} are $\lambda = 1$, $\mathbf{u}_1 = \begin{bmatrix} 0 \\ 1 \\ -1 \end{bmatrix}$, $\mathbf{u}_2 = \begin{bmatrix} 1 \\ -1 \\ 0 \end{bmatrix}$; $\lambda = 2$, $\mathbf{r} = \begin{bmatrix} 0 \\ 0 \\ 1 \end{bmatrix}$.

A set of eigenvectors and generalized eigenvectors is

$$\mathbf{z}_1 = \begin{bmatrix} 0 \\ 1 \\ -1 \\ 0 \\ 0 \end{bmatrix}, \mathbf{z}_2 = \begin{bmatrix} 1 \\ -1 \\ 0 \\ 0 \\ 0 \end{bmatrix}, \mathbf{z}_3 = \begin{bmatrix} 0 \\ 0 \\ 1 \\ 0 \\ 0 \end{bmatrix}, \mathbf{w}_1 = \begin{bmatrix} 0 \\ 1 \\ -1 \\ 1 \\ -1 \end{bmatrix}, \mathbf{w}_2 = \begin{bmatrix} 1 \\ 1 \\ 0 \\ 0 \\ 1 \end{bmatrix}.$$

The preceding $\mathbf{z}_1$ and $\mathbf{w}_1$ are eigenvectors corresponding to $\lambda = 1$; $\mathbf{z}_2$ is a generalized eigenvector of order 2 where $(H - I)\mathbf{z}_2 = \mathbf{z}_1$; $\mathbf{z}_3$ is an eigenvector corresponding to $\lambda = 2$ while $(H - 2I)\mathbf{w}_2 = \mathbf{z}_3$.

11. For $\lambda = 1$ and $\lambda = 2$ a chain of generalized eigenvectors is

$$\boldsymbol{v}_1 = \begin{bmatrix} 1 \\ -1 \\ 0 \\ 0 \\ 0 \\ 0 \end{bmatrix}, \boldsymbol{v}_2 = \begin{bmatrix} -1 \\ 0 \\ 0 \\ 1 \\ 0 \\ 0 \end{bmatrix}, \boldsymbol{v}_3 = \begin{bmatrix} 0 \\ 0 \\ 1 \\ -1 \\ 0 \\ -1 \end{bmatrix}, \boldsymbol{v}_4 = \begin{bmatrix} 0 \\ 0 \\ -1 \\ -2 \\ -1 \\ 3 \end{bmatrix}; \mathbf{w}_1 = \begin{bmatrix} 0 \\ 2 \\ 0 \\ 0 \\ 0 \\ 0 \end{bmatrix}, \mathbf{w}_2 = \begin{bmatrix} 2 \\ 0 \\ 1 \\ 1 \\ 0 \\ 0 \end{bmatrix},$$

respectively.

4.2 EXERCISES

1. a) $\begin{bmatrix} 0 & -7 & 5 \\ -11 & -3 & -12 \end{bmatrix}, \begin{bmatrix} 12 & -22 & 38 \\ -50 & -6 & -15 \end{bmatrix}, \begin{bmatrix} 7 & -21 & 28 \\ -42 & -7 & -14 \end{bmatrix}$

(b) $-x^2 - 4x,\ -x^2 - 2x + 3,\ -3x^2 + 4x + 8$ (c) $e^x - 2\sin x$, $e^x - 2\sin x + 3\sqrt{x^2+1},\ -2e^x - \sin x + 3\sqrt{x^2+1}$

3. Subsets (a), (c), and (e) are vector spaces; the others are not.

7. Part (b) is proved in the same way as (a). To prove (c), start with $(0+0)\boldsymbol{v} = 0\boldsymbol{v}$. For (e) if $a \neq 0$, then multiply by $1/a$.

9. Subsets (a), (b), (d), and (e) are vector spaces; (c) is not.

11. Subset (a) is a vector space; (b) is not.

4.3 EXERCISES

1. Subsets (b) and (c) are subspaces; (a) and (d) are not.

3. One spanning set is $\{A_1, A_2\}$ where

$$A_1 = \begin{bmatrix} 1 & 1 \\ 0 & -1 \end{bmatrix}, \quad A_2 = \begin{bmatrix} 0 & 0 \\ 1 & 0 \end{bmatrix}.$$

5. For (1b):

$$\begin{bmatrix} 1 & 1 & 0 \\ 0 & 0 & 0 \end{bmatrix}, \begin{bmatrix} -2 & 0 & 1 \\ 0 & 0 & 0 \end{bmatrix}, \begin{bmatrix} 0 & 0 & 0 \\ 1 & 0 & 0 \end{bmatrix}, \begin{bmatrix} 0 & 0 & 0 \\ 0 & 1 & 0 \end{bmatrix}, \begin{bmatrix} 0 & 0 & 0 \\ 0 & 0 & 1 \end{bmatrix}$$

For (1c):

$$\begin{bmatrix} 1 & 1 & -1 \\ 0 & 0 & 0 \end{bmatrix}, \begin{bmatrix} 0 & 0 & 0 \\ 1 & 0 & 0 \end{bmatrix}, \begin{bmatrix} 0 & 0 & 0 \\ 0 & 1 & 0 \end{bmatrix}$$

For (2a): $p(x) = a_0 + a_1x + a_2x^2$ is in W if and only if $2a_0 + 2a_1 + 4a_2 = 0$ or $a_0 = -a_1 - 2a_2$. Thus $p(x)$ in W has the form $p(x) = -a_1 - 2a_2 + a_1x + a_2x^2 = a_1(x-1) + a_2(x^2-2)$. A spanning set for W is therefore $\{x-1, x^2-2\}$.
For (2b): $\{1, x^2 - 4x\}$
For (2d): $\{x, x^2 - 1\}$

11. a) $\begin{bmatrix} 0 & 1 & 0 \\ 0 & 0 & 0 \\ 0 & 0 & 0 \end{bmatrix}, \begin{bmatrix} 0 & 0 & 1 \\ 0 & 0 & 0 \\ 0 & 0 & 0 \end{bmatrix}, \begin{bmatrix} 0 & 0 & 0 \\ 0 & 0 & 1 \\ 0 & 0 & 0 \end{bmatrix}$

(b) $\begin{bmatrix} -1 & 0 & 0 \\ 0 & 1 & 0 \\ 0 & 0 & 0 \end{bmatrix}$, $\begin{bmatrix} -1 & 0 & 0 \\ 0 & 0 & 0 \\ 0 & 0 & 1 \end{bmatrix}$, $\begin{bmatrix} 0 & -1 & 0 \\ 0 & 0 & 1 \\ 0 & 0 & 0 \end{bmatrix}$, $\begin{bmatrix} 0 & 0 & 1 \\ 0 & 0 & 0 \\ 0 & 0 & 0 \end{bmatrix}$

(c) $\begin{bmatrix} 1 & 1 & 0 \\ 0 & 0 & 0 \\ 0 & 0 & 0 \end{bmatrix}$, $\begin{bmatrix} 0 & 0 & 1 \\ 0 & 0 & 1 \\ 0 & 0 & 0 \end{bmatrix}$, $\begin{bmatrix} 0 & 0 & 0 \\ 0 & 1 & 0 \\ 0 & 0 & 1 \end{bmatrix}$

(d) $\begin{bmatrix} 1 & 0 & 0 \\ 0 & 1 & 0 \\ 0 & 0 & 1 \end{bmatrix}$, $\begin{bmatrix} 0 & 1 & 0 \\ 0 & 0 & -1 \\ 0 & 0 & 0 \end{bmatrix}$, $\begin{bmatrix} 0 & 0 & 1 \\ 0 & 0 & 0 \\ 0 & 0 & 0 \end{bmatrix}$

4.4 EXERCISES

1. a) $\begin{bmatrix} 2 \\ 1 \\ 1 \\ 1 \\ 2 \\ 1 \end{bmatrix}$ (b) $\begin{bmatrix} 1 \\ 0 \\ 0 \\ 0 \\ 1 \\ 0 \end{bmatrix}$ (c) $\begin{bmatrix} 3 \\ 4 \\ 2 \\ 1 \\ 0 \\ 8 \end{bmatrix}$ 5. $[C_1]_B = \begin{bmatrix} 1 \\ 0 \\ 1 \\ 1 \end{bmatrix}$, $[C_2]_B = \begin{bmatrix} 1 \\ 1 \\ 0 \\ 1 \end{bmatrix}$, $[C_3]_B = \begin{bmatrix} 1 \\ 1 \\ 0 \\ 0 \end{bmatrix}$

7. a) The coordinate vectors are $\begin{bmatrix} 0 \\ -1 \\ 0 \\ 1 \end{bmatrix}$, $\begin{bmatrix} -1 \\ 0 \\ 1 \\ 0 \end{bmatrix}$, $\begin{bmatrix} 4 \\ 1 \\ 0 \\ 0 \end{bmatrix}$;

the vectors are linearly independent in $\mathscr{P}_3$.

(b) The coordinate vectors are $\begin{bmatrix} -1 \\ 2 \\ 1 \\ 0 \end{bmatrix}$, $\begin{bmatrix} 2 \\ -5 \\ 1 \\ 0 \end{bmatrix}$, $\begin{bmatrix} 0 \\ -1 \\ 3 \\ 0 \end{bmatrix}$;

the vectors are linearly dependent in $\mathscr{P}_3$.

(c) The coordinate vectors are $\begin{bmatrix} 0 \\ 0 \\ -1 \\ 1 \end{bmatrix}$, $\begin{bmatrix} 0 \\ -1 \\ 1 \\ 0 \end{bmatrix}$, $\begin{bmatrix} -1 \\ 1 \\ 0 \\ 0 \end{bmatrix}$, $\begin{bmatrix} -1 \\ 0 \\ 0 \\ 1 \end{bmatrix}$;

the vectors are linearly independent in $\mathscr{P}_3$.

(d) The coordinate vectors are $\begin{bmatrix} 1 \\ 0 \\ 0 \\ 1 \end{bmatrix}$, $\begin{bmatrix} 1 \\ 0 \\ 1 \\ 0 \end{bmatrix}$, $\begin{bmatrix} 1 \\ 1 \\ 0 \\ 0 \end{bmatrix}$, $\begin{bmatrix} 1 \\ 0 \\ 0 \\ 0 \end{bmatrix}$;

the vectors are linearly independent in $\mathscr{P}_3$.

9. The coordinate vectors are clearly linearly independent:

$$\begin{bmatrix}1\\0\\0\\0\end{bmatrix}, \begin{bmatrix}1\\1\\0\\0\end{bmatrix}, \begin{bmatrix}-2\\0\\1\\0\end{bmatrix}, \begin{bmatrix}1\\3\\3\\1\end{bmatrix}.$$

11. One basis is $\begin{bmatrix}1&0\\0&0\end{bmatrix}, \begin{bmatrix}0&1\\1&0\end{bmatrix}, \begin{bmatrix}0&0\\0&1\end{bmatrix}$.

4.5 EXERCISES

1. b) One basis for V_1 is $\begin{bmatrix}1&0&0\\0&0&0\\0&0&0\end{bmatrix}, \begin{bmatrix}0&0&0\\1&0&0\\0&0&0\end{bmatrix}, \begin{bmatrix}0&0&0\\0&1&0\\0&0&0\end{bmatrix}, \begin{bmatrix}0&0&0\\0&0&0\\1&0&0\end{bmatrix},$

$\begin{bmatrix}0&0&0\\0&0&0\\0&1&0\end{bmatrix}, \begin{bmatrix}0&0&0\\0&0&0\\0&0&1\end{bmatrix}$. A basis for V_2 can be obtained by using the transpose of each basis vector for V_1. (c) $\dim(V) = 9$, $\dim(V_1) = 6$, $\dim(V_2) = 6$.

3. $V_1 \cap V_2$ is the set of all (3×3) diagonal matrices, and the dimension of this subspace is 3.

5. If $p(x) = a_0 + a_1x + a_2x^2 + a_3x^3 + a_4x^4$, then $p(x)$ is in W if and only if

$$\begin{aligned} 2a_0 + 2a_2 + 2a_4 &= 0 \\ 2a_0 + 8a_2 + 32a_4 &= 0. \end{aligned}$$

These constraints require that $a_2 = -5a_4$ and $a_0 = 4a_4$; so $p(x)$ in W must have the form $p(x) = 4a_4 + a_1x - 5a_4x^2 + a_3x^3 + a_4x^4$. A natural basis for W is $\{x^4 - 5x^2 + 4, x, x^3\}$; so $\dim(W) = 3$. **7.** A natural basis for W is $\begin{bmatrix}0&1&0\\-1&0&0\\0&0&0\end{bmatrix}, \begin{bmatrix}0&0&1\\0&0&0\\-1&0&0\end{bmatrix},$

$\begin{bmatrix}0&0&0\\0&0&1\\0&-1&0\end{bmatrix}$; so $\dim(W) = 3$. **11.** a) $A = \begin{bmatrix}1&-1&1\\0&1&-2\\0&0&1\end{bmatrix}$, $\begin{bmatrix}5\\2\\1\end{bmatrix} = A\begin{bmatrix}8\\4\\1\end{bmatrix}$;

so $x^2 + 4x + 8 = (x+1)^2 + 2(x+1) + 5$ **13.** $A^{-1} = \begin{bmatrix}1&1&1\\0&1&2\\0&0&1\end{bmatrix}$ (a) $A^{-1}[\mathbf{w}]_B$

$= A^{-1}\begin{bmatrix}2\\-3\\7\end{bmatrix} = \begin{bmatrix}6\\11\\7\end{bmatrix}$; so $p(x) = 7x^2 + 11x + 6$ (b) $p(x) = -x^2 + 2x + 4$

(c) $p(x) = x + 5$ (d) $p(x) = -x^2 - 2x + 8$

4.6 EXERCISES

5. Suppose $a_1\mathbf{v}_1 + a_2\mathbf{v}_2 + \ldots + a_k\mathbf{v}_k = \boldsymbol{\theta}$, and form the inner product of this and $\mathbf{v}_i$. Observe that this product gives $a_i\langle \mathbf{v}_i, \mathbf{v}_i \rangle = \langle \mathbf{v}_i, \boldsymbol{\theta} \rangle$. To show $\langle \mathbf{v}_i, \boldsymbol{\theta} \rangle = 0$, consider $\langle \mathbf{v}_i, \boldsymbol{\theta} + \boldsymbol{\theta} \rangle = \langle \mathbf{v}_i, \boldsymbol{\theta} \rangle$. **7.** $p_0(x) = 1$, $\langle p_0, p_0 \rangle = 3$; $p_1(x) = x - \dfrac{\langle p_0, x \rangle}{3} p_0(x) = x - 1$, $\langle p_1, p_1 \rangle = 2$; $p_2(x) = x^2 - \dfrac{\langle p_1, x^2 \rangle}{2} p_1(x) - \dfrac{\langle p_0, x^2 \rangle}{3} p_0(x) = x^2 - 2x + 1/3$.

11. c) $T_2(x) = 2x^2 - 1$, $T_3(x) = 4x^3 - 3x$, etc.

4.7 EXERCISES

7. $T(p) = T((x+1)^2 + 3(x+1) - 5) = x + 3(x^3 - 2x) - 5x^4$, or $T(p) = -5x^4 + 3x^3 - 5x$. Similarly $T(q) = -3x^4 + 7x^3 - 13x$. **9.** In Example 4.25, $T(a_2x^2 + a_1x + a_0) = 4a_2 + 2a_1 + a_0$; so $T(p) = 0$ when $a_0 = -2a_1 - 4a_2$ or when $p(x)$ has the form $p(x) = a_2(x^2 - 4) + a_1(x - 2)$. A basis for $\mathcal{N}(T)$ is $\{x - 2, x^2 - 4\}$, and T is not one to one. In Example 4.28, $\mathcal{N}(T) = \{\boldsymbol{\theta}\}$; and T is one to one. In Example 4.29, $\mathcal{N}(T) = U$; and T is not one to one. In Example 4.27, $\mathcal{N}(T) = \boldsymbol{\theta}_V$; and T is one to one.

13. $0 \leqslant \dim(\mathcal{R}(T)) \leqslant 3$; so $2 \leqslant \dim(\mathcal{N}(T)) \leqslant 5$. T cannot be one to one. **15.** In Example 4.26, $\mathcal{R}(T) = R^1$. In Example 4.28, $\mathcal{R}(T) = V$. In Example 4.29, $\mathcal{R}(T) = \{\boldsymbol{\theta}_V\}$. In Example 4.27, $\mathcal{R}(T) = R^p$. In Example 4.30, $\mathcal{R}(T) = R^1$. In Example 4.31, $\mathcal{R}(T) = C[a, b]$.

4.8 EXERCISES

3. $(T+S)(p) = r$ where $r(x) = (x+2)p(x) + p'(0)$; $(H \circ T)(p) = r$ where $r(x) = p(x) + (x+2)p'(x) + 2p(0)$; $S \circ (H \circ T)(p) = r$ where $r(x) = 2p'(0) + 2p''(0)$. To see that T is one to one, suppose $T(p)$ is the zero vector where $p(x) = a_3x^3 + a_2x^2 + a_1x + a_0$. Then $T(p) = \theta$ means $(x+2)p(x) = \theta$, or $a_3x^4 + (2a_3 + a_2)x^3 + (2a_2 + a_1)x^2 + (2a_1 + a_0)x + 2a_0 = \theta$. But this result can hold only if p is the zero vector; so $\mathcal{N}(T) = \{\theta\}$ and T is one to one. A basis for $\mathcal{R}(T)$ is $\{T(1), T(x), T(x^2), T(x^3)\} = \{x+2, x(x+2), x^2(x+2), x^3(x+2)\}$. If q is in $\mathcal{R}(T)$, then $T^{-1}(q) = p$ where $p(x) = q(x)/(x+2)$. **5.** For $p(x) = a_4x^4 + a_3x^3 + a_2x^2 + a_1x + a_0$, we see that $H(p) = 4a_4x^3 + 3a_3x^2 + 2a_2x + a_1 + a_0$. Thus $H(p) = \theta$ if and only if $a_4 = a_3 = a_2 = 0$ and $a_1 = -a_0$, or $p(x)$ has the form $p(x) = a_1x - a_1$. A basis for $\mathcal{N}(H)$ is $\{x - 1\}$, and the nullity of H is 1. A spanning set for $\mathcal{R}(H)$ is $\{H(x^4), H(x^3), H(x^2), H(x), H(1)\} = \{4x^3, 3x^2, 2x, 1, 1\}$; so $\mathcal{R}(H) = \mathcal{P}_3$; and the rank of H is 4. **7.** T is not one to one since $\mathcal{N}(T)$ contains all the constant polynomials. T is not onto since $T(p) = x^2$ has no solution.

4.9 EXERCISES

For T: $Q = \begin{bmatrix} 2 & 0 & 0 & 0 \\ 1 & 2 & 0 & 0 \\ 0 & 1 & 2 & 0 \\ 0 & 0 & 1 & 2 \\ 0 & 0 & 0 & 1 \end{bmatrix}$, For S: $Q = \begin{bmatrix} 0 & 1 & 0 & 0 \\ 0 & 0 & 0 & 0 \\ 0 & 0 & 0 & 0 \\ 0 & 0 & 0 & 0 \\ 0 & 0 & 0 & 0 \end{bmatrix}$. For H: $Q = \begin{bmatrix} 1 & 1 & 0 & 0 & 0 \\ 0 & 0 & 2 & 0 & 0 \\ 0 & 0 & 0 & 3 & 0 \\ 0 & 0 & 0 & 0 & 4 \end{bmatrix}$.

3. For $T: Q = \begin{bmatrix} 1 & 0 & 0 \\ 0 & 2 & 0 \\ 0 & 0 & 3 \end{bmatrix}$. For $T^{-1}: Q = \begin{bmatrix} 1 & 0 & 0 \\ 0 & 1/2 & 0 \\ 0 & 0 & 1/3 \end{bmatrix}$. 5. $Q = \begin{bmatrix} 3 & 0 & 0 \\ 0 & 3 & 0 \\ 0 & -1 & 3 \\ 0 & 0 & 0 \end{bmatrix}$

7. $Q = \begin{bmatrix} 1 & 0 & 0 \\ 0 & 3 & 6 \\ 0 & 1 & 4 \end{bmatrix}$ 9. $Q = \begin{bmatrix} 0 & 0 & 1 & 1 \\ 1 & 0 & 1 & 0 \\ 0 & 1 & 0 & 0 \\ 0 & 0 & 0 & 3 \end{bmatrix}$ 13. $Q = \begin{bmatrix} 3 & -2 & 0 \\ 1 & 3 & -4 \\ 0 & 0 & 5 \\ 0 & 0 & 1 \end{bmatrix}$

4.10 EXERCISES

3. $P = \begin{bmatrix} \frac{5}{3} & \frac{2}{3} \\ -\frac{4}{3} & -\frac{1}{3} \end{bmatrix}$ 5. $P = \begin{bmatrix} 1/2 & -\frac{1}{3} \\ -1/2 & \frac{2}{3} \end{bmatrix}$; $[\mathbf{a}]_C = P\begin{bmatrix} 2 \\ 0 \end{bmatrix}$, or $[\mathbf{a}]_C = \begin{bmatrix} 1 \\ -1 \end{bmatrix}$, and therefore $\mathbf{a} = \mathbf{w}_1 - \mathbf{w}_2$. Similarly $\mathbf{b} = \frac{10}{3}\mathbf{w}_1 - \frac{11}{3}\mathbf{w}_2$, $\mathbf{c} = 3\mathbf{w}_1 - 2\mathbf{w}_2$, $\mathbf{d} = \frac{7}{3}\mathbf{w}_1 - \frac{5}{3}\mathbf{w}_2$.

7. $P = \begin{bmatrix} -1 & 1 & 2 & 3 \\ 1 & 0 & 0 & -3 \\ 0 & 0 & 1 & 0 \\ 0 & 0 & 0 & 1 \end{bmatrix}$; $[p]_C = P\begin{bmatrix} 2 \\ -7 \\ 1 \\ 0 \end{bmatrix}$ or $[p]_C = \begin{bmatrix} -7 \\ 2 \\ 1 \\ 0 \end{bmatrix}$; and therefore $p(x) = -7x + 2(x+1) + (x^2 - 2x)$. Similarly $q(x) = 13x - 4(x+1) + (x^3 + 3)$ and $r(x) = -7x + 3(x+1) - 2(x^2 - 2x) + (x^3 + 3)$.

9. With respect to the natural basis the matrix representation for T is

$$Q = \begin{bmatrix} 1 & 1 & 0 \\ 0 & 2 & 4 \\ 0 & 0 & 3 \end{bmatrix}.$$

The matrix Q can be diagonalized by

$$S = \begin{bmatrix} 1 & 1 & 2 \\ 0 & 1 & 4 \\ 0 & 0 & 1 \end{bmatrix}.$$

The matrix representation for T is diagonal with respect to the basis $C = \{1, x+1, x^2 + 4x + 2\}$. The transition matrix between B and C is

$$P = \begin{bmatrix} 1 & -1 & 2 \\ 0 & 1 & -4 \\ 0 & 0 & 1 \end{bmatrix}.$$

For (a), $[p]_C = P[p]_B$; so $[p]_C = \begin{bmatrix} -13 \\ 3 \\ 1 \end{bmatrix}$ and $[T(p)]_C = \begin{bmatrix} -13 \\ 6 \\ 3 \end{bmatrix}$. Therefore $T(p) = 3(x^2 + 4x + 2) + 6(x+1) - 13$. Similarly for (b), $T(p) = 3(x^2 + 4x + 2) - 8(x+1) + 7$; and for (c), $T(p) = 6(x^2 + 4x + 2) - 22(x+1) + 11$.

5.2 EXERCISES

1. $\det(A) = -5$, $\det(B) = -31$, $\det(C) = 0$, $\det(D) = 2$ **3.** a) $A_{11} = -2$, $A_{21} = -1$, $A_{23} = 3$, $A_{33} = 1$ (b) $A_{11} = -2$, $A_{21} = -8$, $A_{23} = 11$, $A_{33} = -4$ (c) $A_{11} = -2$, $A_{21} = 7$, $A_{23} = -7$, $A_{33} = 3$ **5.** a) $\det(A) = 8$ (b) $\det(A) = 14$ (c) $\det(A) = -35$
7. $\det(A) = -9$, $\det(B) = 2$ **11.** $\det(A) = 24$, $\det(B) = 6$

5.3 EXERCISES

1. a) $\begin{vmatrix} 1 & 0 & 0 \\ 2 & -4 & -1 \\ 1 & -3 & 0 \end{vmatrix} = -3$ (b) $\begin{vmatrix} 2 & 0 & 0 \\ 0 & 2 & 3 \\ 1 & -1 & 3 \end{vmatrix} = 18$ (c) $\begin{vmatrix} 0 & 1 & 0 \\ 3 & 1 & 0 \\ 2 & 0 & 3 \end{vmatrix} = -9$

5. a) $[\mathbf{A}_1, \mathbf{A}_2, \mathbf{A}_3, \mathbf{A}_4] \to [\mathbf{A}_1, \mathbf{A}_4, \mathbf{A}_3, \mathbf{A}_2] \to [\mathbf{A}_1, \mathbf{A}_4, \mathbf{A}_2, \mathbf{A}_3]$; the original determinant is 6. (b) $[\mathbf{A}_1, \mathbf{A}_2, \mathbf{A}_3, \mathbf{A}_4] \to [\mathbf{A}_3, \mathbf{A}_2, \mathbf{A}_1, \mathbf{A}_4] \to [\mathbf{A}_3, \mathbf{A}_2, \mathbf{A}_4, \mathbf{A}_1] \to [\mathbf{A}_3, \mathbf{A}_4, \mathbf{A}_2, \mathbf{A}_1]$; the original determinant is -48. (c) $[\mathbf{A}_1, \mathbf{A}_2, \mathbf{A}_3, \mathbf{A}_4] \to [\mathbf{A}_2, \mathbf{A}_1, \mathbf{A}_3, \mathbf{A}_4] \to [\mathbf{A}_2, \mathbf{A}_3, \mathbf{A}_1, \mathbf{A}_4] \to [\mathbf{A}_2, \mathbf{A}_4, \mathbf{A}_1, \mathbf{A}_3]$; the original determinant is -12.

7. a) $\det(A) \to \begin{vmatrix} 1 & 0 & 0 & 0 \\ 2 & 1 & 1 & -5 \\ 2 & -4 & 4 & -3 \\ 0 & 1 & 6 & 2 \end{vmatrix} \to \begin{vmatrix} 1 & 0 & 0 & 0 \\ 2 & 1 & 0 & 0 \\ 2 & -4 & 8 & -23 \\ 0 & 1 & 5 & 7 \end{vmatrix} = 171$

(b) $\det(A) \to \begin{vmatrix} 2 & 0 & 0 & 0 \\ 1 & 1 & 2 & 3 \\ 1 & 1 & 2 & 4 \\ -1 & 4 & 0 & 1 \end{vmatrix} \to \begin{vmatrix} 2 & 0 & 0 & 0 \\ 1 & 1 & 0 & 0 \\ 1 & 1 & 0 & 1 \\ -1 & 4 & -8 & -11 \end{vmatrix} = 16$

(c) $\det(A) \to \begin{vmatrix} 1 & 0 & 0 & 0 \\ 0 & 1 & 4 & 1 \\ 2 & -1 & -1 & -2 \\ 2 & 0 & -3 & 0 \end{vmatrix} \to \begin{vmatrix} 1 & 0 & 0 & 0 \\ 0 & 1 & 0 & 0 \\ 2 & -1 & 3 & -1 \\ 2 & 0 & -3 & 0 \end{vmatrix} = -3$

5.4 EXERCISES

1. a) $A \to \begin{bmatrix} 1 & 0 & 3 \\ 2 & 1 & 1 \\ 4 & 3 & 1 \end{bmatrix} \to \begin{bmatrix} 1 & 0 & 0 \\ 2 & 1 & -5 \\ 4 & 3 & -11 \end{bmatrix} \to \begin{bmatrix} 1 & 0 & 0 \\ 2 & 1 & 0 \\ 4 & 3 & 4 \end{bmatrix}$; $\det(A) = -4$.

(b) $B \to \begin{bmatrix} 1 & 0 & 0 \\ 2 & 0 & 1 \\ 2 & -3 & 1 \end{bmatrix} \to \begin{bmatrix} 1 & 0 & 0 \\ 2 & 1 & 0 \\ 2 & 1 & -3 \end{bmatrix}$; $\det(B) = 3$.

(c) $C \to \begin{bmatrix} 2 & 0 & 0 \\ 1 & 2 & 2 \\ -1 & 3 & 3 \end{bmatrix} \to \begin{bmatrix} 2 & 0 & 0 \\ 1 & 2 & 0 \\ -1 & 3 & 0 \end{bmatrix}$; $\det(C) = 0$.

3. a) $x_1 = 2$, $x_2 = 2$, $x_3 = 1$ (b) $x_1 = -1$, $x_2 = 0$, $x_3 = 3$ **11.** $\det(A^5) = [\det(A)]^5 = 243$

5.5 EXERCISES

1. a) $\det(A) \to \begin{vmatrix} 1 & 2 & 1 \\ 0 & -1 & 0 \\ 0 & 6 & 2 \end{vmatrix} \to \begin{vmatrix} 1 & 2 & 1 \\ 0 & -1 & 0 \\ 0 & 0 & 2 \end{vmatrix}$; $\det(A) = -2$.

(b) $\det(A) \to \begin{vmatrix} 1 & 2 & 1 \\ 0 & 3 & 1 \\ 2 & -2 & 2 \end{vmatrix} \to \begin{vmatrix} 1 & 2 & 1 \\ 0 & 3 & 1 \\ 0 & -6 & 0 \end{vmatrix} \to \begin{vmatrix} 1 & 2 & 1 \\ 0 & 3 & 1 \\ 0 & 0 & 2 \end{vmatrix}$; $\det(A) = -6$.

(c) $\det(A) \to \begin{vmatrix} 1 & 0 & 1 \\ 0 & 2 & 4 \\ 0 & 2 & -2 \end{vmatrix} \to \begin{vmatrix} 1 & 0 & 1 \\ 0 & 2 & 4 \\ 0 & 0 & -6 \end{vmatrix}$; $\det(A) = \quad -12$.

7. For $P_1 = \begin{bmatrix} 0 & 1 & 0 \\ 1 & 0 & 0 \\ 0 & 0 & 1 \end{bmatrix}$, $\quad P_1A = \begin{bmatrix} 1 & 2 & 4 \\ 0 & 1 & 3 \\ 2 & 2 & 1 \end{bmatrix}$. Next for $L_1 = \begin{bmatrix} 1 & 0 & 0 \\ 0 & 1 & 0 \\ -2 & 0 & 1 \end{bmatrix}$,

$L_1(P_1A) = \begin{bmatrix} 1 & 2 & 4 \\ 0 & 1 & 3 \\ 0 & -2 & -7 \end{bmatrix}$. Setting $L_2 = \begin{bmatrix} 1 & 0 & 0 \\ 0 & 1 & 0 \\ 0 & 2 & 1 \end{bmatrix}$,

$L_2(L_1P_1A) = \begin{bmatrix} 1 & 2 & 4 \\ 0 & 1 & 3 \\ 0 & 0 & -1 \end{bmatrix} = U.\ Q = L_2L_1P_1 = \begin{bmatrix} 0 & 1 & 0 \\ 1 & 0 & 0 \\ 2 & -2 & 1 \end{bmatrix}$.

9. $\operatorname{Adj}(A) = \begin{bmatrix} 0 & 1 & -1 \\ -2 & 1 & 0 \\ 1 & -1 & 1 \end{bmatrix}$; $\quad \operatorname{Adj}(B) = \begin{bmatrix} -1 & -1 & 1 \\ -3 & 2 & -2 \\ 3 & -2 & -3 \end{bmatrix}$;

$\operatorname{Adj}(C) = \begin{bmatrix} -4 & 2 & 0 \\ 1 & 0 & -1 \\ 1 & -2 & 1 \end{bmatrix}$. $\quad$ **11.** $A^{-1} = \operatorname{Adj}(A)$; $\quad B^{-1} = -\frac{1}{5}\operatorname{Adj}(B)$; $\quad C^{-1} = -\frac{1}{2}\operatorname{Adj}(C)$.

6.1 EXERCISES

1. $a = (1/2 + 1/8)8 = 5$, $b = 6.75$, $c = 38$ **3.** $a = 2857$, $b = 145.75$, $c = 51.1875$
5. $a = .111 \times 2^4$, $b = .111011 \times 2^5$, $c = .11011 \times 2^3$, $d = .10000110111 \times 2^8$
7. $a = .E \times 16$, $b = .1\,D8 \times 16^2$, $c = .6C \times 16$, $d = .86E \times 16^2$ **11.** $a + b = .20242 \times 10^2$, $ab = .63306 \times 10^0$, $a(b+3c) = .30379 \times 10^3$, $c + (d+d) = .50001 \times 10^1$, $(c+d) + d = .50000 \times 10^1$

GLOSSARY OF SYMBOLS

Below is a list of frequently used symbols and the pages where they are defined.

$A = (a_{ij})$	matrix, 12
$[A \mid b]$	augmented matrix, 13
R^n	Euclidean n-space, 27
$A = [\mathbf{A}_1, \mathbf{A}_2, \ldots, \mathbf{A}_n]$	column form of a matrix, 31
$\mathcal{O}$	zero matrix, 38
A^{T}	matrix transpose, 39
$\lVert \mathbf{x} \rVert$	norm of a vector, 41, 225
$\boldsymbol{\theta}$	zero vector, 45, 202
I	identity matrix, 57
$\mathbf{e}_k$	natural unit vectors in R^n, 58
A^{-1}	inverse of the matrix A, 59
$\mathcal{N}(A)$	null space of A, 81
$\mathcal{R}(A)$	range space of A, 81
$\dim(W)$	dimension of W, 95, 220
$\det(A)$	determinant of A, 122, 251
$\mathcal{P}_n$	set of polynomials of degree n or less, 204
$\mathrm{Sp}(Q)$	span of Q, 210
$[\mathbf{w}]_B$	coordinate vector, 215
$\langle \mathbf{u}, \mathbf{v} \rangle$	inner product, 224
$T \circ S$	composition of transformations, 236

INDEX